FRACTURE OF COMPOSITE MATERIALS

FRACTURE OF COMPOSITE MATERIALS

Proceedings of the Second USA-USSR Symposium,
held at Lehigh University, Bethlehem, Pennsylvania USA
March 9-12, 1981

edited by

G.C. SIH
Institute of Fracture and Solid Mechanics
Lehigh University, Bethlehem, Pennsylvania, U.S.A.

V.P. TAMUZS
Institute of Polymer Mechanics
Academy of Sciences of the Latvian SSR
Riga, U.S.S.R.

1982

MARTINUS NIJHOFF PUBLISHERS
THE HAGUE / BOSTON / LONDON

Distributors

for the United States and Canada
Kluwer Boston, Inc.
190 Old Derby Street
Hingham, MA 02043
USA

for all other countries
Kluwer Academic Publishers Group
Distribution Center
P.O. Box 322
3300 AH Dordrecht
The Netherlands

Library of Congress Cataloging in Publication Data

USA-USSR Symposium on Fracture of Composite
Materials (2nd : 1981 : Lehigh University)
Proceedings of Second USA-USSR Symposium on
Fracture of Composite Materials.

1. Composite materials--Congresses. 2. Fracture
mechanics--Congresses. I. Sih, G. C. (George C.)
II. Tamuzs, V. P. (Vitauts P.) III. Title.
TA418.9.C6U73 1981 620.1'186 82-7969
ISBN 90-247-2699-9 AACR2

ISBN 90-247-2699-9 (this volume)

PRINTED IN THE NETHERLANDS

CONTENTS

PREFACE

The Second USA-USSR Symposium on *Fracture of Composite Materials* took place at Lehigh University, Bethlehem, Pennsylvania, during 9-12 March, 1981. This bilateral program between the U.S. and Soviet Union was organized by Professor George C. Sih of the Institute of Fracture and Solid Mechanics at Lehigh University and Dr. Vitauts P. Tamuzs of the Institute of Polymer Mechanics of the Academy of Sciences of the Latvian SSR in Riga. The First Symposium was held in 1978 at Jurmala near the coast of Riga Bay.

The primary reasons for initiating this series of Symposia were to disseminate present knowledge, to promote interchange of ideas, and to stimulate additional studies on the development of composite materials between the U.S. and USSR. Both countries have a vested interest in developing the capability to assess and utilize the attractive mechanical properties of composites so that they can be tailor-made to meet specific design requirements. Despite the increasing number of published papers and articles, there is no communication more effective than on a person-to-person basis. It is with this objective in mind that a small group of engineers and scientists from the U.S. and USSR have planned to meet every two years to report recent progress on composite material research. The size of this group is approximately sixty (60) participants. The presentation involves about forty (40) technical papers which are published in a volume. The first Proceedings are already in print and can be obtained from Sijthoff and Noordhoff located in Alphen aan den Rijn, The Netherlands. The Russian version has been printed by the publishing house, "Zinātne", in Riga, USSR.

The ten members of the Soviet delegation at the Symposium were headed by the corresponding member of the USSR Academy of Sciences, Professor V. V. Bolotin. The Executive Secretary of the delegation was Professor V. P. Tamuzs. A U.S.A. organizing committee was formed to select topics for discussion and speakers. The members were:

G. C. Sih (Chairman)	S. C. Chou
C. W. Smith	R. E. Rowlands
T. T. Chiao	C. C. Chamis

The main theme of the Symposium is the fracture of composite materials possessing geometric discontinuities or inherent material defects. Because of the complexity of composite fracture behavior which is characteristic of heterogeneous materials, many of the analyses have to resort to simplifying assumptions. Generally speaking, the *basic concept* of linear elastic fracture mechanics is applicable when a composite system exhibits brittle fracture behavior. This, however, does not imply that all composites can be characterized by the single crack critical stress intensity factor formula $K_{1c} = \sigma\sqrt{\pi a}$ or the equivalent of G_{1c}, often referred to as the Griffith theory. It should be remembered that the Griffith equation applies only to self-similar cracks and assumes that all the released energy is converted to the extension of a single crack. These idealized assumptions obviously deviate far from the fracture modes in composites which may involve fiber breaking, matrix cracking, debonding, etc. The burden is on the practitioner to apply fracture mechanics correctly and to perform the appropriate stress analysis. The concept of fracture toughness when applied to composites should be interpreted with care. It applies strictly to a homogeneous isotropic material and is considered to be independent of loading and crack geometry. There is little to be gained if the measured fracture toughness parameter changes with the direction of loading and/or fiber reinforcement. For in that case, information on the failure load will suffice. The objective is to predict the allowable load on a composite from a knowledge of the toughness values of its constituents.

For unidirectional composites in which cracking occurs predominantly in the matrix, it suffices to know the fracture toughness of the matrix for predicting failure. In the cross-ply laminates, material damage consists of delamination as well as cracking and analytical modeling of the damage zones becomes difficult. Oversimplifications can significantly distort the physical interpretation of the results. Experimental studies on delamination in laboratory specimens containing free edges on all sides may not represent the condition in structure application. It is essential that the fracture modes in the actual structures be realistically simulated in the laboratory specimens.

Another area of major concern is a knowledge of the mechanical properties of the constituents from which the overall mechanical response of the composite can be determined. Such an approach encounters enormous difficulties and uncertainties because the properties of the fiber-to-matrix interface cannot be easily assessed. The dependence of fiber strength on initial defects is another source of difficulty. Stochastic models for estimating the strength of unidirectional fiber composites are proposed. Considerations are given to composite damage in the form of fiber breakage, debonding, etc. Since the main life of composites up to fracture is of dispersed nature, the statistical models including both the dispersed stage of damage accumulation and the transition to macrocrack propagation play an important role in formulating the fracture prediction methods.

Non-destructive evaluation of material damage can greatly assist the understanding of how composites respond to mechanical loading. Parameters representing inherent material properties need to be distinguished from those describing apparent changes in material behavior caused by physical damage such as defects or cracks. This area deserves the utmost attention in future research, both in theory and experiment. Lacking in particular is the predictive capability of composite failure.

Clearly, this Second USA-USSR Symposium did not cover the whole problem of composite fracture. Each of the topics discussed could be the topic of one or several symposia. Nevertheless, as a forum for reporting new findings, exchanging ideas, establishing directions of future research and making new friends, this Symposium has served well.

A number of the U.S. organizing committee members accompanied the Soviet delegates on a one-day excursion to Atlantic City. The theory of stochastic processes was applied once more to Blackjack, this time with less success. The bus ride, however, provided an additional opportunity to discuss technical problems.

The organizing committee wishes to acknowledge all those who contributed to the presentation and preparation of technical papers. Their efforts have made the Symposium a great success. Special thanks are also due to Robert Bolton, David Chu, Robert Kolkka, Michael Lieu, Erdogan Madenci, Peter Matic, Thomas

Moyer, Spence Reid and Stone Shih who were responsible for making the detailed arrangements. Mrs. Barbara DeLazaro and Mrs. Constance Weaver spent a great deal of their time organizing and coordinating numerous activities associated with the Symposium. Their assistance in typing the manuscripts is also gratefully acknowledged. Acknowledgements also go to Valentina Kiseljova and Marite Smalka who have contributed to the preparation of the Soviet papers.

March, 1981
Lehigh University, USA
Institute of Polymer Mechanics, USSR

G. C. Sih
V. P. Tamuzs
Editors

REMARKS ON SYMPOSIUM SERIES

The first USSR-USA Symposium on Fracture of Composite Materials was held at the Baltic resort site of Jurmala, USSR, during September of 1978. This second Symposium is being held at Lehigh University in the United States of America. The topic on the fracture mechanics of composites is chosen for the USSR-USA scientific cooperation because of the important role composites play in the design of modern engineering structures. Composites offer high strength in addition to low weight. The manufacture of composites requires comparatively low labor and waste, and is desirable as far as pollution is concerned when compared with steel making. Composite structures are destined for many future applications.

There is no doubt in my mind that the topic of this Symposium is also valuable from the scientific point of view. I personally believe that the "mechanics of fracture" and "mechanics of composites" are the two foremost branches of modern applied mechanics. Fracture of composite materials includes a combination of the two and should pave the way for the development of applied mechanics.

The choice of holding this Symposium at Lehigh University is also appropriate for the School of Engineering of this University is associated with a number of renowned researchers and scholars in the field of mechanics of fracture. Among them are Professor G. R. Irwin who taught at Lehigh for many years and Professor P. C. Paris who is a graduate of this University. Professors G. C. Sih, R. S. Rivlin and F. Erdogan are still engaged in research and teaching at Lehigh. These colleagues have made substantial contributions to the mechanics of fracture.

The first USSR-USA Symposium on Fracture of Composite Materials was most successful as evidenced by the 1979 Proceedings published both in Russian and English. The Proceedings represent one of the most up-to-date comprehensive studies on the mechanics of fracture of composites. It also provided an excellent opportunity for the Soviet and American scientists to exchange ideas and research findings, and to focus attention on future problems.

The success of this second Symposium is clearly demonstrated by the contents of this book which includes the progress in this field for the past two and one-half years. Among the topics covered are: development of stochastic models of fracture of composites with application to predicting the scale effects in composite structures; influence of environmental conditions on composite strength, integrity and durability; design of multilayered composite shells; and others. It is unfortunate that some of the papers presented by the American authors are not made available in written form. Many of the stimulating discussions also did not go on record.

The next USSR-USA Symposium on Fracture of Composite Materials is planned to take place in USSR. It is hoped that up to that time many new findings and achievements will be made in this interesting and rapidly developing field of applied mechanics.

March, 1981

Mechanical Engineering Research Institute
USSR Academy of Sciences

V. V. Bolotin
Head of USSR Delegation

Professor G. C. Sih, Director of Institute of Fracture and Solid Mechanics, Lehigh University, USA, addresses the audience on Fracture of Composite Materials.

Professor V. V. Bolotin, Head of USSR delegation, USSR Academy of Sciences, discusses a stochastic model of fracture of composite materials.

Technical session at the Sinclair auditorium.

Question period during the Symposium
on Fracture of Composite Materials.

Participants attending a technical session.

Question raised by Professor V. V. Panasyuk at center.

Members of the USSR delegation look over the choice of drinks.

Cocktail hour in the Neville Lounge at the University Center.

Co-Chairman, Dr. G. C. Sih, briefs on the objectives of the USA-USSR Symposium.

Banquet at the Asa Packer Dining Hall.

Dr. W. Deming Lewis, President of Lehigh University, presents a plaque to the Head of the USSR delegation, Professor V. V. Bolotin.

Professor G. C. Sih receives a plaque from Dr. V. Tamuzs (center) and Professor V. V. Bolotin with Dr. I. Knets standing by.

Members of the USA-USSR delegation on Fracture of Composite Materials. From left to right front row: L. V. Nikitin, G. A. Teters, C. W. Smith, V. Tamuzs, V. V. Bolotin, G. C. Sih, V. V. Panasyuk, E. M. Wu, S. T. Mileiko, V. V. Vasil'ev and back row: R. E. Rowlands, I. Knets, F. W. Crossman, S. C. Chou, V. K. Khlebnikov.

Group picture of the Second USA-USSR Symposium on Fracture of Composite Materials, Lehigh University, Bethlehem, Pennsylvania, March 9-12, 1981.

SECTION I
ANALYTICAL MODELING

STOCHASTIC MODELS OF FRACTURE OF UNIDIRECTIONAL FIBER COMPOSITES

V. V. Bolotin

USSR Academy of Sciences
Moscow 111250, USSR

ABSTRACT

Stochastic models of fracture of unidirectional fiber composites based upon the combination of dispersed damage accumulation and macrocracks nucleation and its further growth are considered. The case of continuous linear elastic fibers and an ideal elastic-plastic matrix are assumed. Both fiber breakage and composite delamination due to damage of the matrix and/or interface are included in the analysis. Macrocracks perpendicular to the fibers are interpreted as three-dimensional regions that account for the fiber strength scale effect.

Two types of final failure under tension along the fibers are considered: the loss of integrity, i.e., the simultaneous development of multiple cracks throughout the specimen, and the more common failure mode crack growth. The results on the unidirectional fiber composite are discussed in connection with the linear theory of fracture mechanics. Within the framework of the stochastic models, the Griffith-Irwin and Paris-Erdogan equations are recovered only when severely restricted assumptions are imposed on the distribution of fiber strength.

INTRODUCTION

The fracture of fibrous composites differs essentially from that of conventional materials. It is well-known that linear fracture mechanics was developed for a single-phase or quasi-homogeneous material. In the case of fibrous composites, additional parameters representing the microstructure are necessary and more than one mode of fracture may be present. Hence, the Griffith-Irwin criterion based on fracture toughness may have limited application for composite materials.

There are two approaches to overcome the aforementioned difficulties. The first is to generalize the concept of linear fracture mechanics by introducing additional scale parameters. The second is to develop special microstructural models taking the microstructure effect of the composite into account according to observed fracture phenomena. In this paper, the second approach is chosen. Considered are high performance composites with relatively weak matrices and high fiber volume fraction. The main cause of transverse macrocracking is assumed to be composite delamination as more fibers are debonded from the matrix. This increases the fiber strength scale effect.

From the viewpoint of microstructural modeling, dispersed damage accumulation, nucleation of microcracks and their growth up to critical sizes are governed by the same mechanism. In Figure 1, a unidirectional fibrous composite specimen is subjected to tension along the fiber direction. During the early

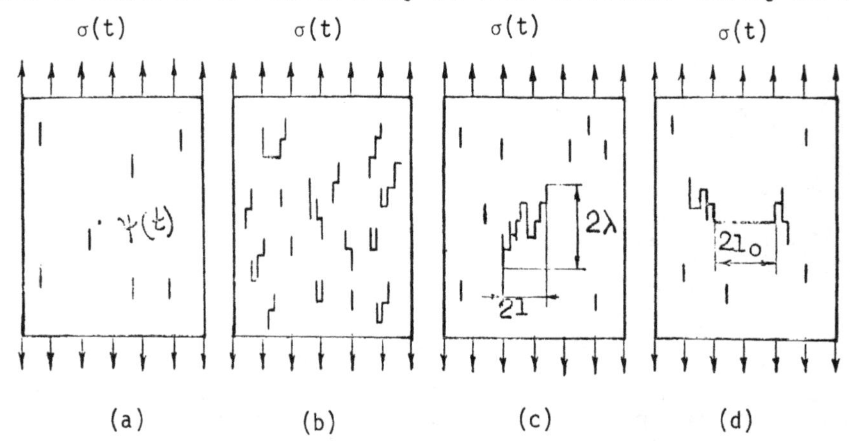

Fig. (1) - Types of fracture: (a) dispersed damage accumulation; (b) loss of integrity; (c) macrocrack nucleation and propagation; (d) propagation of an artificially initiated crack

stages, Figure 1(a), dispersed damage accumulation dominates. At time t_*, the damage density reaches a critical level and the character of the process changes qualitatively. The simultaneous development of multiple cracks throughout the specimen corresponds to a loss of structural integrity, Figure 1(b), while the initiation and subcritical growth of one or more macrocracks lead to disperse damage accumulation, Figure 1(c). In the former case, final fracture occurs at $t \approx t_*$ and the latter at $t \approx t_{**}$ when the macrocrack reaches its critical size.

It is necessary to distinguish the difference between failure due to crack nucleation as a natural process and that of an artificial crack inserted in a virgin specimen, Figure 1(d). In the latter case, damage accumulation starts from the intense stress concentration near the crack tip. As the crack spreads, delamination near the crack tip region tends to relax the stress concentration and dispersed damage spreads throughout the specimen. The failure pattern resembles that of a specimen with a natural crack. The above comment applies, in principle, to all materials of a nonhomogeneous nature, particularly for those high performance composites whose mechanical properties are strongly anisotropic.

A normal procedure [1-4] is to consider the specimen as a set of microstruture elements. Each element is a fiber segment with a portion of the matrix attached to it, and its length is twice that of the transfer length λ_e. All fibers are supposed to be initially continuous and have circular sections of equal radius r. The volume fiber fraction v_f is constant in the specimen with volume V. The fiber material is linear elastic up to fracture with a Young's modulus E_f. The matrix material is ideally elastic-plastic with Young's modulus E_m, the shear modulus G_m and ultimate shear stress τ_m. The latter can be interpreted either as the yield stress of the matrix, or the friction stress of the damaged interface, or the ultimate stress including all inelastic behavior together with matrix damage. The matrix is assumed to be relatively weak, i.e., $E_m/E_f << 1$, $\sigma_m/\sigma_f << 1$ where σ_m and σ_f are ultimate stresses. Time effects for the matrix are negligible. In particular, the characteristic time of delamination is small in comparison with the characteristic life time of fibers.

DISPERSED DAMAGE ACCUMULATION

At least two damage rates should be considered for unidirectional fibrous composites. The first ψ_1 is the ratio of the number of microcracks, i.e., breakage of a single microstructure element, to the total number of these elements. The second ψ_2 characterizes damage of the matrix and/or fiber-matrix interface. This rate is assumed to be equal to the ratio of the summed length of damaged interface to the complete length of all fibers in the volume V. Hence, damage accumulation is described by a vector (two-dimensional) process, whose components are $\psi_1(t)$ and $\psi_2(t)$. If the matrix is deformed elastically, then $\psi_2 \equiv 0$. This case was considered in [1]. Instead of the damage rate ψ_2, a third rate $\psi_3 = \psi_1+\psi_2$ can be introduced. This rate is equal to the ratio of the summed transfer length to the complete length of all fibers.

The following analysis will follow the rigor in [2]. Estimates by orders of magnitude will be widely used. In particular, multipliers whose order of magnitude is equal to unity will be omitted together with those terms that are small. It is of special importance to those expressions involving the fiber volume fraction v_f. Assumed is that $v_f \sim (v_f)^{-1/2} - 1 \sim 1$ and the multipliers of v_f are set to unity. For example, $\sigma \sim \sigma_f$ where σ is the nominal stress and σ_f the fiber stress. As a rule, delamination of the composite and scale strength effect of fibers overshadow the effective crack tip stress concentrations. The limit theorems in probabilistic theory are used to evaluate the macroscopic parameters with "almost sure" estimates identified by their medians with quantiles of probability $1-e^{-1} = 0.632...$, etc. Most of the assumptions are made to obtain analytical results and can be omitted if more exact estimates are required.

The length of an element is equal to $2\lambda_e$;

$$\lambda_e \sim r\left(\frac{E_f}{2G_m}\right)^{1/2} \tag{1}$$

If the stress level is relatively high, the matrix and the interface in the vicinity of a fiber break becomes nonelastic. The transfer length in this case

$$\lambda_p \sim r\,\frac{\sigma}{2\tau_m} \tag{2}$$

depends on the nominal fiber stress denoted simply by σ. Equation (2) follows from equation (1) by letting $\sigma = \sigma_Y$ where σ_Y can be found from the equality λ_e

$= \lambda_p$. Further, an estimate relating the elastic and nonelastic cases can be made by using the brackets:

$$\lambda \sim \{r(\frac{E_f}{2G_m})^{1/2}, r\frac{\sigma}{2\tau_m} \tag{3}$$

If the stresses vary in time and the condition $\sigma<\sigma_Y$ is violated, it is reasonable to count the total number of elements and assume it is a constant with the order of magnitude $N \sim V/V_r$ where $V_r \sim 2\pi r^2\lambda_e$.

The second damage rate is introduced as

$$\psi_2(t) = \int_0^t \Lambda(t,\tau)d\psi_1(\tau) \tag{4}$$

The kernel in equation (4) is

$$\Lambda(t,\tau) = \begin{cases} 0, & \sigma(t,\tau) < \sigma_Y \\ \frac{\sigma(t,\tau)}{\sigma_Y} - 1, & \sigma(t,\tau) \geq \sigma_Y \end{cases} \tag{5}$$

with $\sigma(t,\tau) = \sup\sigma(\tau_1)$ at $\tau_1 \varepsilon[\tau,t]$.

When the damage density is small, interaction between the cracks is negligible. Let the damage accumulated in each element be governed by

$$\frac{d\phi}{dt} = f(\sigma,\phi,\psi_1,\psi_2,s) \tag{6}$$

Here, $\phi(t)$ is a damage rate for the element in consideration, and s is a strength parameter assumed to be a random value with the distribution function $F_s(s;\lambda)$. The latter depends on the active fiber length 2λ. Initially, when the microcrack have yet emerged, $\lambda = \lambda_e$.

The time-to-failure τ of the element at a given $\sigma(t)$, $\psi_1(t)$ and $\psi_2(t)$ is a solution of the inverse boundary problem of equation (6) with the boundary conditions $\phi(o) = 0$ and $\phi(\tau) = 1$. Equation (6) is valid for constant stress or long-time acting slowly varying stresses, as well as cyclic stress with slowly varying amplitudes. In the latter case, t is the "slow" time and $\sigma(t)$ is the envelope for the cyclic loading function.

The processes $\psi_1(t)$ and $\psi_2(t)$ are step-wise continuous. Further, assume that $\psi_1 N >>1$, $\psi_1 << 1$ and $\psi_2 << 1$. An asymptotic distribution for a smoothed process $\psi_1(t)$ was presented in [1]. As the scatter of this process is very small, the semi-probabilistic estimates are valid:

$$\psi_1(t) \sim F_\tau(t), \quad \psi_2(t) \sim \int_0^t \Lambda(t,\tau)dF_\tau(\tau) \tag{7}$$

where $F_\tau(\tau)$ is the elements life distribution function.

A SPECIAL CASE

The solution simplifies if the right-hand side of equation (6) does not depend on ψ_1 and ψ_2. Consider in detail the following special case of equation (6)

$$\frac{d\phi}{dt} = \frac{1}{t_c}\left(\frac{\sigma}{s_c}\right)^n \tag{8}$$

Here, t_c is the characteristic time, and n the power exponent of the fiber life-stress curves under constant loading. If the cyclic loading with slowly varying amplitudes is considered, t_c may be interpreted as the characteristic cycle duration, σ the stress amplitude, and n the power exponent of fiber fatigue curves.

Let s be the strength parameter in the Weibull's distribution

$$F_s(s;\lambda) = 1 - \exp\left[-\frac{\lambda}{r}\left(\frac{s}{s_c}\right)^\alpha\right] \tag{9}$$

where s_c is a characteristic fiber strength. The exponent α is much greater than unity such that the fiber strength variance w_s can be approximated as $w_s \approx \pi/(\alpha\sqrt{\sigma})$ [5]. If the order of magnitude of t_c is the same as that of the duration of standard short-time tests, the constant s_c takes the meaning of the characteristic short-time stress for fiber specimens whose length are equal to 2r. In fact, if sufficiently large test lengths are used, extrapolation is required for estimating s_c.

Let σ = const < σ_Y. Using equations (7) and (9), it is found from equation (8) that

$$\psi_1(t) = 1 - \exp\left[-\frac{\lambda_e}{r}\left(\frac{\sigma}{s_c}\right)^\alpha\left(\frac{t}{t_c}\right)^\beta\right] \tag{10}$$

where the exponent $\beta = \alpha/n$ characterizes the fiber life variance. The result for $\sigma = \dot{\sigma}t < \sigma_Y$ and $\dot{\sigma}$ = const is

$$\psi_1(t) = 1 - \exp\left\{-\frac{\lambda_e}{r}\left(\frac{\dot{\sigma}t}{s_c}\right)^\alpha\left[\frac{t}{(n+1)t_c}\right]^\beta\right\} \tag{11}$$

Typical diagrams for $\psi_1(t)$ are plotted in Figure 2 by means of equations (10) and (11). The broad lines correspond to constant loading, and the narrow lines to loading with constant velocities. The following numerical data are used in Figure 2: $n=4$, $\lambda_c/r = 10^2$, $\sigma/s_c = 10^{-1}$ and $\dot{\sigma}t_c/s_c = 10^{-3}$.

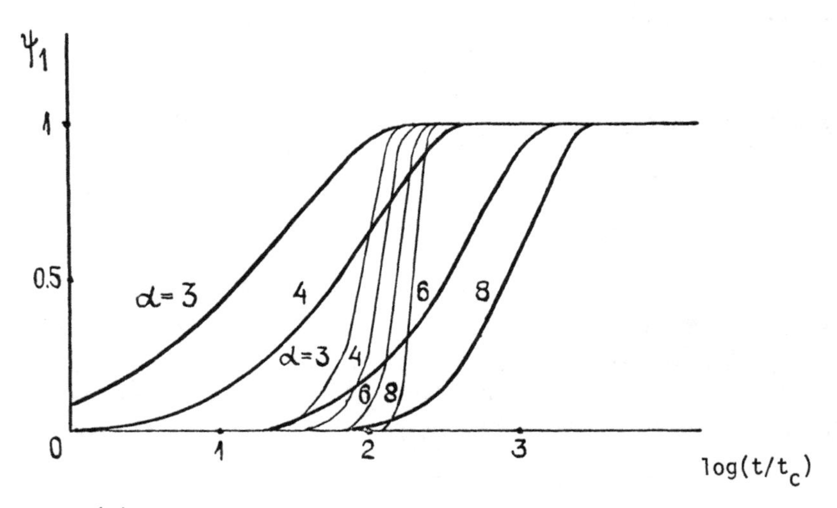

Fig. (2) - The first damage rate versus time: broad lines - under constant loading; narrow lines - under loading with constant velocity

The diagrams for the quantity $L = \log(I/I_0)$ are plotted in Figure 3 where $I \sim \sigma^2\psi_1$, and I_0 is a constant. The value L characterizes the logarithmic level of the elastic energy release per time unit due to the stream of microcracks. This value may be interpreted in the first approximation as the level of acoustic emission in standard non-destructive tests [6].

If the conditions $\sigma<\sigma_Y$ does not hold during the loading process, the second damage rate ψ_2 becomes significant. The diagrams for two damage rates are plotted in Figure 4. Numerical data $n=\alpha=4$, $\lambda_e/r = 10^2$, $\sigma_Y/s_c = 0{,}2$ are assumed. Values of $\dot{\sigma}t_c/s_c$ for curves 1, 2 and 3 are equal to 10^{-3}, 10^{-4} and 10^{-5}, respectively.

Although the diagrams in Figures 2 to 4 are presented for values of ψ_1 and ψ_2 close to unity, the suggested model becomes invalid at relatively low damage level. In addition, the diagrams in Figure 4 do not take into account energy released by the cracks.

FAILURE DUE TO LOSS OF INTEGRITY

For the elastically deformed matrix $\psi_2 = 0$, $\psi_1 = \psi_3$, the damage rate ψ_3 is assumed to be responsible for the failure resulting if the loss of integrity. The failure condition is

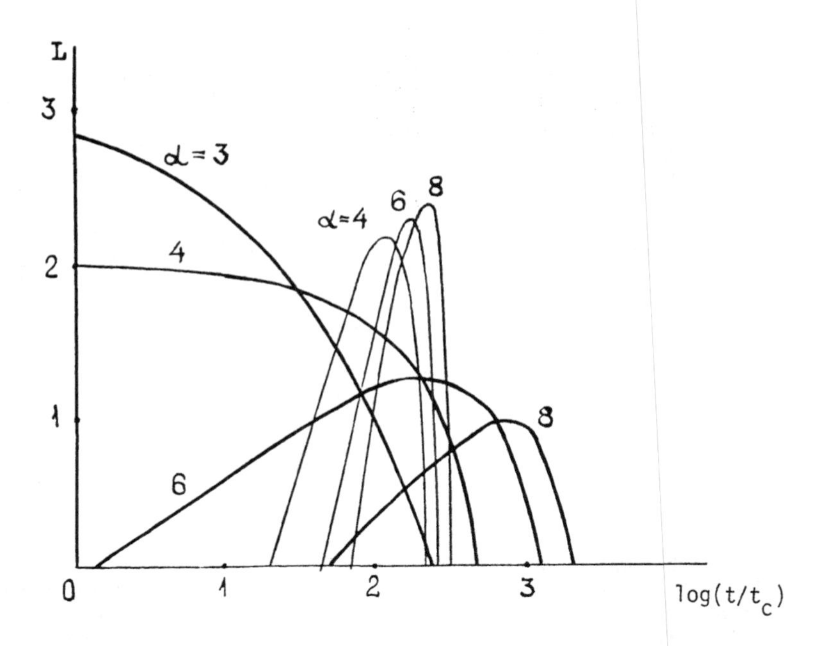

Fig. (3) - Logarithmic level of released energy versus time: broad lines - under constant loading; narrow lines - under loading with constant velocity

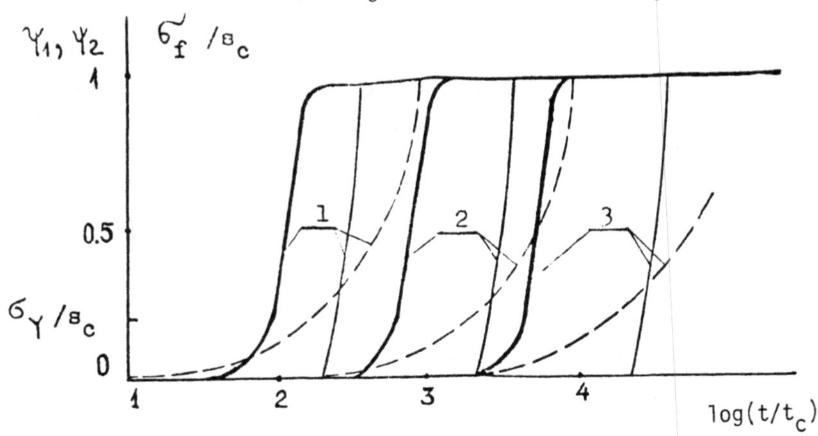

Fig. (4) - Two damage rates under loading with constant velocity versus time; broken lines - stresses growth

$$\psi_3 = \psi^* \tag{12}$$

The critical value ψ^* depends on the composite properties and, generally on the stress level under consideration.

The upper bound $\psi^* = 1$ corresponds either to the breaking of all microstructure elements or complete delamination of all fibers. A more realistic estimate can be obtained by assuming that the loss of integrity occurs if almost each broken element has at least one broken neighbor. If this coupling of elementary cracks takes place throughout the specimen, a number of new vacant neighbors occu and the probability of more microcracks to emerge becomes rather near to unity.

Let n^* be the total number of neighbors. If the transfer length of fiber is $\lambda = \lambda_e \psi_3/\psi_1$, the probability of coupling for almost each broken element is

$$1 - (1-\psi_1)^{n^*\psi_3/\psi_1} \sim n^*\psi_3$$

Hence,

$$\psi^* \sim (n^*)^{-1} \qquad (13)$$

In the case of a hexagonal array, the number of side neighbors $n^* = 6$ and the total number of all neighbors, including skew located ones, $n^* = 20$. Experimenta data show that the loss of integrity of most unidirectional fiber composites occurs when the volume fraction of the broken elements reaches approximately 10%. The more detailed analysis requires the consideration of different variants of elementary cracks stress concentration and increasing of debonded fiber length. The growth of nominal stresses due to damage is also significant as far as the loss of integrity is concerned.

Let the criterion in equation (12) be applied to estimate the ultimate stress σ^* for the special case considered in the previous section. For short-time loading $t^* \sim t_c$, the damage rate ψ_3 will be obtained by replacing λ_e in equation (10) by λ. By an order of magnitude where $1-e^{-1} = 0.632 \text{ --- } \sim 1$, equation (12) becomes

$$\frac{\lambda}{r} \left(\frac{\sigma}{s_c}\right)^{\alpha} \sim \psi^*$$

The ultimate stress is

$$\sigma^* \sim \left\{ s_c(\psi^*)^{1/\alpha} \left(\frac{2G_m}{E_f}\right)^{1/2\alpha}, \; s_c(\psi^*)^{1/(\alpha+1)} \left(\frac{2\tau_m}{s_c}\right)^{1/(\alpha+1)} \right\} \qquad (14)$$

where the first term in the braces corresponds to the case of elastic matrix, and is similar to the Rosen's estimate [3]. The second term corresponds to the case of non-elastic matrix. As α is usually large in comparison with unity, ψ^* has a weak influence on the ultimate stress. Equation (14) shows very clearly how the characteristic fiber strength, its variance, and the elastic and inelastic properties of matrix affect the ultimate stress of the composite.

NUCLEATION OF MACROCRACKS

In this paper, a macrocrack nucleus is considered as a bundle of n_* broken fibers, and the transverse size of a nucleus is thus $l_* \sim rn_*^{1/2}$. In the case of a hexagonal array, $n_* = 7$ corresponds to an internal crack. Macrocrack nucleation occurs when at least one broken element will be found in the volume V with n_*-1 broken neighbors. Hence, the probability distribution function for the time-to-nucleation t_* is

$$F_*(t_*) = 1 - [1 - F_{\tau^*}^{n_*-1}(t_*;\lambda)]^{\psi_1(t_*)N} \tag{15}$$

Here $F_{\tau^*}(t;\lambda)$ is the probability of a secondary break in the time interval [0,t]. Strictly speaking, various sequences of secondary breaks should be considered, as it is done for example in [4,7]. After each break, the debonded fiber length 2λ varies making the computations somewhat cumbersome. It is assumed in equation (15) that all secondary breaks occur simultaneously, and 2λ is an averaged debonded fiber length. Putting into equation (15) $N = V/V_r$, $F_{\tau^*}(t;\lambda) \approx \psi_3(t)$, and assuming that $\psi_1 N >> 1$ with $\psi_3 << 1$, the asymptotic distribution of time-to-nucleation is

$$F_*(t_*) \sim 1 - \exp[- \frac{V}{V_r} \psi_1(t_*)\psi_3(t_*)^{n_*-1}] \tag{16}$$

For those composites in which the first nucleus occurs only on the surface of the specimen [5], it is necessary to replace V by S and V_r by S_r in equation (16), where S and S_r are squares of the specimens and microstructure element's surface, respectively. The number n_* is assumed to be 3 or 4.

Contrary to the failure due to the loss of integrity, the macrocrack nucleation process is highly scale-sensitive. In order to study the scale effect in more detail, consider again the special case mentioned earlier. Let σ = const, and the matrix be elastic up to failure. Using equations (10) and (16) yield

$$F_*(t_*) \sim 1 - \exp[- \frac{V}{V_r} (\frac{\lambda_e}{r})^{n_*} (\frac{\sigma}{s_c})^{\alpha n_*} (\frac{t_*}{t_c})^{\beta n_*}] \tag{17}$$

The effective concentration factor $\kappa^{\alpha(n_*-1)}$ can be included in equation (17) as a multiplier in the brackets.

The ratio of mathematical expectations of the times-to-nucleation for equal nominal stress level σ and for various specimen volumes V_1 and V_2 is

$$\frac{E[t_{*,1}]}{E[t_{*,2}]} = (\frac{V_2}{V_1})^{1/\beta n_*} \tag{18}$$

Similarly, the ratio of mathematical expectations of the stresses $\sigma_*^{\prime}$ for generating macrocrack nucleation for the same time is interval

$$\frac{E[\sigma_{*,1}]}{E[\sigma_{*,2}]} = \left(\frac{V_2}{V_1}\right)^{1/\alpha n_*} \tag{19}$$

The exponents α and β in equations (18) and (19) of the fibers strength scale effect are multiplied by the number n_* of fibers forming a macrocrack nucleus. It reflects the well-known fact that fiber composite materials are much more sensitive to the scale effect than the corresponding fibers. Using equations (18) and (19), the number n_* can be evaluated from the experimental data.

MACROCRACK PROPAGATION

The macrocracks will be taken as three-dimensional domains accounting for fiber delamination and breakage of the randomly situated fibers that develop into a macrocrack. Consider, for example, the analogy of a penny-shape crack in linear fracture mechanics. The internal crack in the composite has a brush-like structure, occupying the domain visualized schematically as a circular cylinder with a radius ℓ and length 2λ as shown in Figure 5.

$$\lambda \sim \{(\ell r)^{1/2} \left(\frac{E_f}{2G_m}\right)^{1/2}, \ \ell \frac{\sigma}{2\tau_m}\} \tag{20}$$

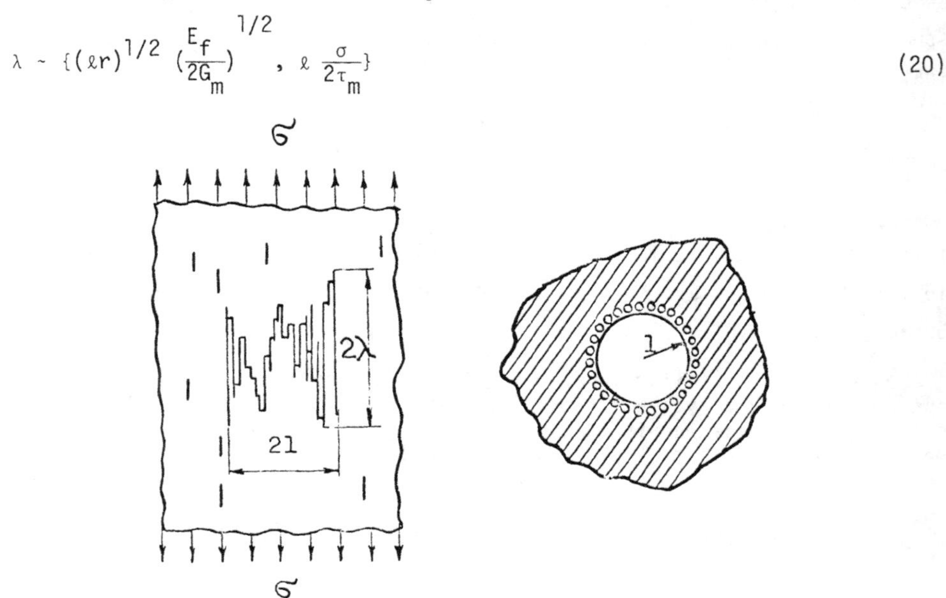

Fig. (5) - The suggested model of the internal macrocrack in unidirectional fiber composites

Equation (20) coincides with equation (3) when $\ell \sim r$ and can be obtained in a similar manner, i.e., using the simplest self-consistent field approach. The transfer from the first term in equation (20) to the second one takes place at $\sigma = \sigma_Y(\ell)$, where $\sigma_Y(\ell)$ is determined by equating the two terms in equation (20) as $\sigma_Y(r) = \sigma_Y$.

Let t be the moment when the crack is of size ℓ. If the number of fibers in the crack front is sufficiently large, this set of fibers may be sufficiently well represented as a sample taken from an ergodic general ensemble. In fact, the properties of the fibers are well mixed. Denoting by $\Delta\tau$ the life of randomly chosen fiber on the crack front, the semi-stochastic estimate for the average velocity of the crack propagation is obtained:

$$\frac{d\ell}{dt} \sim \frac{2r}{E[\Delta\tau]} \tag{21}$$

Let $\Delta\tau << t_{**} - t_{*}$, and the right-hand side of equation (6) be independent of ϕ. This gives

$$\Delta\tau(t) \approx \frac{1 - \frac{\lambda(t)}{\lambda_e} \int_0^t f[\sigma(\tau), s(\lambda_e)]d\tau}{f[\kappa\sigma(t), s(\lambda)]} \tag{22}$$

where κ is the effective concentration factor. Equation (22) is valid until the dτ nominator becomes positive. Otherwise, $\Delta\tau(t) \equiv 0$. The integral in equation (22) is equal to the damage rate accumulated in the microstructure element before it has reached the crack front. Stress concentration at the next row of fibers is not taken into account. If this damage is neglected, the averaging procedure in equation (22) may be performed by means of the distribution function $F_s(s;\lambda)$. It follows that

$$\frac{d\ell}{dt} \sim 2r \left[\int_0^\infty \frac{p_s(s;\lambda)ds}{f(\kappa\sigma, s)}\right]^{-1} \tag{23}$$

where $p_s(s;\lambda)$ is the corresponding density of probability.

For the special case given by equations (8) and (9), it is found by using equation (20) that

$$\frac{d\ell}{dt} \sim \left\{\frac{2r}{t_c} \left(\frac{\kappa\sigma}{s_c}\right)^n \left(\frac{\ell}{r}\right)^{1/2\beta} \left(\frac{E_f}{2G_m}\right)^{1/2\beta}, \frac{2r}{t_c} \left(\frac{\kappa\sigma}{s_c}\right)^n \left(\frac{\ell}{r}\right)^{1/\beta} \left(\frac{\sigma}{2\tau_n}\right)^{1/\beta}\right\} \tag{24}$$

where $\Gamma(1 + 1/\beta) \sim 1$ is assumed. Equation (24) has the same structure as the Paris-Erdogan equation in linear fracture mechanics. Obviously, equation (24) coincides with the Paris-Erdogan equation only in the case $\alpha=1$, and this condition holds both for the elastic and non-elastic matrices. If this condition is fulfilled, the exponent of the strength intensity factor is equal to n if the matrix is elastic, and to 2n if the matrix is non-elastic. Here, n is the power exponent of the life-stress curves for the fibers.

The case $\alpha=1$ corresponds to the exponential distribution of fiber strength. It seems that the value $\alpha=1$ is not arbitrary: when $\alpha=1$, mechanical properties of a composite are most closely related to those of conventional materials. It will be shown later that the value $\alpha=1$ also applies to other conditions where linear fracture mechanics can be applied. Unfortunately, the exponent α is equal to 5 and more for fibers used in composite technology. Therefore, the case $\alpha=1$ is only of academic interest.

CONDITIONS OF MACROCRACKS INSTABILITY

The critical crack size ℓ_{**} as a function of ℓ will be determined when most of the fibers at the crack front with the probability close to unity is broken. Consider an internal crack whose size is small in comparison with those in the specimen of Figure 5. Neglecting premediate damage of the fibers, the instability condition in the form $\kappa\sigma > s(\lambda)$ is assumed. The effective concentration factor κ generally, depends on the size ℓ. The value ℓ_{**} is obtained from the equation $F_s(\kappa\sigma;\lambda)$, where λ is evaluated by equation (20).

For the special case governed by equations (8) and (9), it follows that

$$\frac{\ell_{**}}{r} \sim \{(\frac{s_c}{\kappa\sigma})^{2\alpha} \frac{2G_n}{E_f}, (\frac{s_c}{\kappa\sigma})^{\alpha} \frac{2\tau_m}{\sigma}\} \qquad (25)$$

The crack is stable until $\ell < \ell_{**}(\sigma)$, where σ is a current nominal stress. The Griffith-Irwin criterion $\sigma^2\ell$ = const follows from equation (25) only if $\alpha=1$ both for the elastic and non-elastic matrices.

CONCLUSION

In this paper, it has been shown within the framework of the suggested stochastic model that a very strong assumption must be made in order to recover the Griffith-Irwin criterion and the Paris-Erdogan equation. The exponential probability distribution of the fiber strength is required both in elastic and non-elastic cases, as well as both for the Griffith-Irwin criterion and for the Paris-Erdogan equation. This result is somewhat unexpected and nevertheless may have a profound meaning. As the strength distribution of fibers used in the composite technology differs strongly from the exponential distribution, the prospects of applying linear fracture mechanics to unidirectional fiber composites seem to be pessimistic.

Finally, the conditions governing the type of failure will be discussed, i.e. the loss of integrity or crack propagation until instability. The larger is the size of the specimen (or a total number of microstructure elements), the higher is the probability that nucleation of one or several macrocracks takes place. It follows from equations (12) and (16) that when

$$V << V_r(\psi^*)^{-n_*}$$

the loss of integrity is more probable. For example, assume a specimen reproduce a microstructure element in the ratio 100:1, and that a nucleus can appear at each point of the volume V. Then at $\psi_1 \sim \psi_3$, $\psi^* \sim 10^{-1}$ and $n_* = 7$, the loss of integrity is awaited. For the larger specimens, the macrocrack initiation becomes more probable.

As far as crack instability is concerned, the situation is more complicated. It is illustrated in Figure 6 where sketches of two probability distribution curves are plotted. The left curve corresponds to time-to-nucleation, t_*, and

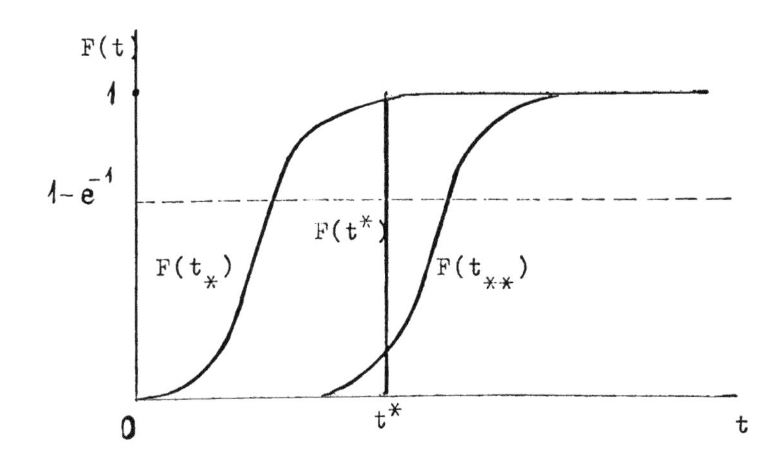

Fig. (6) - Probability distribution functions: t_* is the time to the first crack initiation, t^* is the time to failure due to the loss of integrity, and t_{**} is the time to failure due to macrocrack instability

the right curve corresponds to the time-to-crack instability, t_{**}. Roughly speaking, in the case of constant nominal stresses, the latter curve is shifted against the first one, and the shift distance is equal to the time required for the crack growth until the size ℓ_{**}. Opposite to crack initiation, the loss of integrity is an almost deterministic event. The corresponding distribution function for the time t^* is presented in Figure 6 by the step-wise line. It is evident that the loss of integrity can occur even after one or several macrocracks have been initiated in the specimen.

REFERENCES

[1] Bolotin, V. V., "Stochastic Models of Cumulative Damage in Composite Materials", Engineering Fracture Mechanics, Vol. 8, pp. 103-113, 1976.

[2] Kelly, A., Strong Solids, Clarendon Press, Oxford, 1973.

[3] Rosen, B., "Mechanics of Reinforcement of Composites", Fiber Composite Materials, ASM, Metals Park, Ohio, pp. 37-76, 1965.

[4] Tamuzs, V. P., "Dispersed Fracture of Unidirectional Composite", Fracture of Composite Materials, Sijthoff and Noordhoff, Alphen aan den Rijn, pp. 13-24, 1979.

[5] Bolotin, V. V., Statistical Methods in Structural Mechanics, Holden-Day, Inc., San Francisco, 1969.

[6] Latishenko, V. A. and Matiss, I. G., "Methods and Means for Non-Destructive Study of the Damageability of Composite Materials", Fracture of Composite Materials, Sijthoff and Noordhoff, Alphen aan den Rijn, pp. 321-328, 1979.

[7] Argon, A., "Statistical Aspects of Fracture", Composite Materials, Vol. 5, Fracture and Fatigue, Academic Press, New York and London, 1974.

TIGHT BOUNDS FOR THE PROBABILITY DISTRIBUTION OF THE STRENGTH OF COMPOSITES*

D. G. Harlow

Drexel University
Philadelphia, Pennsylvania 19104

and

S. L. Phoenix

Cornell University
Ithaca, New York 14853

ABSTRACT

Using the chain-of-bundles probability model as the basic mathematical model for composite strength, a sequence of tight but conservative bounds for the probability distribution for the strength of composites is considered. This sequence of bounds numerically converges rapidly to a limiting distribution function from which the probability distribution for composite strength can be obtained easily via an application of the "weakest link rule". A detailed example is given to illustrate these principles and additional conclusions.

INTRODUCTION

In previous studies [1-4], we have considered the commonly accepted chain-of-bundles probability model for the statistical strength of composite materials. The favorable features of this micro-mechanical model are that it incorporates variability in fiber strength, local load redistribution on fibers immediately adjacent to broken fibers, and propagating sequences of fiber fractures which form catastrophic cracks. In our first study [1], we reviewed past statistical analyses, paying particular attention to the nature of load redistribution around failed fibers. We discussed some fundamental analytical differences between _equal load sharing_ for loose bundles, and _local load sharing_ in composite materials; the latter load sharing being a consequence of the binding matrix. Except for some approximate analyses, the chain-of-bundles model under local load sharing had proven, for the most part, to be intractable.

In the sequel [2], we obtained some precise numerical results for a version of the model under a Weibull distribution for fiber strength, but, because of

*This research was supported by the United States Department of Energy under contract DE-AS02-76-ER04027.

the computational complexity, we were limited to composites with fewer than 10 fibers. Nevertheless, a striking convergence was observed in the results, and this led to an important conjecture about the statistical behavior of composite strength. One consequence of the conjecture was that under a Weibull distribution for the strength of single fibers, a composite specimen has strength which approximately follows another Weibull distribution; the shape and scale parameters for this latter Weibull distribution were in agreement with those observed experimentally, and were different from those assumed for the individual fibers. While the numerical results were quite convincing, proof of the conjecture was believed not to be imminent.

In a third paper [3], we turned our attention to obtaining conservative analytical bounds on the probability distribution for composite strength for application to composites of any size. The bounds were based on the occurrence of two or more adjacent broken fibers in the composite, a necessary though not sufficient condition for composite failure. Although the bounds approached a Weibull distribution as composite size increased, they were judged to be rather loose because of the use of certain inequalities in their derivation, and they became even more conservative as composite size or variability in fiber strength increased. Nevertheless, the results supported the earlier conjecture [2] on the approximate Weibull behavior of composite strength. But the major innovation in that study was the use of a recursive technique for generating probabilities for larger bundles from those of smaller bundles, thus permitting analysis for any number of fibers in the composite.

In the fourth paper [4], the recursive technique underwent further development. Again, bounds were obtained based on the probability of occurrence of two or more adjacent failed fibers in the composite, but for such events the new results were exact. When the variability in fiber strength was low, or the size of the composite was small, the bounding distributions were found to be virtually identical to the conjectured exact distributions; otherwise, the bounds became increasingly conservative with increasing size or variability. Nevertheless, for large composite materials, each bound approached a Weibull distribution, again in support of the earlier conjecture [2].

In this study, we advance further by developing distributions based on the occurrence of k or more adjacent broken fibers in the composite for $k = 1,2,3,..$ Analytically, we will find that for each k, a characteristic distribution function $W^{[k]}(x)$ arises, and as k increases, these distributions converge extremely rapidly to a limiting distribution function $W(x)$. Furthermore, this limit $W(x)$ will appear to be numerically identical to the limiting distribution arising in our earlier conjecture [2]; its importance is that the probability distribution for the strength of a composite can be calculated from it by a simple application of the "weakest link rule". Graphs of this limiting distribution $W(x)$ are provided for this purpose for cases of practical interest.

There is little doubt that the sequence of bounding distribution functions obtained in this paper converges in an appropriate sense to the true distribution function for composite strength. Nonetheless, certain difficult mathematical questions remain having to do with the convergence of certain boundary and error terms as composite size increases. However, numerical results reveal regular and rapidly convergent behavior for these terms.

The most important practical conclusions of this study are as follows:

1. Composite materials essentially have a weakest link structure, that is, the strength distributions of composite specimens differing in volume will be related by the weakest link rule of probability theory. However, the strength distribution of a unit volume of the material is not that of its weakest fiber, but rather is governed by the interaction of the local load redistribution around fiber breaks and the underlying probability distribution for fiber strength.

2. The strength of composite materials approximately follows a Weibull distribution even when the distribution for single fibers exhibits marked deviations from Weibull behavior in the upper tail. For fibers one-centimeter long, the length often used in tension tests, it is the middle and lower tail of the fiber strength distribution which is important. Thus, the practice of fitting a Weibull distribution to fiber tension test data is supported.

3. The shape parameter for the (approximate) Weibull distribution for the strength of a composite is quite insensitive to the initial variability in fiber strength, and typically lies in the range from 20 to 30. Thus, the variability in composite strength likewise changes little. However, the mean for this Weibull distribution diminishes dramatically as the variability in fiber strength increases provided that the mean strength of a certain characteristic length in the material is held fixed.

SUMMARY OF THE MODEL

We begin with a breif review of the basic features and notation of the model. More elaborate descriptions appear in [1-4].

The chain-of-bundles model. The composite material is a parallel structure of n fibers in the form of a tape of length ℓ. The structure is partitioned into a series of m short sections or bundles of length δ, so that $\ell = m\delta$. Thus, the fibrous material is a chain of m bundles, each containing n fiber elements of length δ.

A total load $\mathcal{L}$ is applied to the composite material, but to compare results for composites of different sizes, we obtain results in terms of the applied load $L = \mathcal{L}/n$. Thus, L is the nominal load per fiber in the cross-section of the composite. The strength will be the maximum applied load L that the structure supports, and since the bundles are arranged in series, this will be the strength of the weakest bundle (again on a per fiber basis).

The mn fiber segments in the composite are assumed to have independent and identically distributed strengths with common distribution function $F_\delta(x)$, $x \geq 0$. Many of our results will hold for quite general forms of $F_\delta(x)$; however, to gain insight in applications, we will consider a Weibull distribution for fiber element strength of the form.

$$F_\delta(x) = 1 - \exp\{-\delta(x/x_0)^\rho\}, \quad x \geq 0 \tag{1}$$

where $\rho > 0$ is the shape parameter, $x_0 > 0$ is the scale parameter associated with unit fiber length ($\delta=1$), and δ is the actual fiber element length. It is more convenient to write (1) as

$$F_\delta(x) = 1 - \exp\{-(x/x_\delta)^\rho\}, \quad x > 0 \tag{2}$$

where the positive constants x_δ and ρ are the usual Weibull scale and shape parameters respectively. From (1) and (2), we have

$$x_\delta = x_0 \; \delta^{-1/\rho} \tag{3}$$

so that x_δ depends on δ, and thus, is not a fiber parameter external to the composite material.

Within each bundle, the nonfailed fiber elements share the load according to a local load sharing rule. If the applied load is L, a surviving fiber element carries load $K_r L$ where

$$K_r = 1 + r/2, \quad r = 1,2,\ldots, \tag{4}$$

and r is the number of consecutive failed fiber elements immediately adjacent to the surviving element (counting on both sides). Meanwhile, a failed fiber element carries no load. We call the K_r load concentration factors.

The bundles that we consider are referred to as linear incomplete bundles; they are linear because the fiber elements form a linear array, and they are incomplete because the above load sharing rule is inconsistent with the total load summing to $\mathcal{L} = nL$. The latter difficulty clearly arises when k consecutive broken fiber elements occur at a bundle edge, and no fiber element is available on the outside to support the shed load (k/2)L. This boundary irregularity is easily circumvented by adjusting the load sharing rule at the bundle boundaries, say, by returning this shed load (k/2)L to the nearest surviving interior fiber element. Then, the total load would be $\mathcal{L} = nL$, and we would have what are to be referred to as complete bundles. However, in this paper we work with incomplete bundles because the recursive analysis, whereby survival probabilities for larger bundles are expressed in terms of those for smaller bundles, would be unduly cluttered by such boundary adjustments. Of course, the behavior of complete bundles and the effects of various boundary conditions are ultimately of interest especially when the number of fibers n in the composite is small.

The basic probability distributions. We let $G_n(x)$, $x \geq 0$ be the distribution function for the strength of a single bundle, and note that $G_n(\cdot)$ is determined totally by $F_\delta(\cdot)$, n and the constants K_r of the local load sharing rule. Historically, the determination of $G_n(x)$ has been the difficult task.

We let $H_{m,n}(x)$, $x \geq 0$ be the distribution function for the strength of the composite material. Because the composite is a series arrangement of statistically and structurally independent bundles, and its strength is that of its weakest bundle, we easily have

$$H_{m,n}(x) = 1 - [1 - G_n(x)]^m, \quad x \geq 0 \tag{5}$$

by the weakest link rule. In applications, m will be large ($10^4 \leq m \leq 10^7$) so that $G_n(x)$ must be known accurately in its extreme lower tail for (5) to apply.

Description of the bounding distributions. We consider a single bundle, and rather than focus on the event of total failure of the bundle, we focus on the event that k or more broken elements occur in a sequence somewhere in the bundle of n fiber elements, where $k \leq n$. The notion is that such failure sequences form cracks transverse to the composite axis, and the longer the crack, the less likely such a crack will remain stable when the load is further increased a small amount.

Indeed, for brittle materials various laws in fracture mechanics relate applied load and maximum permissible crack length; cracks exceeding this length grow catastrophically and fracture the material. Furthermore, the maximum stable crack size diminishes as the applied load increases.

Failure events and important distribution functions for a single bundle, we define the event $A_n^{[k]}(x)$ as

$$A_n^{[k]}(x) = \{\text{k or more broken fiber elements are adjacent somewhere in the bundle when the applied load is } L = x\}, \tag{6}$$

for k = 1,2,3,... The complementary event $\bar{A}_n^{[k]}(x)$ is, of course,

$$\bar{A}_n^{[k]}(x) = \{\text{no sequence of broken fiber elements consists of k or more such elements under the applied load } L = x\}. \tag{7}$$

The key quantities to be computed are

$$G_n^{[k]}(x) = \Pr\{A_n^{[k]}(x)\}, \quad x \geq 0 \tag{8}$$

for k = 1,2,..., however, the labor will be in computing

$$Q_n^{[k]}(x) = \Pr\{\bar{A}_n^{[k]}(x)\} = 1 - G_n^{[k]}(x), \quad x \geq 0 \tag{9}$$

which is the probability that no sequence of k broken fiber elements will occur anywhere in the bundle under load x. Clearly, $G_n^{[k]}(x)$, $x \geq 0$ will be a proper distribution function if $F_\delta(x)$ is, but we want to relate $G_n^{[k]}(x)$ to the distribution function $G_n(x)$ for bundle failure.

Now for $k < n$, the event $A_n^{[k]}(x)$ is necessary but not sufficient for bundle failure (a crack is necessary but not sufficient for fracture). Furthermore, the event $A_n^{[k]}(x)$ is at least as likely as event of failure, and since, the event $A_n^{[k_1]}(x)$ contains the event $A_n^{[k_2]}(x)$ if $0 < k_1 < k_2 \leq n$, then

$$G_n(x) = G_n^{[n]}(x) \leq G_n^{[n-1]}(x) \leq \dots \leq G_n^{[2]}(x) \leq G_n^{[1]}(x), \quad x \leq 0 \tag{10}$$

Now, the behavior of certain of the above quantities is known from past studies. $G_n^{[1]}(x)$ is just the distribution function for the strength of the weakest fiber element in the bundle, and by the weakest link rule, we compute this most trivial bound as

$$G_n^{[1]}(x) = 1 - [1 - F_\delta(x)]^n, \quad x \geqq 0 \tag{11}$$

Notice that none of the load sharing constants K_r are involved in (11), and consequently, $G_n^{[1]}(x)$ cannot yield anything but a crude bound on $G_n(x)$. In [4], we studied $G_n^{[k]}(x)$ for k = 2. We found that $G_n^{[2]}(x)$ was seemingly a good estimate of $G_n(x)$ in certain cases. On the other hand, the behavior of $G_n(x)$ has been studied in [2] for circular bundles, but only for $n \leqq 9$ fiber elements. The thrust here is to expand our understanding by studying $G_n^{[k]}(x)$ for arbitrary but fixed k and any n. Ultimately, the goal is to uncover totally the behavior of $G_n(x) = G_n^{[n]}(x)$ for increasing n. This task must be deferred to a future publication. Here, we will be satisfied with an understanding of $G_n^{[k]}(x)$, and with answer to the question "How large must k be for $G_n^{[k]}(x)$ to be virtually the same as $G_n(x)$ in applications?"

Once $G_n^{[k]}$ is obtained, we easily generate from (5) the sequence of bounding distribution functions

$$H_{m,n}^{[k]}(x) = 1 - [1 - G_n^{[k]}(x)]^m, \quad x \geqq 0 \tag{12}$$

for k = 1,2,... . Indeed, we have the inequalities

$$H_{m,n}(x) = H_{m,n}^{[n]}(x) \leqq H_{m,n}^{[n-1]}(x) \leqq \cdots \leqq H_{m,n}^{[1]}(x) \tag{13}$$

for bounding the distribution function $H_{m,n}(x)$ for the strength of the composite material. Ultimately, it is the behavior of these bounds which is of interest.

DISCUSSION OF RESULTS

In [2], we considered the transformation

$$W_n(x) = 1 - [1 - G_n(x)]^{1/n}, \quad x \geq 0 \tag{14}$$

which amounts to a weakest link scaling in reverse of $G_n(x)$ back to single fiber size. The results of a careful numerical study led to the conjecture that $W_n(x)$ converges to some limiting distribution function $W(x)$, $x \geq 0$ rapidly as $n \to \infty$, and that the true distribution function for composite strength $H_{m,n}(x)$ is approximated extremely accurately as follows:

$$H_{m,n}(x) = 1 - [1 - W(x)^{mn}, \quad x \geq 0 \tag{15}$$

On the other hand, we have shown in [4,5] that the bounding distribution $H_{m,n}^{[k]}(x)$ has the form

$$H_{m,n}^{[k]}(x) = 1 - [1 - W^{[k]}(x)]^{mn} [\pi^{[k]}(x) + o_n^{[k]}(x)]^m, \quad x \geq 0 \tag{16}$$

where i) $W^{[k]}(x)$ is a characteristic distribution function that satisfies $W^{[k]}(x) \to W(x)$ as $k \to \infty$ for some limiting distribution function $W(x)$, $x \geq 0$, ii) $\pi^{[k]}(x) \to 1$ as $x \to 0$ and is typically near unity, and iii) $o_n^{[k]}(x) \to 0$ extremely rapidly as $n \to \infty$. Indeed even for relatively small values of n, $H_{m,n}^{[k]}(x)$ is indistinguishable from

$$\mathcal{H}_{m,n}^{[k]}(x) = 1 - [1 - W^{[k]}(x)]^{mn}, \quad x \geq 0 \tag{17}$$

This is a useful result in itself, and as we increase k, it supports the validity of the above conjecture. In [5], the basis of a mathematical proof is given for this result, but a few details still remain for the proof to be complete. This limiting distribution function $W(x)$, $x \geq 0$ evidently cannot be expressed in terms of the usual classical functions in any simple or straightforward way. However, we have an algorithm for its computation to any desired level of accuracy and supply graphs for most cases of practical importance.

From (13), we know that $H_{m,n}^{[k]}(x)$ and $\mathcal{H}_{m,n}^{[k]}(x)$ will serve as upper bounds on the distribution function $H_{m,n}(x)$ for the strength of the composite. But the ultimate question is "How close is $H_{m,n}(x)$ to the "limit" distribution function.

$$\mathcal{H}_{m,n}(x) = 1 - [1 - W(x)]^{mn}, \quad x \geq 0$$

where $W(x)$, $x \geq$ is the limiting distribution function referred to above?" An answer to this question is extremely important because the above formula is so easy to use, being of the simple weakest link form. Moreover, in all the cases that we have studied, $W(x)$ here is <u>numerically identical</u> to the limit $W(x)$ of the conjecture of Harlow and Phoenix [2] to within the accuracy of our computations. Thus, if $H_{m,n}(x)$ converges to $\mathcal{H}_{m,n}(x)$ in some appropriate sense as the number of fibers n increases, the conjecture would finally be proven.

From a practical point of view, the important consequences of the analytical results are as follows:

1. $\mathcal{H}_{m,n}^{[k]}(x) = 1 - [1 - W^{[k]}(x)]^{mn}$ becomes an upper bound on $H_{m,n}(x)$ as n increases, and in typical cases, this bound will appear to tighten extremely rapidly as k increases.

2. $\mathcal{H}_{m,n}(x) = 1 - [1 - W(x)]^{mn}$ is an accurate representation of $H_{m,n}(x)$ in typical cases when n is not small, say greater than eight.

The numerical results were obtained under the Weibull distribution (2) for the strength of fiber elements. All graphs use Weibull coordinates since these are most convenient for use with the weakest link rule. (For a discussion of the advantages of Weibull coordinates, the reader is referred to [4]).

<u>Convergence of $W^{[k]}(x)$ to $W(x)$ as k increases</u>. In Figure 1, we have plotted $W^{[k]}(x)$ for $1 \leq k \leq 10$, and for the typical value $\rho = 5$ of the Weibull shape parameter. The convergence of $W^{[k]}(x)$ to $W(x)$ appears to be complete at $k = 10$ for $x > 0.17\ x_\delta$ and for probabilities above 10^{-14}; thus, the result for $W(x)$ is useable in the predictive formula $H_{m,n}(x) = 1 - [1 - W(x)]^{mn}$ when the number of fiber elements in the composite is as large as 10^{13}.

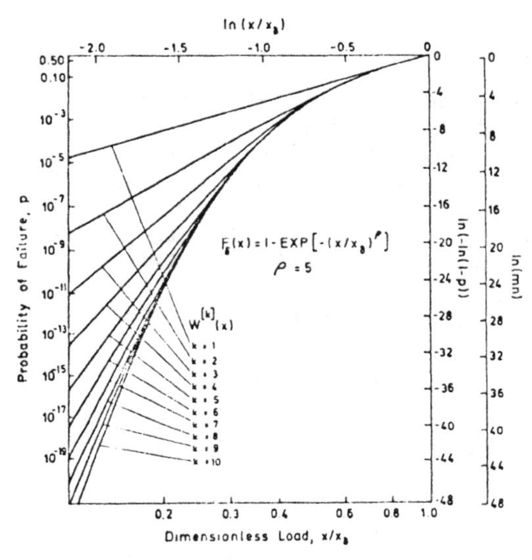

Fig. (1) - Convergence of the characteristic distribution function $W^{[k]}(x)$ to its limit $W(x)$ as k increases

Fig. (2) - Comparison of $W^{[6]}(x)$ for a wide range of the Weibull shape parameter ρ. The divergence point is approximately $x/x_\delta = 1/k_6 = 0.25$ for all ρ

Also in Figure 1, we notice that the upper tails of the $W^{[k]}(x)$ are all tangent to $W^{[1]}(x) = F_\delta(x) = 1 - \exp\{-(x/x_\delta)^\rho\}$, the Weibull distribution for the strength of single fiber elements. The interpretation is that at loads x above x_δ/K_1, the composite will fail if one element fails, and adjacent elements are unlikely to withstand the overload K_1x. However, the lower tail of $W^{[k]}(x)$ is tangent to a Weibull distribution $F^{[k]}(x)$

$$F^{[k]}(x) = 1 - \exp\{-(x/x_\delta^{[k]})^{k\rho}\}, \quad x \geq 0 \tag{18}$$

where $k\rho$ is now the shape parameter, and $x_\delta^{[k]}$ is a scale parameter. In [4], we gave the expression for $x_\delta^{[k]}$ for k = 2, and Smith in a recent Ph.D. thesis [6], has obtained $x_\delta^{[k]}$ for higher k using a different mathematical technique Indeed, Smith shows that $H_{m,n}^{[k]}(x)$ behaves as $F^{[k]}(x[mn]^{1/(k\rho)})$ as n grows large, a result which extends our Result 4 of [4]. Our graphical results would indicate that for a given value of k, this asymptotic Weibull distribution is reasonably accurate only over a limited range of the load x spanning the critical value x_δ/K_k, since the graphical distance between W(x) and $F^{[k]}(x)$ is least there. Otherwise, the asymptotic Weibull distribution will be overly conservative.

In Figure 2, we have plotted $W^{[6]}(x)$ and the limit W(x) for a wide range of ρ, the Weibull shape parameter. For $\rho \geq 5$, $W^{[6]}(x)$ and W(x) are virtually identical when $x/x_\delta > 1/K_6 = 1/4$. But, for $\rho = 3$, which represents a large variability in fiber strength, the overload K_6x on a survivor must significantly exceed x_δ for failure to be almost certain. Consequently, the divergence point for x/x_δ is slightly higher than $1/K_6$. While the divergence points for a given k remain approximately fixed as ρ increases, the probability range over which $W^{[k]}(x)$ and W(x) are identical, increases by <u>orders of magnitude</u> as ρ increases. Thus, useable results for a wide range of probabilities become easier to compute as the variability in fiber strength diminishes.

<u>Behavior of $G_n^{[k]}(x)$ and $H_{m,n}^{[k]}(x)$ as n is increased</u>. As was explained earlier [2,4], the advantage of using Weibull coordinates is that the distribution function $W_n = 1 - [1 - W]^n$ is just the distribution function W. Translated vertically the actual distance ln(n) on the graph where ln is the natural logarithm. Thus, on Figure 3, $\mathcal{G}_n^{[k]}(x) = 1 - [1 - W^{[k]}(x)]^n$ is just $W^{[k]}(x)$ translated vertically ln(n), and a scale has been provided for this purpose. For $\rho \geq 3$, we have found that $\mathcal{G}_n^{[k]}(x)$ and $G_n^{[k]}(x)$ are graphically indistinguishable for $n \geq 4$ and $k \leq n$. The major consequence is that in applications $\mathcal{H}_{m,n}(x) = 1 - [1 - W(x)]^{mn}$ is virtually identical to $H_{m,n}(x)$, the true distribution function for composite strength. On Figure 2, $\mathcal{H}_{m,n}(x)$ is simply W(x) shifted vertically the amount ln(mn).

AN EXAMPLE

In Figure 3, we have plotted various distribution functions for a composite specimen with $mn = 10^6$ fiber elements. Results are shown for the Weibull distribution (2) with ρ = 5, 10, and 25. These parameter values are believed to be typical. The graphs for $H_{m,m}(x)$ were generated from graphs of W(x), some of which are shown in Figure 2, as described earlier; the operation is to shift the W(x) graphs vertically the actual distance ln 10^6.

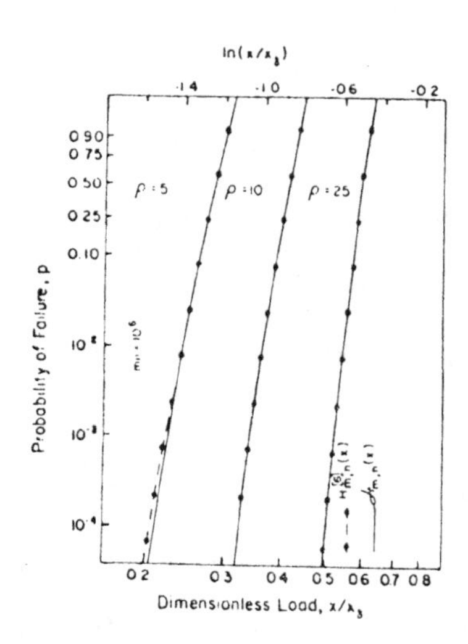

Fig. (3) - Distributions for the strength of a composite specimen of typical size

Now when mn = 10^6 and ρ = 5 for the regular Weibull distribution, $H_{m,n}^{[k]}(x)$ and $H_{m,n}(x)$ are virtually identical when k = 6 but not when k = 5. Thus, for this volume of composite and rather large variability in fiber strength, k = 6 is sufficient for applications. In the other two cases, k = 3 is sufficient when ρ = 10 and k = 2 is sufficient when ρ = 25. Clearly, the value of k which is sufficient for applications will increase as composite of volume or variability in fiber strength increases.

From Figure 3, we see that the curves are almost linear indicating that the strength of a composite approximately follows a Weibull distribution. The shape parameters for these (approximately) Weibull distributions are quite insensitive to the initial variability in fiber strength. When the shape parameter for the fiber strength is ρ = 5, 10, or 25 then the shape parameter for the composite strength is approximately ρ_c = 30, 30, or 50 respectively. Thus, variability in composite strength changes little, and it is significantly reduced.

This example with the volume of a typical laboratory specimen and the conclusions listed earlier, also indicate how Figure 2 can be used for different size specimens.

REFERENCES

[1] Harlow, D. G. and Phoenix, S. L., J. Comp. Mater., 12, 195-214, 1978.

[2] Harlow, D. G. and Phoenix, S. L., J. Comp. Mater., 12, 314-334, 1978.

[3] Harlow, D. G. and Phoenix, S. L., I. J. Fracture, 15, 321-336, 1979.

[4] Harlow, D. G. and Phoenix, S. L., I. J. Fracture, to appear 1980.

[5] Harlow, D. G. and Phoenix, S. L., I. J. Fracture, to appear 1981.

[6] Smith, R. L., "Limit Theorems for the Reliability of Series-Parallel Load-Sharing Systems", Ph.D. Thesis, Cornell University, May, 1979.

TWO APPROACHES IN FRACTURE MECHANICS OF COMPOSITES

S. T. Mileiko, F. Kh. Suleimanov and V. C. Khokhlov

Academy of Sciences of the USSR
Moscow, USSR

INTRODUCTION

The fracture mechanics of composites suggests two ways to calculate the limit of an element with a crack or similar defect. The first approach is based on micromechanical models of a composite material (see, for example [1]). Alternately, a homogeneous anisotropic media with some effective fracture properties can be considered in the framework of the macromechanical approach (for example [2]). These two approaches have their own advantages and disadvantages and ideally they should complement each other.

In this paper, two examples of fracture models are considered, one being of the micromechanical type and the other being of a macromechanical nature.

The first example is a macrocrack in an elastic composite with plane fibres containing regular defects spaced. It is easy to determine a load-dependent boundary of a region characterized by fibre multicracking in front of the macrocrack. A limit region of cracking can be found if a definite value of energy dissipation can be prescribed to each macrocrack. Such a model can give the qualitative information relating the dependence of fracture toughness to the main parameters of the composite structure.

The second example introduces a characteristic size as a fracture parameter of the composite, where its fracture toughness is determined by the damage accumulation within a region of stress concentration. The main assumption is that in a composite any crack oriented in a "tough" direction can be considered as an elliptical hole with a characteristic radius of curvature. Such an approach makes it possible to estimate a limit load of the elements with holes, cracks and other defects having just a moderate volume of experimental data.

MICROMECHANICAL MODELS

Micromechanical models aim at two goals. First, they are to be used to optimize a material structure because they give dependencies of effective fracture characteristics upon structural parameters of composites (namely, fibers and matrix properties, interface properties, component's volume fractions, fiber diameters, etc.). They also should be used to choose technological parameters of composites fabrication. Second, models of composite structures should be seen as a basis for phenomenological constitutive relations of composites. At the present time, however, micro- and macromodels are related only qualitatively. Micromodels can provide a way to construct phenomenological relationships but they do not give quantitative data. The problem of the calculation of elastic moduli is certainly the only exception, but even in this case, rigorous values of elastic moduli of a composite media are obtained only for regular distribution of fibers and defects [3,4].

Therefore, the main purpose in the consideration of micromodels remains the first one mentioned above. There are some examples of the effective use of micro models. Mention can be made of a model of failure of a unidirectional composite [1,5] which has been developed for practical use in the case of a metal matrix composite. The model provides insight into ways of decreasing the strength scatter and increasing the mean strength of a composite. An effective way appears to be the use of the matrix of a fibrous composite structure. Here the effect of a large increase in fracture toughness of a metal matrix - metal fiber composite for the case of small volume fractions of high strength/low ductility wires in a low strength/high ductility matrix produces the desired effect [1]. This results in a non-monotonic curve (curve I in Figure 1) and shows the dependence

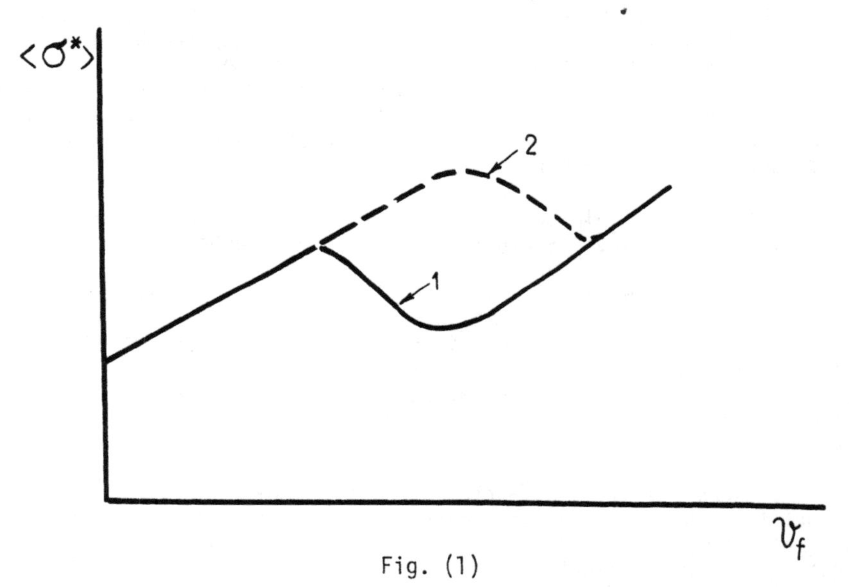

Fig. (1)

of the mean strength $\langle\sigma^*\rangle$ of the composite upon the volume fraction v_f of brittle fibers shifts to curve 2 [6]. The possibilities of controlling the composite strength appear to be large enough, as can be seen in Figure 2, which gives results for composite boron-steel-aluminum.

In [1], an experimental observation, the increase of fracture toughness K^* of a brittle fiber - metal matrix composite with the increase of fiber volume fraction was noted. A model of a composite with a macrocrack which can give $K^*(v_f)$ dependencies of the type observed in experiments [7] will now be discusse

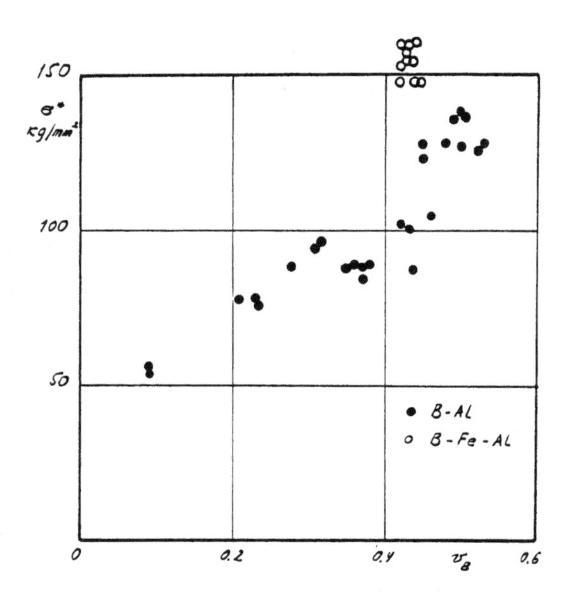

Fig. (2) - The dependence of strength of boron-aluminum and boron-steel-aluminum composites upon volume fraction of boron fibers. The volume fraction of steel wires is 6-7 percent

A. A Macrocrack in a Model Composite

A model should be developed which could be used to estimate the possible contribution of fiber multi-cracking in front of a crack to the fracture toughness of a composite. Irwin's relationship is assumed to be valid. Namely,

$$K_*^2 = CG \tag{1}$$

where G is the effective surface energy, and C is a constant with the dimension of stress depending on the elastic moduli of a composite. Then, let G be written as

$$G = G^0 + \Delta G(\sigma_0) \tag{2}$$

Here, G^0 is determined by linear summation of the values of G of the composite components, and $\Delta G(\sigma_0)$ is determined by the microcrack's density and depends on applied stress σ_0. Note that ΔG is analogous to Orowan's plastic correction to the surface energy by Griffith. It should likewise be noted that, generally, the value of ΔG can also be determined by other processes of local failure (delamination, pull out and so on).

Therefore, if $\Delta G(\sigma_0)$ can be obtained, then the value of K_* will be determined from the solution of the equation:

$$[G(\sigma_0)C]^{1/2} = \lambda\sigma \cdot \sqrt{\ell} \tag{3}$$

where ℓ is the crack length, λ is a factor depending on the body shape. A diagram in Figure 3 illustrates the situation. An obvious complication of the situation due to the dependence of the stress intensity factor K on microcracking is also traced.

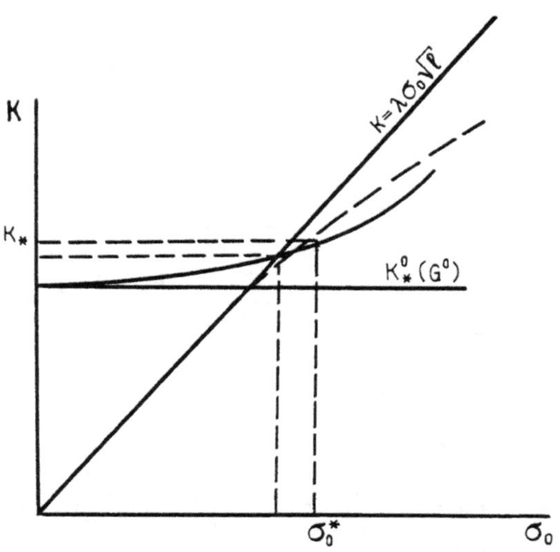

Fig. (3) - The graphical illustration of the solution of equation (3). Curve $K(\sigma_0)$ can deviate from a linear dependence because of microcracking in front of the crack and corresponding decreasing of effective elastic moduli of the material within the process zone

Consider now a plane semi-infinite model of the non-homogeneous media, Figure 4. The model contains alternating layers of two kinds. One has regularly spaced defects such that a microcrack arises when the stress at any point of the defect reaches σ^*. Let the microcrack length always be d. This means that the microcrack cuts a layer completely but never penetrates neighboring layers of the other kind. (The local plastic deformation of the latter layers or the local interface delamination prevents the penetration). For simplicity, the case of isotropic layers with equal elastic moduli will be considered. This case is not too extraneous since the alumina fiber-molybdenum matrix composite described in [8] is elastically almost isotropic.

At the applied stress σ_0, the boundary of the microcracking region in which the tensile stress σ^* at all the defects must be obtained. The defects are numbered by two indices (Figure 4) - the first shows a defect number along the fiber and the second shows a fiber number. The even fibers bear only the

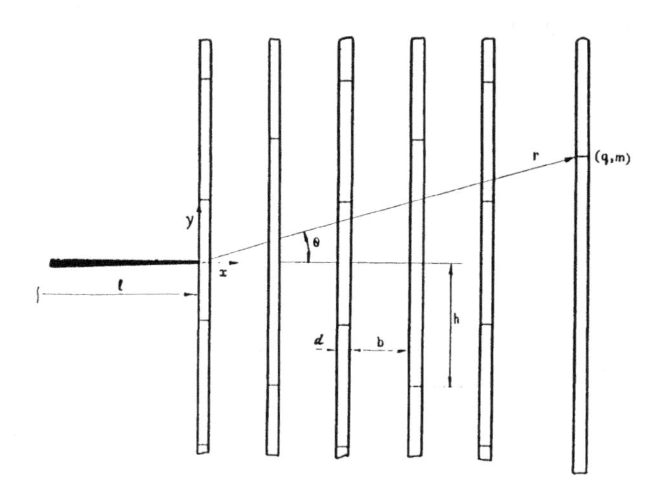

Fig. (4) - A plane model of a composite

even defects and vice versa, this being the result of a particular system of regularly distributed defects. Now, it is convenient to assume all the defects are contained within a finite region which surely includes the region of microcracking. Each microcrack generates an elastic field of its own, and each microcrack is influenced by the macrocrack field and the field of only the direct neighbor microcracks. So the stress at point (q,m) of a fiber is determined as:

$$\sigma_y^{(q,m)} = \sigma(r,\theta) + \Delta\sigma(q,m) + \sigma_o \tag{4}$$

where

$$\sigma(r,\theta) = \frac{\sigma_o\sqrt{\ell}}{\sqrt{2\pi r}} f(\theta)$$

$$r = \sqrt{m^2b^2+q^2h^2}$$

$$\theta = \tan^{-1}\left(\frac{qh}{mb}\right)$$

and $\Delta\sigma(q,m)$ is the stress generated by the presence of microcracks at points $(q\pm1,\ m\pm1)$. Index y on σ here and further on is dropped.

Now we introduce dimensionless values:

$$\bar{\sigma} = \sigma/\sigma^*,\ \mu = d/b,\ \bar{\ell} = \ell/b$$

and rewrite (4), dropping the bar notation,

$$\sigma(r,\theta) = \sigma(q,m) = \frac{\sigma_0\sqrt{\ell}}{\sqrt{2\pi(1+\mu)}}\,\phi(q,m) \tag{5}$$

where

$$\phi(q,m) = \frac{(1+\xi)^{3/2}(3\xi-2)}{2\sqrt{2}\xi}$$

$$\xi = (1+\eta^2)^{1/2},\quad \eta = \frac{q}{m}\,\frac{h}{b+d} = \frac{q}{m}\,\chi\, v_f$$

$$v_f = \frac{d}{b+d},\quad \chi = h/d$$

The condition for some defect (k,s) to be located within the region of microcracking is:

$$\psi(k,s) = \sigma(k,s) + \Delta\sigma(k,s) + \sigma_0 \geq 1 \tag{6}$$

The equality in (6) means the defect lies at the boundary of the microcracking region.

The value of $\sigma(k,s)$ is given by equation (5), and this is a bit higher than should be within the microcracking region and lower outside this region because it does not take into account the decreasing of the elastic moduli of the media due to microcracking. The value of σ_0 should be included in (6) because the microcracking zone may be large enough.

For arbitrary defects, the value of $\Delta\sigma$ is given by sum

$$\Delta\sigma(q,m) = \kappa_{ij}\Delta\sigma_{q+i,m+j}(q,m)$$

where $i,j = \pm 1$, $\kappa_{ij} = 1$ if point $(q+i,m+j)$ lies within the microcracking zone, and $\kappa_{ij} = 0$ if the point lies outside this zone.

The value of $\Delta\sigma_{q+i,m+j}(q,m)$ can be obtained taking into consideration the non-homogeneous stress field of the microcracks. But it will lead to complications, so we will assume these fields to be locally homogeneous. The mutual influence of only the direct neighbor microcracks, which has been already assumed helps in the assumption mentioned above. Usually, $d<\ell$ and it also supports the assumption. So:

$$\Delta\sigma_{q+i,m+j}(q,m) = \frac{\sigma_0\sqrt{\ell}}{2\pi}\sqrt{\frac{\mu}{1+\mu}}\,\phi(q+i,m+j)\phi(i,j)$$

where i,j = ±1.

If a point lies within the microcracking region, then inequality (6) gives a series of inequalities

$$\psi^t(k,s) \geq 1 \tag{7}$$

and each one is determined by its own set of values. This is because the position of the boundary of the cracking zone is not known. Determining the boundary demands the consideration of all the possible sets of $\kappa^t_{ij} = 0$, $\kappa^t_{rs} = 1$.

The calculation of variants with increasing value of σ_0 has been done by a computer and this gives a picture of microcracking in front of the macrocrack. Two examples are shown in Figure 5. Note that the assumption of the

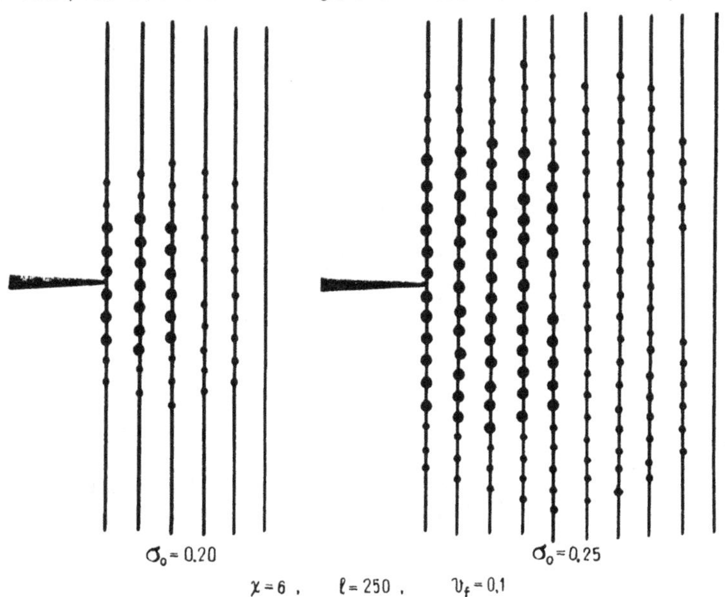

Fig. (5) - Microcracking in the model composite at the tip of the crack ($\chi = 6$, $\ell = 250$, $v_f = 0.1$). The large points are obtained neglecting mutual interaction of microcracks

mutual influence of the closest neighboring microcracks essentially only expands the region of cracking comparatively with the region obtained in the case $\Delta\sigma = 0$ (no mutual microcracks influence at all).

The calculation also gives the dependence of the largest number n of cracks along one "fiber" upon the value of σ_0. These dependencies are presented in [7].

Now it is interesting to get a limit size of the microcracking region, namely such a size when the region stops to grow up in y-direction and goes on to grow in x-direction only. This means the start of macrocrack propagation [1] or K reaching the value of K_*. A possible method of solution to this problem is to prescribe some value γ of dissipated energy to each microcrack. There is no strict way to apply this procedure. For example, the strain distribution in the cracking region (without taking into account the cracking itself) can be found, and then assume:

$$\gamma = \frac{\varepsilon_y}{\varepsilon_y^o} G_m$$

where G_m is the effective surface energy of the matrix, and ε_y^o is the largest matrix strain in the cracking region at the critical condition ($K = K_*$). Such a procedure is therefore iterative in nature.

Additional rough estimations are obtained in [7], where G is taken in form:

$$G = G_m(1+\nu n) \tag{8}$$

where $\nu<1$ is a constant. Equation (8) means that $G_f = 0$, and a fully plastic energy dissipation takes place at microcracks closest to the surface of the propagating macrocrack, while the other cracks dissipate the energy with the intensity given by ν. Then equation (3) is reduced to:

$$[G_m(1+\nu n(\sigma_o))]^{1/2} = \frac{\sigma_o\sqrt{\ell}}{E^{1/2}} \tag{9}$$

where E is Young's modulus of the isotropic composite. This equation gives the values of K_* and σ_o^*, Figure 3.

Some qualitative conclusions follow the calculation (using, for example, values of $\nu = 0.1$, $\sigma^* = 150$ kg/mm^2) with the results shown in Figure 6. They are as follows:

1. Fracture toughness K_* of a composite can go down with increasing fiber volume fraction v_f ($\chi = 26$ in Figure 6a) as well as have a maximum at some value of v_f ($\chi = 8$, $\chi = 10$).

It corresponds to the behavior of some composites in the experiment. In [1], the value of K_* grows up with fiber volume fraction in the case of boron-aluminum composites (χ is small). In experiments by Cooper and Kelly [9] with tungsten-copper composites (χ is large), the value of K_* goes down with fiber volume fraction. Within the framework of the model, the value of K_* can grow up to infinity at small enough χ. This corresponds to a composite which does not feel the presence of a crack up to some length.

2. Fracture toughness K_* goes to a constant value when the crack length ℓ increases. This constant value of K_* can be taken as a characteristic value

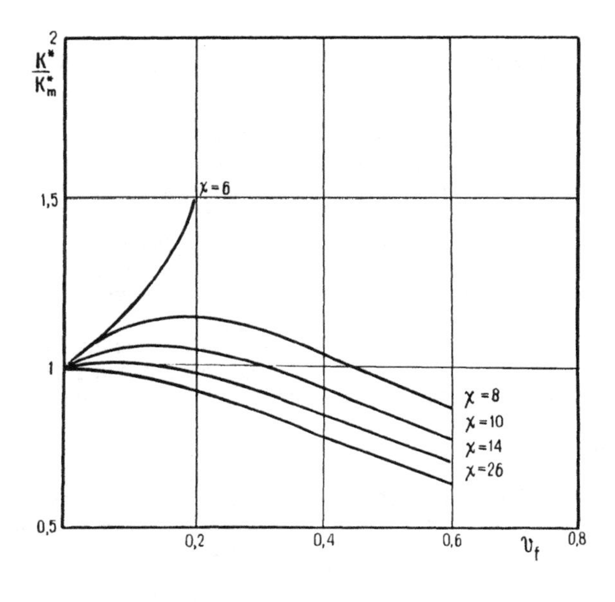

$K_m^* = 300 kgf/mm^{3/2}$, $\ell = 250$

Fig. (6a)

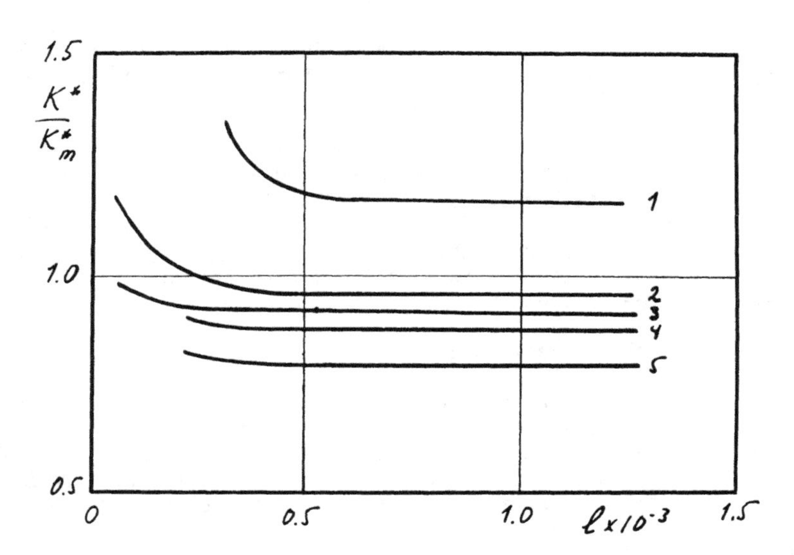

Fig. (6b)

Fig. (6) - The dependence of fracture toughness of a composite upon the structural parameters. a) $K_m^* = 300$ kgf/mm^2, $\ell = 250$; b) $K_m^* = 300$ kgf/mm^2, curves 1-3: $v_f = 0.2$, curves 4-5: $v_f = 0.4$, $\chi = 6$ (curve 2), $\chi = 14$ (curves 1,4), $\chi = 26$ (curves 3,5)

for a given structure (Figure 6b). However, a necessary crack length may be too large to have it in an experiment.

3. Fracture toughness K_* of a composite is influenced by the matrix fracture toughness, Figure 6 . If a composite structure is considered as an amplifier of the toughness [1], then this amplifier appears to be nonlinear.

4. Fracture toughness of a composite depends strongly on the influence of fiber strength characteristics on the critical length (the mean distance between fiber defects to be broken in a particular experiment).

5. Since fracture toughness of a composite depends upon many factors, there appear to exist many possibilities to control fracture toughness of composites. At the same time, it explains the large scatter in the experimental data.

PHENOMENOLOGICAL APPROACH

The development of the macromechanical approach is a result of the natural desire to transfer classical fracture mechanics to the analysis of the fracture of composites. But if plastic deformation is the only thing which damages the pure elastic structure of mechanics when linear fracture mechanics applies to the analysis of cracks in metals, we find many more obstacles to the application of linear fracture mechanics to non-homogeneous solids. Local failures of various types in the vicinity of the crack tip, doubtful validity of the concept of effective elastic moduli, and many other factors resist such an application of linear fracture mechanics. Despite this, all the attempts to go through these obstacles are similar to Irwin's correction of crack length, namely the introduction of some characteristic length. The situation here is well-known so the results already obtained and to be obtained shall not be discussed as they can be found in review papers, for example [10].

One problem should be mentioned. It is the estimation of the limit load of a composite plate with a hole. The difficulties here are the same as in the case of a crack where they arise because of a change of the material structure in the stress concentration zone. So, almost all attempts to solve this problem which start with the well-known paper by Whitney and Nuismer [11] are based on the introduction of a characteristic length in a failure criterion. This size is to be determined from an experiment.

A. A Model of a Blunt Crack in a Composite [12]

Let a usual sharp crack of length 2ℓ exist in a composite body with no external load. During loading, a fracture process zone arises at the crack tip. Assume that the influence of this zone on the crack results in the transformation of the crack into an elliptical hole and the radius of curvature of this hole at the "crack" tip reaches a characteristic value for a given material size, say r^*, at fracture. Then assume that at the same time the local strength σ^* of the material also is reached at the same point. Neglecting a change of elastic moduli of the material in the fracture process zone and using the well-known solution of the problem of an orthotropic plate in tension with an elliptical hole [13], we obtain for the limit load (or stress $\tilde{\sigma}$)

$$\sigma^*/\tilde{\sigma} = 1 + \frac{p+q}{pq}\sqrt{\frac{\ell}{r^*}} \qquad (10)$$

where p and q are the roots of a characteristic equation which depends on the elastic moduli of a material as:

$$p,q = \left[\frac{E_x}{2G_{xy}} - \nu_{yx} \pm \sqrt{\left(\frac{E_x}{2G_{xy}} - \nu_{yx}\right)^2 - \frac{E_x}{E_y}}\right]^{1/2} \tag{11}$$

Here, E_x, E_y, G_{xy}, ν_{yx} are the usual elastic moduli.

Equation (10) contains two parameters σ^* and r^* to be determined in an experiment. The dependence of $1/\tilde{\sigma}$ on $\sqrt{\ell}$ is given by a straight line. The intersection of this line with the $1/\tilde{\sigma}$ - axis gives the value of σ^*, and the slope of this line determines the value of r^*.

Therefore, the first check on the validity of the assumptions made above is the linearity of the experimental dependence of $1/\tilde{\sigma}$ on $\sqrt{\ell}$. The bulk of such experimental data showing this linearity is given elsewhere [11]. Two representative illustrations of this dependency are shown in Figure 7. It is interesting to note that the value of σ^* obtained in such a way is always larger than the mean strength of a specimen without a notch. This is a result of the scale effect.

Now it is possible to give a more general sense to the assumption made above and then support it by experimental data. Assume now that a defect of arbitrary shape exists with the radius of curvature at a dangerous point less than r^*. Also, assume the defect transforms under the load into an elliptical hole with the radius of curvature r^* and at failure the stress at the dangerous point reaches the value of σ^*. Thus, plates with defects of equal lengths in the direction normal to the applied tensile load and with radii of curvature less than r^* will fail at the same load. The limiting load is given by equation (10).

In the case of a plate with a central circular hole of radius r, the dependence of $1/\tilde{\sigma}$ on $\sqrt{r}$ should be obtained (schematically drawn in Figure 8). Here, interval BC corresponds to a straight line [10] and the slope of this line is related to the value of r^*. Point A corresponds to the strength of a specimen without a notch, such that

$$r_o = \left[\frac{pq}{p+q}\left(\frac{\sigma^*}{\sigma_o} - 1\right)\right]^2 r^* \tag{12}$$

The horizontal part of the curve at $r>r^*$ corresponds to the elastic behavior of the plate and the fracture occurs when the stress at the tip of the hole reaches value σ^*.

Part AB needs also some comments. Here (at $r<r_o$), the rupture does not necessarily occur in the vicinity of the hole. The local strength here may be higher than the local stress when rupture condition is fulfilled elsewhere in the specimen. When writing a previous paper [11], the authors saw this possibility; however, they had not observed such a type of fracture. A special experiment has been carried out and the result is shown in Figure 9.

Value σ_o' (point A) depends on specimen size. If the elastic moduli of the material are known, then the slope of part AB gives the value of σ_o^* and there-

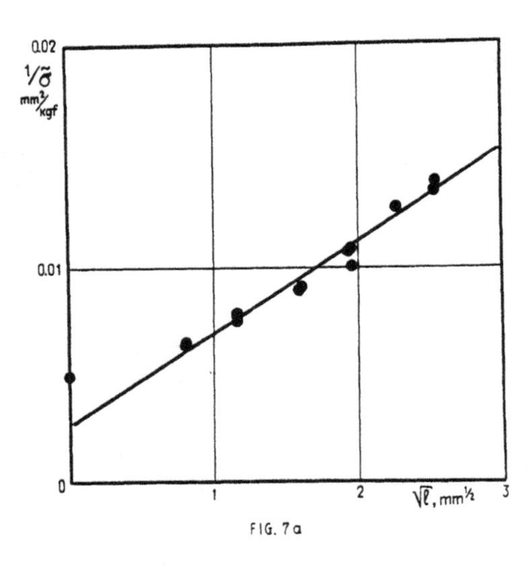

FIG. 7a

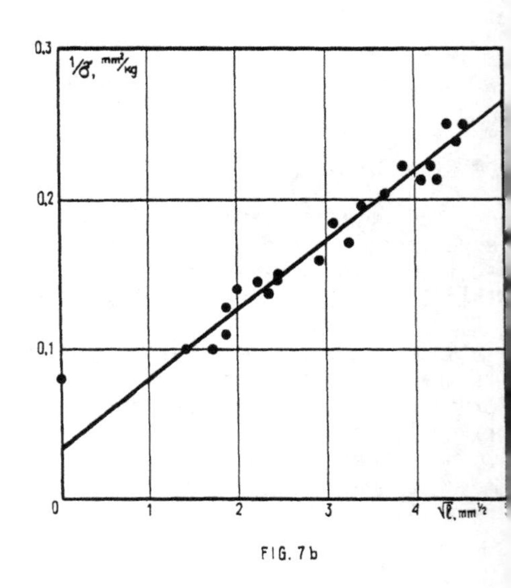

FIG. 7b

Fig. (7) - The dependencies of $1/\tilde{\sigma}$ upon $\sqrt{\ell}$ for composite plates with a central crack. The values of $\tilde{\sigma}$ is obtained taking into account a final width of a plate, $\tilde{\sigma} = \tilde{\sigma}' \sqrt{(w/\pi\ell)\text{th}(\pi\ell/w)}$ where $\tilde{\sigma}'$ is the experimental value, w is the width of a specimen. Materials characteristics are given in Table 1. a) Unidirectional boron-aluminum (experimental data by Averbuch and Hahn [14]); b) Glass fibers mat reinforced plastic (experimental data by Owen and Cann [16])

fore the location of point c. It determines the value of σ_0^*, but the stress concentration factor $k = \sigma^*/\sigma_0^*$ is given by the elastic constants of the material, namely,

$$k = 1 + \frac{p+q}{pq} \tag{13}$$

Therefore, when interpreting experimental data, one needs either to take into account relationship (13) or to accept a possible deviation of the stress concentration factor given by an experiment from that given by the theory. We follow the second way.

Now consider the experimental data. In addition to the examples already discussed elsewhere [11], the results of the tests of graphite fibers - epoxy matrix composites are shown in Figure 10. A very good correspondence can be seen between the real behavior of composites and the model behavior. Moreover, the observed scatter of experimental data does not imply the need for a better model which might lead to more complicated calculations.

TABLE 1 - CHARACTERISTICS OF MATERIALS

Materials, type of the defect	$\frac{p+q}{pq}$	σ^* kgf/mm^2	σ_0 kgf/mm^2	σ_0^* kgf/mm^2	r^* mm	
Boron-aluminum, unidirectional, $v_f = 0.5$, central notch	2.5*	372	200	106.3	2.55	[14]
Boron-aluminum, unidirectional, $v_f = 0.5$, central hole	2.55	118	93	33.2	8.09	[15]
Graphite fibers reinforced plastic, two-directional, central notch, central hole**	-	57.7	41.3	-	-	-
Glass-reinforced plastic, central notch	2.0	29.4	12.5	9.8	2.18	[16]

* Estimation

** Specimens due to Dr. T. G. Sorina, elastic characteristics have not been determined.

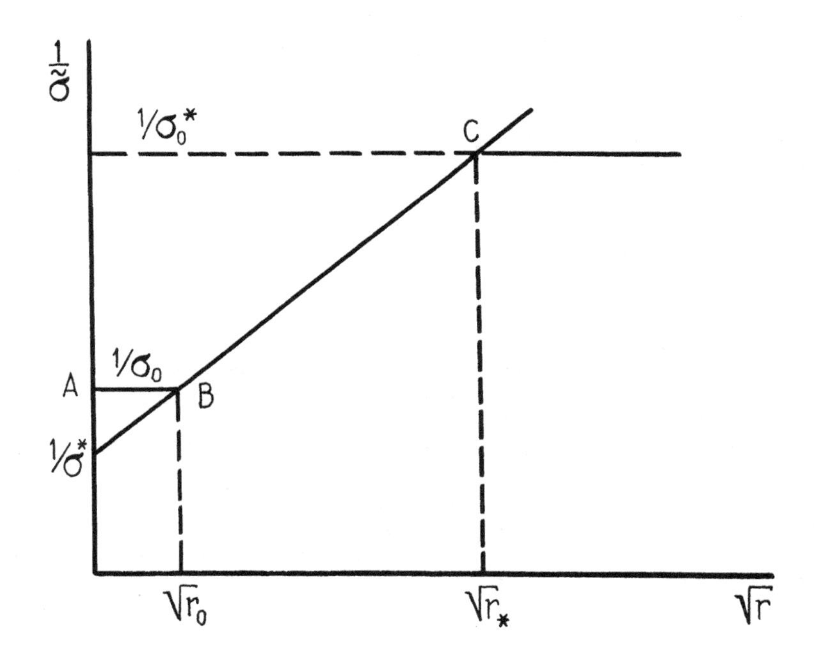

Fig. (8)

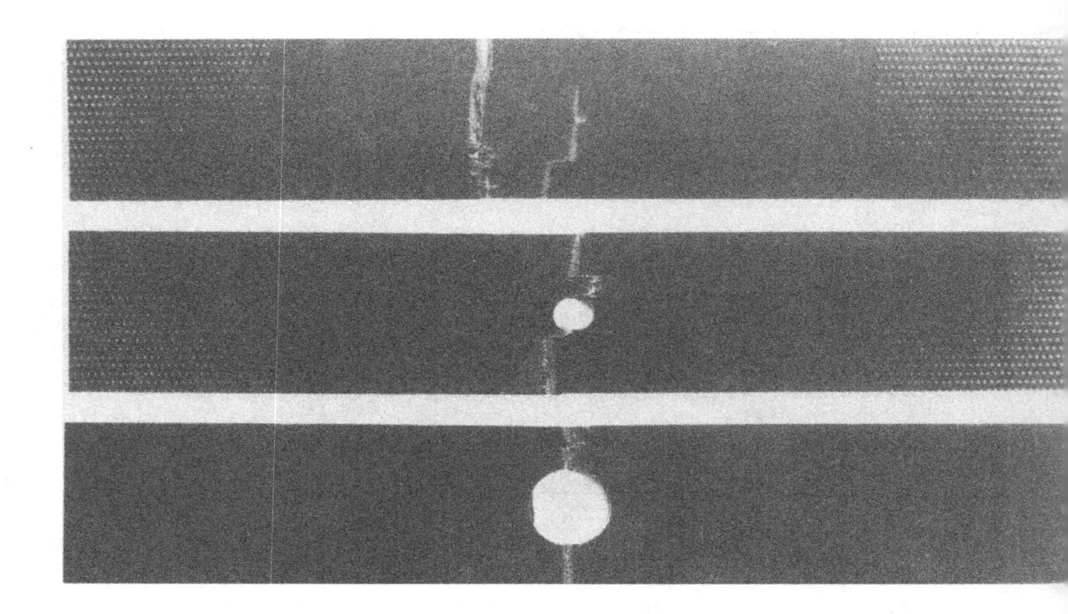

Fig. (9) - GrFRP specimens after testing. When the diameter of the hole decreases,the probability of the fracture far away from the hole increases

It is interesting to note that about 20 years ago, Kuhn [17] made an attempt to avoid operations with the singularity at the tip of a crack by analyzing the behavior of elliptical holes in metal plates. He considered the usual solution

$$\sigma^*/\tilde{\sigma} = 1 + \alpha c^* \sqrt{\ell} \qquad (14)$$

where α is the constant giving the correction due to the finite size of the plate to be valid when the radius of the hole is going to zero. Then an analysis of experimental data leads to the determination of the constants c^* and α. One should note that a fundamental assumption of the authors is directly opposite to that by Kuhn, namely we have assumed that in a composite body cracks do not exist.

The model discussed above does not pretend to give a sufficient description of all possible cases of fracture of a composite element with macrodefects, but the comparison of this model with experimental data leads to the conclusion that it corresponds well enough to real fracture processes which occur on the structural (micromechanical) level.

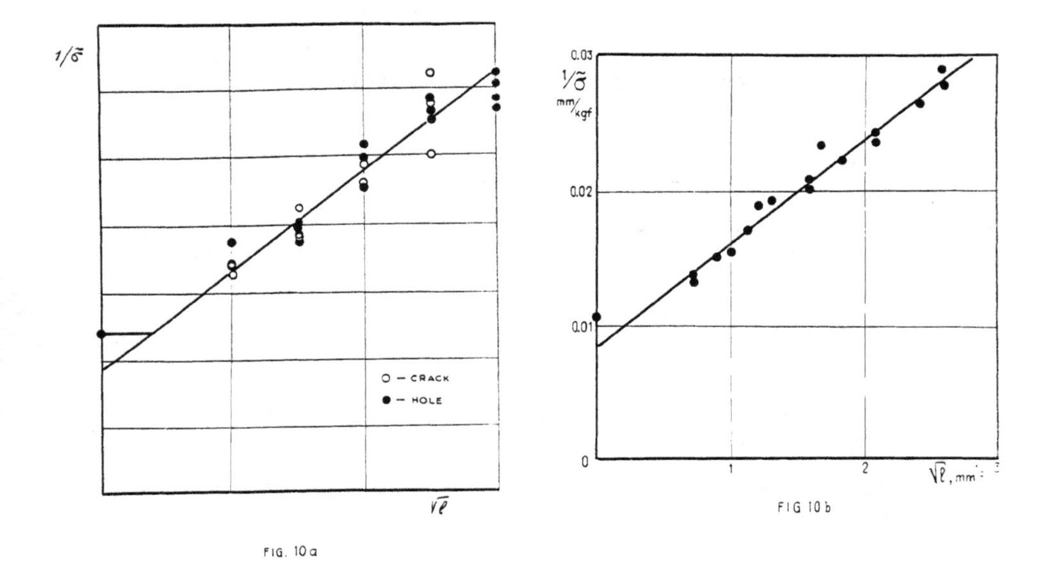

Fig. (10) - The dependencies of $1/\tilde{\sigma}$ upon $\sqrt{r}$ for composite plates with central circular hole. The value of $\tilde{\sigma}$ is obtained taking account a final width of a plate,

$$\tilde{\sigma} = \tilde{\sigma}'[2 + (1 - 2\frac{r}{w})^3]/3(1 - 2\frac{r}{w})$$

where $\tilde{\sigma}'$ is the experimental value, w is the width of a specimen. Materials' characteristics are given in Table 1. a) Graphite fiber reinforced plastic (0/90°); b) Unidirectional boron-aluminum (experimental data by Broockman and Sierakowski [15])

ACKNOWLEDGEMENT

The authors thank Dr. T. G. Sorina for specimens of graphite fiber reinforced plastic.

REFERENCES

[1] Mileiko, S. T., "Micro- and Macrocracks in Composites", in: Fracture of Composite Materials, Sijthoff and Noordhoff, pp. 3-12, 1979.

[2] Sih, G. C., "Fracture Mechanics of Composite Materials", ibid., pp. 111-130.

[3] Vanin, G., "Interaction of Cracks in Fibrous Media", ibid., pp. 131-144.

[4] Vanin, G., "Stress Fields and Moduli of Fibrous Composites", in: Fibrous Composite Materials, "Naukova Dumka", Kiev, (in Russian), 1970.

[5] Mileiko, S. T., Sorokin, N. M. and Zirlin, A. M., "The Strength of Boron-Aluminium, a Composite with Brittle Fibres", Polymer Mechanics, No. 5, pp. 840-846 (in Russian), 1973.

[6] Mileiko, S. T. et al, Published by Patent Offices of England (N1558743 - 08.11.79) and France (N2416270 - 31.08.79).

[7] Mileiko, S. T. and Suleimanov, F. Kh., "A Model of the Macrocrack in a Composite", Mechanics of Composite Materials, 1981 (in press, in Russian).

[8] Mileiko, S. T. and Kazmin, V. I., "The Strength of Supphire Fibres and Supphire-Molybdenum Composites", Mechanics of Composite Materials, No. 4, p. 723 (in Russian), 1979.

[9] Cooper, G. A. and Kelly, A., "Tensile Properties of Fibre Reinforced Metals: Fracture Mechanics", J. Mech. Phys. Solids, Vol. 15, No. 4, p. 279, 1967.

[10] Hahn, H. T., "Fracture Behavior of Composite Laminates", in: Proc. of International Conference on Fracture Mechanics and Technology, Hong Kong, pp. 285-296, 1977.

[11] Whitney, J. M. and Nuismer, R. J., "Stress Fracture Criteria for Laminated Composites Containing Stress Concentration", J. Composite Materials, Vol. 8, pp. 253-265, 1974.

[12] Mileiko, S. T., Khokhlov, V. Kh. and Suleimanov, F. Kh., "The Fracture of a Composite Containing a Macrodefect", Mechanics of Composite Materials, No. 2 (in Russian), 1981.

[13] Lechnizky, S. G., "Elasticity Theory of Anisotropic Bodies", "Nauka", Moscow, 1977 (in Russian).

[14] Averbuch, J. and Hahn, H. T., "Crack-Tip Damage and Fracture Toughness of Boron/Aluminum Composites", J. Composite Materials, Vol. 13, pp. 82-107, 1979.

[15] Broockman, E. C. and Sierakowski, R. L., "Fracture of Unidirectional B-Al Composite with a Central Hole", Fibre Science and Technology, Vol. 12, No. 1, pp. 1-10, 1979.

[16] Owen, M. J. and Cann, R. J., "Fracture Toughness and Crack-Growth Measurements in GRP", J. Mater. Science, No. 8, pp. 1982-1996, 1979.

[17] Kuhn, P., in: Fracture, H. Liebowitz, ed., Vol. 5, Academic Press, 1966.

COMPUTER SIMULATION OF VARIOUS FRACTURE MECHANISMS IN FIBROUS COMPOSITE MATERIALS

I. M. Kopyov, A. S. Ovchinsky and N. K. Bilsagayev

A. A. Baikov Metallurgy Institute, Academy of Science
Moscow, USSR

ABSTRACT

Strength properties and failure appearance of composites are predicted by computer simulation of fracture processes. Four fracture mechanisms have been simulated: fiber breaks by loading, fiber breaks by overloading, debonding of broken fiber ends from matrix, matrix cracks adjoining the location of fiber breaks. The "switch on" criteria of the micromechanisms are obtained by analysis of the stress redistribution induced by the breaks in a composite. Some dynamic effects are taken into consideration. Simulation results agree satisfactorily with experimental data of a C-Aℓ composite prepared by the forced transfusing method. The fiber-to-matrix bond strength influence on composite strength and failure appearance is investigated.

INTRODUCTION

The prediction of composite strength properties is most efficiently handled by a semi empirical approach which combines theoretical models of fracture processes with experimentally observed behavior. Detailed fractography and structural analysis of fracture surfaces allows elementary break sequences, break interactions, and their development with time or load to be modelled. In this way, fracture developments on different material structural levels can be generalized. The final composite fracture is preceded by structural breaks forming defects within the material structural elements. These defects can take the form of fiber breaks, fiber debonding and interfibrous matrix cracks. These defects and their interaction have been simulated by computer for a brittle fiber reinforced composite subjected to tension loading.

A COMPOSITE MODEL

The fibers in the composite model are dispersed hexagonally and represented as chains of ℓ_c-length segments whose endpoints are fiber defects - possible places of fiber breaks. It is assumed that the fiber arrangement distributes all the defects within a row of planes separated by a distance of ℓ_c, Figure 1. Earlier work [1] described a failure process which was carried out in each plane in succession. Later models [2,3] were developed that randomly scattered the defects in the material volume.

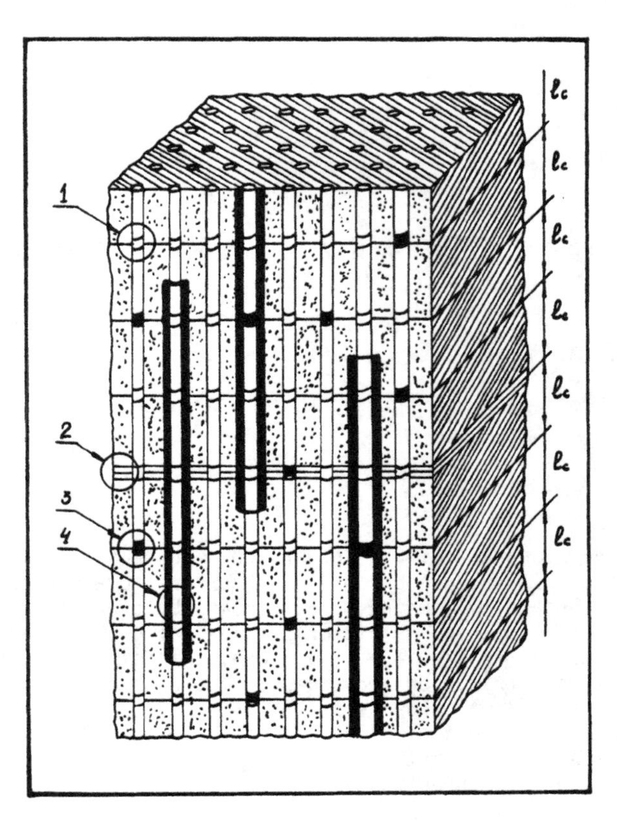

Fig. (1) - A quasi-voluminous model of a composite: 1 - fiber defects brought in defects planes ℓ_c - distanced; 2 - the model defect plane of the material; 3 - local fiber break places; 4 - a fiber segment debonded from matrix

This paper suggests a quasi-voluminous model that traces the bulk fracture behavior of the material by observation of a single plane within the defected ones. This single plane, the model plane, is to be penetrated by debonded zones of fibers that have broken within other planes, Figure 2. Therefore, these debonded fibers are not subject to the external load within the model plane. Assuming that all the planes are equivalent statistically, the number of defects in the model plane that are not subject to the external loading can be calculated as the summary length of debonded zones produced by fiber breaks in the model plane, measured in ℓ_c.

There are initial parameters of the model that reflect determined component and structure properties: fiber (E_f) and matrix (E_m) elastic moduli, matrix yield stress (σ_{my}) and hardening modulus (E_{my}), fiber volume ratio (V_f) and statistical properties: mean fiber ($\bar{\sigma}_{fb}$), matrix ($\bar{\sigma}_{mb}$) and bonding ($\bar{\tau}_{jb}$) strengths, scattering and scaling parameters.

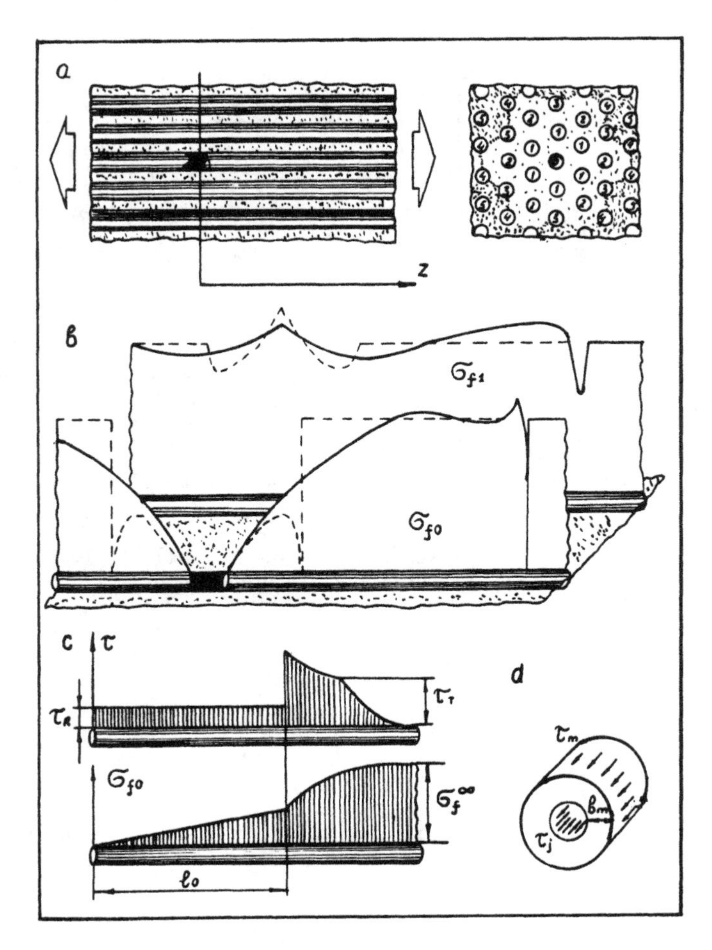

Fig. (2) - The stress redistribution model: a) An elementary composite bulk with a fiber break and fiber indexation; b) Stress distribution diagrams of a broken (σ_{f0}) and neighboring (σ_{f1}) fibers at different time; c) friction and shear stress and tensile stresses of a broken fiber end; d) A circular matrix element adjoining to a broken fiber place, under shear stresses τ_j and τ_m

Using those parameters, the computer forms a row of two-dimensional arrays, hexagonally arranged, representing a model-plane and containing fiber strengths (σ_{fb}), local matrix strengths (σ_{mb}) and local bonding strengths (τ_{jb}) distributed randomly according to Weibull's function, load levels on fibers, load transferring ratios and others.

The fracture simulation begins with some fiber ruptures. The breaks produce shear stresses on the fiber-matrix boundary, increase the matrix stress intensity and overload neighboring fibers. Due to a stress redistribution following the breaks, the fracture mechanism will be examined.

FRACTURE MICROMECHANISMS WORK-UP CRITERIA

The stress redistribution analysis is carried out by means of a model [4] taking into consideration the stress distribution along z-axis, Figure 2, in the broken fiber, in neighboring fibers and in the boundary. Some dynamic effects, Figure 2b, have been investigated [5,6]. The analysis has a given volume and elasticity ratio depending on a) the fiber critical length ℓ_c, b) the overloading ratio of a fiber neighboring to the broken one (k_f), c) the maximum shear stress on the broken fiber boundary (τ_{max}), d) the maximum matrix stress intensity ($\sigma^i_{m\ max}$) at elastic or elastoplastic deformation of matrix.

When $\tau_{max} = \tau_{jb}$ or $\sigma_{im\ max} = \sigma_{mb}$, the two latter values can be formulated as a beginning criteria of micromechanisms. This means the minimal fiber stress, in case of the fiber break, would be followed by debonding (σ_{fj}) or matrix crack (σ_{fm}). Thus we have three micromechanism initiation criteria: $\sigma_f > \sigma_{fb}$ (fiber break), $\sigma_f > \sigma_{fj}$ (debonding), $\sigma_f > \sigma_{fm}$ (matrix crack), where σ_f is a local stress of a fiber.

An analysis of the stress redistribution dynamics allows some important effe that have been used in the simulation to be distinguished:

1. Overloading waves arising and running along fibers neighboring to the broken one can break the fibers some distance from the primary fiber break plane (this excuses the planar distribution of fiber defects because the waves will find a defect to break inevitably).

2. The dynamic overloading ratio is two times more than the static one (it is used in the stress transferring simulation).

3. The debonded fiber section is loaded by friction τ_R, Figure 2b, (debonding length can be estimated as $\ell_0 = d_f(\sigma_f - \sigma_{fj})/4\tau_R$).

4. Increasing the debonding length decreases the overloading ratio (it is used in simulation of micromechanisms interactions).

Balance of the matrix element adjoining a broken fiber end allows the depth of a circular matrix crack to be evaluated, Figure 2d. It is assumed that the shear deformation induced by the break is superposed on the uniform tensile deformation in matrix. When the matrix deforms after the fiber break elastically, the depth $b_m = d_f(\sigma_f/\sigma_{fm} - 1)/2$.

COMPUTER SIMULATION OF FRACTURE PROCESSES

The composite load is simulated by an incremental increase in deformation followed by the uniform axiomatic operation sequence:

1. Calculation of fiber stresses and comparing with fiber strengths.

2. If a fiber defect broke, its stress (σ_0) is to be taken off and stored.

3. Calculation of debonding and matrix cracking criteria and comparing in order to determine which process will follow the fiber break.

4. Calculation of the debonded zone length (respective matrix crack depth).

5. Calculation of a new load transferring coefficient (using the debonded zone length.

6. When debonding process begins, the influence of neighboring planes is simulated by means of randomly subtracting the load from the proper number of fiber defects in the model plane.

7. Fiber defect stresses in defects neighboring the broken one are to be increased by adding a $k_f\sigma_0$ value [1] (load transferring).

8. If a neighboring defect is unable to take that overload (being previously broken), then the overload is to be redistributed as in item 6 among neighbors of this previously broken defect (adding a $k_f^{\prime}k_f\sigma_0$). Thus, the transferred value reduces (k_f<1/6) step by step, and such a transferring process is finite.

9. When a neighboring defect, having taken an overload, breaks, it means it breaks through the first overloading step and items 2-8 are to be repeated for that broken defect. Sequently, breaks of the second and so on overloading steps are possible. Either such a process is limited in a part of a composite defect plane or it leads to an "avalanche-like" failure of all or almost all the fibers in the plane (a "planar avalanche").

10. Calculation of the strength of the fibers, matrix and composite by means of the summary number of fiber breaks, load subtractions, matrix cracking, debonding lengths. This allows a σ-ε diagram to be drawn [7].

11. The composite deformation increases and the operation sequence repeats.

Every new load transferring or overloading step is performed after completing all the previous operations on all the defects of the plane. The plane contains 30x30 defects.

SIMULATION RESULTS

The simulation enables one to reproduce qualitatively different fracture mechanisms, Figure 3, of a C-Aℓ composite prepared by a forced transfusing method [8]. Due to different levels of physic-chemical interaction of components, three characteristical fracture appearances had been observed:

1. A bundle breaks with large debonding lengths and fiber separation; having a low composite strength as a result of a low bonding strength.

2. A "splinter"-like break having a grown complicated fracture appearance and the maximal composite strength.

3. A brittle smooth break caused by some degradation of initial component properties during their interaction in preparing and, partly, by a high bonding strength.

Using the component and bonding properties of these interaction levels [8,9] as entry-parameters of the simulation, a satisfactory correlation of experimental data with computated strength values has been obtained.

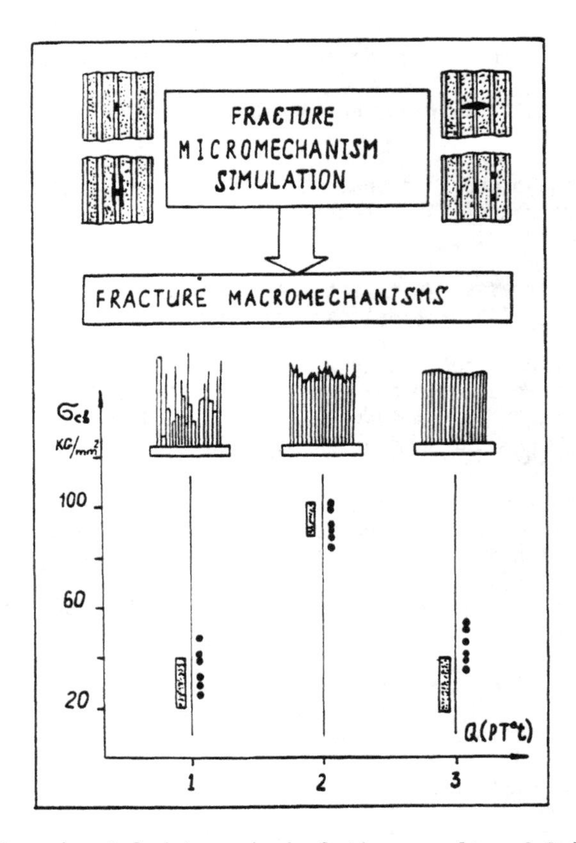

Fig. (3) - Experimental data and simulation results of C-Aℓ composite strength at three characteristical component interaction levels (1,2,3 - fracture-characterizing points [8], $Q(p,T^{O},t)$ - a measure of energetical affection in composite preparing process)

As an example of the ability to study the influence of some factors on the composite strength, the bonding strength dependence (fixing other parameters) is shown, Figure 4.

At low bonding strengths, single fiber breaks being accompanied by large debonding lengths lead quickly to elimination of the load received by the majority of fibers in the plane (a volume avalanche). Realization of the high fiber strength is impossible. It results in a low strength composite. When the bonding strength rises approaching the matrix strength, the composite strength increases till some stage and levels off, sometimes slightly decreases. The fracture mechanism changes. Matrix cracking replaces debonding. It results in a planar avalanche-like process of fiber overloading fractures. Due to the statistical scatter of properties, an intermediate bonding strength level gives both fiber debonding and matrix cracking in the model plane. These two mechanisms hinder each other in growth of the avalanche-like process. When fibers have retained a satisfactorily high strength level, the maximal composite strengt comes to realization.

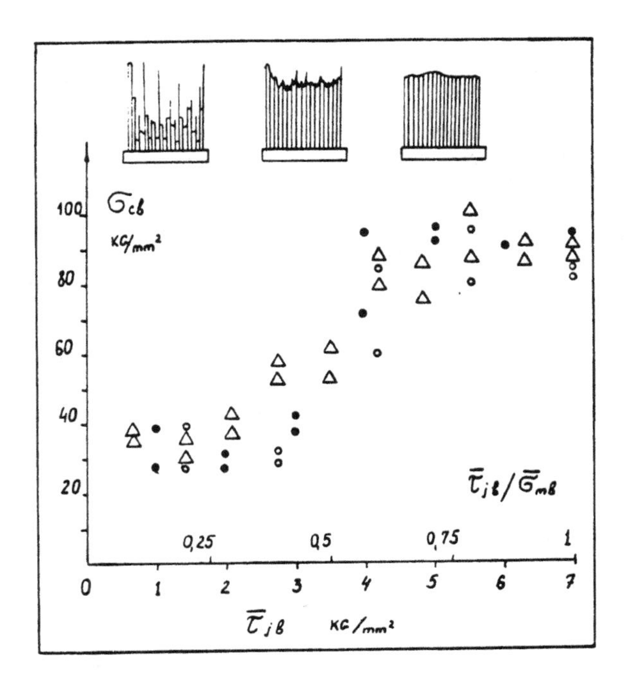

Fig. (4) - Fracture appearance and bonding strength dependence of composite strength (● - $\bar{\sigma}_{mb}$ = 7 kG/mm^2, o - $\bar{\sigma}_{mb}$ = 5 kG/mm^2, Δ - $\bar{\sigma}_{mb}$ = 10 kG/mm^2, (relative scale - $\bar{\tau}_{jb}/\bar{\sigma}_{mb}$)

The fracture process simulation allows optimization of the proportion of component properties, and bonding strength when these parameters are dependent reciprocally. This reflects a specific physico-chemical interaction of components during preparation.

REFERENCES

[1] Ovchinsky, A. S., Nemtsova, S. A. and Kopyov, I. M., "Mathematic Simulation of Composite Material Fracture Processes of Composite Reinforced by Brittle Fibers", Mekhanika Polimerov, No. 5, pp. 800-808, 1976.

[2] Kopyov, I. M., Ovchinsky, A. S. and Bilsagayev, N. K., "Computer Simulation of Fracture Processes in Composites Having Defects of Bonding Strength", Mekhanika Kompozitnykh Materialov, No. 2, pp. 217-221, 1979.

[3] Kopyov, I. M., Ovchinsky, A. S. and Bilsagayev, N. K., "Computer Simulation of Composite Fracture Process with Bonding Strength Defects Between Components", In: Fracture of Composite Materials, Sijthoff and Noordhoff, Alphen aan den Rijn, pp. 55-61, 1979.

[4] Ovchinsky, A. S., Kopyov, I. M., Sakharova, Ye. N. and Moskvitin, V. V., "Stress Redistribution at a Brittle Fiber Break in Metallic Composite Materials", Mekhanika Polimerov, No. 1, pp. 19-29, 1977.

[5] Sakhsrova, Ye. N. and Ovchinsky, A. S., "Dynamics of Stress Redistribution in a Broken Fiber of a Composite Material", Mekhanika Kompositnykh Materialov, No. 1, pp. 57-64, 1979.

[6] Sakharova, Ye. N. and Ovchinsky, A. S., "Dynamics of Stress Redistribution in a Breaking of a Composite Material Fiber", No. 4, pp. 608-615, 1980.

[7] Ovchinsky, A. S., Kopyov, I. M. and Bilsagayev, N. K., "A Method for Drawing of a Composite Material Deformation Diagram According to Statistical Strength Distribution of Reinforcing Fibers", Mekhanika Polimerov, No. 6, pp. 58-64, 1975.

[8] Salibekov, S. Ye., Zabolotsky, A. A., Turchenkov, V. A., Kontsevich, I. A. and Fadyukov, Ye. M., "Factors Having Influence on Structure and Properties Formation in $A\ell$ - Carbon Composite Materials, Poroshkovaya Metallurgiya, No. 2, pp. 58-64, 1977.

[9] Portnoy, K. I., Zabolotsky, A. A. and Turchenkov, V. A., "On Evaluation of Component Interaction and Compatibility in Fibrous Composite Materials", Poroshkovaya Metallurgiya, No. 10, pp. 64-71, 1978.

APPLICATION OF THE CUBIC STRENGTH CRITERION TO THE FAILURE ANALYSIS OF COMPOSITE LAMINATES

R. C. Tennyson, G. E. Wharram and G. Elliott

University of Toronto
Downsview, Ontario, Canada, M3H 5T6

ABSTRACT

Failure analyses are presented based on the application of quadratic and cubic forms of the tensor polynomial lamina strength criterion to various composite structural configurations in a plane stress state. Solutions are given for complex biaxial load conditions, including several cases of multi-mode failure. Some experimental data are also provided to support these calculations.

INTRODUCTION

In the design of composite laminates for load bearing applications, one of the major difficulties is that of selecting a suitable strength criterion. Although many criteria have been developed [1], it is evident that by far the majority of them have been of the quadratic form. However, it would appear that the most general failure criterion proposed is the tensor polynomial,

$$f(\sigma_i) = F_i\sigma_i + F_{ij}\sigma_i\sigma_j + F_{ijk}\sigma_i\sigma_j\sigma_k + \cdots$$
$$= 1$$

which can be shown to encompass all other failure criteria which are currently available [2]. This general description of a strength surface was advocated as early as 1966 by Malmeister [3] and developed extensively by Tsai and Wu [4] in quadratic and higher order forms [2]. Because of the preference for using quadratic models in the literature, the question that naturally arises concerns the nature of the load cases for which a higher order strength equation is required.

In [5,6], it was demonstrated that the cubic criterion was necessary to adequately predict both the initial and ultimate strengths of pressurized tubes. A multi-mode analysis was required in several instances for various laminate configurations to account for redistribution of the load after initial ply failure. Although many investigators have shown that excellent agreement can be obtained between test data and predictions based on a quadratic model, this is generally restricted to certain load and laminate configurations. Again, it was illustrated in [6] that the cubic and quadratic formulations are in agreement for simple load cases such as uniaxial tension and compression, although pure shear (positive) strength predictions for off-axis laminates do differ substantially.

Thus, the purpose of this report is to present further results on the development of the cubic strength criterion, its application to specific problems and comparisons with quadratic model theory and test data. In addition, the capability of this model to accurately predict multi-mode failures of laminates is also described. Furthermore, the effect of nonlinear shear stress/strain behaviour is included to give some insight as to its role in determining the strength of laminates.

A. Laminate Strength Analysis

In order to utilize a failure equation, one is first confronted with the calculation of the lamina principal stresses. For a plane stress state, the kth lamina in-plane stresses corresponding to the structural axes (x,y) can be determined for any ply orientation (θ) from Hooke's law,

$$\begin{bmatrix} \sigma_x \\ \sigma_y \\ \tau_{xy} \end{bmatrix}_k = [\overline{Q}_{ij}]_k[A'][N] + [\overline{Q}_{ij}]_k[B'][M] + Z_k\{[\overline{Q}_{ij}]_k[C'][N] + [\overline{Q}_{ij}][D'][M]\} \quad (1)$$

where Z_k denotes the distance of the ply from the mid-plane. Thus, the principal lamina stresses are given by

$$\begin{bmatrix} \sigma_1 \\ \sigma_2 \\ \sigma_6 \end{bmatrix}_k = [T]_k \begin{bmatrix} \sigma_x \\ \sigma_y \\ \tau_{xy} \end{bmatrix}_k \quad (2)$$

where σ_6 denotes the shear stress in the principal material axes plane, 1-2, where

$$[T]_{\pm\theta} = \begin{bmatrix} m^2 & n^2 & \pm 2mn \\ n^2 & m^2 & \mp 2mn \\ \mp mn & \pm mn & (m^2-n^2) \end{bmatrix} \quad (3)$$

$m = \cos\theta$, $n = \sin\theta$

For a symmetric laminate subject only to in-plane loading, the principal stresses become

$$\begin{bmatrix} \sigma_1 \\ \sigma_2 \\ \sigma_6 \end{bmatrix}_k = [T]_k[\overline{Q}_{ij}]_k[A_{ij}]^{-1} \begin{bmatrix} N_x \\ N_y \\ N_{xy} \end{bmatrix} \quad (4)$$

Nonlinear Effects. In treating nonlinear stress/strain behaviour, only shear was considered in the manner of Hahn and Tsai [7], i.e.,

$$\varepsilon_6 = \tilde{S}_{66} + \tilde{S}_{6666}\sigma_6^3 \qquad (5)$$

Again, for symmetric laminates only, equation (4) can be used by replacing the $[\overline{Q}_{ij}]_k$ and $[A_{ij}]$ matrices with

$$[\overline{Q}']_k = [\overline{Q}_{ij}]_k + g_k(\varepsilon_x,\varepsilon_y,\gamma_{xy})[f(\theta)]_k$$

$$[A'] = \sum_{k=1}^{\bar{n}} [\overline{Q}']_k(h_k - h_{k-1}) \qquad (6)$$

where $g_k(\varepsilon_x,\varepsilon_y,\gamma_{xy})$ is the real root of

$$y^3 + \frac{3}{\tilde{S}_{66}}\,y^2 + \left[\frac{3}{\tilde{S}_{66}^2} + \frac{S_{66}}{\tilde{S}_{6666}} \cdot \frac{1}{\varepsilon_6^2}\right]y + \frac{1}{\tilde{S}_{66}^3} = 0 \qquad (7)$$

$$\varepsilon_6^2 = \frac{(1-\cos 4\theta)}{2}(\varepsilon_x-\varepsilon_y)^2 - (\sin 4\theta)(\varepsilon_x-\varepsilon_y)\gamma_{xy} + \frac{(1+\cos 4\theta)}{2}\gamma_{xy}^2 \qquad (8)$$

$$[f(\theta)]_k = \frac{1}{2}\begin{bmatrix} (1-\cos 4\theta) & -(1-\cos 4\theta) & -\sin 4\theta \\ & (1-\cos 4\theta) & \sin 4\theta \\ \text{symmetric} & & (1+\cos 4\theta) \end{bmatrix} \qquad (9)$$

An iterative procedure is then used to solve for the failure loads.

B. Lamina Cubic Failure Criterion

The general form of the tensor polynomial failure criterion proposed by Malmeister, Tsai and Wu is

$$F_i\sigma_i + F_{ij}\sigma_i\sigma_j + F_{ijk}\sigma_i\sigma_j\sigma_k + \ldots$$

$$= f(\sigma) \begin{cases} <1 & \text{no failure} \\ =1 & \text{failure} \\ >1 & \text{exceeded failure} \end{cases} \qquad (10)$$

for i,j,k = 1,2,3,...,6. F_i, F_{ij} and F_{ijk} are strength tensors of the 2nd, 4th and 6th rank, respectively. If one restricts the analysis to a plane stress state and considers only a cubic formulation as being representative of the failure surface, then equation (10) reduces to [5,6],

$$F_1\sigma_1 + F_2\sigma_2 + F_{11}\sigma_1^2 + F_{22}\sigma_2^2 + F_{66}\sigma_6^2 + 2F_{12}\sigma_1\sigma_2 + 3F_{112}\sigma_1^2\sigma_2 + 3F_{221}\sigma_2^2\sigma_1$$
$$+ 3F_{166}\sigma_1\sigma_6^2 + 3F_{266}\sigma_2\sigma_6^2 = 1 \qquad (11)$$

where it has been assumed that certain material property symmetry conditions hold true and odd order terms in σ_6 vanish. This requires the lamina to exhibit identical strength for both positive and negative shear (relative to the principal axes). The principal strength tensor components are defined by

$$F_1 = \frac{1}{X} - \frac{1}{X'} \qquad F_2 = \frac{1}{Y} - \frac{1}{Y'} \qquad F_6 = \frac{1}{S} - \frac{1}{S'}$$

$$F_{11} = \frac{1}{XX'} \qquad F_{22} = \frac{1}{YY'} \qquad F_{66} = \frac{1}{SS'} \tag{12}$$

and the interaction terms (F_{12}, F_{112}, F_{221}, F_{166}, F_{266}) are described in [2,5].

In general, assume the principal stresses can be described in terms of some variable load parameter λ such that

$$\sigma_1 = C_{11}\lambda + C_{12} \qquad \sigma_2 = C_{21} \;\; + C_{22} \qquad \sigma_6 = C_{31}\lambda + C_{32} \tag{13}$$

where C_{12}, C_{22} and C_{32} represent some additional constant load components. Hence substituting equation (13) into equation (11), one obtains

$$a\lambda^3 + b\lambda^2 + c\lambda + d = 0 \tag{14}$$

where

$$a = 3(F_{112}C_{11}^2C_{21} + F_{122}C_{11}C_{21}^2 + F_{166}C_{11}C_{31}^2 + F_{266}C_{21}C_{31}^2)$$

$$\begin{aligned} b = {} & 3F_{112}(C_{11}^2C_{22} + 2C_{11}C_{12}C_{21}) + 3F_{122}(C_{12}C_{21}^2 + 2C_{11}C_{21}C_{22}) \\ & + 3F_{166}(C_{12}C_{31}^2 + 2C_{11}C_{31}C_{32}) + 3F_{266}(C_{22}C_{31}^2 + 2C_{21}C_{31}C_{32}) \\ & + F_{11}C_{11}^2 + F_{22}C_{21}^2 + F_{66}C_{31}^2 + 2F_{12}C_{11}C_{21} \end{aligned}$$

$$\begin{aligned} c = {} & 3F_{112}(2C_{11}C_{12}C_{22} + C_{12}^2C_{21}) \\ & + 3F_{122}(2C_{12}C_{21}C_{22} + C_{11}C_{22}^2) + 3F_{166}(C_{11}C_{32}^2 + 2C_{12}C_{31}C_{32}) \\ & + 3F_{266}(C_{21}C_{32}^2 + 2C_{22}C_{31}C_{32}) + 2F_{11}C_{11}C_{12} + 2F_{22}C_{21}C_{22} \\ & + 2F_{66}C_{31}C_{32} + 2F_{12}(C_{11}C_{22} + C_{21}C_{12}) + F_1C_{11} + F_2C_{21} \end{aligned} \tag{15}$$

$$\begin{aligned} d = {} & 3F_{112}C_{12}^2C_{22} + 3F_{122}C_{12}C_{22}^2 + 3F_{166}C_{12}C_{32}^2 \\ & + 3F_{266}C_{22}C_{32}^2 + F_{11}C_{12}^2 + F_{22}C_{22}^2 + 2F_{12}C_{12}C_{22} \\ & + F_{66}C_{32}^2 + F_1C_{12} + F_2C_{22} - 1.0 \end{aligned}$$

Once the coefficients, equations (15), are calculated for a given ply orientation and structural configuration subject to prescribed loads, the solution of equation (14) yields the desired failure condition in terms of λ.

Evaluation of Strength Parameters. The principal strength parameters defined by equation (12) can readily be determined from simple tension, compression and pure shear tests [2]. However, to evaluate the interaction parameters, recourse to the method described in [5] is made. This approach is based on a biaxial strength test and four constraint equations. The latter conditions can be derived by setting the discriminant of the cubic polynomial coefficients to zero, thus requiring the failure equation (14) to yield three real roots, two of which must be equal. This satisfies the physical consideration of having only two distinct roots for two collinear loading paths. As reported in [5], this procedure was employed utilizing internal pressure loading of $\pm\theta$ laminates having 45°, 50°, 55° and 60° orientations. Note that no experiments were used except for the one biaxial tension test. Solutions were obtained for a graphite/epoxy material, Table 1, and the planar surfaces plotted in Figures 1, 2 and 3.

TABLE 1(a) - SUMMARY OF PRINCIPAL STRENGTH PARAMETERS

Graphite/Epoxy (3M SP-288 T300)

F_1	F_{11}	F_2	F_{22}	F_6	F_{66}
$(KSI)^{-1}$	$(KSI)^{-2}$	$(KSI)^{-1}$	$(KSI)^{-2}$	$(KSI)^{-1}$	$(KSI)^{-2}$
-2.482×10^{-3}	4.239×10^{-5}	1.035×10^{-1}	3.936×10^{-3}	0	8.009×10^{-3}

TABLE 1(b) - SUMMARY OF INTERACTION STRENGTH PARAMETERS FOR CUBIC MODEL

	F_{12}	F_{112}	F_{221}	F_{166}	F_{266}
	$(KSI)^{-2}$	$(KSI)^{-3}$	$(KSI)^{-3}$	$(KSI)^{-3}$	$(KSI)^{-3}$
Constraint Equation	-4.424×10^{-4}	5.170×10^{-7}	-5.985×10^{-6}	-4.054×10^{-6}	-2.268×10^{-4}
Modified	-3.855×10^{-4}	3.116×10^{-7}	-5.713×10^{-6}	-4.054×10^{-6}	-2.268×10^{-4}

It can be readily noted that, for the cubic model, a difficulty arises in the $(-\sigma_1,-\sigma_2)$ stress quadrant, Figure 1, which appears open-ended. This results from the fact that complex roots can occur in the solution of equation (14) since the double root constraint was only imposed locally on the failure surface. Small changes in the 1-2 interaction coefficients, Table 1(b), were found to close this plane as shown in Figure 1. Clearly, a biaxial compression/compression test would be desirable. However, it should be cautioned that complex solutions can occur and care must be exercised in defining the failure surface in such load regions.

C. Applications of Strength Criteria

General Biaxial Stress States. Comparisons between quadratic and cubic strength predictions were presented in [6] for tension, compression and shear loading of simple laminates. Although some differences were found, by far the

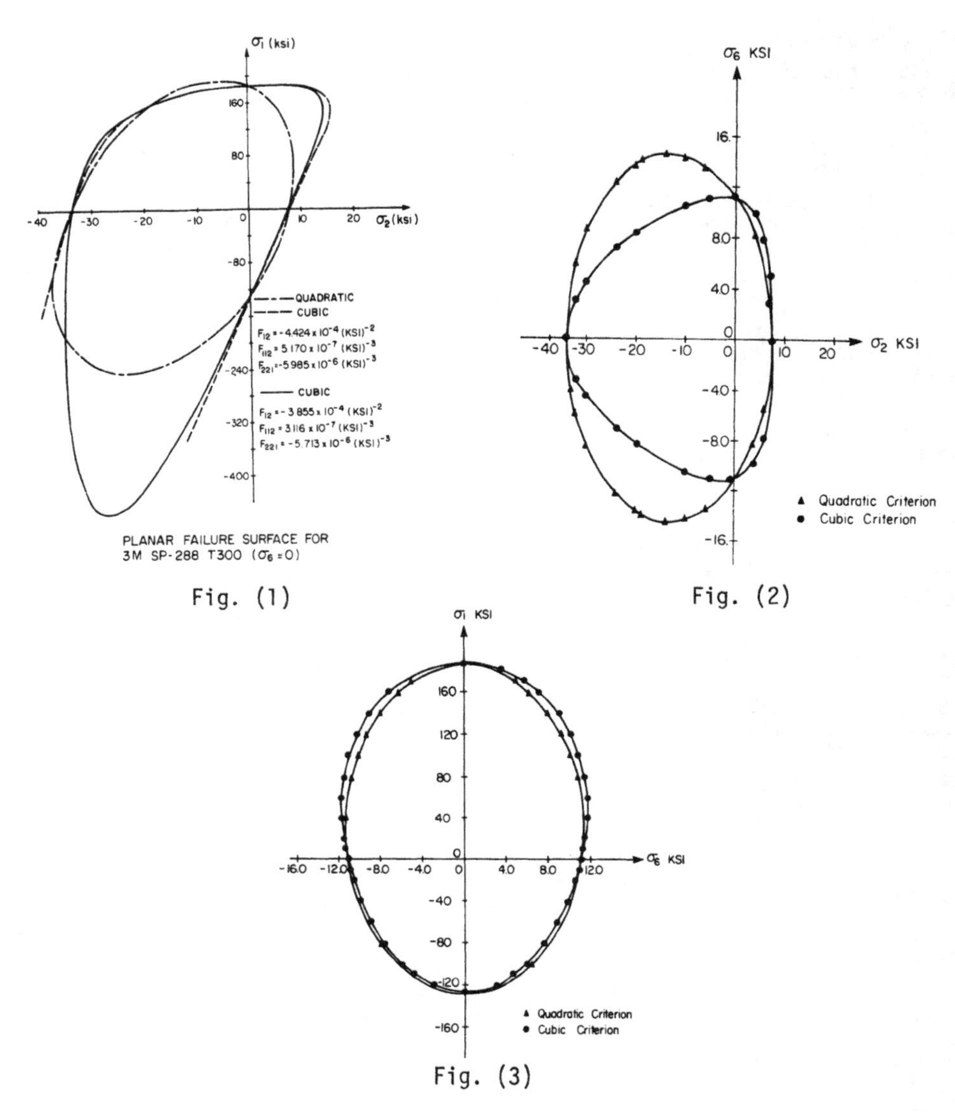

Fig. (1)

Fig. (2)

Fig. (3)

most important case examined was that of biaxial tension for $(\pm\theta)_s$ laminates. Let us now consider the same symmetric balanced construction and solve for the failure loads for any biaxial stress ratio $\alpha = \pm N_x/\pm N_y$ with $N_{xy} = 0$. Using the modified interaction strength parameters, Table 1(b), failure curves were obtained for varying θ and are shown in Figures 4-7. From these results, one can then plot maximum strengths and optimum fiber angles as a function α, as presented in Figures 8-11. Included in these latter graphs are the quadratic predictions for comparative purposes. It is of interest to note that in Figure 8, some complex root solutions were obtained for a limited number of cases. However, the general trend of the remaining data can be used to provide some estimate of the failure loads.

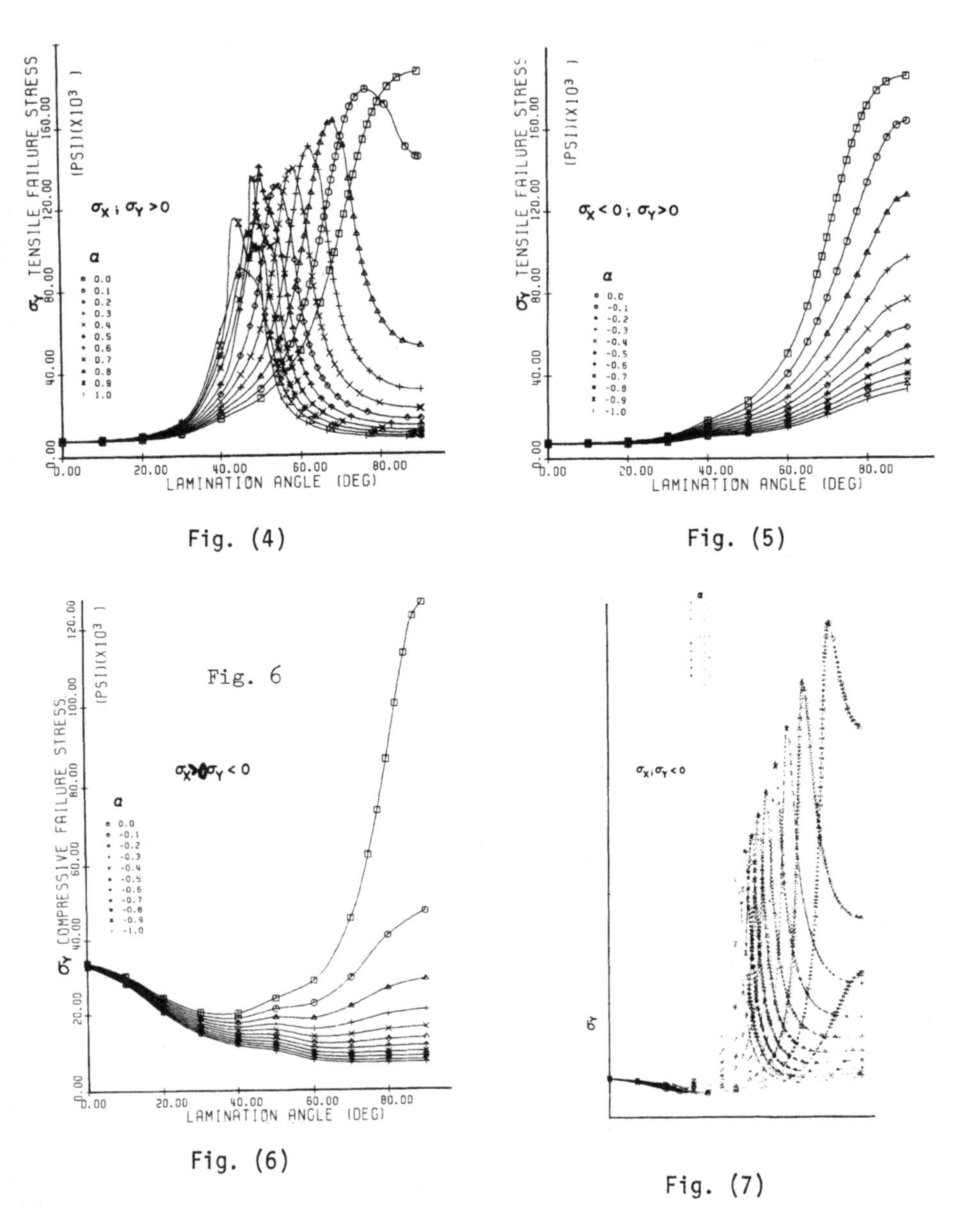

Fig. (4)

Fig. (5)

Fig. (6)

Fig. (7)

It can be seen that for $+\alpha$, large differences in failure loads are predicted, Figures 8 and 9, whereas $-\alpha$ solutions are virtually the same. However, it should be emphasized that the biaxial compression results are preliminary and must await experimental configuration.

Limited experimental confirmation of the cubic strength criterion was obtained from internal pressure tests conducted on $(\pm\theta)_s$ graphite/epoxy tubes [5].

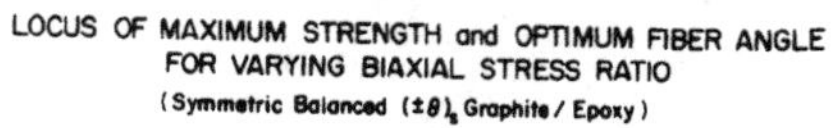

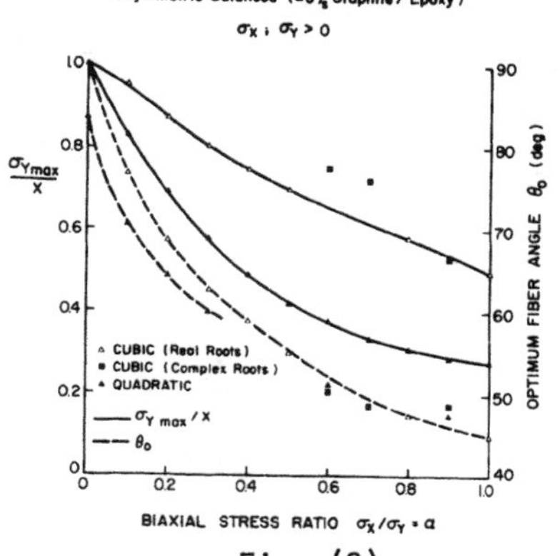

Fig. (8)

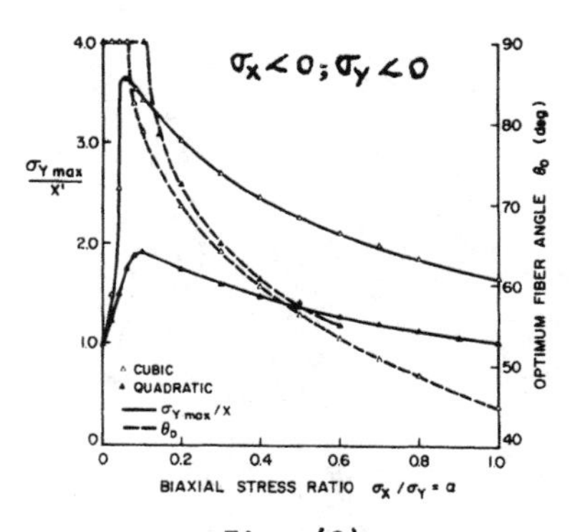

Fig. (9)

LOCUS OF MAXIMUM STRENGTH and OPTIMUM FIBER ANGLE
FOR VARYING BIAXIAL STRESS RATIO
(Symmetric Balanced (±θ)s Graphite/Epoxy)
σX < 0 ; σY > 0

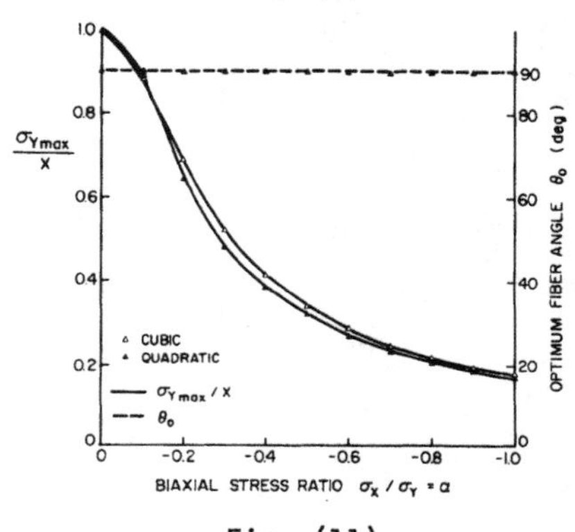

Fig. (11)

LOCUS OF MAXIMUM STRENGTH and OPTIMUM FIBER ANGLE
FOR VARYING BIAXIAL STRESS RATIO
(Symmetric Balanced (±θ)s Graphite / Epoxy)
σX > 0 ; σY < 0

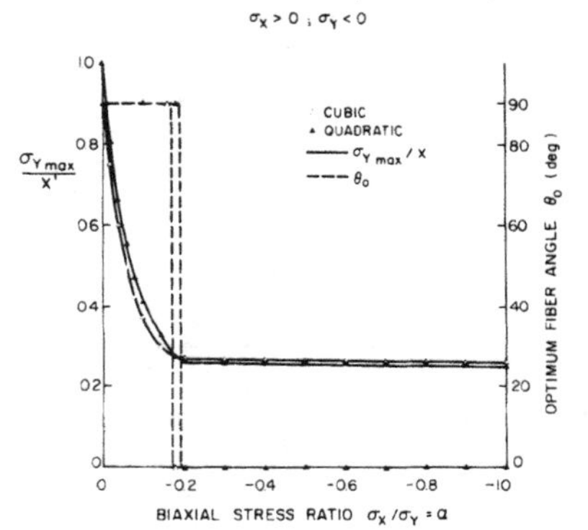

Fig. (10)

A comparison of the test data with both quadratic and cubic predictions is shown in Figure 12. It is clear that agreement with the cubic model is quite good while the quadratic formulation substantially underestimates the strength, particularly in the optimum fiber angle range.

The effect of nonlinear shear behaviour was also investigated for similar pressure loaded cylinder conditions. Figure 13 demonstrates that the nonlinear corrections are small.

Multi-Mode Failure and Combined Loading. Up to this point, we have considered the strength of laminates where failure of all laminae comprising the structure occurred simultaneously. Multi-mode failure of a laminate arises from the progressive failures of laminae that occur at various stages of loading. Since the failure mode of each ply can differ, assumptions have to be made re-

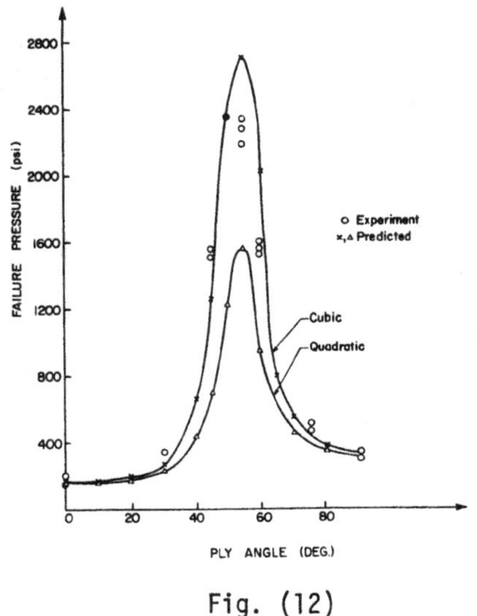

Fig. (12)

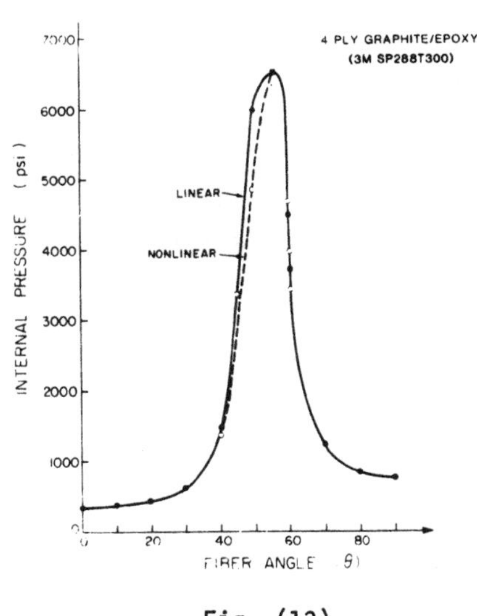

Fig. (13)

garding the effect of the "failed" ply (or plies) on the stiffness of the remaining structure. For the configurations reported in [6], matrix failure was encountered and a modified stiffness matrix was calculated including only the E_{11} contributions from the "failed" plies.

To demonstrate the application of a lamina failure criterion in predicting both initial and ultimate strengths, let us consider graphite/epoxy tubes of $(0, \pm 60)_S$ symmetric construction. Figure 14 presents a comparison of quadratic and cubic model predictions for internal pressure loading ($\alpha = 0.5$). Again, quadratic theory substantially underestimates both initial and ultimate failure loads.

In the second experiment, a pre-torque shear stress of ~19.8 KSI was applied and the tube subsequently loaded by internal pressure until it fractured. The load/strain curves are shown in Figure 15 where it can be seen that the agreement between test and predicted results is excellent. In this instance, quadratic theory predicts only one failure load at a value well below the onset of actual initial ply failure, see Table 2.

The third test involved pre-loading the tube with an internal pressure of 1.10 KSI and then torque loading it to failure. Figure 16 contains comparisons of the experimental torque/shear strain data with predictions based on the cubic criterion, which are in quite good agreement. It can be observed that the test sample exhibited nonlinear behaviour up to first failure and then followed a different response slope up to final fracture. Although theory predicts a second failure mode, it is too close to the ultimate strength and was not observed experimentally. Note that quadratic estimates in this case exceed the cubic values, see Table 2, and this appears to be typical when failure due to torsion loading is involved.

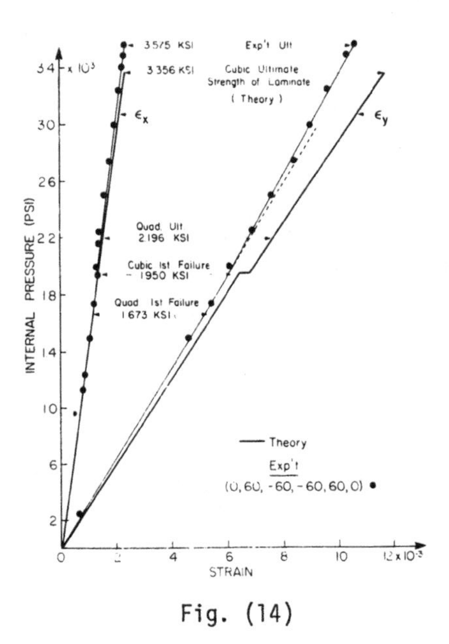

Fig. (14)

Fig. (15)

TABLE 2

	Cubic (Quad.)			
	Strength Predictions		Experiment	
Load Case	First Failure	Ultimate	First Failure	Ultimate
Internal Pressure	1.95 (1.67) KSI	3.36 (2.20) KSI	2.00 KSI	3.58 KSI
Internal Pressure + Pre-Torque of 19.8 KSI shear stress	1.96 (1.63) KSI	2.20 (1.63) KSI	1.90 KSI	2.38 KSI
Torsion + Internal Pressure Pre-Load of 1.1 KSI	5060 (5190) in-lbs	8330~8550 (8910) in-lbs	4900 in-lbs	7480 in-lbs

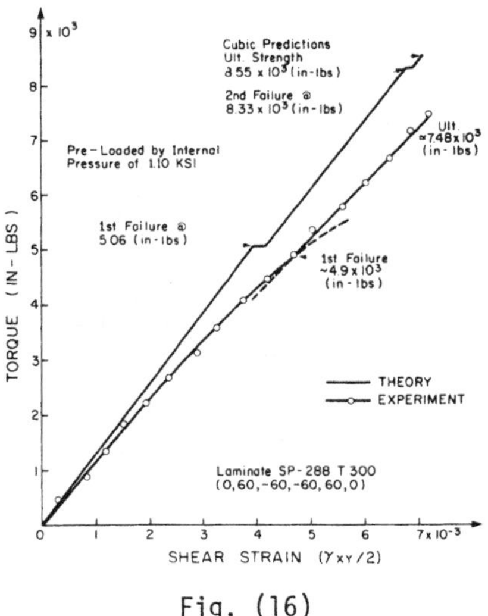

Fig. (16)

CONCLUSIONS

Experimental and analytical results have been presented demonstrating the necessity of incorporating a higher order strength criterion than the quadratic form, particularly for biaxial and combined loading. The cubic polynomial form of the strength criterion has been shown to provide good correlation with test data for the range of configurations studied, although care must be exercised for those cases where complex solutions are obtained. There is no doubt that significant differences between the two formulations can exist both in terms of failure stresses and in the selection of optimum ply orientations. Of some concern are the higher failure loads which are predicted by the quadratic theory for certain load conditions. Clearly, additional experiments must yet be performed to better define the interaction coefficients and provide a larger data base before the cubic model can be used with complete confidence for design purposes.

ACKNOWLEDGEMENTS

The authors wish to acknowledge the financial assitance of both the Natural Sciences and Engineering Research Council of Canada (Grant No. A2783) and the National Aeronautics and Space Administration, Langley Research Center, (Grant No. NSG-7409, Sup. No. 2), for their support of this program

REFERENCES

[1] Sandhu, R. S., "A survey of failure theories of isotropic and anisotropic materials", AFFDL Tech. Rept. AFFDL-TR-72-71.

[2] Wu, E. M., "Phenomenological anisotropic failure criterion", Treatise on Composite Materials, Broutman, Krock and Sendeckyj, eds., Academic Press, 1973.

[3] Malmeister, A. K., "Geometry of theories of strength", Mekhanika Polimerov, Vol. 2, No. 4, 1966.

[4] Tsai, S. W. and Wu, E. M., "A general theory of strength for anisotropic materials", J. Composite Materials, Vol. 5, 1971.

[5] Tennyson, R. C., MacDonald, D. and Nanyaro, A. P., "Evaluation of the tensor polynomial failure criterion for composite materials", J. Composite Materials, Vol. 12, 1978.

[6] Tennyson, R. C., Nanyaro, A. P. and Wharram, G. E., "Application of the cubic polynomial strength criterion to the failure analysis of composite materials", J. Composite Materials Supplement, Vol. 14, 1980.

[7] Hahn, H. T. and Tsai, S. W., "Nonlinear elastic behavior of unidirectional composite laminae", J. Composite Materials, Vol. 7, 1973.

NOMENCLATURE

A_{ij}, B_{ij}, D_{ij}	$= \int_{-h/2}^{h/2} (\overline{Q}_{ij})_k (1, z, z^2) dz$
$[A']$	$= [A^*] - [B^*][D^*]^{-1}[C^*]$
$[B']$	$= [B^*][D^*]^{-1}$
$[C']$	$= -[D^*]^{-1}[C^*]$
$[D']$	$= [D^*]^{-1}$
$[A^*]$	$= [A]^{-1}$
$[B^*]$	$= [A]^{-1}[B]$
$[C^*]$	$= [B][A]^{-1}$
$[D^*]$	$= [D] - [B][A]^{-1}[B]$
E_{11}, E_{22}	= orthotropic moduli of elasticity measured in the 1 and 2 directions, respectively
G_{12}	= shear modulus of elasticity measured in 1-2 plane
h	= total laminate thickness
K	= curvature
M_x, M_y	= bending moment resultants
M_{xy}	= twisting moment resultant
N_x, N_y	= normal laminate stress resultants acting along x,y axes, respectively
N_{xy}	= shear laminate stress resultant acting in x-y plane
$\overline{Q}_{11}$	$= Q_{11}m^4 + 2(Q_{12}+2Q_{66})n^2m^2 + Q_{22}n^4$
$\overline{Q}_{22}$	$= Q_{11}n^4 + 2(Q_{12}+2Q_{66})n^2m^2 + Q_{22}m^4$
$\overline{Q}_{12}$	$= (Q_{11}+Q_{22}-4Q_{66})n^2m^2 + Q_{12}(m^4+n^4)$
$\overline{Q}_{66}$	$= (Q_{11}+Q_{22}-2Q_{12}-2Q_{66})n^2m^2 + Q_{66}(m^4+n^4)$
$\overline{Q}_{16}$	$= (Q_{11}-Q_{12}-2Q_{66})nm^3 + (Q_{12}-Q_{22}+2Q_{66})n^3m$
$\overline{Q}_{26}$	$= (Q_{11}-Q_{12}-2Q_{66})n^3m + (Q_{12}-Q_{22}+2Q_{66})nm^3$
Q_{11}	$= E_{11}/(1-\nu_{12}\nu_{21})$
Q_{22}	$= E_{22}/(1-\nu_{12}\nu_{21})$

Q_{12} = $\nu_{21}E_{11}/(1-\nu_{12}\nu_{21})$

Q_{66} = G_{12}

S,S' = positive and negative lamina shear strengths measured in 1-2 plane

X,X' = tensile and compressive lamina strengths measured in 1 direction

Y,Y' = tensile and compressive lamina strengths measured in 2 direction

Subscripts

x,y = orthogonal in-plane structural axes

1,2 = lamina material axes parallel and orthogonal to fiber reinforcement

Greek Symbols

γ = shear strain

ε = normal strain

θ = fiber orientation relative to x-axis

ν = Poisson's ratio

σ = normal stress

τ = shear stress

APPLICATION OF IMPULSE FUNCTIONS TO THE RELIABILITY OF FIBER REINFORCED COMPOSITES

I. T. Komozin and I. N. Preobrazhenskiy

Zaporozhye Industrial Institute
USSR

INTRODUCTION

A vast amount of bibliography has been devoted to study the macroscopic behavior of composites due to the influence of the microstructure. From a practical viewpoint, this is of limited interest since the macroscopic behavior can be easily determined by experiments. The state of stress and strain in a structural element can differ significantly if the loading conditions in the experiment are not properly simulated.

BASIC FORMULATION

Consider a two-component medium. Based on the distribution of reinforcements, the reliability of composites can be discussed without considering the traditional approach of boundary-value problem solving.

Let a function

$$\Gamma(x-\xi) = \begin{cases} 1\ , & \text{if } x>\xi \\ 1/2, & \text{if } x=\xi \\ 0\ , & \text{if } x<\xi \end{cases} \tag{1}$$

possess the properties as shown in equation (1). An integral form of the function can be written as

$$\int_0^x \left[\frac{d}{dx}\Gamma(x-\xi)\right]dx = 1,\ \xi\varepsilon(0,x) \tag{2}$$

In accordance with equations (2), it follows that

$$\int_0^x f(x)\left[\frac{d}{dx}\Gamma(x-\xi)\right]dx = f(\xi) \tag{3}$$

Now, consider a generalization of the function in equation (1) in R^2

$$\Gamma(x-\xi,y-\eta) = \begin{cases} 1\ , & \text{if } (x,y)\varepsilon(x>\xi,y>\eta) \\ 1/2, & \text{if } (x,y)\varepsilon D \\ 0\ , & \text{if } (x,y)\varepsilon(x>\xi,y>\eta) \end{cases} \tag{3}$$

where

$$D = \{(\xi,y),\ y>\eta;\ (x,\eta),\ x>\xi\}$$

Note that equation (3) can be further generalized in R^2 to read as

$$\int_0^x\int_0^y f(x,y)\ [\frac{d}{dx}\ \Gamma(x-\xi;y-\eta)]dxdy = \int_0^y f(\xi,y)dy(\overset{\rightarrow}{\leftarrow}) \tag{4}$$

A linear combination of the functions in equations (3) lead to an impulse function in R^2, i.e.,

$$\delta(x,y) = \Gamma(x-x_1;y-y_1) - \Gamma(x-x_1;y-y_2) - \Gamma(x-x_2;y-y_1) + \Gamma(x-x_2;y-y_2) \tag{5}$$

which is defined as

$$\delta(x,y) = \begin{cases} 1\ , & \text{if } (x,y)\varepsilon D_1 \\ 0\ , & \text{if } (x,y)\notin D_1 \\ 1/2, & \text{if } (x,y)\varepsilon \end{cases} \tag{6}$$

where

$$D_1 = (x_1<x<x_2;\ y_1<y<y_2)$$

Analogous to equation (5), an impulse function in R^3 can be written:

$$\begin{aligned} j(X,Y,Z) = {} & \Gamma(X-X_1,Y-Y_1,Z-Z_1) - \Gamma(X-X_2,Y-Y_1,Z-Z_1) - \Gamma(X-X_1,Y-Y_1,Z-Z_1) \\ & + \Gamma(X-X_2,Y-Y_2,Z-Z_1) + \Gamma(X-X_2,Y-Y_1,Z-Z_2) + \Gamma(X-X_1,Y-Y_2,Z-Z_2) \\ & - \Gamma(X-X_1,Y-Y_1,Z-Z_2) - \Gamma(X-X_2,Y-Y_2,Z-Z_2) \end{aligned} \tag{7}$$

such that

$$j(x,y,Z) = \begin{cases} 1/2, & \text{if } (x,y,Z)\varepsilon S \\ 1\ , & \text{if } (x,y,Z)\varepsilon D_2 \\ 0\ , & \text{if } (x,y,Z)\notin D_2 \end{cases} \tag{8}$$

where S represents the surface of the area and $D_2 = (X_1<X<X_2;\ Y_1<Y<Y_2;\ Z_1<Z<Z_2)$. The condition in equation (3) now becomes

$$\int_V f(x,y,z)\left[\frac{d}{dx}\Gamma(x-\xi,\ y-\eta,\ z-\zeta)dV = \int_\eta^y \int_\zeta^z f(\xi,y,z)dydz \right.$$
$$V = \{x\varepsilon(0,x);\ y\varepsilon(0,y);\ z\varepsilon(0,z)\} \qquad (9)$$

Equations (3), (4) and (9) are fundamental as they permit the usage of variational methods.

Let a rectangular area be oriented in the coordinate system xyz and at the angles α_1, α_2, α_3 with respect to the XYZ axes. Then

$$\begin{bmatrix} X \\ Y \\ Z \end{bmatrix} = (A)\begin{bmatrix} x \\ y \\ z \end{bmatrix},\ (a_{ij}) = (A),\ i,j = 1,3 \qquad (10)$$

For a rectangular parallelepipedon referred to the above referenced axes, the function in equation (9) takes the form

$$\begin{aligned}\delta^*(x,y,z) = {}& \Gamma[(A)\begin{bmatrix} x-x_1 \\ y-y_1 \\ z-z_1 \end{bmatrix}] - \Gamma[(A)\begin{bmatrix} x-x_2 \\ y-y_1 \\ z-z_1 \end{bmatrix}] \\ & - \Gamma[(A)\begin{bmatrix} x-x_1 \\ y-y_2 \\ z-z_1 \end{bmatrix}] + \Gamma[(A)\begin{bmatrix} x-x_2 \\ y-y_2 \\ z-z_1 \end{bmatrix}] + \Gamma[(A)\begin{bmatrix} x-x_2 \\ y-y_1 \\ z-z_2 \end{bmatrix}] \\ & + \Gamma[(A)\begin{bmatrix} x-x_1 \\ y-y_2 \\ z-z_2 \end{bmatrix}] - \Gamma[(A)\begin{bmatrix} x-x_1 \\ y-y_1 \\ z-z_2 \end{bmatrix}] - \Gamma[(A)\begin{bmatrix} x-x_2 \\ y-y_2 \\ z-z_2 \end{bmatrix}]\end{aligned} \qquad (11)$$

In many cases, the cross-section of the reinforcements is circular. If the generator of the cylinder is parallel to the axis OZ, the function, characterizing the volume of the cylinder, can be obtained from equation (7) as

$$\begin{aligned} j^*(X,Y,Z) = {}& j(X_1=\mu_1,\ X_2=\mu_2,\ Y_1=\omega_1,\ Y_2=\omega_2,\ Z_1,\ Z_2)_{k=1} \\ & + \sum_{k=2}^{S} j(X_1=\mu_1,\ X_2=\mu_2,\ Y_1=\omega_3,\ Y_2=\omega_4,\ Z_1,\ Z_2) \end{aligned} \qquad (12)$$

where

$$\mu_i = a + (-1)^i \, r\cos kt$$

$$\omega_i = b + (-1)^i \, r\sin kt, \; i = 1,2 \tag{13}$$

$$\omega_3 = b + r\sin(k-1)t; \; \omega_4 = b - r\sin(k-1)t$$

In equations (13), (a,b) are the coordinates of the cylinder axis, $t = \pi/2s$, s is a positive integer governing the pitch of the set of rectangles approximating the area of a circle and r is the cylinder radius.

If the generator of the cylindrical surface and the axis OZ make an angle α_3, the function in equation (11) can be transformed in the same way as equation (12):

$$\delta(x,y,z) = \delta^*\{x_1=\mu_1, \, x_2=\mu_2, \, y_1=\omega_1, \, y_2=\omega_2, \, z_1, \, z_2\}_{k=1} + \sum_{k=1}^{s} \delta^*\{x_1=\mu_1, \, x_2=\mu_2, \, y_1=\omega_3, \, y_2=\omega_4, \, z_1, \, z_2\} \tag{14}$$

Use of the function $\delta(x,y,z)$ can be made in a similar fashion for other orientations of the reinforcements. Subsequently, consider the impulse function in equation (14) is known for the ith element oriented in an arbitrary position of the referenced coordinates. By means of linear superposition, the generalized function for the composite is obtained:

$$\sigma = \sum_{i=1}^{n} \sigma_i(x,y,z) \tag{15}$$

where $\sigma_i(x,y,z)$ is the impulse function for the ith element and n the number of elements in the volume.

A TWO-COMPONENT COMPOSITE

Assume that the components of the bimaterial composite are isotropic. Their elastic constants of the composite are

$$E = (E_a - E_M)\delta + E_M$$

$$\nu = (\nu_a - \nu_M)\delta + \nu_M \tag{16}$$

where E_a, E_M, ν_a and ν_M are the moduli of elasticity and Poisson's ratios of the reinforcements and matrix, respectively.

In the theory of elasticity, the equations of equilibrium for the three-dimensional problem have the same appearance as those for a homogeneous material except that the elastic parameters now vary with the space coordinates:

$$\Omega_i(U,V,W) = \Delta_{,x}\lambda + 2e_{11,x}G + \sigma G(e_{12,y} + e_{13,z})$$

$$+ \Delta\lambda_{,x} + \omega(2e_{11}\sigma_{,x} + e_{12}\sigma_{,y} + e_{13}\sigma_{,z}) + P_1 = 0 \qquad (17)$$

where $\Delta = e_{11} + e_{22} + e_{33}$ is the sum of the relative deformations in the directions of x, y, z and G = G(M) is the modulus of the transverse elasticity, i.e.,

$$G = \frac{E_M+(E_a-E_M)\sigma}{2N}; \ \lambda_{,x} = \sigma_{,x}L(\vec{r}) \qquad (18)$$

The Lamé coefficient, $\lambda(x,y,z)$, is

$$\lambda = \frac{E_M\nu_M+(E_a\nu_a-E_M\nu_M)\sigma}{H} \qquad (19)$$

and

$$\omega = \frac{1}{2}\frac{1}{N^2}\,[(E_a-E)(1+\nu_M) - (\nu_a-\nu_M)E_M]$$

$$L = \frac{1}{H^2}\,\{(E_a\nu_a-E_M\nu_M)(1+\nu_M)(1-2\nu_M) - (\nu_M-\nu_a)(1+2\nu_M+2\nu_a)E_M\nu_M\} \qquad (20)$$

$$N = 1 + (\nu_a-\nu_M)\sigma + \nu_M$$

$$H = (1+\nu_M)(1-2\nu_M) + (\nu_M-\nu_a)[1 + 2(\nu_M+\nu_a)\sigma]$$

The subscripts x, y, z, following a comma, signify differentiation with respect to the particular variable. Integration of equation (17) is made by the Bubnov variational method in the finite-element form [1].

A SYSTEM OF LINEAR EQUATIONS

The solution for the jth-element becomes

$$\begin{bmatrix} u \\ v \\ w \end{bmatrix} \quad \sum_{j=1}^{j=27} N_j \begin{bmatrix} u_j \\ v_j \\ w_j \end{bmatrix} \qquad (21)$$

where N_j is the Lagrangian family of basic functions of the rectangular complex element [2]. The displacement components of the jth node are u_j, v_j, w_j. In order to minimize the error on the solution of equation (21), the basic functions N_j will be orthogonalized with respect to the volumes of all the finite elements:

$$\sum_{e=1}^{n} \int_{v_e} \sum_{j=1}^{27} \begin{bmatrix} u_j \\ v_j \\ w_j \end{bmatrix} \Omega_i(u,v,w)dV_e = 0; \; i = 1,2,3 \tag{22}$$

The method adopted here is described in [3]. From the first of equations (22), the coefficients of the eth-element of the linear system of algebraic equation is found:

$$B_{k,i} = \int_{v_e} N_k\{(\lambda+2G)N_{i,xx} + \sigma G(N_{i,yy}+N_{i,zz}) + \sigma_{,x}(L+2\omega)N_{i,x} + \omega\sigma_{,y}N_{i,y}$$

$$+ \omega\sigma_{,z}N_{i,z}\}dV; \; k,i = 1,2,3,\ldots,27$$

$$(23)$$

$$B_{k,27+i} = \int_{v_e} N_k\{(\lambda+\sigma G)N_{i,xy} + \omega\sigma_{,y}N_{i,x} + L\sigma_{,x}N_{i,y}\}dV$$

$$B_{k,54+i} = \int_{v_e} N_k\{(\lambda+\sigma G)N_{i,xz} + \omega\sigma_{,z}N_{i,x} + L\sigma_{,x}N_{i,z}\}dV$$

The second of the system of equations (22) gives

$$B_{k+27,i} = \int_{v_e} N_k\{(\lambda+G\sigma)N_{i,xy} + L\sigma_{,y}N_{i,x} + \omega\sigma_{,x}N_{i,y}\}dV$$

$$B_{k+27,27+i} = \int_{v_e} N_k\{(\lambda+2G)N_{i,yy} + \sigma G(N_{i,zz}+N_{i,xx}) + \sigma_{,y}LN_{i,y}$$

$$+ \omega(2N_{i,y}\sigma_{,y} + N_{i,x}\sigma_{,x} + N_{i,z}\sigma_{,z})\}dV \tag{24}$$

$$B_{27+k,54+i} = \int_{v_e} N_k\{(\lambda+\sigma G)N_{i,zy} + L\sigma_{,y}N_{i,z} + \omega N_{i,y}\sigma_{,z}\}dV$$

$k,i = 1,2,\ldots,27$

It follows that the third of equations (22) leads to

$$B_{54+k,i} = \int_{V_e} N_k\{(\lambda+G)N_{i,xz} + L\sigma_{,z}N_{i,x} + \omega\sigma_{,x}N_{i,z}\}dV$$

$$B_{54+k,27+i} = \int_{V_e} N_k\{(\lambda+G)N_{i,yz} + L\sigma_{,z}N_{i,y} + \omega N_{i,z}\sigma_{,y}\}dV \qquad (25)$$

$$B_{54+k,54+i} = \int_{V_e} N_k\{(\lambda+2G)N_{i,zz} + 2G(N_{i,xx}+N_{i,yy}) + L\sigma_{,z}N_{i,z} + \omega(2N_{i,z}\sigma_{,z} + N_{i,x}\sigma_{,x} + N_{i,y}\sigma_{,y})\}dV$$

The elements in the column to the right in equation (26) for the eth-element are:

$$d_k = \int_{V_e} N_k P_{1,k} dV; \quad d_{27+k} = \int_{V_e} N_k P_{2,k} dV;$$

$$d_{54+k} = \int_{V_e} N_k P_{3,k} dV; \quad k = 1,2,\ldots,27$$

in which $P_{i,k}$ represents the components of the external forces. Hence for the eth-element:

$$(b_{k,i})\begin{bmatrix} u_i \\ \vdots \\ u_{27} \\ v_1 \\ \vdots \\ v_{27} \\ w_1 \\ \vdots \\ w_{27} \end{bmatrix} = \begin{bmatrix} d_1 \\ d_2 \\ \cdot \\ \cdot \\ \cdot \\ d_{81} \end{bmatrix} \qquad k,i = 1,2,\ldots,81 \qquad (26)$$

The system of equations (26) is solved by the method of Zeidel simple iteration. The advantage lies in its simplicity to complete a cycle of interaction. A relatively small computer storage space is required. An important factor in favor of this method is that rounding-off error is not accumulated which is essential for a large system. A sample example of the integrals will be made. It involves the derivative of the impulse function $\sigma(x,y,z)$ for one rectangular element D_2.

$$\int_{V_e} (L+2\omega) N_{i,x}\sigma_{,x} dV = \iint (L+2\omega)\Big|_{S\left(\substack{x=x_1 \\ x=x_2}\right)} \Big[\text{ɜ}\Big|_{y_1,z_1}^{y_e,z_e} + \text{ɜ}\Big|_{y_2,z_2}^{y_e,z_e}$$

$$- \text{ɜ}\Big|_{y_1,z_2}^{y_e,z_e} - \text{ɜ}\Big|_{y_2,z_1}^{y_e,z_e}\Big] dydz \tag{27}$$

$$\text{ɜ} = \left|N_{x,i}N_k\right|_{x=x_1} - \left|N_{i,x}N_k\right|_{x=x_2} ; \quad S_{x=x_1} = S_{x=x_2}$$

where y_e and z_e are the upper limits of the rectangular finite element. The integrals in equation (27) are then calculated:

$$\int_{y_1}^{y_e}\int_{z_1}^{z_e} \left|L+2\omega\right|_{S(x=x_1)} \left|N_k N_{i,x}\right|_{x=x_1} dydz = \left|L+2\omega\right|_{\sigma=\frac{1}{2}} \int_{y_1}^{y_2}\int_{z_1}^{z_2} \left|N_{i,x}N_k\right|_{x=x_1} dydz$$

$$+ \; L+2\omega\Big|_{\sigma=0} \Big(\int_{y_2}^{y_e}\int_{z_1}^{z_e} \left|N_{i,x}N_k\right|_{x=x_1} dydz$$

$$+ \int_{y_1}^{y_e}\int_{z_2}^{z_e} \left|N_{i,x}N_k\right|_{x=x_1} \Big) dydz \tag{28}$$

The integrals of the second group in equations (23) to (25) containing the functions $\lambda(x,y,z)$ and $G(x,y,z)$ may be resolved as

$$\int_{V_e} \lambda N_k N_i dV = |\lambda|_{\sigma=0} \int_{V_e} N_k N_i dV + \left(|\lambda|_{\sigma=1} - |\lambda|_{\sigma=0}\right) \int_{x_1}^{x_2}\int_{y_1}^{y_2}\int_{z_1}^{z_2} N_i N_k dV \tag{29}$$

Finally, the third group of the integrals including the product of the two impulse functions σG gives

$$\sigma G = \frac{\sigma E_a}{2N} \tag{30}$$

Use has been made of the obvious identity

$$\sigma(x,y,z) = \sigma^2(x,y,z); \; \forall (x,y,z) \varepsilon S$$

Equation (30) thus gives

$$\int_{V_e} \sigma G N_k N_{i,yy} dV = \frac{E_a}{2(1+\nu_a)} \int_{D_2} N_k N_{i,yy} dV \tag{31}$$

Summing up all the elements with the aid of equations (27) to (30), the quantities in equations (23) to (25) may be calculated for the functions in equations (11) and (14) in the same way.

DISCUSSION OF RESULTS

Comparison will be made of the results obtained by the method described here for the case of bending of a hinged orthotropic beam having the dimensions $2\ell = 40$ mm, $2h = 10$ mm and $b = 2h$ aligned along the directions of the coordinate axes x,y,z. Suppose that it is loaded at mid-span by a concentrated load P, which is uniformly distributed along its width. The properties of the reinforcements are: $E_a = 7310$ kgs/mm^2, $\nu_a = 0.35$ and $2r = 2{,}25$ mm. The percentage of reinforcement is assumed [5] to be 32.4% along the OX axis and 3.1% along the OZ axis. The maximum tangential stresses along the width of the beam at $\xi=0$ are given in Table 1. They have been computed according to the known solution [4] taking into account the modulus of the transverse elasticity $G_{13} = 350$ kgs/mm^2 given in [5]. The results obtained by the present method are given in Table 1. With 1/4-symmetry, the beam area is approximated three elements along the length and four elements along the height. In three-dimensions, each element has 81 degrees of freedom. On the border of the area, the conditions are

$$w_x = \frac{\ell}{2} = u_{x=0} = v_{y=0} = 0$$

The distribution of the reinforcements is shown in Figure 1.

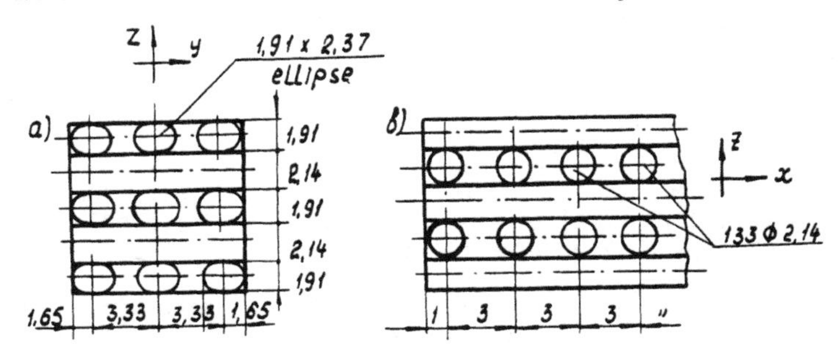

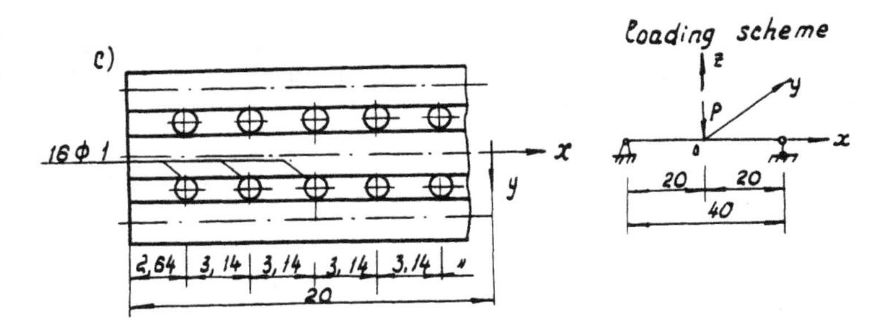

Fig. (1) - The distribution of reinforcing material in the beam.
a) in direction of axis OX; b) Oy; c) Oz

TABLE 1

$\xi = \frac{x}{\ell}$	0.05	0.15	0.25	0.35
$2 \frac{\tau_{xz}^{max}}{p} h^2$	$\frac{2,4134}{2,5362}$	$\frac{0,9604}{0,9901}$	$\frac{0,7803}{0,7771}$	$\frac{0,75}{0,7601}$

It is interesting to note that an impulse representation of the elastic behavior, according to equation (16), leads to the nonuniform character of the deformation field. Their corresponding products, however, yield the continuous character of the stresses and displacements. On the contact surfaces, the reinforcements and matrix are required to satisfy the conditions:

$$\tau_{ija} = \tau_{ijm};\ \sigma_{ija} = \sigma_{ijm};\ u_a = u_m;\ v_a = v_m;\ w_a = w_m\ (i = 1,2,3)$$

Analogous derivation has been obtained for the axisymmetric deformation of cylinders [6].

REFERENCES

[1] Segerland, L., "Application of finite element method", Moscow, "Mir", p. 392, 1979.

[2] Nemchinov, Y. I., "Computations for three-dimensional structures (finite element method)", Kiev, "Budivelnik", p. 228, 1980.

[3] Zenkevich, O., "Finite element method in engineering", Moscow, "Mir", p. 380, 1975.

[4] Tarnopolskiy, Y. M., Zhigun, I. G. and Polyakov, V. A., "Analysis of the distribution of tangential stress for three-pointwise bending of the beams made of composites", "Mekhanika Polimerov", No. 1, pp. 56-62, 1977.

[5] Zhigun, I. G., Dushin, M. I., Polyakov, E. A. and Yakushin, V. A., "Composite materials reinforced by the system of three straight mutually orthogonal fibers", 2, Experimental Studies, Mekhanika Polimerov, No. 3, pp. 1011-1018, 1973.

[6] Byrke, M. S., "Computations for the multi-layered and non-uniform cylinders using step functions", Izvestiya Vysshikh Uchebnykh Zavedeniy, Mashinostroeniye, No. 8, pp. 5-8, 1980.

SECTION II

CRACK AND FRACTURE ANALYSIS

TRANSIENT HYGROTHERMAL AND MECHANICAL STRESS INTENSITIES AROUND CRACKS

G. C. Sih

Lehigh University
Bethlehem, Pennsylvania 18015

and

A. Ogawa*

National Aerospace Laboratory
Tokyo, Japan

INTRODUCTION

One of the main concerns of using advanced fiber-reinforced polymeric matrix composites as aerospace structural components is the degradation of the material due to moisture at elevated temperatures. This is of particular concern to the matrix-dominated composite. The deterioration of material properties cannot be easily understood without a sound analytical modeling of the physical problem. A majority of the investigations [1-3] on this subject assume that temperature and moisture do not interact and that they each obey the simple (Fickian) diffusion theory. As a consequence, experimental data are available only for changes in moisture condition [4] while the temperature environment is taken to be constant. In contrast to the common belief, coupling between temperature and moisture can be extremely important [5,6] when the surface temperature undergoes rapid changes. The difference in the hygrothermal stresses with and without coupling can differ anywhere from 20 to 80% depending on the surface temperature gradient. This casts a new light on the subject and suggests a series of new experiments before reliable predictions on the life of structural components due to service environments could be made.

In two recent publications, the transient hygrothermal stresses around a spherical [7] and circular [8] cavity were determined. The stresses were found to oscillate in time and vary in a complicated nature depending on the boundary condition whether moisture and/or temperature are applied to the cavity. An analysis was also performed in [8] to investigate the possible sites of failure. The hygrothermally and mechanically induced stresses were found to peak at distances away and near the cavity, respectively. Based on the minimum strain en-

---

*This work was completed in the U.S. when Akinori Ogawa held the position of Visiting Scientist at the Institute of Fracture and Solid Mechanics at Lehigh University.

ergy density criterion, actual locations of possible failure were determined and discussed in detail.

The primary objective of this investigation is to develop an analytical method for determining the redistribution of hygrothermal and mechanical stresses due to crack-like defects such that the initiation of failure in composites exposed to service environments could be better understood. These defects are modeled as narrow ellipses. Use is made of the time-dependent finite element method developed in an earlier publication [8]. Different grid patterns are constructed as the aspect ratios of the elliptically shaped crack are altered. Presented graphically are hygrothermal stresses near the ellipse for different time. Time dependent stress intensity factors are also defined to study the possible fracture of the T300/5208 epoxy resin.

MATHEMATICAL MODEL

The thermodynamic treatment for deriving the coupled diffusion equations is given in [9] and will not be repeated here. If T stands for the temperature and C the mass of moisture per unit volume of void space in the solid, then the governing equations take the forms

$$D\nabla^2 C - \frac{\partial}{\partial t}(C - \lambda T) = 0$$

$$\mathcal{D}\nabla^2 T - \frac{\partial}{\partial t}(T - \nu C) = 0 \qquad (1)$$

in which D and $\mathcal{D}$ are the diffusion coefficients with units of area per unit time and λ and ν are the coupling coefficients with units of mass per unit volume per unit temperature and the reciprocal, respectively. The Laplace operator in equations (1) is in two dimensions given by $\nabla^2 = \partial^2/\partial x^2 + \partial^2/\partial y^2$. The domain of interest is that of a multiply-connected rectangular region R having the dimensions: 6 units in height and 8 units in width. A narrow elliptical opening with semi-major axis a of one unit and variable semi-major axis b is centered at the origin of a rectangular coordinate system (x,y) and is shown in Figure 1. Initially for t<0, the region R possesses the following temperature and moisture fields:

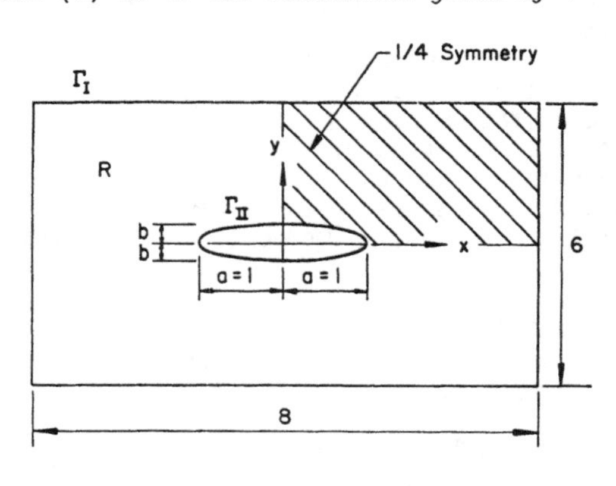

Fig. (1) - A rectangular region with a crack-like opening

$$T(x,y,t) = T_o(x,y)$$

$$C(x,y,t) = C_o(x,y) \qquad (2)$$

For t>0, the temperature and/or moisture on the boundary Γ_I and/or Γ_{II} are changed such that equations (2) become

$$T(x,y,t) = T_0(x,y) + \Delta T(x,y,t)$$

$$C(x,y,t) = C_0(x,y) + \Delta C(x,y,t) \tag{3}$$

Once the boundary conditions at $t = t_0$ are known, equations (1) may be solved numerically for $T(x,y,t)$ and $C(x,y,t)$.

The time dependent two-dimensional finite element method developed in [8] will be applied to evaluate equations (1). The two types of boundary conditions are *sudden moisture change* with

$$\Delta C = \begin{cases} 0 & \text{on } \Gamma_I \\ \Delta C_B & \text{on } \Gamma_{II} \end{cases} \tag{4}$$

$$\Delta T = 0 \text{ on } \Gamma_I \text{ and } \Gamma_{II}$$

and *sudden temperature change* with

$$\Delta C = 0 \text{ on } \Gamma_I \text{ and } \Gamma_{II}$$

$$\Delta T = \begin{cases} 0 & \text{on } \Gamma_I \\ \Delta T_B & \text{on } \Gamma_{II} \end{cases} \tag{5}$$

Since the problem is one-quarter symmetry, there is only the need to consider the grid pattern shown in Figure 2 consisting of 100 nodal points. The elements near the elliptical cavity are smaller in size so as to accommodate the high gradient of the local moisture and temperature. In fact, special consideration is given to the sub-region A enclosed by the corner nodes 6, 27, 48 and 47. Special grid pattern in Figure 3 is constructed for $b/a = 0.067$ or $\rho = b\ /a = 0.0044$. The numerical results will be presented subsequently.

NUMERICAL RESULTS ON DIFFUSION

The systems of equations (1) is solved numerically by subjecting the elliptical cavity to sudden changes in the surface moisture and/or temperature according to the conditions described by equations (4) and/or (5). The moisture and temperature distribution are assumed to be constant in the direction normal to the xy-plane. For the T300/5208 epoxy resin material, the coupling constants are [6] given by $D/\mathcal{D} = 0.1$, $\lambda = 0.5$ and $\nu = 0.5$. Of particular interest is the variation of C and T in the material ahead of elliptical cavity as a function of time.

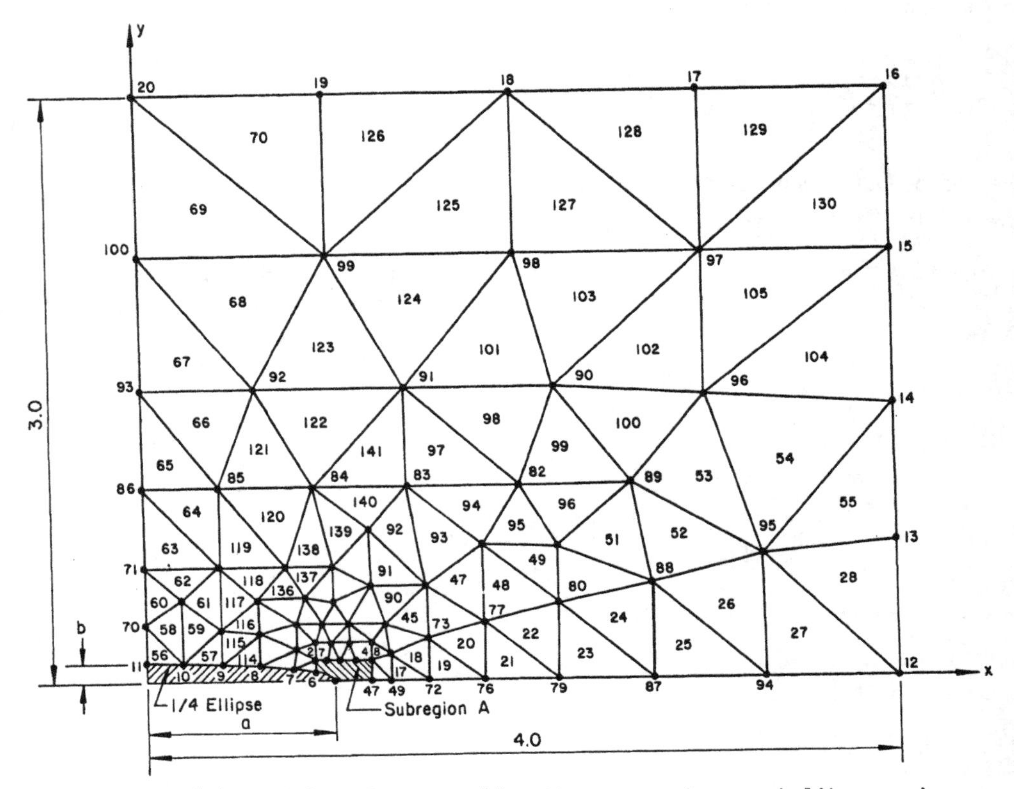

Fig. (2) - Finite element grid pattern around a crack-like opening

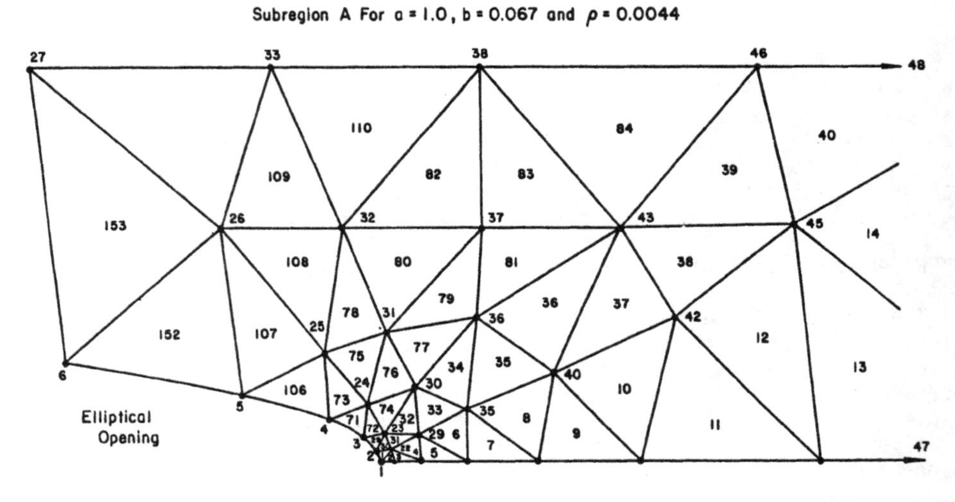

Fig. (3) - Sub-region ahead of an elliptically-shaped crack with b/a = 0.067

Sudden moisture change. Figures 4 and 5 give the results for the case when the surface moisture on the ellipse is suddenly raised to another constant value,

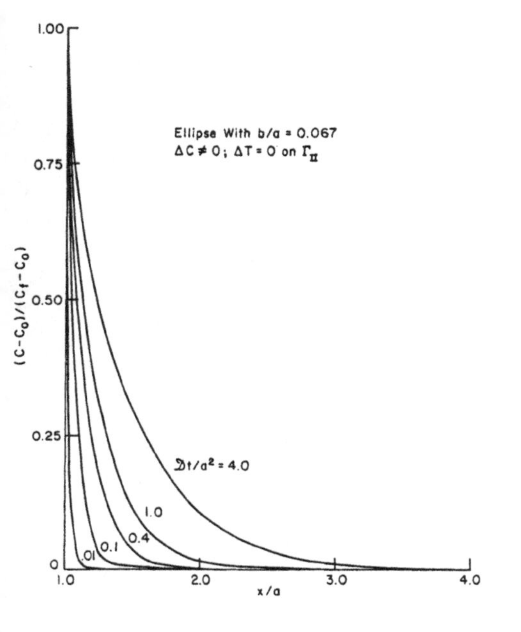

Fig. (4) - Normalized moisture content versus distance for sudden moisture change

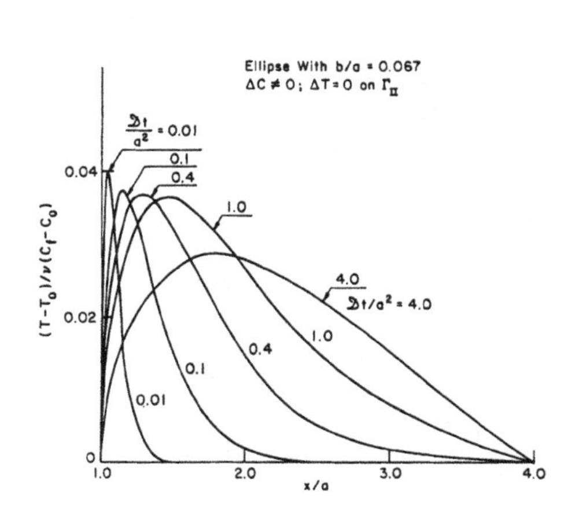

Fig. (5) - Normalized temperature with distance for sudden moisture change

while the surface temperature is unchanged. Refer to the conditions in equations (4). The initial and final condition on the outside boundary of R in Figure 1 is maintained constant at all time. Plots of $(C-C_0)/(C_f-C_0)$ versus x/a for different values of $\mathcal{D}t/a^2$ are shown in Figure 4. The end point of the narrow ellipse is given by x/a = 1.0. Initially, the moisture drops very rapidly in the vicinity of the elliptical cavity and it gradually diffuses into the material as time is increased. The transient behavior of the temperature is exhibited in Figure 5. The quantity $(T-T_0)/\nu(C_f-C_0)$ is seen to peak very sharply at first and the variations become more gradual as time increases. The peaks move into the material with decreasing amplitude and they take lower values as the ellipse becomes more slender.

Sudden temperature change. The disturbances within the solid due to coupling of heat and moisture are considerably more pronounced when the surface temperature on the elliptical cavity undergoes rapid changes, equations (5). Figure 6 gives the variations of $(T-T_0)/(T_f-T_0)$ with x/a and time. For small time, only the material near the ellipse experiences temperature change. The temperature gradient tends to spread over a larger portion of the material as time goes by. Numerical results of $(C-C_0)/\lambda(T_f-T_0)$ versus x/a for the case of sudden temperature change are similar to those shown in Figure 5 except for the factor $\mathcal{D}/D$ which is equal to ten in the present problem, i.e.,

$$\left[\frac{C-C_o}{\lambda(T_f-T_o)}\right]_{\Delta T = \text{const}} = \frac{\mathcal{D}}{D}\left[\frac{T-T_o}{\nu(C_f-C_o)}\right]_{\Delta C = \text{const}} \tag{6}$$

Hence, there is no need to repeat the plot.

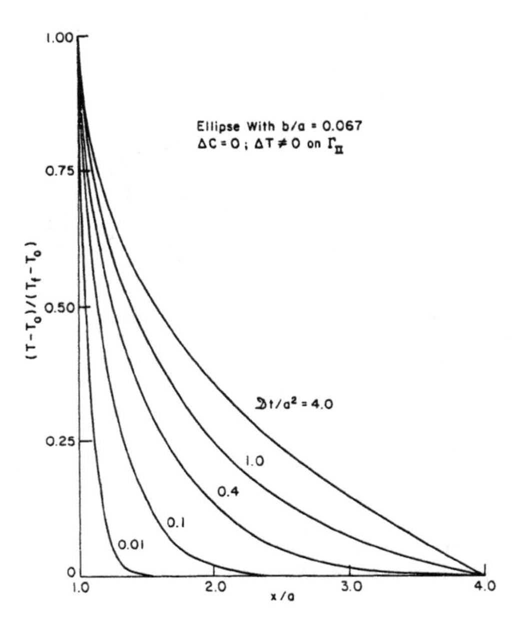

Fig. (6) - Normalized temperature versus distance for sudden temperature change

HYGROTHERMAL STRESSES

For an isotropic and homogeneous material, the hygrothermal stresses may be obtained as

$$\sigma_{ij} = E(\varepsilon_{ij} - \alpha\Delta T\delta_{ij} - \beta\Delta C\delta_{ij}) \tag{7}$$

in which E is the Young's modulus, α the coefficient of thermal expansion and β the coefficient of moisture expansion. The stress and strain components are denoted respectively by σ_{ij} and ε_{ij}. The condition of plane strain with $\varepsilon_z = 0$ will be assumed such that

$$\sigma_z = \nu_p(\sigma_r+\sigma_\theta) \tag{8}$$

with ν_p being the Poisson's ratio. The material properties pertaining to T300 /5208 are given as follows:

$$\alpha = 4.5 \times 10^{-5}\ \text{m/m°C};\quad \beta = 2.68 \times 10^{-3}\ \text{m/m/\% } H_2O$$
$$E = 3.45\ \text{GN/m}^2\ (5 \times 10^5\ \text{psi});\quad \nu_p = 0.34 \tag{9}$$

For clarity sake, the discussion for the cases of moisture change and temperature change will be carried out separately.

Sudden application of moisture. When the moisture on the elliptical cavity is suddenly raised, there results a nonuniform expansion and contraction of material which, in turn, give rise to stresses and strains. Figure 7 shows that the radial stresses are compressive along the line of symmetry ahead of the ellipse. They decrease rapidly at first and then increase steadily towards zero. The corresponding circumferential stress is given in Figure 8. Unlike σ_r, σ_θ starts out

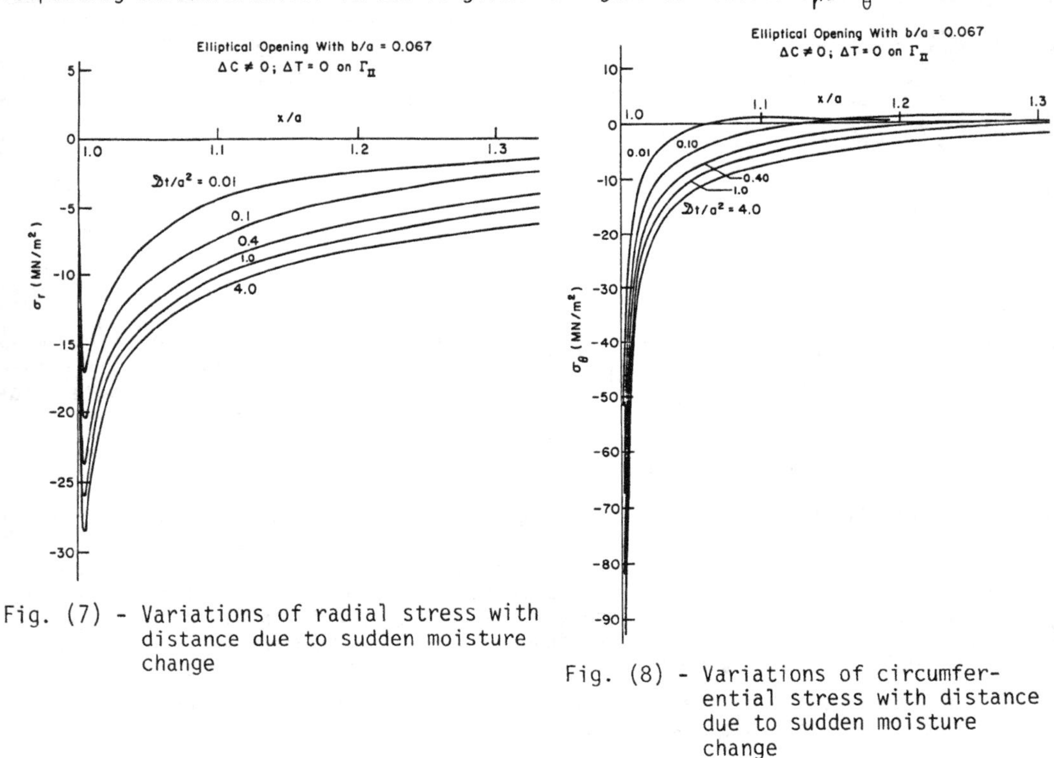

Fig. (7) - Variations of radial stress with distance due to sudden moisture change

Fig. (8) - Variations of circumferential stress with distance due to sudden moisture change

in compression and then becomes tensile for small time.

Sudden application of temperature. The stresses resulting from a sudden change of the surface temperature in the cavity boundary are qualitatively different from those discussed earlier. The radial stresses in Figure 9 is no longer always compressive. For $\mathcal{D}t/a^2 = 4.0$, σ_r first becomes tensile rising to a peak and then becomes compressive. For small time, σ_r oscillates violently near the cavity end and returns to zero at $x/a = 1$. Similarly, the circumferential stress displayed graphically in Figure 10 is also tensile for $\mathcal{D}t/a^2 = 4.0$ attaining its largest value at $x/a = 1.0$ and then decrease in magnitude until it becomes compressive. The opposite trend is observed for small time, i.e., θ_θ is compressive near the cavity and tensile away from the cavity. The transverse normal stress component σ_z may be obtained from equation (8) from the results in Figures 7 to 10 in a straightforward manner and no special treatment is needed.

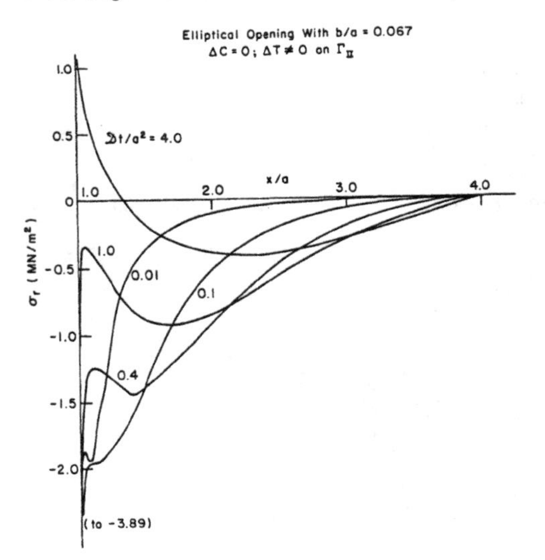

Fig. (9) - Variations of radial stress with distance for sudden temperature change

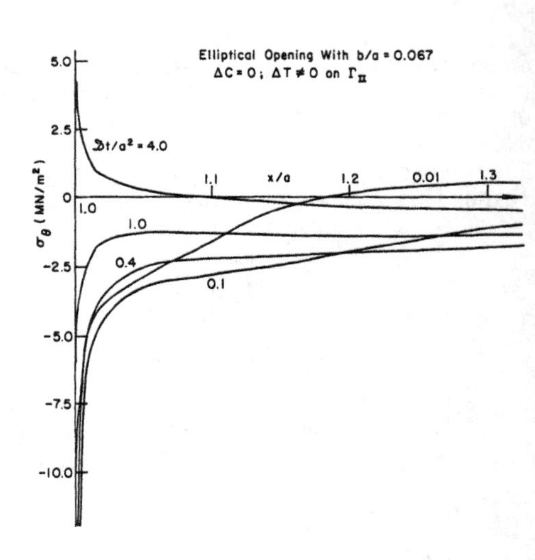

Fig. (10) - Variations of circumferential stress with distance for sudden temperature change

TIME DEPENDENT STRESS INTENSITY FACTOR

It is well-known that mechanical imperfections can interact with the sudden changes in moisture and/or temperature in the material causing it to degrade in strength and/or fracture toughness. A parameter that has been commonly used in fracture mechanics [10] is the stress intensity factor whose critical value for a given material can be related to the energy required to initiate the propagation of a line crack. If the stresses are symmetric with respect to the crack plane, only a single parameter k_1 is needed to describe the intensity of the local stresses. Since the hygrothermal stress field is time dependent, the resulting stress intensity factor k_1 will also fluctuate with time.

In addition to the hygrothermal stresses, mechanical stresses will also assume to be present owing to a uniform static tensile load σ_0 applied normal to the crack which is approximated by a narrow ellipse with radius of curvature ρ. Since the stresses induced by diffusion are not coupled with the mechanical stresses, the combined k_1 factor may be obtained by superposition:

$$k_1 = \sigma_0\sqrt{a}\ [1 \pm \frac{1}{2}\ (\frac{\sigma_\theta}{\sigma_0})\ \sqrt{\frac{\rho}{a}}] \tag{10}$$

In the absence of heat and moisture, $\sigma_\theta = 0$ and equation (10) reduces to the familiar result of $k_1 = \sigma_0\sqrt{a}$.

Figures 11 to 13 display the variations of the normalized stress intensity factor $k_1/\sigma_0\sqrt{a}$ with time for three different values of the applied mechanical

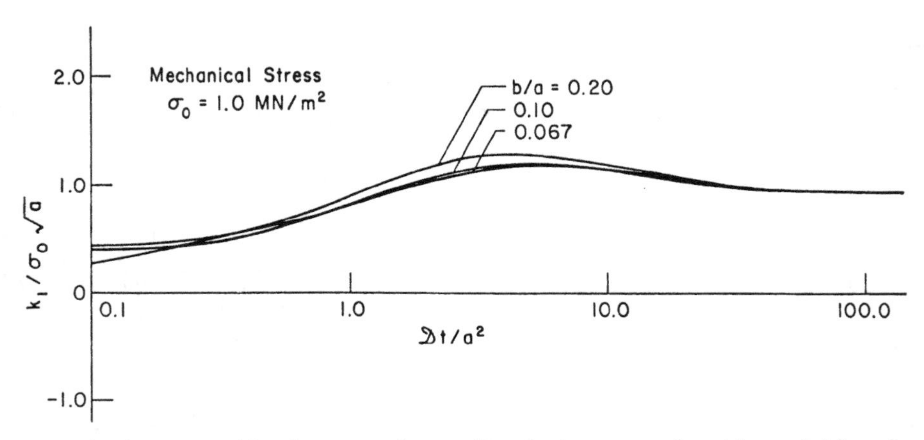

Fig. (11) - Normalized stress intensity factor as a function of time for σ_0 = 1.0 MN/m^2

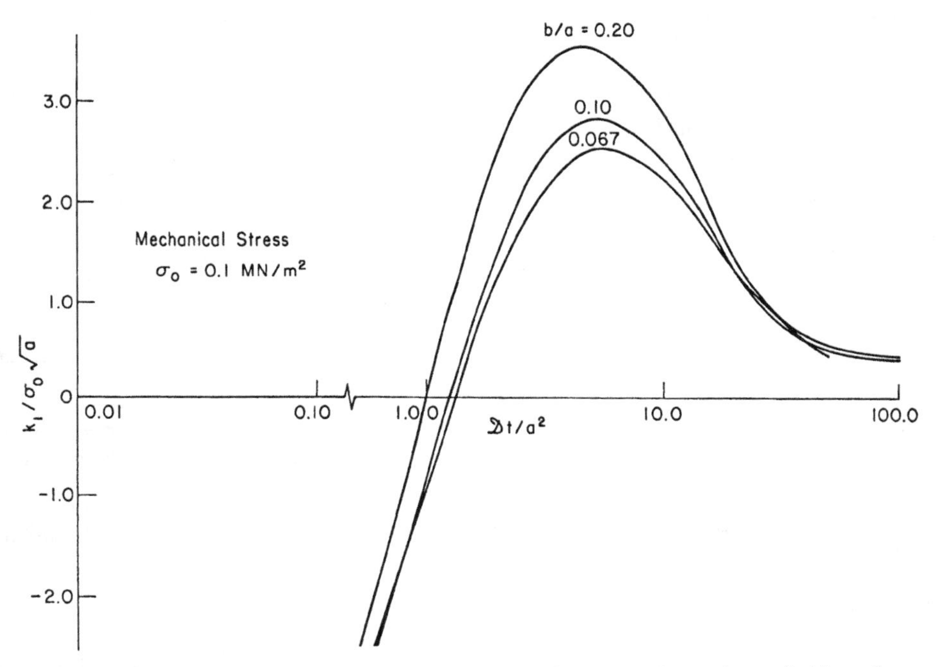

Fig. (12) - Normalized stress intensity factor as a function of time for σ_0 = 0.1 MN/m^2

stress σ_0 = 1.0, 0.1 and 0.01 having the units MN/m^2 while b/a takes the values of 0.20, 0.10 and 0.067. The temperature on the crack surface is raised suddenly

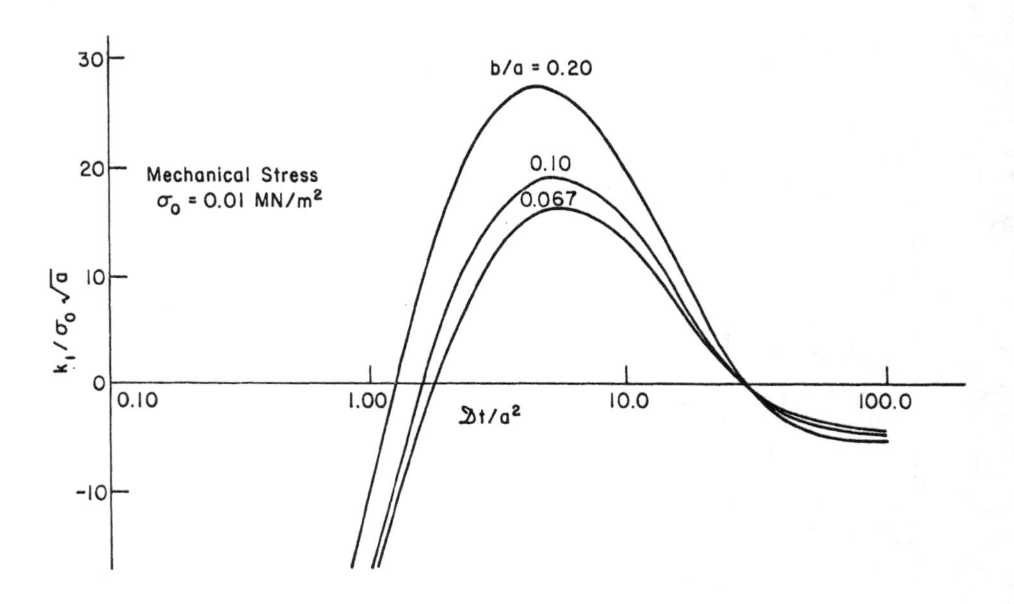

Fig. (13) - Normalized stress intensity factor as a function of time for $\sigma_0 = 0.01$ MN/m^2

introducing additional stresses. When σ_0 is relatively large, Figure 11 shows that $k_1/\sigma_0\sqrt{a}$ is not sensitive to changes in b/a. All the curves rise slowly to a peak at $\mathcal{D}t/a^2 \simeq 4.5$ and then decrease in magnitude. In the limit as $t \to \infty$, $k_1 = \sigma_0\sqrt{a}$ is recovered. The hygrothermal effect becomes more pronounced when the applied mechanical stress σ_0 is reduced in magnitude, Figure 12. For $\sigma_0 = 0.1$ MN/m^2, the maximum value of $k_1 = 3.61\ \sigma_0\sqrt{a}$ occurs at b/a = 0.20 and occurs at $\mathcal{D}t/a^2 \simeq 4.5$. The peaks for k_1 decreases as b/a is decreased and occur at a later time. A similar trend is also observed in Figure 13 when σ_0 is further reduced to 0.01 MN/m^2.

Depending on the gradient of the applied moisture and/or temperature, the hygrothermal stresses alone could lead to failure. The additional mechanical stresses can further aggravate the state of affairs near the crack and cause the crack to run. The results presented in Figures 11 to 13 offer some insight into the fracture toughness requirement for the T300/5208 resin material when both hygrothermal and mechanical disturbances are present.

CONCLUDING REMARKS

The simultaneous diffusion equations for the coupling of heat and moisture have been solved for a region containing a crack-like imperfection. Studied in detail are the influence of the crack tip radius of curvature modeled by a narrow ellipse on the redistribution of the hygrothermal stresses. The stresses near the crack front are found to vary more sharply as the ellipse becomes more slender. Because of the time dependent nature of the diffusion process, the stresses tend to oscillate and can be either compressive or tensile depending on the elapsed time.

The coupling of heat and moisture is particularly significant when the crack boundary temperature is raised suddenly. The maximum intensification of the local stresses occurs at $\mathcal{D}t/a^2 \approx 4.5$. The crack tip radius of curvature ρ affects the local hygrothermal and mechanical stresses in different ways. The former tends to increase with ρ while the latter behaves in the opposite fashion. It is a combination of loading and geometry that must be analyzed for determining the critical condition of crack instability.

ACKNOWLEDGEMENT

The authors wish to acknowledge the financial support of the Army Materials and Mechanics Research Center, Watertown, Massachusetts, under Contract Number DAAG46-79-C-0049 with the Institute of Fracture and Solid Mechanics at Lehigh University.

REFERENCES

[1] Tenney, D. R. and Unnam, J., "Analytical Prediction of Moisture Absorption in Composites", Journal of Aircraft, Vol. 15, No. 3, pp. 148-154, 1978.

[2] Y. Weitsman, "Diffusion with Time-Varying Diffusivity with Application to Moisture-Sorption in Composites", Journal of Composite Materials, Vol. 10, pp. 193-204, 1976.

[3] Carter, H. C. and Kibler, K. C., "Langmuir-Type Model for Anomalous Moisture Diffusion in Composite Resins", Journal of Composite Materials, Vol. 12, pp. 118-131, 1978.

[4] Shen, C. H. and Springer, G. S., "Moisture Absorption and Desorption of Composite Materials", Journal of Composite Materials, Vol. 10, pp. 2-20, 1976.

[5] Hartranft, R. J. and Sih, G. C., "The Influence of Coupled Diffusion of Heat and Moisture on the State of Stress in a Plate", Journal of Polymer Mechanics, Mekhanika Polimerov, USSR (in press).

[6] Sih, G. C., Shih, M. T. and Chou, S. C., "Transient Hygrothermal Stresses in Composites: Coupling of Moisture and Heat with Temperature Varying Diffusivity", International Journal of Engineering Science, Vol. 18, pp. 19-42, 1980.

[7] Hartranft, R. J. and Sih, G. C., "Stresses Induced in an Infinite Medium by the Coupled Diffusion of Heat and Moisture from a Spherical Hole", International Journal of Engineering Fracture Mechanics (in press).

[8] Sih, G. C. and Ogawa, A., "Two-Dimensional Transient Hygrothermal Stresses in Bodies with Circular Cavities: Moisture and Temperature Coupling Effects", Institute of Fracture and Solid Mechanics Technical Report, Lehigh University, July 1980.

[9] Hartranft, R. J. and Sih, G. C., "The Influence of the Soret and DuFour Effects on the Diffusion of Heat and Moisture in Solids", International Journal of Engineering Science (in press).

[10] Sih, G. C. and Macdonald, B., "Fracture Mechanics Applied to Engineering Problems - Strain Energy Density Fracture Criterion", Journal of Engineering Fracture Mechanics, Vol. 6, pp. 361-386, 1974.

STRENGTH EVALUATION OF COMPOSITES WITH SHARP-POINTED INCLUSIONS

V. V. Panasyuk and L. T. Berezhnitskij

SSR Academy of Sciences
Lvov, USSR

ABSTRACT

Fracture mechanics principles have been applied to composites with heterogeneous, sharp-pointed (with cusps on the contour), rigid, non-interacting inclusions contained in an infinite isotropic matrix under conditions of plane and anti-plane strain. An algorithm has been developed to calculate the stress intensity factors k_1, k_2 and k_3 near the tips of the arbitrary inclusions. The algorithm has been obtained on the basis of the modified boundary shape disturbance method. Some particular cases are given to illustrate the effect of the inclusion material rigidity, the defect shape and the external load type on the local stress distribution and ideal bonding seal conditions at the media interface. The possibility and range of application of conventional fracture mechanics criteria to the strength evaluation of composites containing defects have been studied.

INTRODUCTION

When used to evaluate the brittle strength of a composite, fracture mechanics methods serve mainly to obtain the external load values permissible with the solid composed of a certain geometry (shape, the solid and defect configuration) containing defects of a certain type (cracks, inclusions, voids) for given characteristics of the composing materials (shear modulus μ_s; Poisson's ratios ν_s; linear thermal expansion factors α_s; material fracture toughness values $k_{1c}^{(s)}$; etc.) under given external conditions (mechanical,thermal, environmental). At present, numerous studies on the limit equilibrium state of deformable cracked solids are available. But the variety of defects in real working structures is in no way limited to cuts, curvilinear notches with cusps at their surface, etc. In development of the brittle fracture theory for composites, the determination of the stress, strain and limit states of solids containing curvilinear dissimilar inclusions with angular points is of importance. Dissimilar, sharp-pointed inclusions or curvilinear cuts (notches) impregnated with some other elastic material may be treated as a natural model of hollow defects of the crack type conventional in fracture mechanics. The distribution of the stress tensor components and elastic displacement vector in the vicinity of the cusps along the interface is of great interest. A short time ago, the absence of asymptotic ratios of the sort for the stress and displacement characteristics near the impregnated defects made it difficult to apply fracture mechanics methods to the composites brittle strength evaluation.

Let us describe in detail the general approach to the evaluation of elastic displacements, the corresponding stresses, and the stress intensity factors in the vicinity of the inclusion tips (cusps in the contour) in an isotropic plate subjected to tension or compression at infinity. Fracture mechanics methods developed for the case of crack type defects were estimated as to their limitations and applicability to the evaluation of the strength of composites containing defects impregnated with heterogeneous materials.

Let us take an infinite elastic plate with a curvilinear sharp-pointed inclusion of a different isotropic material. It is assumed that at the interface idea bonding conditions are satisfied. Let us introduce a system of Cartesian coordinates xO_y so that its origin will coincide with the geometric center of the inclu sion and its x-axis be directed at one of the cusps. Let such a plate be subject to tension (or compression) at infinity with external stresses N_1 and N_2, the direction of the stresses N_1 making an angle of α with the axis Ox. It is assumed that the inclusion is prevented from rotating, and the difference in the linear thermal expansion factor values of composing materials may be neglected. To eval uate the local stress-strain state near the inclusion, we use an elastic problem solution technique including a preliminary study of the problem for a region bounded by a smooth contour of a more general type and reduction of the contour to a sharp-pointed one [1,2].

Let $\omega(\zeta)$ represent the conformal mapping of the exterior of the unit circle in the ζ-plane upon the exterior of the inclusion in the physical plane be expressed through

$$\omega(\zeta) = R[\zeta + \frac{m}{\zeta} + \varepsilon\omega_o(\zeta)]; \quad \omega_o(\zeta) = \sum_{n=1}^{\infty} f_n\zeta^{-n}; \quad |\zeta| \geq 1 \tag{1}$$

where R and m are constants depending on the shape of inclusion. f_n, generally speaking, are complex factors, $\varepsilon<1$ is a small parameter. In the case of a sharp-pointed inclusion, ε is chosen in such a way that at the points of the unit circle σ_{oj} corresponding to the cusps,

$$1 - \frac{m}{\sigma_{oj}^2} + \varepsilon\omega_o'(\sigma_{oj}) = 0 \tag{2}$$

is satisfied.

The stress-strain problem for a plate with a smooth-shaped (with no cusps) inclusion (1) is solved with the modified boundary shape disturbance method [3,4] and assumption of the precise solution for a plate with a circular [1] or elliptical [5,6] elastic inclusion as a zeroth approximation for the case studied Using a certain number of approximations, we obtain a solution close to the exact one. The parameter ε is assigned values in accordance with condition (2), i.e., the smooth contour is contracted into a sharp-pointed one, and we obtain the solution of the problem.

Complex potentials characterizing the stress-strain state in the inclusion and in the plate are obtained in the following form:

$$\phi_1(z) = \sum_{i=0}^{\infty} \sum_{n=1}^{\infty} \varepsilon^i c_{in} z^n; \quad \psi_1(z) = \sum_{i=0}^{\infty} \sum_{n=1}^{\infty} \varepsilon^i d_{in} z^n$$

$$\phi_2(\zeta) = R[\Gamma\zeta + \sum_{i=0}^{\infty} \sum_{n=1}^{\infty} \varepsilon^i a_{in} \zeta^{-n}]; \quad \psi_2(\zeta) = R[\Gamma'\zeta + \sum_{i=0} \sum_{n=1} \varepsilon^i b_{in} \zeta^{-n}] \tag{3}$$

where c_{in}, d_{in}, a_{in}, b_{in} are coefficients to be evaluated, and constants Γ and Γ' determine the stress state at infinity as follows

$$\Gamma = \frac{N_1+N_2}{4} + i\,\frac{2\mu_2\omega_\infty}{1+\kappa_2}; \quad \Gamma' = -\frac{1}{2}(N_1-N_2)e^{-2i\alpha} \tag{4}$$

ω_∞ is the rotation parameter of an infinitely remote part of the plane, μ_s $(s = 1,2)$ are shear moduli (here and subsequently, indices "1" and "2" refer to the inclusion and the plate material, respectively); $\kappa_s = (3-\nu_s)/(1+\nu_s)$ for plane stress; $\kappa_s = 3-4\nu_s$ for plane strain; ν_s is Poisson's ratio.

On the inclusion contour the conditions of continuity of the displacement and stress vectors crossing the interface in terms of the Muskhelishvili potentials are

$$\bar{\phi}_1(\tfrac{1}{\sigma}) + \frac{\bar{\omega}(\frac{1}{\sigma})}{\omega'(\sigma)}\,\phi_1'(\sigma) + \psi_1(\sigma) = \bar{\phi}_2(\tfrac{1}{\sigma}) + \frac{\bar{\omega}(\frac{1}{\sigma})}{\omega'(\sigma)}\,\phi_2'(\sigma) + \psi_2(\sigma)$$

$$\frac{\kappa_1}{\mu_1}\,\bar{\phi}_1(\tfrac{1}{\sigma}) - \frac{1}{\mu_1}\left[\frac{\bar{\omega}(\frac{1}{\sigma})}{\omega'(\sigma)}\,\phi_1'(\sigma) + \psi_1(\sigma)\right] = \frac{\kappa_2}{\mu_2}\,\bar{\phi}_2(\tfrac{1}{\sigma}) - \frac{1}{\mu_2}\left[\frac{\bar{\omega}(\frac{1}{\sigma})}{\omega'(\sigma)}\,\phi_2'(\sigma) + \psi_2(\sigma)\right] \tag{5}$$

where to simplify the expressions we use the notations

$$\phi_1(\sigma) = \phi_1[\omega(\sigma)]; \quad \phi_1(\sigma) = \sum_{i=0}^{\infty} \sum_{n=1}^{\infty} R\varepsilon^i c_{in}'[\sigma + \frac{m}{\sigma} + \varepsilon\omega_o(\sigma)]^n; \quad c_{in}' = R^{n-1}c_{in}$$

$$\psi_1(\sigma) = \psi_1[\omega(\sigma)]; \quad \psi_1(\sigma) = \sum_{i=0}^{\infty} \sum_{n=1}^{\infty} R\varepsilon^i d_{in}'[\sigma + \frac{m}{\sigma} + \varepsilon\omega_o(\sigma)]^n; \quad d_{in}' = R^{n-1}d_{in}; \tag{6}$$

$$\sigma = \ell^{i\gamma}$$

By substituting (3) and (6) into (5), we obtain a series of coupled boundary conditions corresponding to each power of ε. Complex potentials $\phi_1(\sigma)$ and $\phi_2(\sigma)$ are obtained through equating the terms containing the same degree σ in each of the expansions by ε. Thus, we obtain an infinite system to evaluate a_{in} and c_{in}. Function $\psi_2(\zeta)$ is derived from its value assigned at the contour of the unit radius circle.

$$\psi_2(\zeta) = \frac{(1+\kappa_1)\mu_2}{\mu_2-\mu_1}\bar{\phi}_1(\frac{1}{\zeta}) - \frac{\kappa_2\mu_1+\mu_2}{\mu_2-\mu_1}\bar{\phi}_2(\frac{1}{\zeta}) - \frac{\phi_2'(\zeta)\bar{\omega}(\frac{1}{\zeta})}{\omega'(\zeta)} \tag{7}$$

In the way described, explicit expressions for $\phi_1(z)$, $\psi_1(z)$, $\phi_2(\zeta)$ and $\psi_2(\zeta)$ for plates containing hypotrochoidal (m=0) and oval (m≠0) inclusions have been deduced. To make the postulates obvious, consider the general expressions for the complex potentials for composites containing hypotrochoidal (m=0) inclusions with three (n=2, f_2 = 1) or four (n=3, f_3 = 1) tips. If n=2 in the first approximation,

$$\begin{aligned} \phi_1(z) &= c_{01}z + \varepsilon c_{12}z^2 \\ \phi_2(\zeta) &= R[\Gamma\zeta + \frac{a_{01}}{\zeta} + \frac{\varepsilon a_{12}}{\zeta^2} + \frac{\varepsilon^2 a_{21}}{\zeta} + \varepsilon^3(\frac{a_{32}}{\zeta^2} + \frac{a_{34}}{\zeta^4}) \end{aligned} \tag{8}$$

and in the second approximation,

$$\begin{aligned} \phi_1(z) &= c_{01}z + \varepsilon c_{12}z^2 + \varepsilon^2(c_{21}z + c_{23}z^3) \\ \phi_2(\zeta) &= R[\Gamma\zeta + \frac{a_{01}}{\zeta} + \varepsilon\frac{a_{12}}{\zeta^2} + \varepsilon^2\frac{a_{21}}{\zeta} + \varepsilon^3(\frac{a_{32}}{\zeta^2} + \frac{a_{34}}{\zeta^4}) + \varepsilon^4(\frac{a_{41}}{\zeta} + \frac{a_{43}}{\zeta^3}) \\ &\quad + \varepsilon^5(\frac{a_{52}}{\zeta^2} + \frac{a_{54}}{\zeta^4} + \frac{a_{56}}{\zeta^6})] \end{aligned} \tag{9}$$

If n=3, in the first approximation,

$$\begin{aligned} \phi_1(z) &= c_{01}z + \varepsilon c_{13}z^3 \\ \phi_2(\zeta) &= R[\Gamma\zeta + \frac{a_{01}}{\zeta} + \varepsilon(\frac{a_{11}}{\zeta} + \frac{a_{13}}{\zeta^3}) + \varepsilon^2\frac{a_{21}}{\zeta} + \varepsilon^3(\frac{a_{31}}{\zeta} + \frac{a_{33}}{\zeta^3} + \frac{a_{35}}{\zeta^5}) \\ &\quad + \varepsilon^4(\frac{a_{45}}{\zeta^5} + \frac{a_{49}}{\zeta^9})] \end{aligned} \tag{10}$$

Due to their complicated form, the explicit expressions for c_{in} and a_{in} are not presented here. From (8)-(10), it follows that the non-uniform stress state is common for a curvilinear elastic inclusion of a non-elliptical (non-circular) shape when there exists a uniform stress state at infinity determined by constant values (4). On assumption that in (8) or (9), $\varepsilon = 1/2$, and in formulae (10), $\varepsilon = 1/3$, the corresponding complex stress potentials for a plate containing hypocyclic inclusions with three or four cusps in the contour are obtained. In this case, conditions (4) at infinity are satisfied absolutely and conditions of ideal bonding are satisfied approximately, i.e., they are satisfied along a contour close to the initial one. However, the geometry in the vicinity of the tip (of the cusp) is preserved.

To study the stress-strain state in the vicinity of the defect tip, transform (within the limits of conventional [1] complex expressions of stresses and displacements into the system of local polar coordinates r, $\theta^{(j)}$ with the origin at the tip of the cusp j and the polar axis directed along the tangent to the inclusion surface at its tip. After the necessary transformations have been performed (with $r \ll \ell$) on the basis of (7)-(10), the following expressions for stress tensor components $\sigma_r^{(j)}$, $\sigma_\theta^{(j)}$, $\sigma_{2\theta}^{(j)}$ and displacements vector $v_r^{(j)}$, $v_\theta^{(j)}$ in the vicinity of the inclusion tip in the matrix are obtained:

$$\sigma_r^{(j)} = \frac{1}{4\sqrt{2r}} \{k_1^{(j)} [5\tilde{\kappa}\cos\frac{\theta^{(j)}}{2} + (\tilde{\kappa}-2)\cos\frac{3\theta^{(j)}}{2}] - k_2^{(j)} [5\tilde{\kappa}\sin\frac{\theta^{(j)}}{2} - (\tilde{\kappa}+2)\sin\frac{3\theta^{(j)}}{2}]\} + 0(1)$$

$$\sigma_\theta^{(j)} = \frac{1}{4\sqrt{2r}} \{k_1^{(j)} [3\tilde{\kappa}\cos\frac{\theta^{(j)}}{2} - (\tilde{\kappa}-2)\cos\frac{3\theta^{(j)}}{2}] - k_2^{(j)} [3\tilde{\kappa}\sin\frac{\theta^{(j)}}{2} + (\tilde{\kappa}+2)\sin\frac{3\theta^{(j)}}{2}]\} + 0(1) \qquad (11)$$

$$\sigma_{r\theta}^{(j)} = \frac{1}{4\sqrt{2r}} \{k_1^{(j)} [\tilde{\kappa}\sin\frac{\theta^{(j)}}{2} - (\tilde{\kappa}-2)\sin\frac{3\theta^{(j)}}{2}] + k_2^{(j)} [\tilde{\kappa}\cos\frac{\theta^{(j)}}{2} + (\tilde{\kappa}+2)\cos\frac{3\theta^{(j)}}{2}]\} + 0(1)$$

$$v_r^{(j)} = \frac{\sqrt{2r}}{8\mu_2} \{k_1^{(j)} [(2\kappa_2-1)\tilde{\kappa}\cos\frac{\theta^{(j)}}{2} + (\tilde{\kappa}-2)\cos\frac{3\theta^{(j)}}{2}] - k_2^{(j)} [(2\kappa_2-1) \times \tilde{\kappa}\sin\frac{\theta^{(j)}}{2} - (\tilde{\kappa}+2)\sin\frac{3\theta^{(j)}}{2}]\} + c_1\cos\theta^{(j)} + c_2\sin\theta^{(j)} + 0(z)$$

$$v_\theta^{(j)} = -\frac{\sqrt{2r}}{8\mu_2}\{k_1^{(j)}\left[(2\kappa_2+1)\tilde{\kappa}\sin\frac{\theta^{(j)}}{2} + (\tilde{\kappa}-2)\sin\frac{3\theta^{(j)}}{2}\right] + k_2^{(j)}\left[(2\kappa_2+1)\times\right.$$

$$\left.\times\,\tilde{\kappa}\cos\frac{\theta^{(j)}}{2} - (\tilde{\kappa}+2)\cos\frac{3\theta^{(j)}}{2}\right]\} + c_2\cos\theta^{(j)} - c_1\sin\theta^{(j)} + O(r) \tag{12}$$

$$\tilde{\kappa} = \frac{\mu_2-\mu_1}{\kappa_2\mu_1+\mu_2} = \frac{E_2(1+\nu_1)-E_1(1+\nu_2)}{\kappa_2E_1(1+\nu_2)+E_2(1+\nu_1)} \tag{13}$$

In these expressions, c_1 and c_2 are related to the inclusion rotation and outer loading; j is the number of the tip; (the tips are numbered according to their order, beginning with the one on the positive axis Ox and counting counterclockwise), $k_1^{(j)}$ and $k_2^{(j)}$ are the stress intensity factors depending on the applied load, the parameters determining the configuration of the solid and the form of inclusion, and on the elastic properties (μ_s, ν_s) of the inclusion and matrix materials. The k_1 and k_2 values are expressed as polynomials in the parameter ε

$$k_1^{(j)} = \sum_{n=1}^{N} \varepsilon^n k_{1n}^{(j)};\; k_2^{(j)} = \sum_{n=1}^{N} \varepsilon^n k_{2n}^{(j)} \tag{14}$$

It is important to note that the local stress and displacement distributions, (11) and (12), in terms of the angle θ, do not depend on the number of chosen approximations; it is the same for each case of two, three or more approximations. The degree of accuracy influences only the stress intensity factors.

Thus, the stress singularity near the cusp of an elastic inclusion has the order of $r^{-1/2}$ and the displacement singularity has the order of $r^{1/2}$ as in the case of a crack tip or rigid inclusion. The stress and displacement distribution near the tip of a heterogeneous inclusion is characterized by the same functional ratios (11) and (12) not depending on the type of load and inclusion shape. One should be aware of the fact that in the case of a hollow defect of the crack type the stress distribution by angle θ does not depend on the elastic properties of the material, and in the case of an absolutely rigid inclusion [9-11], this distribution is affected by Poisson's ratio of the matrix. The stress distribution in the vicinity of an elastic heterogeneous inclusion depends on the ratio of the shear moduli for the materials composing the composite and on the Poisson's ratio of the matrix material.

From (11), it follows that:

$$\frac{\partial\sigma_\theta}{\partial\theta} = -\frac{3}{2}\sigma_{r\theta} \tag{15}$$

i.e., there are no tangential stresses in the direction θ_* where the normal stress

σ_θ reaches its maximum (minimum) intensity. This conclusion remains in force for the sharp-pointed defects of different shape regardless of the elastic properties of the inclusion and matrix materials. It is interesting to note that for plane stress in the vicinity of the inclusion at $\theta=0$, the singular stress ratio t $= \sigma_r/\sigma_\theta$ lies within the limits $1 \geq t \geq -\frac{3+\nu_2}{1-\nu_2}$ depending on the modulus of elasticity of the inclusion. To obtain $k_1^{(j)}$ and $k_2^{(j)}$, we use the first two asymptotic expressions in (11) and deduce

$$k_1^{(j)} - ik_2^{(j)} = \frac{2}{\tilde{\kappa}} \frac{\phi_2'(\sigma)}{\sqrt{e^{i\lambda_j}\omega''(\sigma_{oj})}} \tag{16}$$

where λ_j is the angle between the polar axis of the local system $r,\theta^{(j)}$ and axis Ox.

A similar method was applied to the local stress strain study for the case of anti-plane conditions and a sharp-pointed elastic tube-type inclusion in a solid [12]. An asymptotic stress and displacement field in the matrix is evaluated:

$$\begin{aligned} \sigma_{rz}^{(j)} &= \frac{1}{\sqrt{2r}} \left(k_{3,1}^{(j)} \sin \frac{\theta^{(j)}}{2} + k_{3,2}^{(j)} \cos \frac{\theta^{(j)}}{2}\right) + O(1) \\ \sigma_{\theta z}^{(j)} &= \frac{1}{\sqrt{2r}} \left(k_{3,1}^{(j)} \cos \frac{\theta^{(j)}}{2} - k_{3,2}^{(j)} \sin \frac{\theta^{(j)}}{2}\right) + O(1) \end{aligned} \tag{17}$$

$$w = \frac{\sqrt{2r}}{\mu_2} \left(k_{3,1}^{(j)} \sin \frac{\theta^{(j)}}{2} + k_{3,2}^{(j)} \cos \frac{\theta^{(j)}}{2}\right) + O(r) \tag{18}$$

where $k_{3,1}^{(j)}$ and $k_{3,2}^{(j)}$ are the stress intensity factors of the flexible and rigid stress components respectively. If there is a tendency to an absolutely flexible inclusion (void) $\mu_1 \to 0$, $k_{3,2}^{(j)} \to 0$, and $k_{3,1}^{(j)}$ is the stress intensity factor expression for the corresponding hollow defect. If $\mu_1 \to \infty$, $k_{3,1}^{(j)} \to 0$, and $k_{3,2}^{(j)}$ is that for an absolutely rigid inclusion, the solid being subjected to anti-plane strain [13]. For $\mu_1 \to \mu_2$ ($k_{3,1}^{(j)} \to k_{3,2}^{(j)} \to 0$) and the singular component in (17) is eliminated, and the terms O(1) determine the elastic equilibrium of the solid in the vicinity of the inclusion. To determine $k_{3,1}^{(j)}$ and $k_{3,2}^{(j)}$, there is the expression:

$$k_{3,1}^{(j)} + ik_{3,2}^{(j)} = i\mu_2 \frac{F_2'(\sigma_{oj})}{\sqrt{e^{-i\lambda_j}\omega''(\sigma_{oj})}} \tag{19}$$

where $F_2'(\sigma_{0j})$ is the first derivative of the stress function at the cusp during the anti-plane strain of the solid.

The general form of stress intensity factors $k_1^{(j)}$ and $k_2^{(j)}$ near the elastic sharp-pointed inclusions for the given type of load is expressed by:

$$k_1^{(j)} = (N_1+N_2)F_1^{(j)} + (N_1-N_2)F_2^{(j)}\cos 2\alpha_1$$
$$k_2^{(j)} = (N_1-N_2)F_3^{(j)}\sin 2\alpha_1, \; (\alpha_1 = \lambda_j-\alpha) \tag{20}$$

where $F_i^{(j)}$ are functions to be determined which depend exclusively on the defect characteristics, shape, size, and the difference in the elastic constants of the filling material and the matrix. If the expressions (10)-(18) are accurate, the $F_i^{(j)}$ function evaluation is carried out very approximately. Values of $k_1^{(j)}$ and $k_2^{(j)}$ were obtained with provision for the second, third and fourth approximations but they are omitted here for the sake of brevity. Some calculations have been performed on their basis. It can be shown that a substantial contribution to the values of $k_1^{(j)}$ and $k_2^{(j)}$ is due only to the first and second approximations. The accuracy and reliability of $k_1^{(j)}$ and $k_2^{(j)}$ depends not only on the form of the defect (on its difference from a circular or elliptical defect) but also on the difference between the elastic constant of the inclusion and that of the matrix. If two or three approximations are enough for nearly rigid or hollow defects, for composites in the $0.1 \leq \frac{\mu_1}{\mu_2} \leq 10$ range, it is necessary to carry out 20-30 approximations to obtain an accuracy of 3 percent. This becomes possible if we use computers by the algorithm proposed in our report and developed in accordance wit discussed results.

In the limiting cases $\mu_1/\mu_2 \to 0$ (hollow defects of the crack type) and $\mu_1/\mu_2 \to \infty$ (absolute rigid inclusions), from (11), (12), (17) and (18) follow the correct expressions for the stress and displacement distribution. We also derive the formulae for the stress intensity factors near the tip of a crack and a rigid inclusion [2,16,11]. If the inclusion and the matrix have the same shear modulus (the Poisson's ratios may differ), $k_1^{(j)} = k_2^{(j)} = 0$, and, thus, there are no singular terms in the stress values, a fact expected for the uniform plate.

The diagrams presented in Figure 1 demonstrate the effect of the defect filling material on the adjacent stress intensity factors under different types of loading. Here, one can see the alteration in the stress intensity factors near the filled $k_1^{(f)}$ and unfilled (hollow) $k_1^{(h)}$ defects of a hypocycloidal with three tips (curves 1) and an astroidal (curves 2) depending on the parameter $d = (\mu_2-\mu_1$ $(\mu_2+\mu_1)$ for (a) uniaxial tension ($\alpha = \pi/2$, $N_2 = 0$; (b) hydrostatic tension ($N_1=N_2$

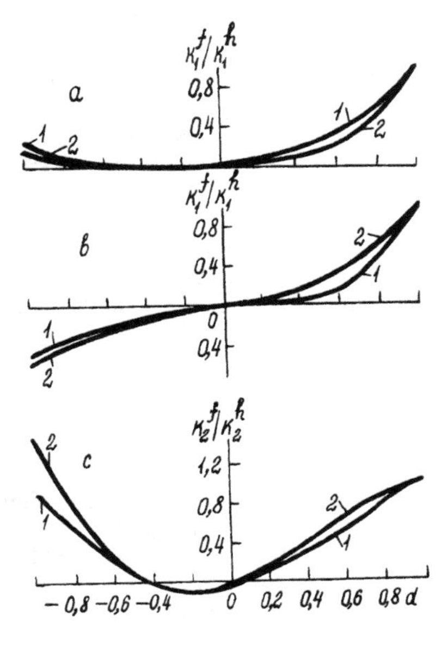

Fig. (1)

and (c) shear ($N_1 = -N_2$, $\alpha = \pi/4$). The calculations were performed for plane stress and $\nu_1 = \nu_2 = 0,3$. For a particular type of loading and shape of inclusion, it is possible to determine the range of values of the parameter causing very small stress intensity factor values. In terms of fracture mechanics, sharp-pointed inclusions in this range do not have much effect on the strength of the composite. If the solution of the problem is obtained with provision for the residual thermal stresses resulting from the difference in linear thermal expansion factors and from the temperature difference, it is evident that this effect on values $k_1^{(j)}$ and $k_2^{(j)}$ is greatest when $\mu_1 > \mu_2$. A considerable nonuniformity of the stress state inside the curvilinear inclusion has been found. In some cases under conditions of the uniaxial tension of composites the stresses is the astroidal inclusion ($\mu_1/\mu_2 = 0,05$) near the cusp were greater than the corresponding stresses in its center by 4 to 5 times.

Expressions (20) may be regarded [15] as a parametric record of the family of ellipses in the $k_1^{(j)}$, $k_2^{(j)}$ plane. Thus, from equation (20), it follows:

$$\frac{[k_1^{(j)} - (N_1+N_2)F_1]^2}{(N_1-N_2)^2 F_2^2} + \frac{k_2^{(j)2}}{(N_1-N_2)^2 F_3^2} = 1 \qquad (21)$$

which is the equation of an ellipse with center at point $[(N_1+N_2)F_1;0]$ and with the axes parallel with the coordinate axes ($k_1 0 k_2$). Thus, in the $k_1^{(j)}$, $k_2^{(j)}$ plane, all possible stress intensity factor values near the sharp-pointed defects in biaxial tension lie on an ellipse with parameters depending on the defect shape and the difference in the elastic constants of the inclusion and matrix materials. The proposed diagrams, by analogy with circular Mohr diagrams, may be applied [15] not only in the analysis of the local stress field intensity in the vicinity of various defects but also in the development of the criteria for the transition of the material near the inclusion tip into its limit state.

It seems quite natural to utilize conventional strength hypotheses and failure criteria expressed through the asymptotic formulae (11), (12), (17) and (18) to evaluate the transition of the matrix material near the defect tip into its limit state. Then, in the long run, the matrix fracture criterion may be expressed in the general form:

$$F(k_1, k_2, k_3, \theta_*, \tilde{\kappa}) = \text{const.} \tag{22}$$

where θ_* are the values of the polar angle θ making the functional take its extreme values. The expression of the functional F is determined by the applied strength hypothesis and comparison with experimental data. In contrast with the case of a crack in close proximity to an inclusion, we observe here a tendency to non-uniform tension-compression. This makes us suspect a more substantial plastic deformation near the inclusion. Consequently, for composites with more plastic matrices, the transition of the matrix material into the limit state near a sharp-pointed inclusion is easily evaluated by maximum tangential stress τ_{max} and energy from transformation parameter W_d.

In [19], the authors give a critical analysis of some conventional failure criteria (including the total strain energy density criterion [20]) asymptotically expressed through the stress tensor components and displacement vectors in the vicinity of an absolutely rigid sharp-pointed inclusion. In the case of anti-plane deformation, the boundary of the plastic zone in the vicinity of the rigid linear inclusion tip has been determined [16]. For an ideal elasto-plastic matrix material, this zone has the shape of a cylindrical circle of radius $r_p = (k_3/\tau)^2/2$, displaced at distance r_p from the tip in the direction of the rigid inclusion where τ_y is the flow limit value in shear. If the plane problem is studied, numerical methods may serve to demonstrate that in contrast with cracks, the plasticity region is displaced in the direction of the rigid inclusion and covers the tip. This makes it possible to obtain the effective solution of the plane elasto-plastic problem on the basis of representing plastic zones by narrow layers of zero thickness at the inclusion-matrix interface.

The results obtained make it possible to estimate the possibility of applying the methods of fracture mechanics to the computation of the brittle strength of composites with sharp-pointed inclusions. In this case, it is necessary to distribute all possible compositions $-1 \leq d \leq 1$ $(d = (\mu_2 - \mu_1)/(\mu_2 + \mu_1))$ into several classes described by the inequalities $-1 \leq -d_1 < -d_2 < -d_3 < 0 < d_4 < d_5 < d_6 \leq 1$. Inclusions out of the range $(-d_1, d_6)$ are regarded as absolutely rigid or hollow defects, and their effect on the strength value is calculated by formulae mentioned in [2,8,17,20]. For the inclusions out of range $(-d_2, d_5)$, it is enough to consider the singular term in the stresses, e.g., $A_\theta/\sqrt{r}$ for soft inclusions and $A_r/\sqrt{r}$

for rigid inclusions. Then, in accordance with fundamental fracture mechanics postulates the local fracture criterion may be expressed by the following formulae as in [8,17,18]:

$$A_{\theta}^{max(\alpha,\theta)} = k_{0\theta} \quad \text{at } d \geq d_5$$
$$A_{r}^{max(\alpha,\theta)} = k_{0r} \quad \text{at } d \leq -d_2 \qquad (23)$$

where $k_{0\theta}$ and k_{0r} are constants of the composite characterizing the crack initiation resistance [17] in directions θ and r. These characteristics are obtained experimentally or expressed through the effective fracture energy density γ for the given material through the matrix strength limit value σ_b [17,21] or through k_{1c} if fracture occurs in the matrix. In the strength evaluation of composites in ranges $(-d_2,-d_3)$ and (d_4,d_5), one should take into account not only singular but also bounded O(1) stresses. And last, in the range $(-d_3,d_4)$, the stress intensity factors near sharp-pointed elastic inclusions, e.g., at $\mu_1/\mu_2 \leq 5$, are very small, thus preventing the application of linear fracture mechanics to the strength analysis of such composites.

REFERENCES

[1] Muskhelishvili, N. I., Some Basic Problems of the Mathematical Elasticity Theory (in Russian), Moscow, 1966.

[2] Panasyuk, V. V., Limit Equilibrium of Cracked Brittle Solids (in Russian), Kiev, 1968.

[3] Lekhnitskij, S. G., Anisotropic Plates (in Russian), Moscow, 1957.

[4] Savin, G. N., Stress Distribution near Holes (in Russian), Kiev, 1968.

[5] Hardiman, N. J., "Elliptic Elastic Inclusion in an Infinite Elastic Plate", Quart. J. Mech. and Appl. Math., 3, pp. 226-230, 1952.

[6] Trush, I. I., Panasyuk, V. V. and Berezhnitskij, L. T., On the Effect of the Inclusion Shape on the Initial Fracture Stage in Two Componential Composites (in Russian). Fiz.-Khim. Mekh. Materialov, 6, pp. 48-53, 1972.

[7] Berezhnitskij, L. T. and Sadivskij, V. M., Stress Distribution near Elastic Inclusions with Cusps in the Contour (in Russian). Fiz.-Khim. Mekh. Materialov, 3, pp. 47-54, 1976.

[8] Panasyuk, V. V., Berezhnitskij, L. T. and Sadivskij, M. V., Stress Intensity Factors and Stress Distribution near Sharp-Pointed Elastic Inclusions (in Russian). Diklady Akad. Nauk Ukr. SSR, 232 (2), pp. 304-307, 1977.

[9] Sih, G. C., "Plane Extension of Rigidly Embedded Line Inclusions", In: Proc. 9th Midwest Mechanics Conference, pp. 61-79, 1965.

[10] Panasyuk, V. V., Berezhnitskij, L. T. and Trush, I. I., Stress Distribution near Defects of the Rigid Sharp-Pointed Inclusion Type (in Russian), Problem Prochnosti (7), pp. 3-9, 1972.

[11] Berezhnitskij, L. T., Delyavskij, M. V. and Panasyuk, V. V., Bending of Thin Plates with Crack-Type Defects (in Russian), Kiev, 1979.

[12] Berezhnitskij, L. T., Panasyuk, V. V. and Sadivskij, V. M., Longitudinal Shear of an Isotropic Solid with Sharp-Pointed Elastic Inclusions (in Russian), Doklady Akad. Nauk. Ukr. SSR, Ser. A95, 5, pp. 413-417, 1977.

[13] Berezhnitskij, M. P. Len', Antiplane Deformation of a Solid with Rigid Inclusions (in Russian). Problemy Prochnosti (8), pp. 10-14, 1975.

[14] Panasyuk, V. V., Berezhnitskij, L. T. and Trush, I. I., Stress Intensity Factors near Rigid Sharp-Pointed Inclusions (in Russian). Problemy Prochnosti (7), pp. 3-7, 1973.

[15] Berezhnitskij, L. T., Graphic Interpretation for Stress Intensity Factors and Evaluation of Brittle Strength in Structural Elements with Crack-Type Defects (in Russian). Fiz-Khim. Mekh. Materialov (5), pp. 27-31, 1977.

[16] Gromyak, R. S., On Plastic Zone in the Rigid Inclusion Vicinity under Antiplane Strain (in Russian). Fiz.-Khim. Mekh. Materialov (4), pp. 124-126, 1979.

[17] Berezhnitskij, L. T., Panasyuk, V. V. and Trush, I. I., On Local Fracture in Brittle Solids with Sharp-Pointed Rigid Inclusions (in Russian). Problemy Prochnosti (10), pp. 8-14, 1973.

[18] Sotkilava, O. V. and Cherepanov, G. P., Some Problems of the Nonuniform Elastic Theory (in Russian). Prikl. Mat. Mekh. 38, pp. 539-550, 1974.

[19] Berezhnitskij, L. T. and Gromyak, R. S., On the Evaluation of the Limit State of a Matrix in the Vicinity of a Sharp-Pointed Rigid Inclusion (in Russian). Fiz.-Khim. Mekh. Materialov (2), pp. 39-47, 1977.

[20] Sih, G. C., "Strain Energy Density Factor Applied to Mixed Mode Crack Problems", Int. Journal of Fracture, 10, (3), pp. 305-321, 1974.

[21] Panasyuk, V. V., Berezhnitskij, L. T. and Gromyak, R. S., On Effect of Material Structure on Local Fracture near Sharp-Pointed Rigid Inclusions (in Russian). Doklady Akad. Nauk. Ukr. SSR., Ser. A, (2), pp. 1096-1101, 1976.

LOCAL FRACTURE IN FIBROUS MEDIA

G. A. Vanin

Ukranian Academy of Sciences
Kiev 57, USSR

Local fracture in the structure of fibrous medium precedes the formation of the main crack and stimulates its accelerated growth. The study of the conditions of growth and arrest of local cracks, taking into account the inhomogeneous structure of the material, makes it possible to define more precisely the redistribution of stresses in neighboring regions and determine the occurrence of new critical states in components of the medium. The formation of a crack at the interface causes not only instability, but also, by changing the stress state in the fiber, contributes to the formation and growth in the fiber of superficial cracks oriented perpendicular to its axis.

Experimentally found initial imperfections in the form of pores and cracks are considerably elongated along the fibers. For that reason, the following assumptions are made:

a. In the initial state, crack borders form surfaces equidistant to interfaces of components or near to them; the projection of a crack border on the surface perpendicular to the normal in the geometrical center of the crack forms an ellipse whose major axis is parallel to the fiber axis and is one order of magnitude larger than the minor axis.

b. Crack propagation proceeds in a self-similar fashion.

c. The crack growth resistance in the matrix and on the interface border at normal separation and shear is specified by two parameters k_1, k_2 and k_{01}, k_{02}, respectively (the index "c" is omitted). For fibers, the parameters k_{a1}, k'_{a1} and k_{a2} are introduced, that characterize, respectively, the tear resistance of fibers or their elements along and across its axis and the resistance to the longitudinal-transverse shear.

One of the problems of the theory consists in deriving the relations between the parameters introduced, taking into account the special features of the structure and the loading type.

The simplifications introduced make possible the two-dimensional analysis of the stress state in a reinforced medium with cracks located far from the ends of the major axis of the ellipse. The evolution of brittle fracture of composite materials taking into account the set of local cracks, as described, is reduced to the problem of finding the explicit dependence of the components of the Z-matrix on parameters that characterize structure, imperfections and initial stresse The Z-matrix relates mean stresses and deformations in representative volume:

$$\langle \varepsilon_{ik} \rangle = Z_{iksn} \langle \sigma_{sn} \rangle \tag{1}$$

When local fractures begin under loading, the indices describing the physical and mechanical properties of the material, i.e., parameters of the Z-matrix, change, leading to additional redistribution of stresses not only in the region of the fibers, but also between layers with differently oriented fibers.

The general stable stress state of the elastic medium, in which fibers are located in parallel and form a regular biperiodical structure is divided into the sum of simplest states. For each state, the characteristics of propagation of local cracks are specified.

The spatially inhomogeneous distribution of fibers within the matrix significantly affects the conditions of local fracture. The fibers of boron have pronounced zones with different physico-chemical properties changing along the fiber's radius.

A characteristic feature of organic and carbon fibers is their considerable anisotropy of properties due to their fibrillar structure, porosity, as well as the presence of predominant sliding planes in carbon.

The polymer matrix in the microregion immediately adjacent to the fiber has anomalous physico-chemical properties caused by the high surface energy of the fibers, the finishing materials, and special features of the hardening process. Since the zone mentioned has a developed interface region, it has a marked influence on the stress distribution and crack growth.

The quantitative evaluation of the effect of local inhomogeneity on the fracture process is possible on the basis of known regularities of changes of physicc chemical characteristic in a small volume. Thus, the direct measurement of the indices of mechanical properties is not practically possible. The experimental check of assumed fracture process models may be carried out indirectly by measurement of integral values.

For the following analysis we introduce the notion of simple and mixed fracture. Simple fracture corresponds to the growth of a single crack in the components of the medium, and mixed fracture corresponds to the simultaneous crack growth in each component and on the component borders.

We consider the material in the case of the simple cohesive fracture of the matrix, when $k_{02}>k_2$, $k_{01}>k_1$ with a biperiodical system of identical cracks havin border radius. The solution of the mixed two-dimensional problem is constructed using the previously [2]. Accordingly, the unknown solution for the matrix is constructed by way of the superposition of functions describing the local field near the fibers and having pointwise symmetry only, and functions specifying the interaction between inclusions taking into account the translational symmetry of the structure. The procedure may be usefully employed for the solution of problems for two- and three-dimensional regions [3]. Harmonic and biharmonic interference functions for arbitrary biperiodical structure are expressed by elliptical functions.

The variables are defined as: Γ - interphase border, $L = \ell+\ell_0$ - cylindrical surface that covers the fiber, the region ℓ_0 of which specifies the discontinuity (the crack), $\bar{u}$ - displacement vector, T_n - limit value of stresses with normal $\bar{n}$. When frictional connections and overlying borders of the cut are absent, the unknown functions in each mesh should satisfy the following conditions:

$$[T_{na}\bar{n}]^+ = [T_n\bar{n}]^-,\ \bar{u}_a^+ = \bar{u}^-,\ (\bar{x}\varepsilon\Gamma),$$
$$[T_n\bar{n}]^+ = [T_n\bar{n}]^-,\ \bar{u}^+ = \bar{u}^-,\ (\bar{x}\varepsilon\ell), \qquad (2)$$

$$[T_n\bar{n}]^+ = 0,\ [T_n\bar{n}]^- = 0,\ (\bar{x}\varepsilon\ell_0)$$
$$\sigma_{ik}(\bar{x}) = \sigma_{ik}(\bar{x}+\bar{\omega}) \qquad (3)$$

Here, the values related to the fiber are denoted by the index "a", and those related to the matrix are without index. Indices plus and minus mean that the limit value is specified along the positive and negative direction of the normal. According to the condition (3), the components of the stress tensor should be invariant to operations of translational and rotational symmetry for the structure chosen.

The relations cited are supplemented by the system of equalities that relate the mean stresses with the local stresses.

$$<\sigma_{ik}> = \frac{\omega_i}{V} \int_{B_i} \sigma_{ik} dB$$

where V - representative volume, B_i - area of facet perpendicular to x_i-axis, ω_i - dimensions of ith side. With $<\sigma_{ik}>$ given, components of the Z-matrix are specified according to equalities

$$<\varepsilon_{ik}> = \frac{1}{V} \sum_{j=1}^{N} \int_{V_j} \frac{\partial U_j}{\partial <\sigma_{ik}>} dN \qquad (4)$$

Here U_j and V_j are the energy and volume of the jth component, and N is the number of components.

Under longitudinal shear, the fracture resistance of the two-component fibrous medium with cracks stretched along the fibers is determined by three parameters k_{02}, k_{a2} and k_2. The first one characterizes the resistance to crack growth at interphase border under adhesive fracture, while the subsequent parameters specify adhesive fracture toughness of fibers and matrix, respectively.

The inhomogeneous structure of the fibers will be presented intergrally using models of inhomogeneous anisotropic media. We introduce a local system of cylindrical coordinates r,θ with origin on the fiber axis. The properties near fiber center are different from those at its border $r=\Lambda$, so models of media with cylindrical anisotropy are employed [4]. Hooke's law for such a medium has the form

$$\sigma_{1r} = G_r \frac{\partial u_a}{\partial r} + G_{r\theta} \frac{1}{r} \frac{\partial u_a}{\partial \theta}$$

$$\sigma_{1\theta} = G_{r\theta} \frac{\partial u_a}{\partial r} + G_\theta \frac{1}{r} \frac{\partial u_a}{\partial \theta} \tag{5}$$

where the displacement function $u(r,\theta)$ satisfies the equation

$$G_r \frac{\partial^2 u_a}{\partial r^2} + \left(\frac{G_r}{r} + \frac{\partial G_r}{\partial r} + \frac{1}{r} \frac{\partial G_{r\theta}}{\partial \theta}\right)\frac{\partial u_a}{\partial r} + \left(\frac{1}{r^2} \frac{\partial G_\theta}{\partial \theta} + \frac{1}{r} \frac{\partial G_{r\theta}}{\partial r}\right)\frac{\partial u_a}{\partial \theta}$$

$$+ 2G_{r\theta} \frac{1}{r} \frac{\partial^2 u_a}{\partial r \partial \theta} + G_\theta \frac{1}{r^2} \frac{\partial^2 u_a}{\partial \theta^2} = 0 \tag{6}$$

The relation between shear moduli and coordinate θ is analyzed approximately for a quantitative description. In the first approximation, the following assumptions are made:

$$G_r = G_r^o \left(\frac{x}{x_1}\right)^{2g} e^{x-x_1}, \; G_{r\theta} = \kappa G_r, \; G_\theta = \nu^2 G_r, \; x = \alpha r, \; x_1 = \alpha\varepsilon \tag{7}$$

Here, G_r^o, g, α, κ, ν^2 - constants, characterizing the change of elastic properties of the fiber, ε - radius of central part of the fiber with uniform structure. Material of the matrix is modeled by homogeneous isotropic medium with shear modulus G, and the inhomogeneity of the microstructure near the fibers is neglected.

Displacement function u is assumed in the form of the real part of the series

$$u = 2\mathrm{Re}[Cz + \sum_{j=0}^{\infty} C_j \zeta^{(j)}(z)] \tag{8}$$

where $z = x_2 + ix_3 = re^{i\theta}$, $\zeta(z)$ is the Weierstrass zeta-function. The solution of equation (6) is constructed in the form

$$u_a = 2\mathrm{Re} \sum_{n>0} e^{-x+in\theta} x^{q_n - in\kappa} [A_n F(a_n,b_n;x) + B_n U(a_n,b_n;x)] \tag{9}$$

where degenerate hypergeometrical functions are introduced.

$$F(a_n,b_n;x) = \sum_{m=0}^{\infty} \frac{(a_n)_m}{(b_n)_m} \frac{x^m}{m!}$$

$$U(a_n,b_n;x) = \frac{\pi}{\sin\pi\, b_n} \left[\frac{F(a_n,b_n;x)}{\Gamma(1-q_n)\Gamma(b_n)} - x^{1-b_n} \frac{F(1-q_n,2-b_n;x)}{\Gamma(a_n)\Gamma(2-b_n)}\right]$$

We assume further

$$q_n = - g + \sqrt{g^2+n^2(\nu^2-\kappa^2)}$$

$$a_n = 1 + 2g + q_n, \; b_n = 1 + 2g + 2q_n$$

The coefficients of expansion in series (8) and (9) are determined from conditions of the interaction of components (2) and from the solution of the problem with mixed boundary conditions for auxiliary piecewise-holographic functions $X(z)$, $Y(z)$

$$X^+(\tau) - X^-(\tau) = 0, \; Y^+(\tau) - Y^-(\tau) = 0, \; (\tau\varepsilon\ell_o)$$

$$[\sigma X(\tau) + Y(\tau)]^+ - [\sigma X(\tau) + Y(\tau)]^- = 0, \; (\tau\varepsilon\ell) \tag{10}$$

$$[\rho X(\tau) - Y(\tau)]^+ + [\bar{\rho} X(\tau) - Y(\tau)]^- = 0, \; (\tau\varepsilon\ell)$$

where τ - coordinate on L; σ, ρ - complex parameters. Expansions $X(z)$, $Y(z)$ in power series z express explicitly the dependence of unknown coefficients on parameters of the problem. The dependence of components of the Z-matrix on the elastic constants and defect geometry is a cumbersome expression, so particular cases are presented.

Numerical analysis of the solution points out the predominant contribution of the first term of series (8) and (9) to the value of the elastic constants and volume content of the fibers $\xi < 0.6$. The subsequent terms of the expansion introduce a significant change in the critical load value, as they reveal for a hexagonal structure the existence of a stability limit, caused by the strong interaction of fibers as they become more tightly packed. Terms of the series have a similar type of structure, so for simplification and clearness of the following analysis, only the first terms in the expansions will be retained.

For materials in which isotropic fibers have a longitudinal shear rigidity increasing towards the lateral surface, it is assumed that:

$$G_r = G_r^o e^x, \; G_\theta = G_r, \; G_{r\theta} = 0 \tag{11}$$

The crack at a distance R from the center is opened by the angle 2θ and its midportion has the coordinate $Re^{i\theta_o}$, where it is assumed (Figure 1)

$$2\theta = \theta_b - \theta_a,\ 2\theta_o = \theta_b + \theta_a$$

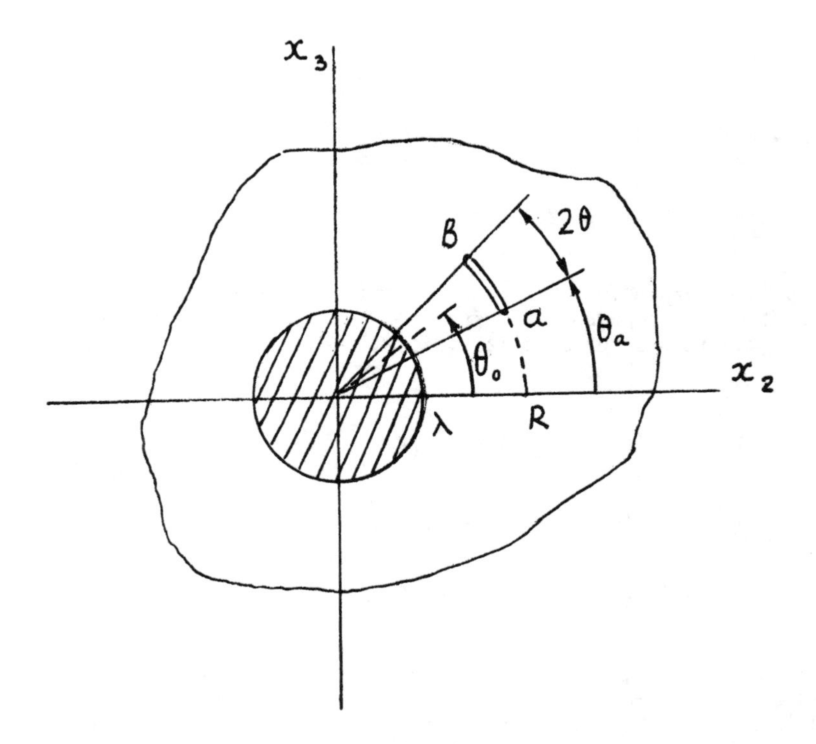

Fig. (1) - Location of local coordinate system in the medium with a crack ab

The generalized Hooke's law for a fibrous medium assumes the form

$$\langle\gamma_{12}\rangle = \frac{1}{G_{12}}\langle\sigma_{12}\rangle + \frac{\mu_{23}}{G_{13}}\langle\sigma_{13}\rangle$$

$$\langle\gamma_{13}\rangle = \frac{\mu_{23}}{G_{13}}\langle\sigma_{12}\rangle + \frac{1}{G_{13}}\langle\sigma_{13}\rangle \tag{12}$$

where the first terms in the expansion of components of the Z-matrix have the form

$$G_{12} = G\,\frac{h^2(\theta)-\xi^2\sin^4\theta G_p^2}{L(\theta)+8\xi\sin^2\theta\cos 2\theta_o G_p(G_a+b)} + \ldots$$

$$G_{13} = G\,\frac{h^2(\theta)-\xi^2\sin^4\theta G_p^2}{L(\theta)-8\xi\sin^2\theta\cos2\theta_o G_p(G_a+G)} + \cdots$$

$$\mu_{23} = \frac{8\xi\sin^2\theta\sin2\theta_o G_p(G_a+G)}{L(\theta)-8\xi\sin^2\theta\cos2\theta_o G_p(G_a+G)} + \cdots \tag{13}$$

Here, $f = R^2/\lambda^2$, $x_o = \alpha\Lambda$, $G_a = G_r^o e^{x_o}\,\dfrac{e^{x_o}-x_o-1}{1+x_o e^{x_o}-e^{x_o}}$

$$L(\theta) = (G_a+G)^2\{16 + f^2\xi^2[\sin^4\theta - 4(1-\cos\theta)^2]\} + 2f\xi^2(G_a-G)^2\sin^2\theta(4+\sin^2\theta)$$

$$+ \xi^2(G_a-G)^2[\sin^4\theta - 4(1+\cos\theta)^2],$$

$$h(\theta) = 4(G_a+G) - 2f\xi(1-\cos\theta)(G_a+G) + 2\xi(1+\cos\theta)(G_a-G),\ G_p = G_a$$

$$- G + f(G_a+G)$$

In the limiting cases ($\theta\leq\theta\leq\pi$), the above relations yield both the values of elastic constants for reinforced medium without cracks and the case of complete cohesive separation of the fiber with part of the matrix.

With weak fiber to matrix adhesion, where $k_{02}<<k_2$, but $k_{a2}>k_{02}$, k_2 is obtained from the above relations by means of the limiting transition $f\rightarrow I$. The relations are valid for integral parameters in the case of materials with interface cracks. Critical shear stresses are determined for the case of symmetrically located crack ($\theta_o = 0$), when

$$\langle\sigma_{13}\rangle = 0$$

If the next stage of crack growth is assumed to occur on the same cylindrical surface, and is restricted to a known angle, the stress concentration at the crack tip is specified by the relation:

$$\sigma_{1n} = \frac{K(\theta,\theta_o)}{\sqrt{2\pi\rho}} \tag{14}$$

where ρ is the distance from the crack tip at the point of symmetry and $K(\theta,\theta_o)$ is a function describing the stress intensity. As a result, the mean critical stresses will be [5]:

$$\langle\overset{*}{\sigma}_{12}\rangle = \frac{k_2 f}{\sqrt{2\pi\lambda}}\cdot\frac{h(\theta)-G_p\xi\sin^2\theta}{2G_p\cos\frac{\theta}{2}\sqrt{\sin\theta}} \tag{15}$$

where

$$G_p = G_a - G + f(G_a+G)$$

According to energy criterion of Griffith's, we find

$$\langle\sigma_{12}^*\rangle = \sqrt{\frac{8\lambda\gamma}{\frac{\partial}{\partial\theta}\left(\frac{F}{G_{12}}\right)}} = \sqrt{\frac{\gamma G}{\lambda}} \cdot \frac{h(\theta)-G_p\xi\sin^2\theta}{2\cos\frac{\theta}{2}\sqrt{\sin\theta}\sqrt{G_p(G_a+G)}} + \ldots \quad (16)$$

where γ is the surface energy. The relations (14) and (15) agree within the parameters depending on ratio of moduli of elasticity. This last feature may be of importance in predicting the limiting stresses for various composite media and for prescribed γ and k_2 values.

With an increase of the index α, in accordance with relation (14), the increa is noted of the shear modulus given above G_a and the stress intensity that leads to a reduction of the critical load, however, according to different laws (15) an (16).

Starting with the equality of critical stresses for a small crack at the interface border in the region adjacent to the fiber, the conditions or equality between the cohesive and adhesive strength of the medium can be approximately evaluated. Assuming the crack is located axisymmetrically ($\theta_0 = 0$) at $R = 1,1\lambda$, when local derivations in mechanical properties of the matrix are suppressed, we have

$$\frac{k_{02}}{k_2} = \frac{fG_a\cos\frac{\theta_1}{2}\sqrt{\sin\theta_1}}{G_p\cos\frac{\theta}{2}\sqrt{\sin\theta}} \cdot \frac{h(\theta)-G_p\xi\sin^2\theta}{2[G_a(1+\xi\cos\theta_1)+\eta G]-\xi G_a\sin^2\theta_1} \quad (17)$$

For glass-reinforced plastics with volume content of the matrix, $\eta = 0.3$; also with $\xi = 0.7$; $f = 1, 21$; $G/G_a = 0.04$, $\theta_1 = \pi/180$, $\theta = \lambda\theta_1/R$, we find

$$k_{02} \sim 1.1k_2$$

This means that the crack growth resistance in adhesive fracture should be 10% higher than in cohesive fracture. The relation above is not a strict one in that it may be made more precise by taking into account the subsequent terms in the solution, and the conditions at the interface border. Assuming a constant radius crack trajectory in the matrix in a cohesive fracture, it is of interest to analyze the changes in the shear moduli given. The data in Figures 2 and 3 illustrate in an approximate manner the effect of a reduction of these parameters in the presence of the one symmetrically located crack ($\theta_0 = 0$) with growth θ. The curves 1 and 2 in Figure 2 indicate the change G_{13}/G at a fiber volume content of $\xi = 0.5$, when $f = 1$ and $f = 1.21$. Curves 3 and 4 show the decrease of the ratio G_{12}/G under analogous conditions. Similar data, but with $\xi = 0.7$ and with re-

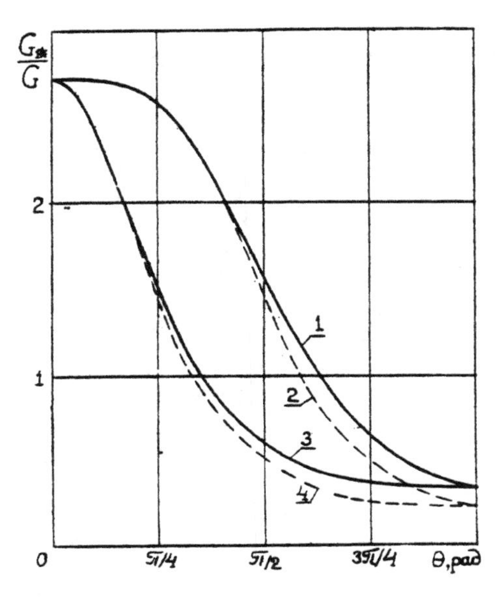

Fig. (2) - Change of shear moduli ratio with crack growth at volume content of fibers $\xi = 0.5$

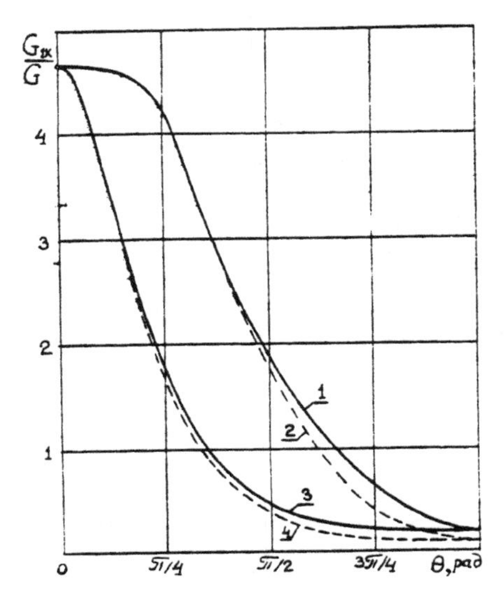

Fig. (3) - Change of shear moduli ratio with crack growth at volume content of fibers $\xi = 0.7$

spective notation retained, are presented in Figure 3. In Figure 4, curves 1 and 2 indicate the change at $\xi = 0.7$ in the side effect μ_{23} depending on location of the center of the crack θ_0 at its opening angle $\theta = \pi/4$ for f = 1.21 and f = 1. Curves 3 and 4 illustrate the change in the same data, but with $\xi = 0.5$.

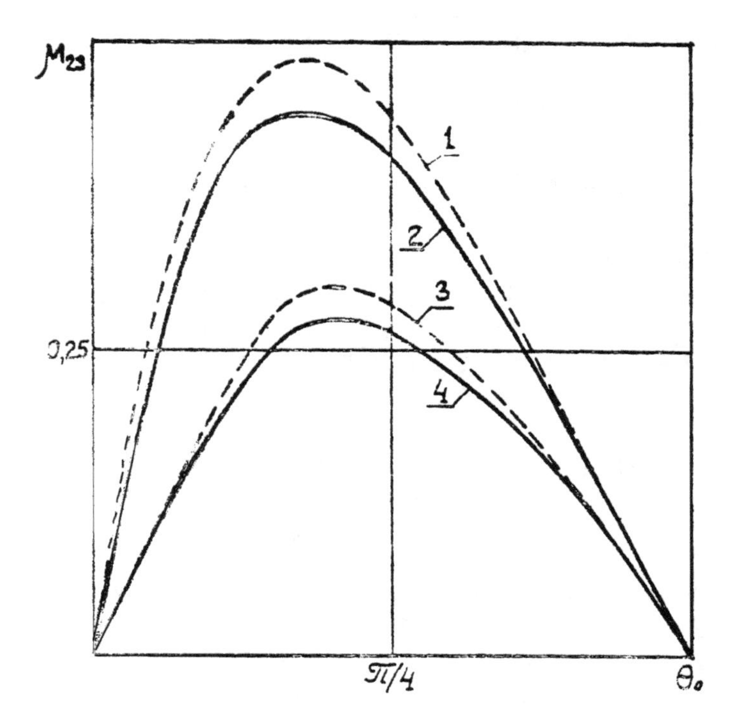

Fig. (4) - Change of crossover effect μ_{23} relative to location of the crack and its geometrical center

While the inhomogeneous isotropic structure of the fibers results in changes in elastic properties and in stress concentrations near the crack in accordance with relations (13) and subsequent ones, the anisotropy of the fibers introduces qualitative changes in the stress distribution and in the relation between the moduli given and component properties. The following case is considered:

$$G_r = G_r^o = \text{const},\ G_\theta = \nu^2 G_r,\ G_{r\theta} = \kappa G_r$$

Assume that the fiber's core with radius ε is isotropic. In order to avoid cumbersome equations, the case of an arbitrarily located crack on the interface border at a medium with $\varepsilon/\lambda \ll 1$ is presented:

$$G_{12} = G\,\frac{4[(1+\xi\cos\theta)G_a+\eta G]^2-\xi^2G_a^2\sin^4\theta}{L(\theta)+4\xi G_a(G_a+G)\sin^2\theta\cos2\theta_o} + \ldots$$

$$G_{13} = G\,\frac{4[(1+\xi\cos\theta)G_a+\eta G]^2-\xi^2G_a^2\sin^4\theta}{L(\theta)-4\xi G_a(G_a+G)\sin^2\theta\cos2\theta_o} + \ldots \qquad (18)$$

$$\mu_{23} = \frac{4\xi G_a(G_a+G)\sin^2\theta\sin2\theta_o}{L(\theta)-4\xi G_a(G_a+G)\sin^2\theta\cos2\theta_o} + \ldots$$

where it is assumed

$$G_a = \sqrt{G_r G_\theta - G_{r\theta}^2}$$

$$L(\theta) = 4[(1-\xi^2\cos^2\theta)G_a^2 + 2(1+\xi^2\cos\theta)G_aG + (1-\xi^2)G^2] + \xi^2G_a^2\sin^4\theta$$

The remaining notations are previous ones. From the comparison of (18) with those previously presented [1], it follows that the parameter

$$\sqrt{G_r G_\theta - G_{r\theta}^2}$$

plays the role of the reduced shear modulus with arbitrary cylindrical anisotropy under shear and with an isotropic inclusion of very small radius. The stresses at the tip of the symmetrically located crack when only $<\sigma_{12}>$ acts are described by a series, the first term of which is equal to

$$\sigma_{1n} = \sum_{m>0} \sqrt{\left(\frac{2\sin\theta}{\rho}\right)^{1-2g_m}} K_m(\;, g_m) = \sqrt{\left(\frac{2\sin\theta}{\rho}\right)^{1-2g_r}} \times$$

$$\times \frac{2G_a(1+2g_r)<\sigma_{12}>\cos\left(\frac{\pi}{4} - \frac{g_r\pi}{2}\right)\sin\left(g_r + \frac{1}{2}\right)\theta}{2[(1+\xi\cos\theta)G_a+\eta G]-\xi G_a\sin^2\theta} + \ldots \tag{19}$$

Here, ρ is the distance from the crack tip along the circular arc

$$\pi g_m \simeq \frac{4GG_aG_{r\theta}}{(G_a+G)G_r^2}\left(\frac{\varepsilon}{\lambda}\right)^{\Omega_m}, \quad \Omega_m = 2mG_a/G_r \tag{20}$$

The occurrence of side effects in Hooke's law for a fiber due to the presence in relation (5) of the constant $G_{r\theta}$, changes the characteristic of the stress state at the crack borders. This conclusion is validated upon examination of the exact equation for g_m:

$$\pi g_m = \operatorname{arctg} \frac{G_aG_{r\theta}(\varepsilon/\lambda)^{\Omega_m}}{G_\theta(G_a+G_{r\theta})+G_{r\theta}^2(\varepsilon/\lambda)^{\Omega_m}}$$

$$- \operatorname{arctg} \frac{G_aG_{r\theta}(G_a-G)(\varepsilon/\lambda)^{\Omega_m}}{(G_a+G)(G_a+G_\theta)G_\theta+(G_a-G)G_{r\theta}^2(\varepsilon/\lambda)^{\Omega_m}} \tag{21}$$

It follows that the deviation mentioned vanishes only when $G_{r\theta} = 0$. These findings suggest that on crack borders in an isotropic matrix, fiber cracks occur that distort the symmetry of the stress-strain relations in the fiber.

The critical stress state in the general case of the longitudinal shear is specified by the previously derived relation [1], developed from the energy criterion

$$\frac{\langle\overset{*}{\sigma}_{12}\rangle^2}{a^2} + \frac{\langle\overset{*}{\sigma}_{13}\rangle^2}{b^2} + \frac{\langle\overset{*}{\sigma}_{12}\rangle\langle\overset{*}{\sigma}_{13}\rangle}{c^2} = 1 \tag{22}$$

In the case that the crack starts its movement at the point θ_a, it should be assumed

$$a^2 = \frac{8\lambda_\gamma}{\left|\left(\frac{\partial}{\partial\theta} - \frac{\partial}{\partial\theta_o}\right)\frac{F}{G_{12}}\right|}, \quad b^2 = \frac{8\lambda_\gamma}{\left|\left(\frac{\partial}{\partial\theta} - \frac{\partial}{\partial\theta_o}\right)\frac{F}{G_{13}}\right|}, \quad c^2 = \frac{4\lambda_\gamma}{\left|\left(\frac{\partial}{\partial\theta} - \frac{\partial}{\partial\theta_o}\right)\frac{\mu_{23}F}{G_{13}}\right|}$$

When the crack starts moving at the tip θ_b, the following holds

$$a^2 = \frac{8\lambda_\gamma}{\left(\frac{\partial}{\partial\theta} + \frac{\partial}{\partial\theta_o}\right)\frac{F}{G_{12}}}, \quad b^2 = \frac{8\lambda_\gamma}{\left(\frac{\partial}{\partial\theta} + \frac{\partial}{\partial\theta_o}\right)\frac{F}{G_{13}}}, \quad c^2 = \frac{4\lambda_\gamma}{\left(\frac{\partial}{\partial\theta} + \frac{\partial}{\partial\theta_o}\right)\frac{\mu_{23}F}{G_{13}}}$$

Equation (22) in coordinates $\langle\overset{*}{\sigma}_{12}\rangle$, $\langle\overset{*}{\sigma}_{13}\rangle$ is represented by an ellipse, the location of which is specified by the value of the derivative of the components of the Z-matrix.

In the case of longitudinal shear, as follows from above, the components of the Z-matrix specify completely the changes in mechanical characteristics of the composite medium and the occurrence of the critical stress state.

REFERENCES

[1] Vanin, G. A., "Interference of Cracks in Fibrous Media", Fracture of Fibrous Materials, pp. 38-45, 1979.

[2] Vanin, G. A., "Longitudinal Shear of Multicomponent Medium with Defects", Prikl. Mekhanika, No. 8, pp. 35-41, 1977.

[3] Vanin, G. A., "Volumetric Elastic Expansion of Medium with Hollow Spherical Inclusions", Prikl. Mekhanika, No. 7, pp. 127-129, 1980.

[4] Lekhicky, S. G., "Theory of Elasticity of an Isotropic Body", p. 416, 1977.

[5] Barenblatt, G. I., "Mathematical Theory of Equilibrium Cracks Forming in the Process of Brittle Fracture", Zhrn. Prikl. Mech. i. Tekn. Fiziki, No. 4, pp. 65-70, 1961.

[6] Cherepanov, G. I., "Mechanics of Brittle Fracture", p. 640, 1974.

SURFACE NOTCHES IN COMPOSITES*

G. P. Sendeckyj

Air Force Wright Aeronautical Laboratories
Wright-Patterson Air Force Base, Ohio 45433

ABSTRACT

Simple mechanics of materials analyses for predicting the static strength of surface notched (scratched) resin-matrix composite laminates are presented. The strength predictions are compared with and are found to be in excellent agreement with the available experimental data.

INTRODUCTION

The surface scratch (notch) is a common service induced defect in airframe structures. In metallic structures, it can either grow into a through-the-thickness crack which can lead to failure or directly cause the failure of the structure. Hence, considerable effort has been devoted to developing analysis methods for assessing its severity and predicting its growth characteristics [1,2]. The experience with scratches in metallic structures caused concern about the effect of surface notches on high performance composite laminates. As a result of this concern, a number of investigations [3-10] were conducted to assess the effect of surface notches on composite structures. These studies showed that surface notches in laboratory size test specimens had a severe effect on the residual strength of the laminate. Careful experimental observations [3-6] indicated that delaminations and matrix cracking, accompanied by local buckling of the delaminated plies, occur near the surface notch prior to failure. Hence, rigorous analyses of the surface notched laminate would require modeling of the delamination zones. Even though such analyses are possible [11,12] using the finite element method, they are extremely time consuming. Sendeckyj [3] suggested that an elementary mechanics of materials analysis could be used to predict the strength reduction caused by surface notches in composite laminates. In his analysis, Sendeckyj assumed that the surface notched laminate could be modeled as a homogeneous beam of variable eccentric cross-section under axial loading. The theoretical results were shown to be in good agreement with experimental data for some graphite-epoxy laminates and in poor agreement for others.

In the present paper, refined mechanics of materials analyses are presented that take into account the ply-by-ply heterogeneity of the laminate. The strength

*This paper is based on inhouse work performed at the Flight Dynamics Laboratory of the Air Force Wright Aeronautical Laboratories under the Solid Mechanics Project funded by the Air Force Office of Scientific Research.

predictions, based on the refined analyses, are shown to be in excellent agreement with the available experimental data for surface notched graphite-epoxy laminates. The strength predictions provide a lower bound for glass-epoxy surface notched laminates and, in general, are in poor agreement with the data. This lack of agreement for glass-epoxy laminates is due to the extreme delamination/matrix cracking observed prior to failure which results in the violation of one of the basic assumptions of the analyses.

MECHANICS OF MATERIALS STRENGTH PREDICTION

A. General Description of the Models

As pointed out in [3], the occurrence of delamination zones near the surface notch prior to failure and the tendency of plies with fibers perpendicular to the loading direction to fail at low strain levels suggest that the strength of surface notched laminates can be predicted by using elementary mechanics of materials models. In these models, the surface notched specimen is assumed to behave like a beam of variable cross-section under axial load. A free body is isolated from the beam by taking two parallel cuts (or sections) perpendicular to the direction of loading. One of the cuts is taken through the surface notch which is assumed to be perpendicular to the load, while the other section is take through a region far removed from the notch. The strains at the cut containing the surface notch are assumed to vary linearly, while those at the other cut are assumed to be constant. The tractions, corresponding to these strain distributions, transmitted across the cut surfaces are computed and overall equilibrium conditions are enforced on the free body, resulting in a system of algebraic equations that can be solved for the strains at the surface notched cross-sectior in terms of the far field strain. The strains at the surface notched section are then compared with the static ultimate strains of the various plies. If the calculated strains exceed the ultimate strains for a given ply, the ply is assumed to have failed and the stresses in it acting on the notched section are set equal to zero and overall equilibrium conditions are again imposed leading to a new strain distribution at the notched section. The new strain distribution is compared to the ultimate strains of the plies. This process is repeated until all plies fail within the surface notched section or no changes occur in the strains at the notch. The highest far field strain value that does not lead to failure of the notched section is then used to calculate the failure stress of the notch laminate. The actual details of the calculations depend on the assumed behavior of the plies making up the laminate.

In the present paper, analyses are developed for three models of laminate behavior. These models, increasing in complexity, are:

Model I: The laminate is modeled as being homogeneous, with a ply-by-ply application of the maximum strain failure criterion. This is basically the mode used in [3].

Model II: The laminate is modeled as being heterogeneous, with a ply-by-ply application of the maximum strain failure criterion. Moreover, Poisson's ratio effects are neglected.

Model III: The laminate is modeled as being heterogeneous, with a ply-by-ply application of the maximum strain failure criterion. Poisson's ratio effects are taken into account in an approximate manner.

B. Basic Assumptions

The mechanics of materials predictions of the strength of surface notched laminates, presented herein, are based on the following assumptions:

1. Plane sections remain plane.

2. All plies at an angle α ($\neq 0°$ or $90°$) with respect to the loading direction are assumed to behave like a ($\pm\alpha°$) laminate; that is, they have the same elastic moduli and failure strains in the loading direction and perpendicular to it as the ($\pm\alpha°$) laminate.

3. All plies have linear stress-strain curves to failure.

4. Ply failure is governed by a maximum strain failure criterion.

5. If in a particular ply the ultimate strain in the direction of loading or perpendicular to it is exceeded, the ply is assumed to carry zero load in that direction while supporting loads in the transverse direction. Moreover, it has a zero Poisson's ratio after failure.

Assumption 1, basic to the mechanics of materials strength prediction method, is reasonable whenever the extent of delamination and matrix cracking damage observed to occur prior to failure is small. If the extent of matrix damage is large as in the case of surface notched glass-epoxy laminates [6] and fatigued graphite-epoxy laminates [5], the assumption is no longer valid and the method is not applicable. In these cases, the mechanics of materials analysis provides a lower bound on the strength.

While not absolutely necessary, assumptions 2 and 3 are made to simplify the analysis. Assumption 2 eliminates the need for resolving the stresses and strains within each ply into components parallel and perpendicular to the fiber directions. Assumption 3, the standard assumption of linear laminated plate theory, is the most serious one since the basic ply and, hence, the laminate exhibits nonlinear stress-strain behavior [13,14]. It causes the stresses corresponding to a given strain state to be in error. As a result, the predicted strengths can be in error. The error introduced by this assumption can be bounded by using the tangent and secant moduli of the affected plies in the computations. Use of the tangent moduli will give an upper bound on the strength of the surface notched laminate, while use of the secant moduli will give a lower bound.

Assumption 5 deals with the modeling of the post-first-ply failure behavior of the laminate. It presupposes that failure of plies with low ultimate strain capability can occur without precipitating laminate failure. Moreover, it indicates how the failed plies are modeled in subsequent computations. Basically, it states that once the ultimate strain in a given direction in a ply is exceeded, the ply ceases to carry loads in that direction while still carrying the loads in the perpendicular direction. Operationally, this is equivalent to assuming that the Young's modulus of the ply in the failed direction and Poisson's ratios are zero after ply failure. It should be noted that this is not the only possible model of the post-first-ply failure behavior of the laminate.

C. Derivations

Consider a midplane symmetric laminate of uniform width, w, and constant thickness. As shown in Figure 1, assume that the laminate contains a centrally

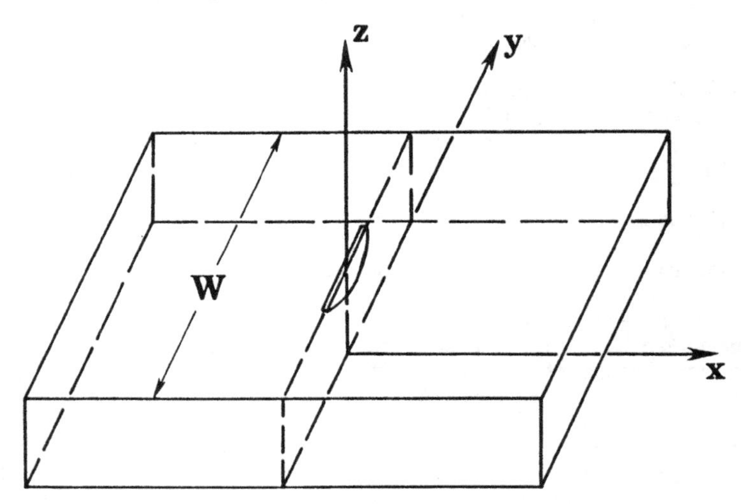

Fig. (1) - Laminate containing a centrally located surface notch

located (in the y,z plane) surface notch of arbitrary symmetric profile. Assume that the laminate consists of N layers. The layers can consist either of multiples of plies or portions of a ply, depending on the degree of computational accuracy required. Assume that the surface notch can be replaced by an "equivalent" surface notch having the stepped rectangular profile shown by the dotted line in Figure 2. The depth of each step of the equivalent notch is equal to the

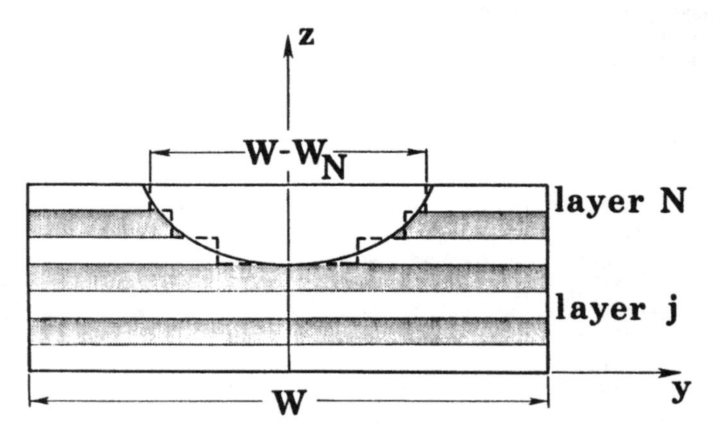

Fig. (2) - Cross-section view showing scheme for modeling arbitrary surface notch geometry

thickness of the layer containing the step, while the width ($w-w_n$, where w_n is the unnotched width of layer n) of each step is chosen so that the net section area of the layer is equal to the net section area of the layer with the original notch geometry.

Assume that the strain distribution on the notched section (x=0 plane) is given by

$$\left.\begin{aligned} e_{ix} &= a + bz \\ e_{iy} &= c + dz \end{aligned}\right\} \quad z_i \leq z \leq z_{i+1},\ i = 1,2,\ldots,N \tag{1}$$

where e_{ix} and e_{iy} are the x- and y-strain components in the ith layer; $z_1 = 0$ and z_{i+1} is the distance from the x,y plane to the top of the ith layer; and a, b,c and d are constants to be determined. The strain distribution on a section far removed from the notch ($x = x_o \neq 0$ plane) is given by

$$e_{ix} = f,\ e_{iy} = g,\ z_i \leq z \leq z_{i+1},\ i = 1,2,\ldots,N \tag{2}$$

where f is the known axial far field strain component and g is to be solved for far field transverse strain component.

Assuming generalized plane stress in the z-direction, the stress states corresponding to the strain states defined by equations (1) and (2) are

$$\left.\begin{aligned} \sigma_{ix} &= Q_{ixx}(a+bz) + Q_{ixy}(c+dz) \\ \sigma_{iy} &= Q_{ixy}(a+bz) + Q_{iyy}(c+dz) \end{aligned}\right\} \quad z_i \leq z \leq z_{i+1} \text{ on } x = 0 \tag{3}$$

$$\left.\begin{aligned} \sigma_{ix} &= Q_{ixx}f + Q_{ixy}g \\ \sigma_{iy} &= Q_{ixy}f + Q_{iyy}g \end{aligned}\right\} \quad z_i \leq z \leq z_{i+1} \text{ on } x = x_o \neq 0 \tag{4}$$

where

$$\begin{aligned} Q_{ixx} &= E_{ix}/(1-\nu_{ixy}\nu_{iyx}) \\ Q_{iyy} &= E_{iy}/(1-\nu_{ixy}\nu_{iyx}) \\ Q_{ixy} &= \nu_{ixy}Q_{iyy} = \nu_{iyx}Q_{ixx} \end{aligned} \tag{5}$$

where

E_{ix} = longitudinal (x-direction) Young's modulus of ith layer,

E_{iy} = transverse (y-direction) Young's modulus of ith layer, and

ν_{ixy} = major Poisson's ratio of the ith layer.

It should be noted that assumption 2 has been imposed in writing equations (3) and (4). If this assumption were not imposed, then shear-extension coupling terms would have to be included in constitutive equations (3) and (4).

As pointed out in the general description of the models section, the five unknown constants (a,b,c,d and g) are obtained by imposing the following stress equilibrium conditions:

$$\int \sigma_y(0,y_o,z)dz = 0 \tag{6}$$

$$\int \sigma_y(0,y_o,z)zdz = 0 \tag{7}$$

$$\int \sigma_y(x_o,y_o,z)dz = 0 \tag{8}$$

$$\iint \sigma_x(0,y,z)dydz = \iint \sigma_x(x_o,y,z)dydz \tag{9}$$

$$\iint \sigma_x(0,y,z)zdydz = \iint \sigma_x(x_o,y,z)zdydz \tag{10}$$

In the transverse equilibrium conditions, equations (6) and (7) on the notched cross-section, y_o is any value of y that does not intersect the notch boundary. In equilibrium conditions, equations (8) through (10), x_o is any value of $x \neq 0$. Moreover, the integrals in equations (9) and (10) are evaluated over the cut section (x=0) with the notch area omitted.

Upon substituting equations (3) and (4) into equations (6) through (10) and performing the indicated integrations, we get

$$A_1a + A_2b + A_3c + A_4d = 0 \tag{11}$$

$$A_2a + A_5b + A_4c + A_6d = 0 \tag{12}$$

$$A_7a + A_8b + A_9c + A_{10}d = A_{11}f + A_{15}g \tag{13}$$

$$A_8a + A_{12}b + A_{10}c + A_{13}d = A_{14}f + A_{16}g \tag{14}$$

$$A_{17}f + A_{18}g = 0 \tag{15}$$

where

$$A_1 = \sum_{i=1}^{N} Q_{ixy}(z_{i+1}-z_i) \tag{16}$$

$$A_2 = \sum_{i=1}^{N} Q_{ixy}(z_{i+1}^2-z_i^2)/2 \tag{17}$$

$$A_3 = \sum_{i=1}^{N} Q_{iyy}(z_{i+1}-z_i) \tag{18}$$

$$A_4 = \sum_{i=1}^{N} Q_{iyy}(z_{i+1}^2-z_i^2)/2 \tag{19}$$

$$A_5 = \sum_{i=1}^{N} Q_{ixy}(z_{i+1}^3-z_i^3)/3 \tag{20}$$

$$A_6 = \sum_{i=1}^{N} Q_{iyy}(z_{i+1}^3-z_i^3)/3 \tag{21}$$

$$A_7 = \sum_{i=1}^{N} w_i Q_{ixx}(z_{i+1}-z_i) \tag{22}$$

$$A_8 = \sum_{i=1}^{N} w_i Q_{ixx}(z_{i+1}^2-z_i^2)/2 \tag{23}$$

$$A_9 = \sum_{i=1}^{N} w_i Q_{ixy}(z_{i+1}-z_i) \tag{24}$$

$$A_{10} = \sum_{i=1}^{N} w_i Q_{ixy}(z_{i+1}^2-z_i^2)/2 \tag{25}$$

$$A_{11} = w \sum_{i=1}^{N} Q_{ixx}(z_{i+1}-z_i) \tag{26}$$

$$A_{12} = \sum_{i=1}^{N} w_i Q_{ixx}(z_{i+1}^3-z_i^3)/3 \tag{27}$$

$$A_{13} = \sum_{i=1}^{N} w_i Q_{ixy}(z_{i+1}^3-z_i^3)/3 \tag{28}$$

$$A_{14} = w \sum_{i=1}^{N} Q_{ixx}(z_{i+1}^2 - z_i^2)/2 \tag{29}$$

$$A_{15} = w \sum_{i=1}^{N} Q_{ixy}(z_{i+1} - z_i) \tag{30}$$

$$A_{16} = w \sum_{i=1}^{N} Q_{ixy}(z_{i+1}^2 - z_i^2)/2 \tag{31}$$

$$A_{17} = \sum_{i=1}^{N} Q_{ixy}(z_{i+1} - z_i) \tag{32}$$

$$A_{18} = \sum_{i=1}^{N} Q_{iyy}(z_{i+1} - z_i) \tag{33}$$

Note that assumption 3 was used in deriving equations (11) through (15). Moreover while the definitions of constants A_{17} and A_{18} are identical to those for A_1 and A_3, different rules are followed in their computation after first-ply failure occurs. These definitions are repeated for the sake of clarity in the discussion of the iterative computation procedure.

D. Computation Procedure

Starting with equilibrium equations (11) through (15), the strength of the surface notched laminate is computed as follows:

1. Assume a value of the far field longitudinal strain, $e_{ix} = f$.

2. Solve equations (11) through (15) for the far field transverse strain, $e_{iy} = g$, and the strain distribution, defined by equation (1), at the surface notched cross-section (x=0) of the laminate.

3. Apply the maximum strain failure criterion on a layer by layer basis; that is, check whether

$$\begin{aligned} a + bz_{i+1} &> \hat{e}_{ix} \\ c + dz_{i+1} &> \hat{e}_{iy} \end{aligned} \qquad \text{for } i = 1,2,\ldots,N \tag{34}$$

where $\hat{e}_{ix}$ and $\hat{e}_{iy}$ are the ultimate longitudinal and transverse strains in the ith layer, respectively. If no layers fail in the notched cross-section, i.e.,

$$
\left.
\begin{array}{l}
a + bz_{i+1} \leq \hat{e}_{ix} \\
c + dz_{i+1} \leq \hat{e}_{iy}
\end{array}
\right\} \quad \text{for every } i = 1,2,\ldots,N \tag{35}
$$

assume a higher value of f and repeat steps 2 and 3. If all the layers fail in the longitudinal direction, that is,

$$a + bz_{i+1} > \hat{e}_{ix} \qquad \text{for every } i = 1,2,\ldots,N \tag{36}$$

assume a lower value of f and repeat steps 2 and 3. If $(a+bz_{i+1}) > \hat{e}_{ix}$ or $(c+dz_{i+1}) > \hat{e}_{iy}$ for at least one but not all of the layers, go the step 4 in which the failed layers are unloaded.

4. If $f > \hat{e}_{ix}$ for some of the layers, set the Q_{ixx} and Q_{ixy} for those layers equal to zero in equations (26), (29) and (32). If $g > \hat{e}_{iy}$ for some of the layers, set the Q_{ixy} and Q_{iyy} for those layers equal to zero in equations (30) through (32). If $(a+bz_{i+1}) > \hat{e}_{ix}$ for some of the layers, set the Q_{ixx} and Q_{ixy} for those layers equal to zero in equations (16), (17), (20), (22)-(25), (27) and (28). If $(c+dz_{i+1}) > \hat{e}_{iy}$ for some of the layers, set the Q_{ixy} and Q_{iyy} for those layers equal to zero in equations (16)-(21), (24), (25) and (28).

5. Repeat steps 2 through 4 until there is no change in the values of the constants A_i $(i = 1,2,\ldots,18)$.

6. If $(a+bz_{i+1}) < \hat{e}_{ix}$ for at least one layer, the surface notched laminate can carry additional load. Hence, increase f, set the Q_{ixx}, Q_{ixy} and Q_{iyy} equal to their original values given by equations (5), and repeat steps 2 through (6) until f becomes sufficiently large so that $(a+bz_{i+1}) > \hat{e}_{ix}$ for every layer of the laminate. The largest value of f such that $(a+bz_{i+1}) \leq \hat{e}_{ix}$ for at least one layer is the ultimate far field strain in the surface notched laminate. The far field stress, corresponding to this value of f and the associated value of g, is the ultimate strength of the surface notched laminate.

It should be noted that the derivations and computation procedure presented above are for Model III. Model II is obtained by setting $Q_{ixy} = 0$ $(i = 1,2,\ldots,N)$ in the equations. Model I is obtained by further setting $E_{ix} = E_x$ $(i = 1,2,\ldots,N)$ in Model II.

THEORY-EXPERIMENT COMPARISON

A limited amount of experimental data is available on the effect of surface notches on the tensile strength of composite laminates [3,5-10]. These data will now be compared with the theoretical predictions based on the mechanics of materials models.

A. Sendeckyj et al Graphite-Epoxy Data [3,5]

Sendeckyj et al [3,5] determined (a) the tensile static strengths and (b) the tensile residual strengths after tension dominated fatigue loading of $[(0/\pm45/0)_S]_3$ graphite-epoxy specimens containing surface notches with circular arc profiles. The specimens were 1.5 inches wide. The surface notches were centrally located and had the geometry shown in Figure 3.

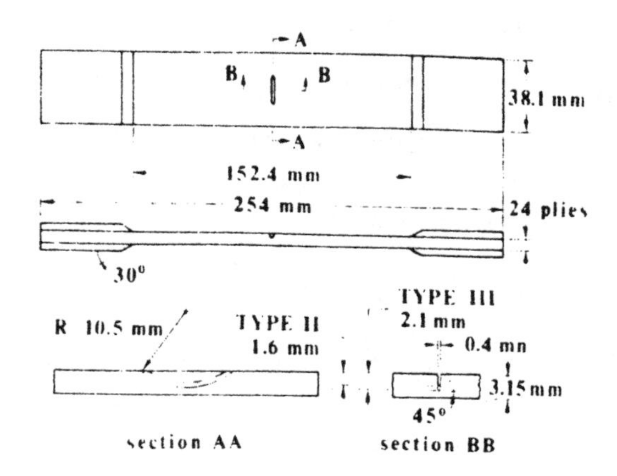

Fig. (3) - Specimen geometry used in [3]

The static test results are compared with theoretical predictions, based on Model I, in Table 1, which presents the specimen and notch dimensions, and ratios of notched to unnotched strengths. The predicted strength ratios were ob-

TABLE 1 - THEORY-EXPERIMENT COMPARISON FOR CIRCULAR ARC SURFACE NOTCHES IN $[(0/\pm45/0)_S]_3$ GRAPHITE-EPOXY LAMINATE

w	t	R	d	Strength Ratio	
(in.)	(in.)	(in.)	(in.)	Model I	Experiment
1.500	0.124	0.828	0.041	0.766	0.752
1.500	0.124	0.828	0.041	0.766	0.764
1.500	0.124	0.414	0.062	0.741	0.782
1.500	0.124	0.414	0.062	0.741	0.716
1.500	0.124	0.414	0.083	0.668	0.659
1.500	0.124	0.414	0.083	0.668	0.661

tained by assuming that the failure strains of the 0° and ±45° plies are equal. As can be seen from the table, the predicted strength ratios are in excellent agreement with the experimentally determined ones. It should be noted that Models II and III give identical predictions for this case.

B. Verette and Labor Graphite-Epoxy Data [7,10]

Verette and Labor [7] tested a series of 2 inch wide, 16 ply thick graphite-epoxy specimens containing 1 inch long centrally located surface notches. The notches had a rectangular profile and were either 3 or 6 plies deep. The experimental results [7,10] are compared with theoretical predictions, based on Model II, in Table 2. The table gives the laminate stacking sequences, notch

TABLE 2 - COMPARISON OF PREDICTED AND MEASURED STRENGTH RATIOS FOR DATA OF [7]

Laminate	Notch Depth	Strength Ratio	
	(plies)	Experiment	Model II
$(0/\pm45/0/\pm45)_S$	3	0.63	0.65
$(0_2/\pm45/0_2/90/0)_S$	3	0.69	0.67
$(0/\pm45/90)_{2S}$	3	0.55	0.65
	6	0.49	0.48

depths, and predicted and experimentally determined strength ratios. Models II and III were used in deriving the predicted strength ratios and it was assumed that the failure strains of the 0° and ±45° plies were equal. Since the two models gave essentially identical predictions, only the predictions based on Model II are given in the table.

As can be seen from Table 2, the predicted and measured strength ratios are in excellent agreement for all of the cases but one. The exception is the 3 ply deep surface notch in the $(0/\pm45/90)_{2S}$ laminate. It is suspected that in this case, there was an error in machining of the notches and, hence, the lack of agreement may be disregarded.

C. Im et al Graphite-Epoxy Data [6]

Im et al [6] determined the static tensile strength of (90/0/0/90) graphite-epoxy specimens with two ply deep surface notches with a rectangular profile. Their data is compared with predictions based on Model II in Figure 4. As can be seen from the figure, the predicted strength values (solid line in the figure) are in excellent agreement with the experimental data for the shorter surface notch. For the full width surface notch, the agreement is poor. This is due to the nature of the damage in this case.

DISCUSSION AND CONCLUSIONS

As can be seen from the theory-experiment comparisons presented in the previous section, the present mechanics of materials strength prediction method for surface notched laminates gives excellent results for the graphite-epoxy laminates. In all cases that were analyzed, the experimentally observed delamination zones were small and, hence, one would expect the model and analyses to be valid. If the delaminations are large as in glass-epoxy laminates [6], an underlying assumption of the model is violated and one would expect the predictions to disagree with the experimental data. This is the case for the data presented in [6].

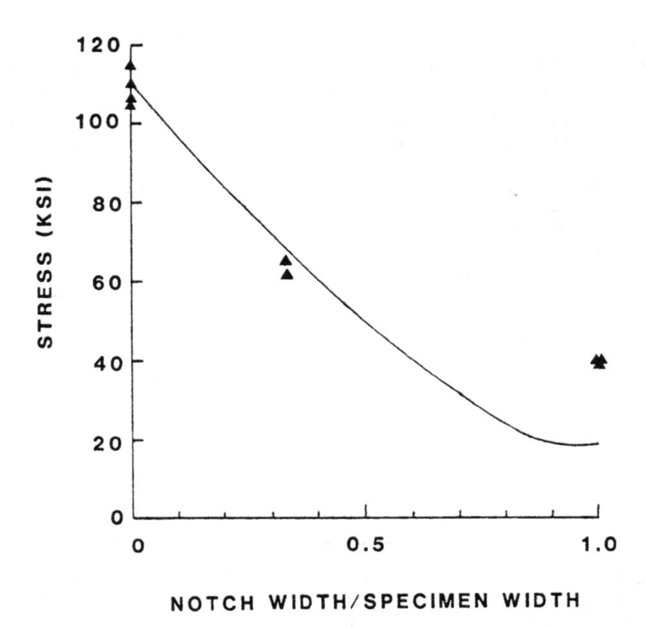

Fig. (4) - Comparison of predicted strength (solid curve) with experimental data (triangles) for (90/0/0/90) graphite-epoxy laminate with two ply deep surface notches

Based on the results presented herein, it is obvious that a simple mechanics of materials analysis is adequate for assessing the effect of surface notches in graphite-epoxy laminates. More work needs to be done on adepting the present re sults to other composite materials.

REFERENCES

[1] Swedlow, J. L., ed., The Surface Crack: Physical Problems and Computationa Solutions, ASME, New York, 1972.

[2] Chang, J. B., ed., Part-Through Crack Fatigue Life Prediction, ASTM STP 687 1979.

[3] Sendeckyj, G. P., "Some observations on fracture behavior of advanced fiber reinforced laminates", Proceedings of the 12th Annual Meeting of the Societ of Engineering Science, University of Texas at Austin Press, pp. 625-634, 1975.

[4] Sendeckyj, G. P., "Fatigue damage accumulation in graphite-epoxy laminates" Failure Modes on Composites - III, T. T. Chiao and D. M. Shuster, eds., Met lurgical Society of AIME, pp. 100-144, 1976.

[5] Sendeckyj, G. P., Stalnaker, H. D. and Kleismit, R. A., "Effect of temperature on fatigue response of surface-notched $[(0/\pm45/0)_S]_3$ graphite-epoxy laminate", Fatigue of Filamentary Composite Materials, K. L. Reifsnider and K. N. Lauraitis, eds., ASTM STP 636, pp. 123-140, 1977.

[6] Im, J., Mandell, J. F., Wang, S. S. and McGarry, F. J., "Surface crack growth in fiber composites", MIT Report, NASA CR-135094, September 1976.

[7] Verette, R. M. and Labor, J. D., "Structural criteria for advanced composites", Northrop Corporation, Technical Report AFFDL-TR-76-142, Volumes I and II, March 1977.

[8] Porter, T. R., "Evaluation of flawed composite structural components under static and cyclic loading", Boeing Aerospace Company, Technical Report NASA CR-135403, February 1979.

[9] Porter, T. R., "Evaluation of flawed composite structure under static and cyclic loading", Fatigue of Filamentary Composite Materials, K. L. Reifsnider and K. N. Lauraitis, eds., ASTM STP 636, pp. 152-170, 1977.

[10] Lo, K. H. and Wu, E. M., "Serviceability of composites surface damage", Fibrous Composites in Structural Design, E. M. Lenoe, D. W. Oplinger and J. J. Burke, eds., Plenum Press, New York, pp. 459-466, 1978.

[11] Anderson, J. M., Hsu, T. M. and McGre, W. D., "Characterization of crack growth in bonded structures", Proceedings of the 12th Annual Meeting of the Society of Engineering Science, University of Texas at Austin Press, pp. 1283-1292, 1975.

[12] Wang, S. S., Mandell, J. F. and McGarry, F. J., "Three-dimensional solution for a through-thickness crack with crack tip damage in a cross-plied laminate", Fracture Mechanics of Composites, ASTM STP 593, pp. 61-85, 1975.

[13] Sendeckyj, G. P., Richardson, M. D. and Pappas, J. E., "Fracture behavior of Thornel 300/5208 graphite-epoxy laminates - Part 1: Unnotched laminates", Composite Reliability, ASTM STP 580, pp. 528-546, 1975.

[14] Sandhu, R. S., "Nonlinear behavior of unidirectional and angle ply laminates", Journal of Aircraft, Vol. 13, No. 2, pp. 104-111, 1976.

SECTION III

MICROMECHANICS

SOME PECULIARITIES OF FRACTURE IN HETEROGENEOUS MATERIALS

V. Tamuzs

L.S.S.R. Academy of Sciences
Riga, USSR

INTRODUCTION

The volume fraction of microcracks plays an essential role in the fracture process of heterogeneous materials. The breaking of a material to separate pieces initiates from the creation of microcracks, their enlargement and coalescence to the formation of macrocracks [1].

At the moment of final fracture of a specimen, a certain amount of microcracks is accumulated in the volume of a material. Their accumulation kinetics, size, and level of damage depend on the structure of a material, loading conditions and specimen size.

Damageability diagnostics is, in fact, the only method by means of which the expected fracture of a particular material can be predicted. Hence, it follows that the classification of the volume fracture regularities is of practical significance. In order to estimate the degree of damage, it is necessary to make a quantitative assessment. Such an assessment might be made through the parameter $\omega = Na^3$ which has been introduced by different authors for estimating the penny-shaped microcrack density in an isotropic material. Here, N is the number of cracks in a unit volume where a is the crack radius. In the case of cracks with different radii, it is necessary to sum up their respective values. V. S. Kuksenko [1] has noted that the value $R = N^{-1/3}$ characterizes the mean distance between the microcrack centers and if the typical linear size of a crack is $\ell \approx 2a$, then the coefficient $K = R/\ell = \frac{1}{2}\,\omega^{-1/3}$ characterizes the ratio of the mean distance between the cracks to their size.

DAMAGE CHARACTERIZATION

In the theoretical calculations, it is convenient to introduce the assumption that a material consists of structural elements with linear size ℓ, each of which can be broken, simulating the development of a crack of the same size. The damage is then characterized by a ratio of the broken elements to their total number, N_0, i.e., in percentage:

$$P = \frac{N}{N_0} \times 100$$

For $N_o \approx 1/\ell^3$, $P \approx K^{-3} \times 100$. In the case of a transversal isotropic material, the size of a structural element in the direction of symmetry is $\ell\alpha$. Then as $N_o \approx 1/\ell^3\alpha$, $K \approx (100 \cdot \alpha/P)^{1/3}$.

The experimental data for the extreme level of damage of various materials [2] are given in Table 1. Nylon-6 and PMMA damage are determined by measurements

TABLE 1 - EXTREME LEVEL OF DAMAGE

Specimen Material	Type of Loading	a(mcm)	$k = \ell^{-1} N^{-1/3}$	$P = \frac{N}{N_o} \cdot 100\%$ ($P \approx K^{-3}$ 100%)
Nylon-6	Tension	0.009	2.5	6
PMMA	Tension	0.17	3.5	2.3
Rock-Solid	Compression	200	5	0.8
PE + PVC	Low Cycle Fatigue	100-150	3.2	3
Epoxy Carbon V_a = 64%	Tension & Bending	10	3.5	2.3
Epoxy Boron V_a = 0.34 Small Specimens	Tension		2.5	6
Epoxy Boron V_a = 0.34 Small Specimens	Low Cycle Fatigue		2.1	10
Epoxy Boron V_a = 0.34 Large Specimens	Tension		4.6	1

based on small-angle X-ray diffraction [1]. The damage of rock-solid and polyester resin filled with PVC particles were estimated by direct crack calculation using a microscope [3,4] after cutting the material. The damage of fibrous composites was determined by counting the fiber breaks which were revealed after burning out the matrix in the broken specimens. In all the investigated cases, the value of P changes from 1 to 10% and it depends sensitively on the size of the specimen.

UNIDIRECTIONAL COMPOSITES

Consider in detail the volume fracture of a unidirectional epoxy carbon composite. The counting of fiber breakage and damage analysis were carried out on specimens tested under static tension and bending at load levels equal to 0.2, 0.4, 0.6, 0.8 and 0.93 of the maximum load. The broken specimen was also tested As it is seen from the results in Figure 1, the breakage of fibers starts at a load level approximately equal to one half of its maximum value. At load level of 0.90 to 0.95 of the maximum value, the breakage turns into an avalanche-like process. Consequently, the fracture of an epoxy-carbon composite even with high

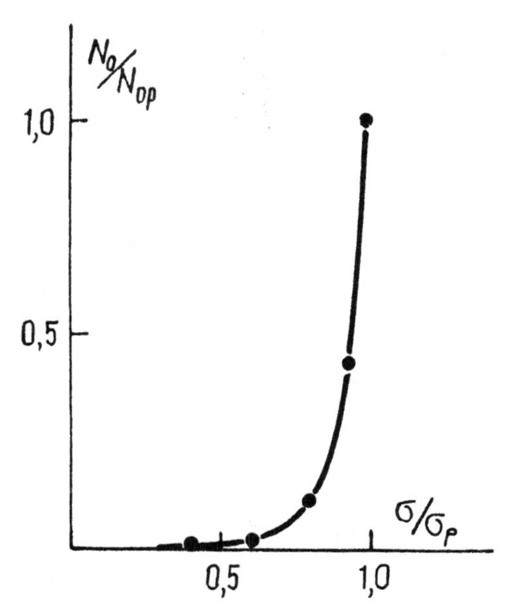

Fig. (1) - Break accumulation kinetics in epoxy carbon composite

volume fraction of fibers is of a cumulative nature.

The length distribution of the broken fiber segments is shown in Figure 2. It can be seen that the density distribution curve has two definite maxima at

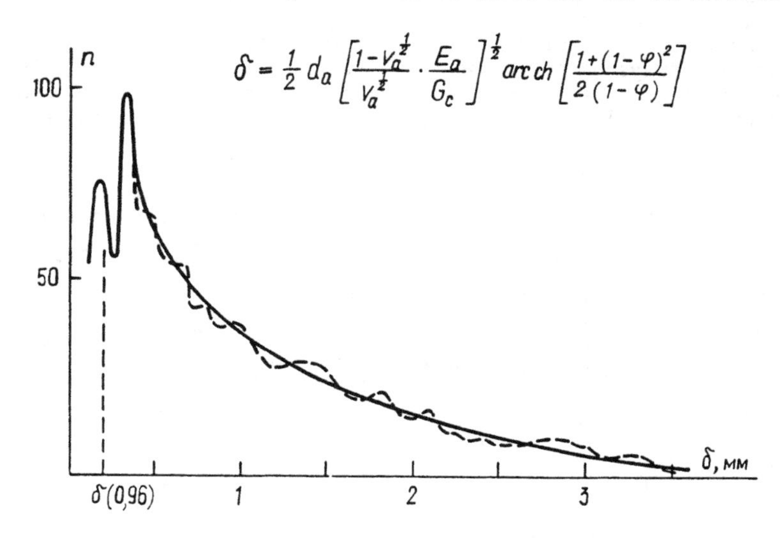

Fig. (2) - Length distribution of broken fiber segments

small lengths. As the fiber damage in segments shorter than the ineffective length δ is hardly probable, it can be assumed that the value of the first maximum corresponds to the size δ, according to the well-known formula:

$$\delta = d_f \left(\frac{E_a}{G_m} \frac{1-V_a^{1/2}}{V_a^{1/2}}\right)^{1/2} \operatorname{arcch}\left[\frac{1+(1-\phi)^2}{2(1-\phi)}\right] \tag{1}$$

Knowing the fiber modulus E_a, matrix modulus G_m, fiber diameter d_f and fiber volume fraction V_a, the value stress reset degree ϕ in a distance δ from fiber breakage can be determined. For all the investigated volume fractions, the value $\phi = 0.96 \div 0.97$ is obtained. Consequently, this value can be applied for the structural element size determination in a unidirectional composite.

THEORETICAL CONSIDERATION

The theoretical analysis of fiber breakage in a unidirectional composite was given in [5,6] for the first time. In [7], the calculation of damage kinetics in a composite subjected to a constant load was carried out. Consider the calculation of volume damage under an increasing load. The following are the main assumptions. Strength of fibers is determined by the Weibull distribution

$$F(\sigma) = 1 - \exp\left[- \frac{\ell}{\ell_0} \left(\frac{\sigma}{\sigma_0}\right)^{\beta}\right] \tag{2}$$

and then the probability of appearance of a single break and groups of 2,3 and more breaks is obtained. The final failure of a specimen takes place when at least one break of an arbitrarily large size appears. In order to simplify the over-stress coefficient calculation in the vicinity of the defect, it is assumed that the crack is always penny-shaped and only those nearest to the defect fibers are overstressed. The overstress along the fiber spreads on the ineffective length. The breakage calculation reduces to the following approximate formulas:

$$P_1 = F(\sigma),\ P_2 = P_1 \cdot P_{2,1},\ P_n = P_{n-1} \cdot P_{n,n-1} \tag{3}$$

where P_1 is the probability of a single break appearance. The function $F(\sigma)$ is determined by equation (2) for $\ell=\delta$. In equation (3), P_n ($n>1$) is the probability of appearance of a group consisting of n broken elements, $P_{2,1}$ is the probability of breaking at least one element next to an already broken one, and $P_{n,n-1}$ is determined in a similar way. The probability $P_{2,1}$ is defined by the following formula:

$$P_{2,1} = 1 - \left[\frac{1-F(K_1,\sigma)}{1-F(\sigma)}\right]^{n_1} \tag{4}$$

where K_1 is the overstress coefficient, $[1-F(K_1,\sigma)]/[1-F(\sigma)]$ is the conditional surviving probability of the overstressed element adjoining the single defect and n_1 is the number of elements subjected to the overstress K_1 and $P_{2,1}$ is the surviving probability of all the overstressed elements. The probability P_{32} is expressed by

$$P_{32} = 1 - \left[\frac{1-F(K_2,\sigma)}{1-F(K_1,\sigma)}\right]^{n_{21}} \left[\frac{1-F(K_2,\sigma)}{1-F(\sigma)}\right]^{n_2-n_{21}} \tag{5}$$

where n_2 is the number of elements surrounding the double defect and n_{21} is the number of the adjoining elements which were subjected to overstress before the break of a second fiber.

The probability of developing at least one defect with size no smaller than n in a specimen with N elements is determined by

$$W_n = 1 - \exp(-P_n N) \tag{6}$$

For sufficiently large n, all the curves for W_m (m>n) merge. This limit curve determines the macrocrack appearance. In contrast to [6], where the final failure for the plane case appears after the breakage of two adjoining fibers, the merging of the curves W_n in the present investigation took place only at n>30 as shown in Figure 3. When $W_{30} = \frac{1}{2}$, the level of maximum load σ on a fiber is reached. Then P_1^{np} and P_2^{np} at the moment of fracture are determined.

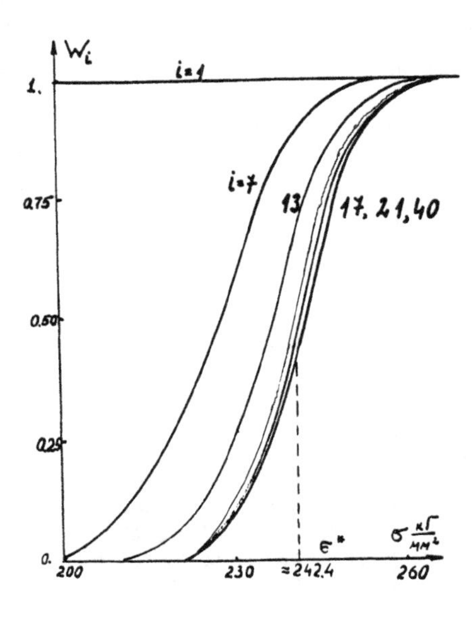

Fig. (3) - Probability of a macrocrack appearance in a composite

DISCUSSION OF RESULTS

The calculated results for the various specimen sizes are presented in Figure 4. The strength dependence on the size of a specimen, i.e., size effect and, what is more essential, the dependence of extreme level of damage on the size of a specimen is seen. This means that for a specimen of a larger size, the final failure takes place at a lower level of the volume damageability level. The obtained theoretical values of extreme damage agree with the experimental data in Table 1.

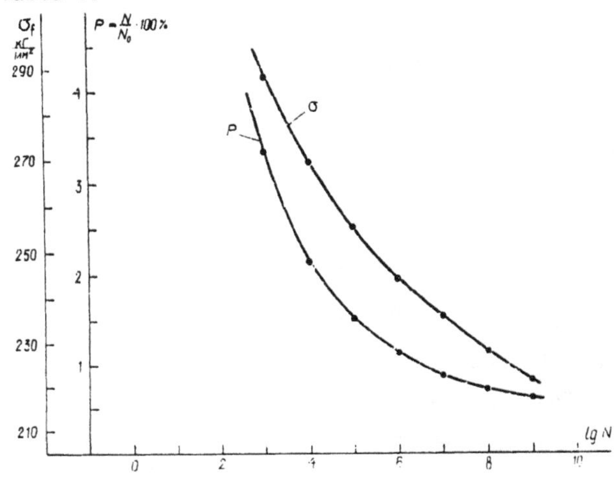

Fig. (4) - Size effect of the extreme damageability

The accumulation of microcracks causes a change in the properties of a material, for instance, the decrease of elasticity modulus. For an isotropic material, refer to [8] for a summary of results and the calculated results for filled polyester resin is in good agreement with the experiment.

For a unidirectional composite, the decrease of modulus can be obtained by using the solution of a penny-shaped crack in a material with transversally isotropic symmetry [9]. It turns out that the decreasing of modulus is proportional to microcrack density P and the ratio $(E_{33}/G_{13})^{1/2}$:

$$\frac{\Delta E}{E_{33}} \approx \frac{8}{3}\sqrt{\frac{E_{33}}{G_{13}}}\, P\left(\frac{m_1}{1+m_1} - \frac{m_2}{1+m_2}\right)\frac{1}{\sqrt{n_1}-\sqrt{n_2}}$$

where values m_1, m_2, n_1 and n_2 are the functions of the material constants [9]. Note that another calculation of modulus change can be found in [10]. This method is based on a direct calculation of unloaded fiber segments in the damaged material. The alteration of mechanical properties of a material can be used for damageability diagnostics and for predicting the moment of fracture. However, such predictions are difficult owing to small changes of the modulus and peculiarities of defect accumulation kinetics under constant load [7]. The difficulty is also attributed to the intense accumulation in the initial loading stage and slow increase of defects in the main life span of the material.

REFERENCES

[1] Tamuzs, V. P. and Kuksenko, V. S., Fracture Micromechanics of Polymer Materials, Zinātne Press, Riga, 1978 (in Russian).

[2] Tamuzs, V. P., "Fracture Peculiarities in Composite Materials", Abstracts of Fourth USSR Conference on Mechanics of Polymer and Composite Materials, Riga, p. 88, 1980 (in Russian).

[3] Mikelson, J. J. and Hohberg, L. J., "Anisotropy of Fatigue Fracture of Filled Amorphous Polymers", Mekhanika Kompozitnikh Materialov, No. 1, pp. 34-41, 1980 (in Russian).

[4] Kuksenko, V. S., Frolov, D. I. and Orlov, L. G., "A Concentration Criterion for Microcrack Enlargement in Heterogeneous Materials", Fracture of Composite Materials, Riga, pp. 25-37, 1979.

[5] Rosen, B. V., "Tensile Failure of Fibrous Composites", AJAA Journal, No. 2, pp. 1985-1994, 1967.

[6] Zweben, C., "Tensile Failure of Composites", AJAA Journal, No. 12, pp. 2325-2331, 1968.

[7] Tamuzs, V. P., "Dispersed Fracture of Unidirectional Composites", Fracture of Composite Materials, Riga, pp. 17-24, 1979.

[8] Malmeisters, A. K., Tamuzs, V. P. and Teters, G. A., The Strength of Polymer and Composite Materials, Riga, 1980 (in Russian).

[9] Kassir, M. K. and Sih, G. C., Three Dimensional Crack Problems, Noordhoff International Publishers, 1975.

[10] Kochetkov, V. A. and Maksimov, R. D., "Stress Redistribution in Polyfiber Composite as a Result of Rupture of Brittle Fibers", Mekhanika Kompozitnikh Materialov, No. 6, pp. 1014-1029, 1980.

PHYSICAL PRINCIPLES OF PREDICTION OF HETEROGENEOUS MATERIAL FRACTURE

V. S. Kuksenko, A. I. Lyashkov, V. N. Saveliev and D. I. Frolov

Academy of Sciences of the USSR
194021 Leningrad, USSR

A general possibility of predicting the fracture of solids is revealed by the concept of fracture as a thermal activation process taking place in a mechanically stressed body and proceeding in time [1,2]. According to this model, the specimen longevity (τ) is related to the uniaxial tensile stress (σ) and the measurement temperature (T) by the expression:

$$\tau = \tau_0 \exp \frac{u_0 - \gamma\sigma}{kT}$$

where u_0 is the fracture activation energy which is close in absolute value to the energy of interatomic bonds, τ_0 is the period of atomic thermal fluctuations in a solid, k is the Boltzmann's constant, γ is a material parameter expressing local overstresses. However, the phenomenological expression for the longevity of a loaded body allows the estimation of the time before fracture ( ) within a logarithmic accuracy only, this being due to the static nature of the effect of thermal fluctuations rupturing interatomic bonds. Moreover, the true values of σ, T, γ are often unknown.

In many cases, a higher degree of accuracy is needed for the estimation of fracture time. New possibilities of fracture prediction are revealed by controlling the course of fracture in a specimen or structure under investigation. The problem is to expose the most general regularities of fracture and criteria of the transition from a stable or stationary phase of fracture to a final unstable one which results in a final macrofracture.

The study of the fracture micromechanics of solids having various structures, including polymer materials [3], allows these general laws of fracture to be found.

The most general feature of fracture of solids is a diffuse, or scattered, accumulation of microcracks in a loaded specimen [3], which determines the further evolution of fracture. Figure 1 shows the characteristic microcrack accumulation curves for various solids. Microcracks in an oriented nylon-6 film under tension (Figure 1a) were recorded by small-angle X-ray scattering method permitting a reliable record of microcracks ranging from tens of angstroms to

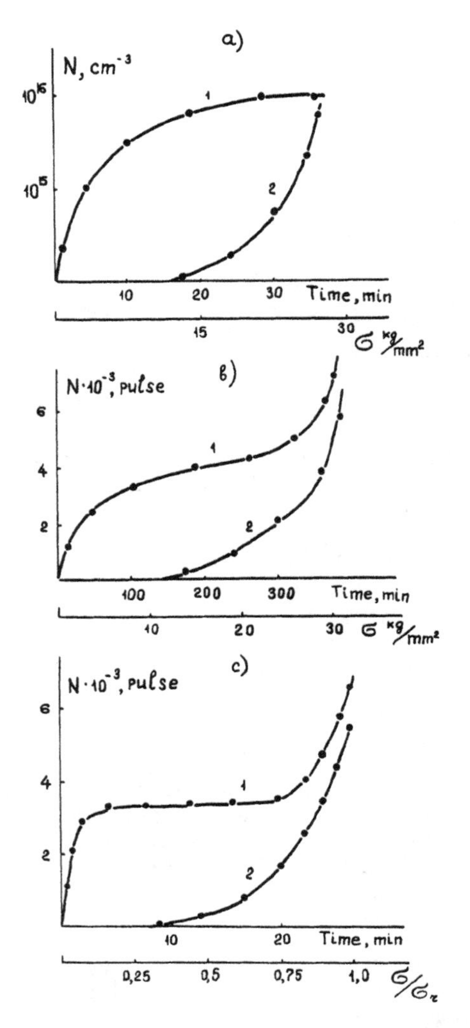

Fig. (1) - Microcrack accumulation curves in solids under the effect of constant load (1) and increasing load (2): a - oriented nylon-6, b - granite, c - glass-reinforced plastic

fractions of a micron in size. Initial nucleus crack sizes here appear to be equal to approximately 200 Å and coincide with structural element sizes, i.e., microfibris. In metal specimens (e.g., steel, Figure 1b) under tension and bending, microcracks were recorded by the acoustic emission method [4,5] which is also a reliable method for rupture recording in composite materials. Figure 1c shows rupture accumulation curves in a one-direction oriented composite. It is important that accumulation curves are qualitatively similar though there is a sharp difference in structure and mechanical properties of materials.

The analysis of the microcrack accumulation curves allows the determination of two stages of the fracture process.

The first stage comprises most of the longevity of a loaded specimen and is related to the accumulation of stable microcracks. In general, for constant loading, this is an attenuating process, though ensembles of microcracks with distances between neighboring cracks comparable to microcrack sizes appear when the concentration of microcracks reaches a certain value. This enables their force interaction, coalescence and enlargement. In its turn, the formation of larger cracks produces increased stresses in certain areas and leads to an intensification of fracture. In these areas, the fracture character is accelerating and not attenuating. This is demonstrated below.

Figure 2a shows microcrack accumulation curves for two loadings: curves 1 and 2 correspond to loads equal to 0.7 and 0.9 of the rupturing load respectively.

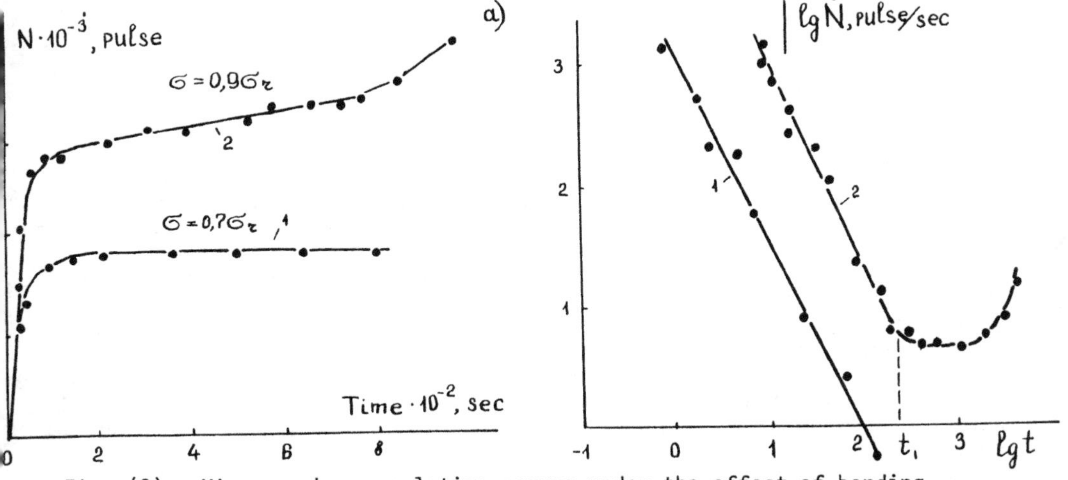

Fig. (2) - Microcrack accumulation curves under the effect of bending of a steel plate by constant loads equal to 0.7 (1) and 0.9 of the rupturing load (2)

Only volume diffuse, microcrack accumulation proceeds with small loadings. The kinetics of accumulation are described by considering thermal fluctuation statistics and the strength distribution of microregions. The kinetics of accumulation are described by a straight line in coordinates $\ell g\dot{N}$-ℓgt (Figure 2b, curve 1), where $\dot{N}$ is the rate of microcrack accumulation, t is the time. The extinction of fracture may result in a complete specimen stabilization.

The behavior of the accumulation curve changes sharply with large loads. In the beginning, the curve is similar to curve 1 (Figure 2b); this can be seen by considering the linear part of curve 2, Figure 2b. The curve then changes sharply; the deviation from a straight line indicates a new stage of fracture or an intensified evolution of the fracture center.

This statement may be confirmed by the following. A component in the microcrack accumulation curve related to a hypothetical center of fracture is selected. This is curve 1, Figure 3. The origin of time is the moment t_1 of Figure 2b. A

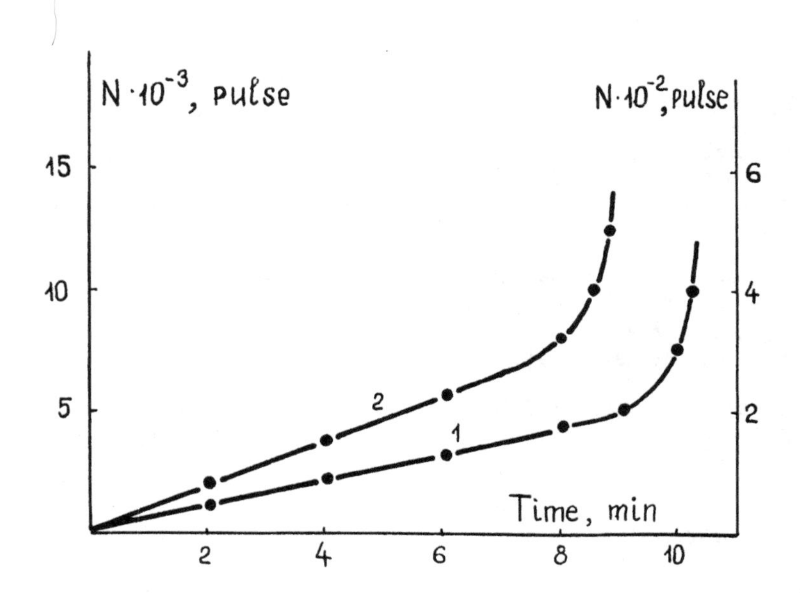

Fig. (3) - Microcrack accumulation curves in fracture centers of granite. 1 - artificially made center of fracture; 2 - natural center of fracture

cut was made in the specimen to follow the crack accumulation curve in an artificially made center of fracture. Under bending, the evolution of fracture is local in character, i.e., it proceeds in the cut peak. The crack accumulation curve describing this case is shown in Figure 3, curve 2.

It can be seen that accumulation curves in the hypothetical or artificial centers (Figure 3, curve 1) and in the natural (Figure 3, curve 2) centers are similar. Thus, the second and final stage of fracture is associated with the evolution of a fracture center resulting from the coalescence and enlargement of microcracks when a certain concentration of stable microcracks in a volume of a loaded specimen is achieved. In this case, the critical concentration of microcracks is a criterion of transition from a stable fracture stage to an unstable, final one.

If the above mentioned model related to the coalescence and enlargement of microcracks upon a certain concentration of stable microcracks is valid, then the enlargement of microcracks should be present in the second stage.

A peculiar composite model - porous glass - is a suitable material to study this transition. Figure 4 shows a micrograph of this material. Uniaxial compression leads to ruptures of the links between pores. The recording of these ruptures by acoustic emission method [6] allows the relation between the sizes of continuum ruptures (ruptures of links) and acoustic signal parameters (i.e., the amplitude and duration of the first signal) to be determined. The loading of the specimen was carried out in a liquid which did not effect the fracture and reduce the mechanical Q-factor of the specimen by approximately two orders of magnitude. This reduced sharply the effect of undesired phenomena associated with multiple reflections. The acoustic signal receiver (compression transducer) was placed in

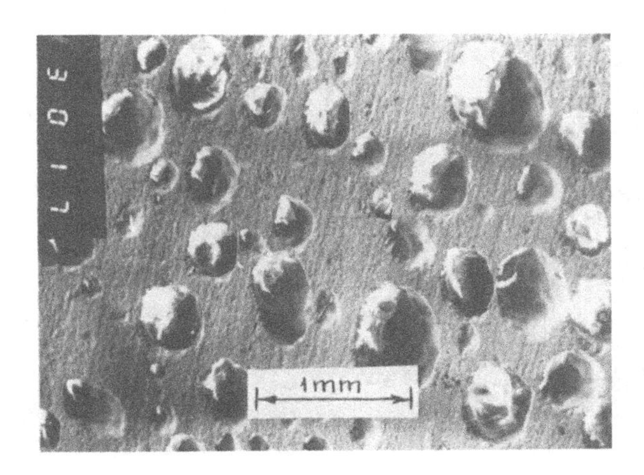

Fig. (4) - Micrograph of a glass microsection

the liquid which removed the distortions due to the superposition of transverse waves. In this case, essentially undistorted signals were recorded (Figure 5a). For comparison, the amplitude (A) of the first emergence and its duration (T) was used. Figure 6 presents the dependence of rupture sizes and acoustic signal parameters. These data allow estimation of the sizes of microcracks being formed under loading by the amplitude and duration of acoustic signals generated upon their formation.

The problem of the transition from the accumulation of initial microcracks to that of larger ones requires discussion. Figure 7 shows an accumulation curve of acoustic signals similar to Figure 5a, indicating the origination of individual ruptures of links only. When a certain concentration of ruptures, N_{cr}, is reached, then signals of a new type appear (their oscillograms are shown in Figure 5b). A special spectral analysis of oscillograms of the first and second types has shown the following. The first type of signals have only one frequency band while signals of the second type have three frequency bands, two of them being close to one another and to the frequency of the first type signal. The frequency of the third band is three times lower, which indicates the formation of a crack of the size three times that of an individual link, according to the dependence of Figure 6, curve 2. The signal structure can be easily explained by the origination of a larger crack as a result of the coalescence of the crack with a new one which was being formed in its neighborhood.

It is important to note that the origination of these enlarged microcracks takes place only if a certain concentration of initial individual ruptures N_{cr} has been accumulated in a specimen. Here, the mean distance between neighboring ruptures is

$$R = N^{-1/3}$$

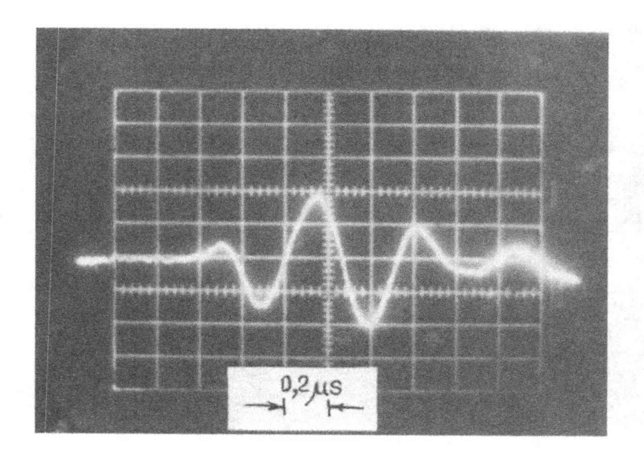

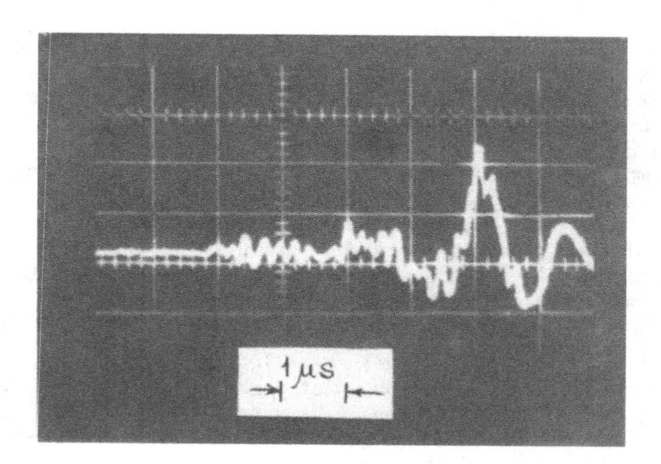

Fig. (5) - Oscillograms of acoustic signals under an individual (a) and co-operative rupture; (b) in a porous glass loaded with uniaxial compression

Considering average sizes of individual ruptures and pore sizes (L) gives the ratio

$$K = \frac{R}{L} \simeq 3$$

With these values of "K", the statistics of the chaotic distribution of ruptures leads to the origination of ensembles of ruptures where the distances be-

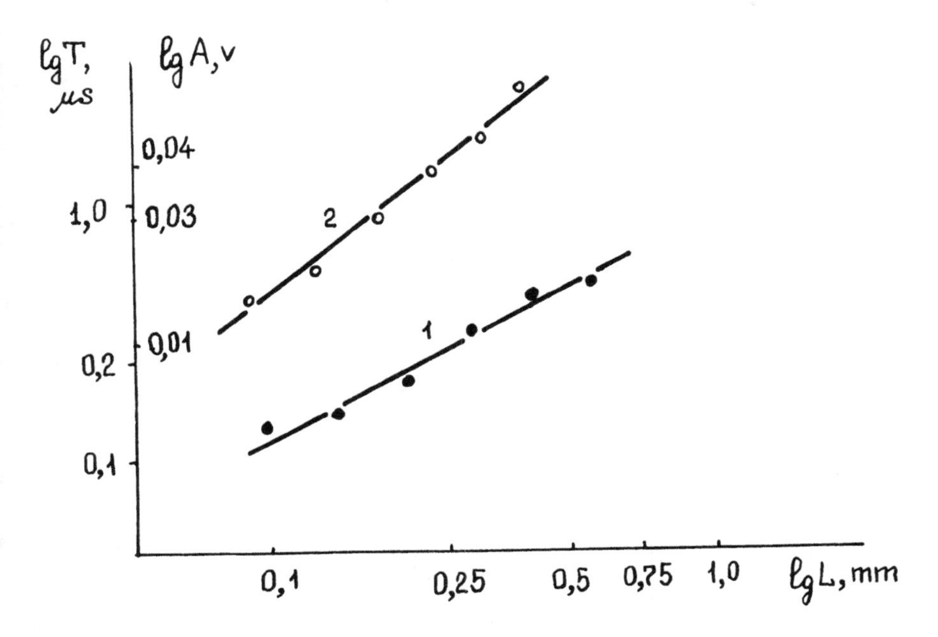

Fig. (6) - Relation between microcrack sizes being formed in porous glasses and parameters of acoustic signals emitted during their generation

tween the cracks are comparable to the sizes of individual ruptures, ensuring their interaction and coalescence.

The enlargement of microcracks during the total time of a constant loading can now be followed. As a parameter characterizing enlargement of cracks, we choose $\bar{A}/\bar{A}_{in}$, where $\bar{A}_{in}$ is the amplitude of acoustic signals not exceeding the maximum amplitude of a stationary stage and $\bar{A}$ is the signal amplitude exceeding $\bar{A}_{in}$; the averaging is performed over a time interval not exceeding 0.1 of the duration of the final stage. Figure 7a shows the dependence in a rock material, diabase, during the total time of the action of loading. It can be seen that during the stable stage (the crack accumulation curve is described by a dash-line) the parameter is constant. It increases sharply not long before the final fracture. There is a relationship between the duration of the final stage (Δt) and total life of a loaded specimen, which is shown in Figure 7b.

The dependencies considered above point to the fact that the final stage is but a particular stage of a total fracture process. Moreover, because of this dependence, the time of macroscopic fracture can be indicated by a systematic enlargement of the cracks or by an intensification of the general process of fracture in the case where the time of the beginning of the final stage is recorded.

In the case when the load does not remain constant, to indicate the time of fracture, we may use the relationship between the amplitude of an acoustic signal

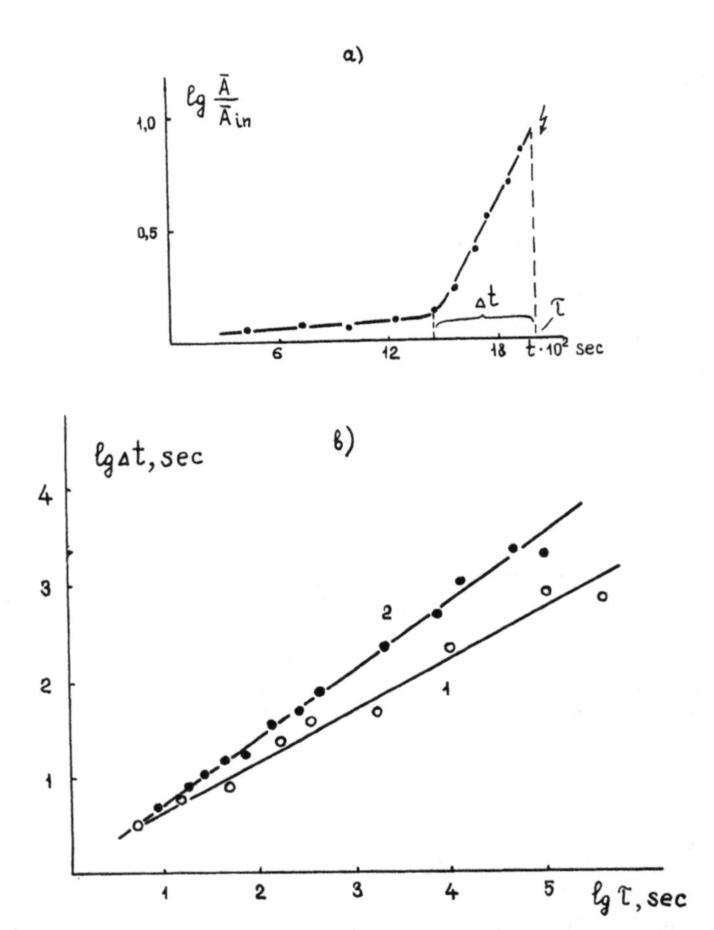

Fig. (7) - Changes in acoustic signal amplitude spectrum in diabase under a constant load (a) and relation between the duration of the final part (Δt) and total longevity of a specimen (b)

and its duration. The ordinate axis may show the size of the largest crack being formed to the given moment of time, Figure 8. When a center of fracture proceeds and still larger cracks appear, then there will be a break in the dependence. The part of the intensified region dependence obtained can be extrapolated in time. The origination of a crack of critical size for the given material and stressed state (Figure 8, dash-line) may be considered the criterion of macrorupture, and its intersection with an experimental curve will give the time of macroscopic rupture.

Thus, the prediction of macroscopic fracture may be based on the general laws of the fracture process and may consist of several stages.

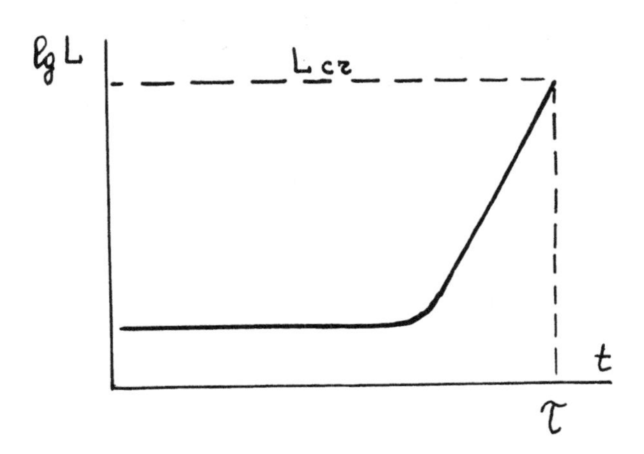

Fig. (8) - Diagram for kinetics of microcrack enlargement in loaded solids

1. Origination of a critical concentration of macrocracks in a loaded specimen (even if it is local) indicates the probability of the transition of fracture into a final stage.

2. The time of origin of an intensified center of fracture may be recorded by a change of the dependence of microcrack accumulation rate (from an attenuating dependence to an accelerated one).

3. The beginning of systematic microcrack enlargement indicates an intensified evolution of fracture centers.

4. The time of macroscopic fracture is estimated from the dependence of the enlargement of microcracks by extrapolating to a critical crack size in a given specimen or structure.

REFERENCES

[1] Zhurkov, S. N., "Kinetic Concept of Durability of Solids", Vestn. AN SSSR, No. 3, p. 46, (in Russian), 1968.

[2] Regel, V. R., Slutsker, A. I. and Tomashevskij, E. E., "Kinetic Nature of Durability of Solids", Nauka, (in Russian), 1975.

[3] Tamuzh, V. P. and Kuksenko, V. S., "Micromechanics of Fracture of Polymer Materials", Zinatne, (in Russian), 1979.

[4] Greshnikov, V. A. and Drobot, Yu. B., "Acoustic Emission", Izd. Standartov, Moscow, (in Russian), 1976.

[5] Kuksenko, V. S., Lyashkov, A. I. and Saveliev, V. N., "Acoustic Emission and Origination and Evolution of Microcracks in Steels", Defektoskopija, No. 6, pp. 57-63, (in Russian), 1980.

[6] Frolov, D. I., Kilkeev, R. Sh., Kuksenko, V. S. and Novikov, S. V., "Relation between Acoustic Signal Parameters and Continuum Rupture Sizes During Fracture of Heterogeneous Materials", Mekhanika Kompozitnykh Materialov, No. 5, pp. 907-911, (in Russian), 1980.

MICROSCOPIC MECHANISMS OF FIBER COMPOSITE DEFORMATION AND FRACTURE

N. A. Pertsev, N. D. Priemsky, A. E. Romanov and L. I. Vladimirov

Academy of Sciences of the USSR
Leningrad 194021, USSR

The wide use of composites has been fostered by the possibility of obtaining improved material properties as compared with the properties of the individual material components. Until recently, the mechanical behaviour of composites has been investigated only on the level of the mechanics of a solid body subjected to deformation [1]. Only lately have papers concerning the study of microscopic processes of deformation and fracture of composites been published.

The present communication considers the sequence of processes which occur during the loading of a composite material with a plastic matrix and plastically undeformable fibers. The main qualitatively distinct stages of the material's plastic deformation are discussed, and the causes of transition from one stage to another are analyzed. On the basis of the theory of dislocations and disclinations, the concrete mechanisms of composite deformation and fracture are calculated. The final purpose of this work is to determine the connection between the plastic deformation of the matrix and the fracture or stratification of the fibers.

In the composite material under analysis, plastic deformation develops only in the matrix and thus leads to accumulation of elementary carriers of deformation (i.e., dislocations) within the matrix. These dislocations are also sources of internal stresses. The movement of dislocations in the matrix is accompanied by their accumulation near the reinforcement which serves as an obstacle for dislocation glide due to its plastic non-deformability. These dislocations may be considered geometrically necessary as has been shown by Ashby [2]. Elastic interaction between moving dislocations and stored ones leading to the hardening of the material. The calculation of this deformation hardening is discussed in [3] for the initial stage of deformation.

The further development of the process consists of the dislocations overcoming the rows of fibers. This can be done either by cutting the reinforcement, or by encircling according to Orowan's mechanism. At this stage of the deformation, the influence of composite heterogeneity on mechanical properties may be manifested in work-hardening due to the shortening of the dislocation sources [4]. The cutting mechanism is energetically favorable only for inclusions of atomic scale [5]. In real composites, the pile-ups of Orowan's loops around inclusions can be observed experimentally (a detailed review is given in [6]), and the number of the loops in each glide plane increases with the growth of deformation. It has been experimentally proven [7] that this phenomenon results in work-hardening up to several percent deformation. Under these conditions, the degree of hardening depends on the change of shape of the loops arrays under the action of applied stress.

To calculate this effect, consider the following model. In the matrix surrounding fibers in planes separated from each other by an equal distance Δ, there are pile-ups of coaxial, circular, prismatic dislocation loops. These pile-ups are formed as a result of the transformation of Orowan's loops (Figure 1). Under

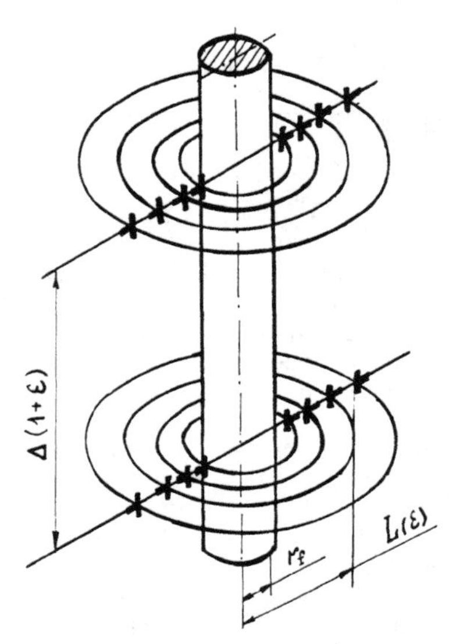

Fig. (1) - Pile-ups of dislocation loops around fibers in a composite. The loops are formed as a result of dislocation encircling of the fibers according to Orowan's mechanism. r_f is the fiber radius; $L(\varepsilon)$ denotes the radius of the external loop in the pile-up

the effect of an external stress, the distance between the planes of pile-ups increases by $\varepsilon\Delta$, where ε is the complete predeformation of the material. The plastic deformation ε^p_{ZZ} of each layer of the material between the neighboring planes is linked with the distribution function ϕ of prismatic loops in the pile-up in the following way:

$$\varepsilon^p_{ZZ}(r) = \frac{b}{\Delta} \int_{r_f}^{r} \phi(r')dr' \tag{1}$$

where b is the Burgers vector of an individual dislocation, r_f is the fiber radius. In this case, the part of the matrix adjoining the fiber is deformed elastically, and the corresponding elastic stress τ^e appears in it (E_m is the elasti modulus of the matrix):

$$\tau^{e}(r) = E_m[\varepsilon - \varepsilon^{p}_{ZZ}(r)] \qquad (2)$$

Finally, the deforming stress τ^p is determined according to the rule of mixtures with regard for the decrease of the area subjected to plastic deformation:

$$\tau^{p} = v_f\tau_f + [v_m - \frac{L(\varepsilon)-r_f}{r_f v_f}]\tau_m + \frac{\sqrt{v_f}}{r_f}\int_{r_f}^{L(\varepsilon)} \tau^{e}(r')dr' \qquad (3)$$

where L is the external radius of dislocation pile-up, v_f and v_m are volume fractions of fibers and matrix respectively, $\tau_f = E_f\varepsilon$ (E_f is the elastic modulus of fibers), τ_m is the matrix resistance to deformation outside a layer of radius $L(\varepsilon)$. In equation (3), the first and the second terms describe the stresses in the fibers and the stresses in the matrix outside the arrays respectively, while the third summand corresponds to the stresses in the matrix inside the layer of the loops. This layer possesses elastic-plastic properties different from those of the pure matrix. The availability of this layer is one of the main reasons for failure of the rule of mixtures when considering composites with dispersion fibers.

The loop distribution function $\phi(r)$ in the planes has been determined [8] from the condition of array equilibrium under the stress $\tau^{e}(r)$. The equation of balance is a singular integral equation of special type. The normalization condition for the function $\phi(r)$ in the singular integral equation helps to determine the dependence on $L(\varepsilon)$. After substitution in equations (1)-(3), $\phi(r)$ and $L(\varepsilon)$ can be determined, and together with the given deformation ε, the hardening caused by a system of Orowan's type loops can be calculated numerically.

At a certain stage of composite deformation the formation of new Orowan's loops stops*; powerful dislocation arrays (consisting of straight dislocations of the same sign) appear close to the fibers (Figure 2). The formation of these dislocation charges is known to be linked with the inhomogeneity of the plastic deformation [5]:

$$\underline{\underline{\Delta\rho}}(r,z) = -\frac{1}{b}\nabla \times \underline{\underline{\varepsilon}}^{p}(r,z)$$

where $\underline{\underline{\Delta\rho}}(r,z)$ denotes the tensor density of dislocation charges and $\underline{\underline{\varepsilon}}^{p}$ is the plastic deformation tensor. In the case of plane deformation caused by dislocations with Burgers vectors of the same type, the number N of dislocations in the array depends on the total deformation as $N = \varepsilon^{p}\Delta/b$, where Δ is again the distance between the slip planes. Such dislocation arrays are formed in composites at a slightly less degree of deformation than in standard materials. This phenomenon is conditioned by the initial inhomogeneity of plastic properties of a fibrous composite. The more irregular the laying of fibers in the material the stronger the phenomenon.

*The pile-ups formed earlier prevent new dislocations from encircling the fibers with increased effective radius $r_f = L(\varepsilon)$.

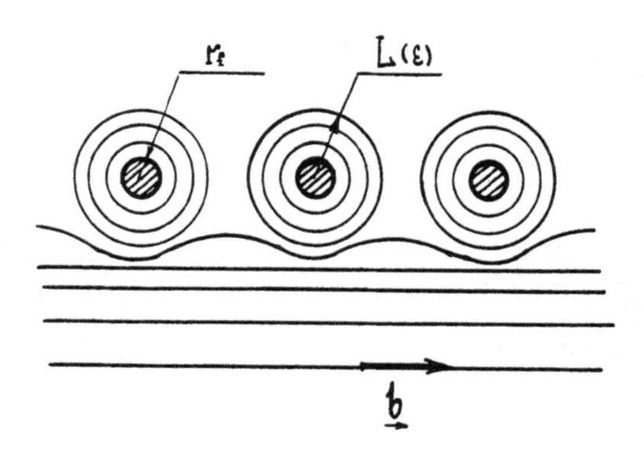

Fig. (2) - Dislocation arrays stopped near fibers with an increased effective radius L(ε)

At this stage of deformation, it is already necessary to take into consideration the possibility of the appearance of collective effects in dislocation ensembles [5] being equivalent to the appearance and movement of partial disclinations. Indeed, this effect is connected with the availability of torques on the areas close to the dislocation pile-ups, i.e., in fact, around the fibers. Analytical expressions for the torques $\vec{M}$ [9] conditioned by super-dislocations modelling the examined arrays have been obtained proceeding from the following expression:

$$\vec{M} = - \int_S \vec{r} \times \vec{\vec{\sigma}}^B(r,z) \cdot d\vec{S} \tag{5}$$

Here $\vec{\vec{\sigma}}^B$ is the super dislocation stress tensor in the coordinate system connected with the fiber axis (Figure 3); the integration is carried out along the surface of area S. In the case of a screw dislocation on the area perpendicular to the fiber axis, the twisting moment $\vec{M}_S$ appears, while the edge dislocation on the same area causes the bending moment $\vec{M}_e$.

The relaxation of the torques on plastic rotations of the matrix is carried out with the help of disclination loops. Due to the plastic non-deformability of the fibers, pairs of disclination loops of different signs always appear. The energy condition of the disclination relaxation can be reproduced in the following way:

$$A = \omega M \geq W_S + W_i + W_\gamma - W_b \tag{6}$$

where A is the work of the initial super dislocation elastic field carried out on the plastic rotation ω of the matrix material; W_S is the self-energy of all

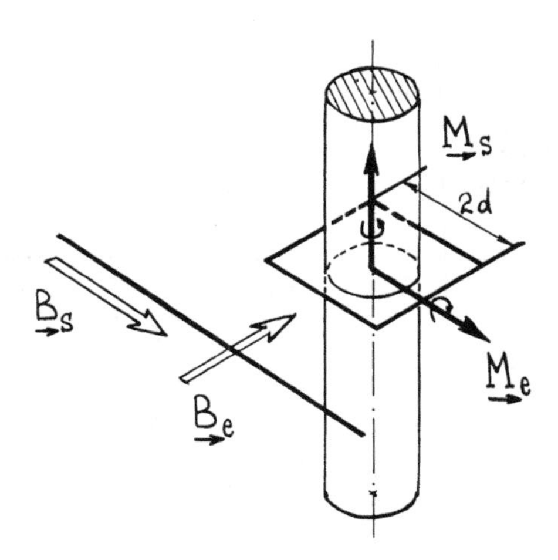

Fig. (3) - Torques on the area perpendicular to the fiber are conditioned by stopped super dislocations. $\underset{\to}{M}_S$ - twisting moment due to screw dislocation with Burgers vector $\underset{\to}{B}_S$, $\underset{\to}{M}_e$ - bending moment connected with edge dislocation ($\underset{\to}{B}_e$)

disclination loops forming a part of the configuration; W_i is the energy of the interaction between the loops; W_γ is the surface energy bound up with the fact that disclinations forming the loops are partial; W_b is the energy of the dislocations available in the relaxation volume before the process starts. The plastic relaxation may cause the loss of strength of the material and the formation of microfracture sites.

Consider the evolution of forming disclination configurations. During the relaxation of the twisting moment M_S, a pair of coaxial twist disclination loops forms around the fiber, which causes the appearance of elastic shear in the fiber-matrix inter layer. When this shear, characterized by total deformation ε and the composite properties r_f and v_f, exceeds the critical value depending on the inter layer strength, peeling takes place. The detailed description of this process has been given in [10]. The alternative possibility is the fracture of the fiber due to the same twisting. This is characteristic of the composites with fibers of low strength.

It should be noted that development of disclination twisting modes in a fibrous composite is localized on individual fibers, because the geometry of deformation does not require its transition from one fiber to another - such plastic rotation does not cause long-range stresses.

Bending disclination modes appearing during relaxation of bending moments M_e from edge super dislocations are characterized by a completely different type of evolution. The geometry of bending leads to the immediate appearance of a zone of tilted material in the matrix (so-called kink band), spreading from one fiber to another. Indeed, the kink bands have been observed experimentally in composit reinforced with carbon or glass fibers [11-13].

The stage of kink band spreading in the fibrous composite requires a more detailed examination as will be shown below, this stage ends in fracture. The fron of incomplete kink band can be represented by the dipole of partial wedge disclin tions with strength ω (Figure 4) [14]. During the movement of the dipole in the

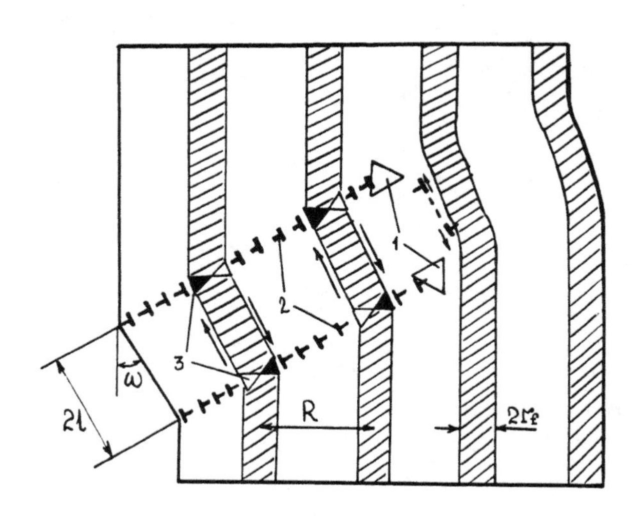

Fig. (4) - Kink band movement in fibrous composite. Here ω is the angle of material rotation inside the band and 2ℓ is the band thickness. R denotes the average distance between fibers in composite

matrix pairs of edge dislocations form according to the mechanism offered for crystals in [15]. When the kink band (the dipole) passes through a row of fibers there must remain (due to the condition of deformation compatibility) special defects around the fibers, corresponding to the bending of fibers and the shear alo the fiber-matrix boundary. These defects are the slipped kinks first described in [16], where it is shown that they consist of two wedge disclination loops of different signs (which are situated in cross-section of fibers), and two glide dislocation loops on the lateral surface of the fiber (Figure 5). In the slipped kink, the Burgers vector of dislocation loops B is connected unambiguously with the strength of disclinations ω by the relation $B = \omega r_f$ ($2r_f$ is the defect size in the direction perpendicular to the vectors $\vec{\omega}$ and $\vec{B}$).

Plastically undeformable fibers obstruct the development of the kink bands and thus create the strengthening at this stage. The disclination dipole (the kink band front) is held up by the fibers and the bends between them. The dipole cuts through one row of the fibers only at an external stress τ, the work A of which, used to displace the dipole through a row, makes up the self energy W of defects which have been left on the fibers. In a corresponding manner, the criti cal stress τ^* of kink band motion in the composite is determined by the following relation:

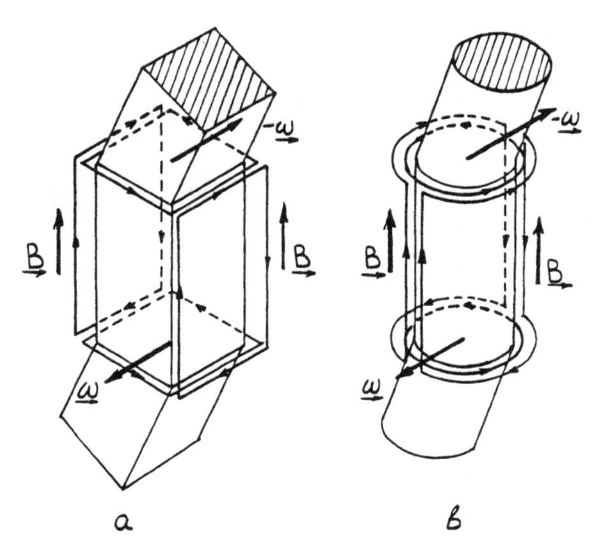

Fig. (5) - Structure of slipped kinks on the fibers of rectangular (a) and circular cross-sections (b). The dimensions of rectangular cross-section are 2c and 2d, the length of the defect along fiber is 2ℓ. $\underset{\rightarrow}{\omega}$ denotes the Frank vector of the disclination loops while $\underset{\rightarrow}{B}$ is the Burgers vector of the dislocation loops

$$A(\tau^{*}) = W \tag{7}$$

The work A may be calculated if the disclination dipole is regarded as the edge super dislocation with Burgers vector $|\underset{\rightarrow}{B}| = 4\ell \mathrm{tg}\,\frac{\omega}{2} \approx 2\ell\omega$:

$$A = 4\tau\omega\ell r_f R \tag{8}$$

Here, $2r_f$ is the fiber diameter (the distance at which the dipole moves overcoming the row of fibers), R is the average distance between the fibers in the composite and 2ℓ is the kink band thickness, (Figure 4).

The self-energy of the slipped kink W_{SK} has been calculated [14] in the approximation of the linear isotropic theory of elasticity for the defect consisting of the rectangular-shaped loops (Figure 5a). The slipped kinks on the circular fibers (Figure 5b) may be replaced by the rectangular-shaped ones with an adequate degree of accuracy by supposing that $2d = 2c = 2r_f$. As the dependence of the energy W_{SK} on the defect dimensions is in its general form rather unwieldy, the deforming stress is calculated at the stage of development of kink bands only for two important particular cases. When the kink band spreads in material with thin fibers, the relation between the sizes of the defect and the fiber is $\ell >> r_f$ and the energy acquires the simple form [14]:

$$W_{SK} = \frac{2G\omega^2 r_f^2 \ell}{\pi} \ln\left(\frac{2\sqrt{2}\, r_f}{r_o}\right) \quad (9)$$

where G is the shear modulus, r_o is the core radius of dislocations and disclinations comprising the defect. Proceeding from (7) and (9), the deforming stress can be determined:

$$\tau^*_{SK} = \frac{G\omega r_f}{2\pi R} \ln\left(\frac{2\sqrt{2}\, r_f}{r_o}\right) \quad (10)$$

In the case of the development of thin (as compared to the diameter of the fibers) kink bands, $\ell << r_f$ and the defect energy can be expressed in the following form (ν is Poisson's ratio):

$$W_{SK} = \frac{2G\omega^2 r_f^3}{3\pi(1-\nu)} \left(\ln \frac{2\ell}{r_o} + 2\right) \quad (11)$$

The corresponding critical stress is equal to:

$$\tau^*_{SK} = \frac{G\omega r_f^2}{6\pi(1-\nu)\ell R} \left(\ln \frac{2\ell}{r_o} + 2\right) \quad (12)$$

In both (10) and (12), the critical stress changes in inverse proportion to the average distance between the fibers. When R is expressed by the volume fraction of fibers in the composite v_f, the strengthening during kink forming is proportional to the square root of the volume fracture of fibers of the prescribed siz

The stress τ^*_{SK} increases when the angle ω of the kink band increases. Due t this dependence, when ω reaches a certain critical value which will be calculate below, the transformation of the slipped kink into other defects becomes energetically profitable. The following variants are possible here (Figure 6): a) Ben fibers with peeling along the border of fiber-matrix. This defect consists of two disclination loops of different signs in cross-sections of the fiber and of shear cracks on the lateral surface. b) Fiber fractured in the cross-section without peeling of the matrix. In this case, tensile cracks appear instead of the disclination loops, while the slipped kink leaves the segments of screw dislocations. c) Fracture of the fiber with peeling of the bent area of the matrix and, correspondingly, complete disappearance of the slipped kink.

In reality, the defect with the least elastic energy will develop. Let us examine the limit case when the kink band passing through the row of fibers is accompanied by fiber peeling complete with fracture. The self-energy of the shear and tensile cracks formed may be written in the following way in the first approximation:

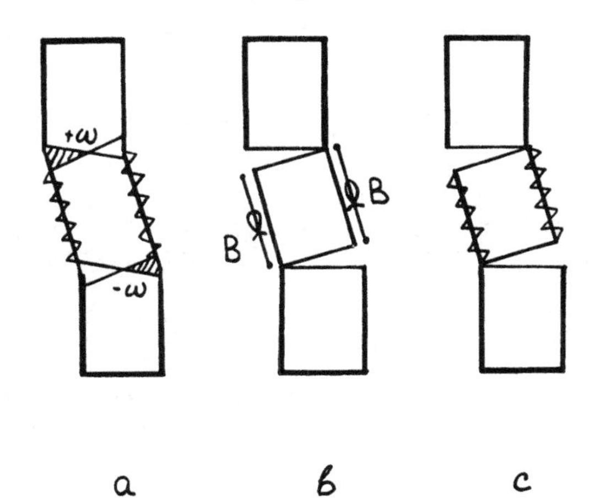

Fig. (6) - Possible variants of slipped-kink transformation caused by fiber fracture or stratification

$$W_c = \gamma_1 8 r_f \ell + \gamma_2 16 r_f^2 \tag{13}$$

Here the specific energy of the interfacial boundary γ_1 is approximately equal to the specific energy of the free surface and may be expressed as $\gamma_1 \approx \gamma_2 \approx Ga/4\pi(1-\nu)$ (a is the interatomic spacing) [5]. By supposing that the kink band passes through the fiber only when the condition analogous to (7) is observed, the critical stress τ_c^* in this case can be determined:

$$\tau_c^* = \frac{Ga}{2\pi(1-\nu)\omega R}\left(1 + \frac{2r_f}{\ell}\right) \tag{14}$$

Since τ_c^* decreases with the increase of the angle ω of the band, a critical angle ω^* of the transition from fiber deformation to fiber fracture exists. In the case of thin fibers, for example, the condition $\tau_{SK} > \tau_c$ indicates that fracture will start when the rotation angle ω obeys the following inequality:

$$\omega > \omega^* = \sqrt{\frac{a}{(1-\nu) r_f \ell n(2\sqrt{2}\, r_f/r_o)}} \tag{15}$$

Numerical estimates show that for micron-thin fibers, the critical angle ω^* is within the range of 1 to 3° for both thin and thick kink bands. This explains the formation of fiber cracks observed experimentally in carbon-fiber composites [13], when kink bands with a rotation angle $\omega \approx 40 \div 60°$ spread. It follows from the condition (15) that bands having such an angle start destroying the fibers from an r_f of several interatomic distances. However, fibers in composites are in a state of strain when kink bands with $\omega \sim 1°$ (crystal matrix) develop in material. The corresponding strengthening is determined from formulae (10) and (12). In the case of a polymeric matrix rotation, the angle is usually on the order of 50° and fiber fracture occurs when the stress τ exceeds the safe stress given by expression (14).

Thus, the formation of fracture sites in a composite is caused by the plastic deformation of the matrix, where the following main stages may be singled out:

1. The formation of powerful dislocation structure consisting of Orowan's loops and impeded by straight-line dislocations near the rows of fibers. This process causes deformation strengthening at the initial stage of deformation.

2. The appearance of rotation (disclination) modes of microplastic deformation relaxing the elastic fields of the preceding dislocation structure.

3. The development of the disclination modes of the macroscopic plastic deformation leading to fracture or peeling of the fibers and therefore to the fracture of the whole material.

REFERENCES

[1] Malmeister, A. K., Tamuzh, V. P. and Teters, G. A., Strength of Polymer and Composite Materials, Riga, 572 p., 1980.

[2] Ashby, M. F., Philos. Magaz., Vol. 21, No. 169, p. 399, 1970.

[3] Newmann, P. and Haasen, P., Philos. Magaz., Vol. 23, No. 182, p. 285, 1971.

[4] Brown, L. M. and Clark, D. R., Acta Met., Vol. 25, No. 5, p. 563, 1977.

[5] Vladimirov, V. I., Einführung in die physikalishe Theorie der Plastizität und Festigkeit, Leipzig, 280 p., 1975.

[6] Chumliackov, Ju. I., Corotaiev, A. D., Bushnev, L. S. and Esipenko, V. F., Fiz. Metall. i Metalloved., Vol. 50, No. 2, p. 367 (in Russian), 1980.

[7] Pajuk, V. A., Corotaiev, A. D., Bushnev, L. S. and Chumliackov, Ju. I., Izvestija VUZov (ser. Physics), No. 3, p. 88 (in Russian), 1974.

[8] Vladimirov, V. I. and Priemsky, N. D., in "Physics of Strength of Composite Materials", Leningrad, p. 19 (in Russian), 1979.

[9] Vladimirov, V. I., Romanov, A. E. and Priemsky, N. D., in "Physics of Streng of Composite Materials", Leningrad, p. 27 (in Russian), 1979.

[10] Vladimirov, V. I., Priemsky, N. D. and Romanov, A. E., Mechanika Comp. Mate No. 5, p. 802 (in Russian), 1980.

[11] Weaver, C. W. and Williams, J. G., J. Mater. Sci., Vol. 10, No. 8, p. 1323, 1975.

[12] Chaplin, C. R., J. Mater. Sci., Vol. 12, No. 2, p. 346, 1977.

[13] Evans, A. G. and Adler, W. F., Acta Met., Vol. 26, No. 5, p. 725, 1978.

[14] Vladimirov, V. I., Pertsev, N. A. and Romanov, A. E., Preprint Phys. - Tech. Institute, No. 683, 45 p., (in Russian), 1980.

[15] Vladimirov, V. I. and Romanov, A. E., Fiz. Tverd. Tela, No. 10, p. 3114 (in Russian), 1978.

[16] Vladimirov, V. I., Pertsev, N. A. and Romanov, A. E., Mechanika Comp. Mater., No. 4, p. 730 (in Russian), 1980.

ANALYSIS OF MICROFRACTURE IN COMPOSITE

L. V. Nikitin and A. N. Tumanov

USSR Academy of Sciences
Moscow, USSR

During the loading of constructions made of fiber reinforced composite, processes of microfracturing may take place in them long before global fracture occurs. These processes are the breakage of fibers and fiber separation from the matrix. They essentially influence the ultimate strength and rigidity of the construction. Determining conditions of microfracture in the composite is especially important for creation of constructions with prescribed properties.

Numerous theoretical and experimental works are devoted to the study of composites. In most cases, very simple models of composites are considered and they are treated by strength of material methods [1-3]. However, the experimental data for the example given in [2] shows a singular trend for the shear stress at the edge of reinforcing element, which is not described by the simple models. On the other hand, it is the intensity of the stress concentration that is responsible for microfracture. A singular stress state is found when a fiber is assumed to be an infinitely thin elastic rod embedded into an elastic plate [4].

The present analysis is based on a rather simple plane model of fibers in an elastic medium. Arising mathematical problems are similar to that of contact problems of load transmission from thin stringers to an elastic plate which were studied in detail abroad [5,6] as well as in the USSR [7-9]. The developed methods of solution are used in the present work. For a single whole fiber, a study close to the present one is carried out in [10].

For a qualitative analysis of microfracture in fiber reinforced composites, consider the plane problem for an unbounded elastic plate with an embedded strip - "fiber" of length ℓ, infinitely small thickness δ, and thickness equal to a whole thickness of the plate h. Let μ and $\kappa = (3-\nu)/(1+\nu)$ be the effective elastic moduli of the composite and E the elastic modulus of a fiber. Microfracture of the composite will be characterized by the rupture stress σ_{cr} of the fiber material or by the rupture force per unit fiber width $P_{cr} = \sigma_{cr}\delta$ and by the critical stress intensity factor K_{cr}, which determines separation of a fiber from the matrix. The rigidity of the reinforcing fibers is much greater than that of the matrix. Therefore, as a first approximation, consider fibers as being absolutely rigid. This makes mathematical treatment of the problem very simple.

First, consider a single rigid fiber embedded in an elastic plate, which is stretched at infinity by the homogeneous stress σ_∞ directed along the fiber. Refer to a Cartesian system of coordinates, xy, in the plate with the origin in the middle of the fiber and x-axis directed along the fiber. Due to symmetry of the problem with respect to the axis y=0, it is possible to consider the upper half-plane only.

Boundary conditions along y=0 are as follows. The displacement v in the y direction due to symmetry vanishes;

$$v=0; \ -\infty<x<\infty, \ y=0 \tag{1}$$

Outside of the fiber, shear stresses are zero

$$\tau_{xy} = 0; \ |x|>\ell, \ y=0 \tag{2}$$

The fiber is assumed to be rigid and bonded with the matrix; thus,

$$u=0; \ |x|<\ell, \ y=0 \tag{3}$$

At infinity, the stress state is uniaxial tension

$$\sigma_x = \sigma_\infty, \ \sigma_y = \sigma_{xy} = 0; \ (x^2+y^2)^{1/2} \to \infty \tag{4}$$

Stresses and displacements in the plane problem of elasticity are expressed in terms of two analytic functions of the complex variable z = x+iy with the help of the Kolosov-Muskhelishvili formulae, which for a half-space can be written in the form

$$\begin{aligned} \sigma_x + \sigma_y &= 4\mathrm{Re}\phi(z) \\ \sigma_y + i\tau_{xy} &= 2\mathrm{Re}\phi(z) + (\bar{z}-z)\phi'(z) + \Omega(z) \\ 2\mu(u'+iv') &= \kappa\phi(z) - \overline{\phi(z)} + (z-z)\overline{\phi'(z)} - \overline{\Omega(z)} \end{aligned} \tag{5}$$

where a prime denotes derivative in respect to z.

The solution of the problem (1) - (4) for an upper half-plane has a form

$$\begin{aligned} \phi(z) &= \frac{(\kappa-1)\sigma_\infty}{8\kappa} + \frac{i(\kappa+1)\sigma_\infty z}{8\kappa(\ell^2-z^2)^{1/2}} \\ \Omega(z) &= \frac{(\kappa-1)^2\sigma_\infty}{8\kappa} + \frac{i(\kappa+1)\sigma_\infty z}{8\kappa(\ell^2-z^2)^{1/2}} \end{aligned} \tag{6}$$

Action of the matrix onto the fiber is described by the shear stress τ_{xy} along the intersection. These stresses are found using (5)

$$\tau_{xy} = \frac{(\kappa+1)^2\sigma_\infty x}{8\kappa(\ell^2-x^2)^{1/2}}; \ |x|<\ell, \ y=0 \tag{7}$$

Since there is no action of any external force on the fiber, the resulting force due to the stress (7) is equal to zero. The shear stresses, (7), however, produces a force P(x) on the fiber cross sections, which is distributed as

$$P(x) = -2\int_x^{\ell} \delta_{xy}(x,o)dx = \frac{(\kappa+1)^2\sigma_\infty(\ell^2-x^2)^{1/2}}{4\kappa} \tag{8}$$

This force reaches its maximum value at the middle of the fiber x=0. The maximum force cannot exceed a certain critical value

$$P_{max} = \frac{(\kappa+1)^2\ell\sigma_\infty}{4\kappa} \leq P_{cr} \tag{9}$$

The shear stress near the fiber edge has a square root singularity with stress intensity factor $K = \lim_{x\to\ell}\sqrt{2(\ell-x)}\tau_{xy}(x,o)$. Similar to the mechanics of brittle fracture, separation of a fiber from matrix will be characterized by the critical value of this coefficient K_{cr}. Then the condition of fiber-matrix interface unity is written as follows

$$K = \frac{(\kappa+1)^2\ell^{1/2}\sigma_\infty}{8\kappa} \leq K_{cr} \tag{10}$$

It can be easily shown that the critical stress intensity factor K_{cr} is related to the energy γ needed to create a unit of separated surface by the equation

$$K_{cr}^2 = \frac{\mu\gamma(\kappa+1)}{\pi\kappa} \tag{11}$$

At the edge of a fiber, normal stresses σ_x and σ_y are singular also

$$\sigma_x \simeq \frac{(\kappa+1)(\kappa+3)\ell^{1/2}\sigma_\infty}{4\sqrt{2}\kappa(\ell-x)^{1/2}}; \quad \sigma_y = -\frac{(\kappa^2-1)\ell^{1/2}\sigma_\infty}{4\sqrt{2}\kappa(\ell-x)^{1/2}}$$

The stress σ_y is negative, so it cannot produce fracture. The stress σ_x makes a trend for local fracture in the form of a tensile crack which is normal to a fiber. It is reasonable to suspect that strength of fiber-matrix interface is weaker than strength of matrix material and a tensile crack might not appear, then local microfracture is characterized by the two parameters P_{cr} and K_{cr} only, and breakage of a fiber or its separation from matrix takes place depending on which of the conditions (9) or (10) is satisfied first. Figure 1 shows the plots σ_∞ versus ℓ for the above mentioned cases

$$\sigma_\infty = \frac{4\kappa P_{cr}}{(\kappa+1)^2\ell}, \quad \sigma_\infty = \frac{8\kappa K_{cr}}{(\kappa+1)^2\ell^{1/2}}$$

It can be seen that if the half-length of a fiber is less than its critical value

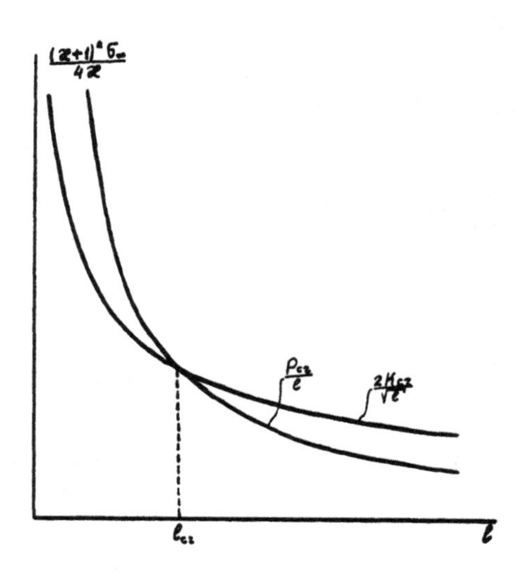

Fig. (1)

$$\ell_{cr} = P_{cr}^2/4K_{cr}^2$$

then, with the growth of the load at infinity, separation of a fiber from the matrix takes place. If, however, the half-length of a fiber is greater than the critical value, then a fiber is broken.

The obtained results allow a choice of fiber length at a given load to be made so that neither fiber breakage nor separation take place. Greater fiber lengths result in lesser rupture loads. Note that this result is obtained for the case when the load is applied far from the fiber edge. Only for this case is it possible to state that the short fibers are optimal for reinforcement.

Consider next what happens after one or another microfracture has taken place. Assume the fiber half-length is less than the critical one. Then for load σ_∞, which is greater than $8\kappa K_{cr}/(\kappa+1)^2\ell^{1/2}$, a fiber is separated from the matrix along or near the edge segments $\ell_1<|x|<\ell$. To find length ℓ_1, it is necessary to solve the problem for a fiber in an elastic matrix accounting for fiber separation. Assume that along the segment of separation sliding, without crack opening, takes place. This suggestion will be justified if along the mentioned segment the stress σ_y is negative. Interaction between a fiber and matrix along the segment of sliding is assumed to be described by the Coulomb law of dry friction with a friction coefficient f. The boundary conditions along this segment have the form

$$v=0,\ \delta_{xy} = -f\sigma_y;\ \ell_1<|x|<\ell,\ y=0 \tag{12}$$

Considering the simplest case of no friction f=0, the segment of sliding can be neglected since the boundary conditions along it are the same as that along the x-axis outside of the fiber. To find the length of separation, it is necessary to use the solution of the Problem (1)-(4) in which ℓ should be replaced by ℓ_1. This results in

$$\ell_1 = \frac{64\kappa^2 K_{cr}^2}{(\kappa+1)^4\sigma_\infty^2} \tag{13}$$

It can be easily checked that the adopted suggestion about absence of an open crack is justified since $\sigma_y<0$ for all $|x|>\ell_1$.

Consider now a fiber half-length greater than the critical one so that for a load σ_∞ exceeding the value $4\kappa P_{cr}/(\kappa+1)^2\ell$, at the fiber middle, fracture takes place. Evidently the boundary conditions (1), (2) and (4) still hold in the case under consideration. But the condition (3) changes. After breakage of a fiber at its center, one can expect that right to the point x=y=0 a broken part of the fiber moves a distance u_0 and left to a distance u_0. At the center, the displacement suffers a jump of $2u_0$. This condition may be written as

$$u' = 2u_0\delta(x); \; |x|<\ell, \; y=0 \tag{14}$$

where $\delta(x)$ is the delta function.

The solution of the problem (1), (2), (4) and (14) has the form

$$\begin{aligned} \phi(z) &= \frac{(\kappa-1)\sigma_\infty}{8\kappa} + \frac{i(\kappa+1)\sigma_\infty z}{8\kappa(\ell^2-z^2)^{1/2}} - \frac{2i\mu u_0\ell}{\pi\kappa z(\ell^2-z^2)^{1/2}} \\ \Omega(z) &= \frac{(\kappa-1)^2\sigma_\infty}{8\kappa} - \frac{i(\kappa+1)^2\sigma_\infty z}{8\kappa(\ell^2-z^2)^{1/2}} + \frac{2i(\kappa+1)\mu u_0\ell}{\pi\kappa z(\ell^2-z^2)^{1/2}} \end{aligned} \tag{15}$$

To find the constant u_0, it is necessary to use the condition of absence of an external force action on each part of the fiber. This gives $u_0 = 0$ which corresponds to the solution without fiber breakage. The displacement u_0 will be non-zero only in the case if infinite force is applied to the fiber. This somewhat paradoxical result may be explained by overidealization of the problem.

In the framework of the assumed idealization, the above paradox can be avoided if the boundary condition (14) is prescribed along the deformed boundary $u_0<|x|<\ell+u_0$ which is admissible in the scope of the linear theory of elasticity. In this case, it is necessary to prescribe some boundary conditions along the segment $|x|<u_0$. For instance, it can be traction free.

It seems more natural to investigate this question considering the problem of two fibers lying along the same line with a gap 2b. The limiting case $b\to0$

must clarify the situation. The posed problem has its own interest as well since it shows mutual fiber influence.

The conditions outside the fibers are the same

$$\tau_{xy} = 0,\ v=0;\ |x|>\ell,\ y=0 \tag{16}$$

The conditions along the fibers are as follows

$$v=0,\ u = \pm u_0;\ b<|x|<\ell,\ y=0 \tag{17}$$

where the plus corresponds to the positive x and the minus to the negative. Along the gap between the fibers, one can assume different conditions. If one assumes that there is no discontinuity in the matrix material, then due to symmetry, the conditions are

$$\tau_{xy} = 0,\ v=0;\ |x|<b,\ y=0 \tag{18}$$

One can assume another situation: the fiber is broken at the center but occupies all segment $|x|<2\ell$ and along the segment $|x|<b$, separation of the fiber from the matrix takes place. In this case, it is possible to consider sliding without an open crack

$$\tau_{xy} = -f\sigma_y,\ v=0,\ \sigma_y<0;\ |x|<b,\ y=0 \tag{19a}$$

or with an open crack

$$\sigma_y = 0,\ \tau_{xy} = 0,\ v>0;\ |x|<b,\ y=0 \tag{19b}$$

The condition (19b) could not be satisfied since at the points $x = \pm b$ where a type of boundary conditions is changed, there appears an oscillating singularity near which the inequality does not hold. To avoid this [11], one can consider a combination of the conditions (19a) and (19b), e.g., assume that close to the center along a segment $|x|<b_1<b$, there is an open crack so that the condition (19b) holds there, whereas along the segment $b_1<|x|<b$, the condition (19a) is satisfied.

Restricting to the simplest case of f=0. It can be shown that the inequality $\sigma_y<0$ holds everywhere along $|x|<\ell$, y=0. The conditions (19) coincide in that case and the both above posed problem can be solved simultaneously.

The solution of the problem (16), (17), (18) with the former conditions at infinity (4) has a form

$$\begin{aligned} \phi(z) &= \frac{(\kappa-1)\sigma_\infty}{8\kappa} + \frac{i(\kappa+1)\sigma_\infty(z^2-\alpha\ell^2)}{8\kappa(z^2-b^2)^{1/2}(\ell^2-z^2)^{1/2}} \\ \Omega(z) &= \frac{(\kappa-1)^2\sigma_\infty}{8\kappa} - \frac{i(\kappa+1)^2\sigma_\infty(z^2-\alpha\ell^2)}{8\kappa(z^2-b^2)^{1/2}(\ell^2-z^2)^{1/2}} \end{aligned} \tag{20}$$

where α denotes the ratio of the complete elliptical integral of the second kind $E(\pi/2,k)$ to the complete elliptical integral of the first kind $K(k)$

$\alpha = E(\pi/2,k)/K(k); \; k = (\ell^2-b^2)^{1/2}/\ell$

The stress intensity factors K_ℓ at the fiber edge $x=\ell$ and K_b at the point $x=b$ may be found from (5) and (20) in the form

$$K_\ell = \frac{(\kappa+1)^2(1-\alpha)\ell^{3/2}\sigma_\infty}{8\kappa(\ell^2-b^2)^{1/2}}; \; K_b = \frac{(\kappa+1)^2(b^2-\alpha\ell^2)\sigma_\infty}{8\kappa b^{1/2}(\ell^2-b^2)^{1/2}} \tag{21}$$

when $b \to 0$, the parameter α behaves as follows:

$\alpha \simeq \ell n^{-1}(\ell/b) \to 0$

From (20), it can be seen that when $b \to 0$, the solution for the broken fiber continuously approaches the solution (6) for the whole single fiber with length 2b, which has no stress concentration at the center. On the other hand, as can be seen from (21), the stress intensity factor K_b approaches infinity when $b \to 0$. Such behavior of K_b clarifies the situation and it is possible to avoid difficulties if it is suggested that the breakage of a fiber is accompanied by an arbitrary small separation of the fiber from the matrix. Existence of a small separation produces a stress intensity factor which is greater for the smaller separation, Figure 2. After fiber breakage, the matrix separation starts. It stops when the length b takes value for which the stress intensity factor is equal to the critical value. This leads to the transcendental equation for the ineffective fiber length

$$K_b = \frac{(\kappa+1)^2(b^2-\alpha\ell^2)\sigma_\infty}{8\kappa b^{1/2}(\ell^2-b^2)^{1/2}} \tag{22}$$

The stress intensity factor at the fiber edge, K_ℓ, drops after fiber breakage. The tensile force in the fiber is reduced also. Its maximum values are at $x = \pm\alpha\ell$ and equal to

$$P_{max} = \frac{(\kappa+1)^2\ell\sigma_\infty}{4\kappa} [\alpha F(\phi,k) - E(\phi,k)] \tag{23}$$

where $F(\phi,k)$ and $E(\phi,k)$ are the elliptical integrals of the first and second kind, respectively, in which ϕ is

$\phi = \text{a2c} \sin(1-\alpha)^{1/2}/(1-b/\ell)^{1/2}$

It is interesting to evaluate the stress concentration near the point of fiber breakage. The most interesting is the stress σ_x since it can overload a neighboring fiber. Asymptotically, the stress decreases along the y-axes as

$$\sigma_x = -\frac{(\kappa^2+6\kappa+5)\ell\sigma_\infty}{8\kappa y \ell n(b/4\ell)}; \; x=0, \; b<<y<<\ell \tag{24}$$

Evidently, fiber separation from the matrix strongly decreases the stress, σ_x, concentration.

Consider now a model in which the fiber is treated analogously to that of a bounded plate with stringers [7,12]. Restricting the analysis to a single fiber the boundary conditions (1), (2) and (4) are unchanged. The condition for fiber equilibrium has the form

$$\frac{\partial P}{\partial x} + 2\tau_{xy} = 0 \tag{25}$$

Assuming the fiber rigidity to be equal to $E\delta$, where E is Young's modulus, from (25) one has

$$E\delta \frac{du}{dx} + 2 \int_{\ell}^{x} \tau_{xy} dx = 0; \ |x|<\ell, \ y=0 \tag{26}$$

This problem (1), (2), (4) and (25) leads to the integrodifferential singula equation for $\tau(x) = \tau_{xy}(x,o)$ given as

$$\frac{2\kappa}{\pi(\kappa+1)} \int_{-\ell}^{\ell} \frac{\tau(t)dt}{t-x} - \frac{4\mu}{E\delta} \int_{\ell}^{x} \tau(x)dx - \frac{(\kappa+1)\sigma_\infty}{4} = 0 \tag{27}$$

When the function $\tau(x)$ is found from (27), the functions $\Omega(z)$ and $\phi(z)$ are dete mined by the equations

$$\Omega(z) = -(\kappa+1)\phi(z) + \frac{(\kappa-1)\sigma_\infty}{4} = \frac{1}{\pi} \int_{-\ell}^{\ell} \frac{\tau(t)dt}{t-z} - \frac{\sigma_\infty}{2} \tag{28}$$

Solution of the equation (26) is assumed in the form of a series of Chebysh polynomials of the first kind $T_n(x)$, with the separated square root singularity [12].

$$\tau(x) = \frac{\ell\sigma_\infty}{(\ell^2-x^2)^{1/2}} \sum_{n=o}^{\infty} B_n T_n(x) \tag{29}$$

Substitution of (29) into (27) after performing integration by parts leads to t following set of equation for unknown coefficients B_n

$$B_m + \frac{4(\kappa+1)\ell\mu}{\pi\kappa\delta E} \sum_{n=1}^{\infty} \frac{[1+(-1)^{m+n}](m^2+n^2)B_n}{n[1-2(m^2+n^2)+(m^2-n^2)^2]} + \frac{(\kappa+1)^2\delta_{1m}}{8\kappa} \tag{30}$$

Due to the absence of external force on the fiber, $B_o = 0$. It can be shown tha equation (30) is quasiregular and may be solved by the method of transaction. Figure 3 shows the solution of (30) by plotting the ratio of the stress intensi factor for an elastic fiber

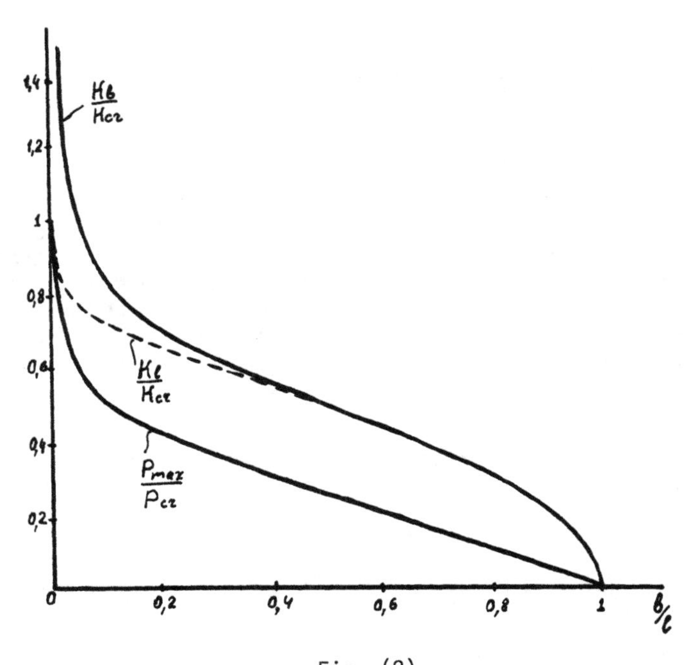
1,4
1,2
1
0,8
0,6
0,4
0,2
0
0,2
0,4
0,6
0,8
1
b/l
Kb/Kcr
Kl/Kcr
Pmax/Pcr

Fig. (2)

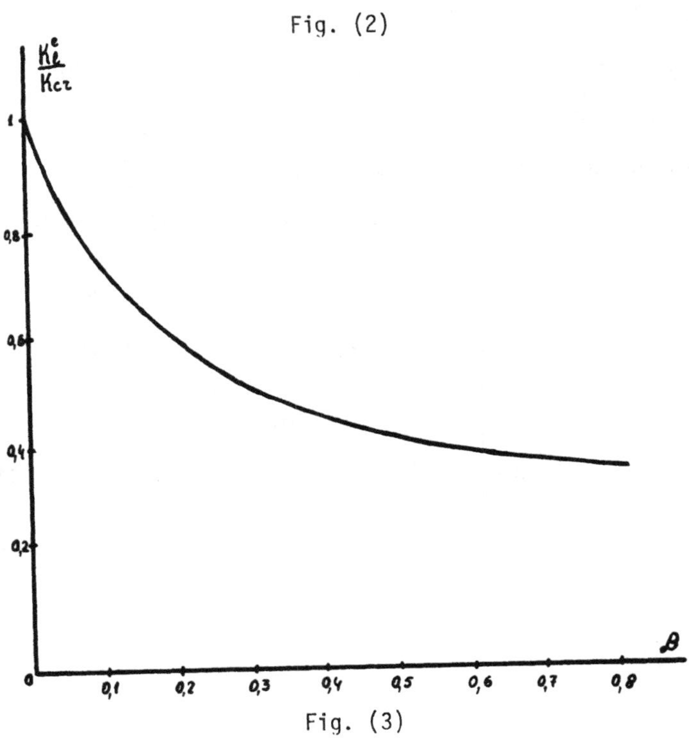
Kl^e/Kcr
1
0,8
0,6
0,4
0,2
0
0,1
0,2
0,3
0,4
0,5
0,6
0,7
0,8
β

Fig. (3)

$$K_\ell^e = \ell^{1/2}\sigma_\infty \sum_{n=1}^{\infty} B_n T_n(\ell)$$

to that for a rigid fiber versus the dimensionless parameter

$$\beta = (\kappa+1)\ell\mu/(\pi\kappa\delta E)$$

which characterizes the fiber rigidity. The results in Figure 3 demonstrate that accounting for fiber elasticity has the effect of decreasing the stress intensity factor. Similarly, it can be shown that fiber elasticity decreases the force in the fiber as well. Thus, while the singular character of the stress is the same for a rigid or elastic fiber, the quantitative results are significantly influenced by the fiber elasticity.

REFERENCES

[1] Mechanics of Composite Materials, Vols. 2 and 5, Academic Press, New York and London, 1974.

[2] Kelly, A., Strong Solids, Clarendon Press, Oxford, 1973.

[3] Kopiev, I. M. and Ovchinskyi, A. S., "Fracture of Fiber Reinforced Materials in Russian, Moscow, Nauka, 1977.

[4] Melan, E., Ing. Arch., 3, No. 2, 1932.

[5] Sternberg, E. and Muki, R., Z. Angew. Math. und Phys., Vol. 21, No. 4, 1970.

[6] Muki, R. and Sternberg, E., Z. Angew. Math. und Phys., Vol. 2, No. 5, 1971.

[7] Arutunyan, N. Ch., Prikladnaya Mathem. Mech., Vol. 32, No. 4, 1968.

[8] Arutunyan, N. Ch. and Mchitaryan, S. M., Prikladnaya Mathem. Mech., Vol. 36, No. 5, 1972.

[9] Kalandiya, A. I., Prikladnaya Mathem. Mech., Vol. 33, No. 3, 1969.

[10] Shiori, J. and Inoue, K., Soviet-Japan Symposium on Composite, Moscow University, 1977.

[11] Comninou, M., J. Appl. Mech., Ser. E, Vol. 44, December 1977.

[12] Kalandia, A. I., "Mathematical Methods of Plane Theory of Elasticity", Moscow, Nauka, 1973.

SECTION IV
UNIDIRECTIONAL COMPOSITES

FRACTURE OF UNIDIRECTIONAL COMPOSITE MATERIALS UNDER THE AXIAL COMPRESSION

A. N. Guz

Academy of Sciences of Ukranian SSR
Kiev, USSR

ABSTRACT

The results of theoretical study of the fracture of unidirectional composite materials under axial compression are discussed. These results are based on the use of the strict three-dimensional linearized stability theory of deformed bodies. A brief analysis is presented of main results for laminated and fibrous composite materials. New results are presented on the stability loss under highly elastic deformations taking into account the interference of two neighboring fibers. The fracture of fibrous materials with metallic matrix is discussed taking into account the plastic deformations.

INTRODUCTION

A loss of stability in a material's structure is considered one of the fracture mechanisms acting in the compression of unidirectional composite materials. The phenomenon mentioned occurs also in the compression of composite materials having insignificant reinforcement in the transverse direction as well as in compressed regions under bending. In one of the first studies in this field [1], an approximate formulation of the stability problem of laminated material was presented, on the basis of a one-dimensional model the work of a binder was considered. In subsequent years, that problem was dealt with in many publications in various approximate formulations.

In the author's papers [2,3] for the problem considered, a formulation was proposed based on the strict three-dimensional linearized theory of stability of deformable bodies. Methods of that theory were presented in the monographs [4-7] and other publications. In subsequent years within the scope of the formulation [2,3] using the apparatus [4-7], a number of problems related to the problem of the fracture of laminated and fibrous materials was studied. The present article deals with an analysis of results, obtained on the basis of the strict formulation [2,3] some new results will be presented. New results, obtained within the framework of the formulation [2,3], are related to the study of the loss of stability under highly elastic and small deformations, taking into account the interaction of two neighboring fibers. These results are also related to the fracture of fibrous materials with metallic matrix under elastic-plastic deformations.

ANALYSIS OF RESULTS

With the use of three-dimensional linearized theory, two approaches were proposed. In a first approach, the material is considered as homogeneous and orthotropic, the effect of the structure is accounted for through reduced constants; that approach which may be considered as a continual one, was proposed originally in [3]. In the second approach, the material is taken to be a piecewise-homogeneous, and for description of the behaviour of filler and binder, the relations of the three-dimensional linearized theory of the stability of deformed bodies are used, which satisfy the conditions of continuity of stress and displacement vectors at interfaces; that approach was originally proposed in [2]. It may be regarded as an approach within the framework of piecewise-homogeneous medium. Now an analysis of the results will be presented, obtained within the scope of two abovementioned approaches. The Cartesian coordinate system is introduced (X_1, X_2, X_3). Reinforcing elements are assumed to be directed along the OX_3 axis, along which the compression is effected.

CONTINUUM APPROACH

The material is considered as homogeneous and orthotropic with reduced elastic constants. The elasticity relations are presented in the form

$$\sigma_{ij} = \delta_{ij} \langle a_{ik} \rangle \varepsilon_{kk} + 2(1-\delta_{ij}) \langle G_{ij} \rangle \varepsilon_{ij};$$
$$\langle a_{ik} \rangle \equiv \langle a_{ki} \rangle; \quad \langle G_{ij} \rangle \equiv \langle G_{ji} \rangle \qquad (1)$$

Before fracture starts (an initial state) under the type of loading considered, an uniaxial, homogeneous stressed state arises

$$\sigma^{o}_{ij} = - p\delta_{i3}\delta_{j3} \qquad (2)$$

At the beginning of the fracture process, disturbances of the stressed state occur; their values in the beginning of the fracture process may be considered as lower than those in the initial state, equation (2). This consideration makes it possible to use for the analysis of disturbances the relations of the linearized elasticity theory for the brittle fracture. We consider the case of the small initial deformations theory, when the initial state is specified according to geometrically linear theory. In this case, taking into account equation (2), we obtain for static problems [4] the following system of equations

$$L_{mj}u_j = 0, \; L_{mj} = \langle a_{mj} \rangle \frac{\partial^2}{\partial X_m \partial X_j} + (1-\delta_{mj}) \langle G_{mj} \rangle \frac{\partial^2}{\partial X_m \partial X_j}$$
$$+ (1-\delta_{im}) \langle G_{im} \rangle \delta_{mj} \frac{\partial^2}{\partial X_i^2} - \delta_{mj}\, p \frac{\partial^2}{\partial X_3^2} \qquad (3)$$

When the process of fracture begins, the microvolume changes should in some way lead to changes in macrovolume. Those changes should arise in properties that are independent of boundary conditions as the process of material fracture is studied and not the effects of testing machines clamps, cross-section form, etc. However, these macrovolume changes should not have a local character only. Evidently, the macrovolume changes are determined by displacements disturbances represented by the equation system (3). It may be reasonably assumed that the beginning of the fracture is associated with the appearance of nonzero solutions of the system (3). These solutions are independent of boundary conditions (an infinite space is considered) and are not of a local nature. Undoubtedly, it is necessary to reject the solutions of homogeneous stressed state type that coincide with initial state of (2) type. Taking into account the above considerations, we conclude that only the presence in the system (3) solutions of hyperbolic equations solutions satisfies the requirements mentioned. Thus, the theoretical ultimate strength within the framework of the continuum approach follows from the condition of the presence of hyperbolic equations within the solutions of the system (3). In [4], it is shown (page 218) that this condition leads to the expression

$$\sigma_{np} = \langle G_{13} \rangle \tag{4}$$

It may be noted that in the approximation (4), the fibrous material is taken to be transverse-isotropic one, hence $\langle G_{13} \rangle \equiv \langle G_{23} \rangle$.

It should be pointed out that owing to the presence of defects in material structure, the shear modulus values in some parts of the material may be considerably less than the mean value of shear modulus. Consequently, for theoretical ultimate strength, one of the following expressions may be assumed

$$\sigma_{np} \leq \langle G_{13} \rangle; \; \sigma_{np} = \min\{\langle G_{13} \rangle\} \tag{5}$$

The proposed continuum approach allows not only to estimate the theoretical ultimate strength, but to predict the nature of the fracture as well. Within the framework of the adopted approach, the beginning of fracture is associated with the appearance of disturbances (changes) in the macrovolume that are determined by the solution of hyperbolic equation. Consequently, it is reasonable to assume that fracture characteristics are determined by propagation of the disturbances. For hyperbolic equations, the disturbances propagate on characteristic surfaces. Taking into account this consideration [4], it was shown (page 220), that in unidirectional composite materials under compression along the reinforcing layers, the fracture occurs in planes and surfaces, which are situated closely to perpendicular ones to the force action line. For the case of materials without defects in the structure in [4], (pages 209-211), it was shown that the critical load value in the state of surface instability is somewhat less than the value (4), which follows from the relation (IV.142) of the monograph [4], (page 211). Thus, the main concepts of the continuum theory discussed here are reduced to following concepts which will be presented separately for materials with some defects and for materials without defects.

For materials with insignificant defects: (1) the theoretical ultimate strength is estimated from expressions (5), (2) the fracture starts from the defect (where the shear modulus is somewhat less than the mean shear modulus) and propagates on the planes, which are close to perpendicular ones to the force action line.

For materials without defects in the structure: (1) the theoretical ultimate strength is estimated from the expression (4), (2) the fracture starts from material surface and propagates on the planes close to perpendicular ones to the force action line.

The continuum approach is discussed at length in [4] and some other publications; in these references, the relation of the theory considered with the internal instability phenomenon was indicated and the results of experimental studies were also given.

Note: In the discussion above and in [4], the main concepts of the continuum approach are presented regarding the brittle fracture. Within the scope of the linearized theory, the continuum approach may be extended to case of nonbrittle fracture by changing the linearized relations for nonelastic models [6]; on that basis, some results can be obtained for strength characteristics and prolonged loading and high temperature.

Consider the laminated and fibrous materials and introduce the notations: G and $G^{(1)}$ - shear moduli of the binder and filler; S and $S^{(1)}$ - concentrations of binder and filler. Only materials without defects in the structure will be considered. Theoretical ultimate strength [4] is estimated on the basis of (4). For laminated materials, we obtain

$$\sigma_{np} = \frac{G^{(1)}}{S^{(1)}G+SG^{(1)}} G \tag{6}$$

For fibrous materials, (6) is used or more exact relation of the form

$$\sigma_{np} = \frac{G^{(1)}(1+S^{(1)})+GS}{G^{(1)}S+(1+S^{(1)})G} G \tag{7}$$

In a similar manner, for hollow fibers (τ_2 and τ_1 - outer and inner fiber radii), we obtain

$$\sigma_{np} = \frac{G^{(1)}(1+S^{(1)})+GS}{G^{(1)}S+(\tau_2^2+\tau_1^2)(\tau_2^2-\tau_1^2)^{-1}(1+S^{(1)})G} G \tag{8}$$

Approach for piecewise-homogeneous medium. In this case for the study of elastic stability of piecewise-homogeneous media, three-dimensional linearized equations [4-7] are used in each component at large (finite) and small precritical deformations. The material is supposed to be free of defects and for that reason, the conditions of the continuity of displacement vector and of stresses on the interface boundary should be satisfied. Main results were obtained for the case of homogeneous precritical state. That condition under the loading type discussed is always satisfied for laminated materials. It is always satisfied exactly for fibrous materials as well, in case the filler and binder are incompressible or in case they are compressible, the value of the Poisson's ratio is the same for both the filler and the binder. The critical loading is assumed to correspond to the theoretical ultimate strength at brittle fracture. The results obtained on the basis of the approach mentioned are presented in [4] for the case of comparatively rigid materials (at small precritical deformations) and in [5] for highly elastic materials (at finite precritical deformations). Some results are presented in publications in periodicals. Main results are given below with reference to fibrous and laminated materials.

Fibrous materials. General solutions were constructed for corresponding linearized equations, these solutions allow to analyze the stability of fibrous materials with fibers of circular cross-section. The procedures are proposed for the investigation of one and two fibers, for infinite row of fibers, for the finite number of fibers and for biperiodical fiber system. It is shown that problem solution is reduced to the analysis of the infinite determinant. The way is shown for proving that these determinants are determinants of normal type and this allows to use approximate procedures for their analysis.

Numerical results were obtained for a single fiber, and these results are related to composite materials with low concentration of filler at small and large precritical deformations. It is established that under axisymmetrical deformations (barrel-like form) the loss of stability does not occur. The loss of stability is observed in the bending mode. Some result of asymptotic character are obtained for low concentration of the filler and for a longwave form of the stability loss.

Laminated materials. General solutions of corresponding linearized equations are constructed. These solutions may be used for stability studies within the framework of the plane axisymmetric and non-axisymmetric three-dimensional problems. Laminated materials with periodic structures were studied under four different (along the OX_1 axis) forms of stability loss. Some results of asymptotic character are obtained for low concentration of the filler and for a longwave form of the stability loss.

Numerical results were obtained for composite materials of periodic structure with small (comparatively by rigid materials) and finite (highly elastic materials) precritical deformations. The critical load values were estimated relating to mechanical and geometric parameters of composite laminated materials with periodic structure. Analysis of results has demonstrated that the stability loss in the structure can occur only when the binder layer thickness considerably exceeds the thickness of filler layer. Moreover, it is shown that when the loss of stability has occurred in the structure of laminated composite material of periodic structure - the critical loading value may be estimated with a sufficient degree of accuracy according to the following calculating procedure. The approximate calculating procedure consists in the analysis of the stability of filler layer connected with two semispaces; the material of the latter coincides with binder's material; for the layer and two semispaces, the three-dimensional linearized stability theory is used. The result obtained allows in the solution of more complex problems not to take account of the mutual interference of layers of the filler using the indicated approximate calculation procedure for layered materials.

Comparative analysis. The continuum approach presented here has made it possible to estimate in a closed form the theoretical ultimate strength and to predict the fracture mode. However, the application limits of that approach could be specified only on the basis of a strict approach within the piecewise homogeneous medium. The last approach even for comparatively simple problems leads to the analysis of characteristic transcendental equations of complex structure. Such an analysis is possible only with the use of numerical methods. However, these approaches can be considered as comparatively strict ones and are based on the use of the three-dimentional linearized stability theory. There are some other procedures based on the approximate representation of the filler and binder behaviour within the limits of the piecewise homogeneous medium. Those theories do not proceed from the strict three-dimensional stability theory. We shall name conventionally these approaches by a general term-approximate approaches. It should be pointed out that within the framework of approximate approaches, many studies were made presented in published articles and reports. All the approximate approaches are much more complicated than the continuum approach presented above, although the validity limits of the approximate approaches could be specified only on the basis of the strict approach within the limits of the piecewise homogeneous medium. That approach is presented above. The use of experimental results for determination of limits of applicability (in particular, the reliability of results) of approximate approaches does not always lead to unique conclusion. This particular feature is associated with the fact that in the problem of the loss of stability in material structure (in the mechanical phenomenon that problem is considered a complex and delicate one) it is not always possible to identify the phenomena. These considerations emphasize the importance of the comparative analysis of approximate approaches (including the continuum approach) and the above mentioned strict approach within the scope of the piecewise-homogeneous medium. It is reasonable to make that analysis primarily for results obtained by utilization of various approaches. An example of such an analysis will be presented below.

In [1], the following theoretical ultimate strength for a laminated material was obtained

$$\sigma_{np} = \frac{G}{S} \tag{9}$$

This result in [1] was obtained within the scope of approximate approach, when the filler was analyzed using the theory of thin rods while the binder was approximated by one-dimensional medium. In the following, the result (9) is extended to fibrous materials. Under some assumptions, the result (9) may be derived from the results of (6) and (7) of the continuum approach. Consider a case when the following conditions are satisfied

$$G^{(1)} >> G; \; S^{(1)} << S \tag{10}$$

From (6), (7) and (10), we obtain for laminated and fibrous materials, respectively

$$\sigma_{np} \approx \frac{G}{S}; \; \sigma_{np} \approx \frac{1+S^{(1)}}{S} G \tag{11}$$

Consequently, under the condition (10), in approximate approach, the results [1] are almost identical with the results (11) of the continuum approach.

We specify the conditions when the expression (9), obtained using approximate approach [1] is identical with the results derived on the basis of piecewise-homogeneous medium [4] approximation when using the three-dimensional linearized elasticity theory. The following notations are introduced: h and $h^{(1)}$ - the thickness of binder and filler layers in laminated material; R and R_{12} - fiber radius and minimal distance between two neighboring fibers; α and $\alpha^{(1)}$ wave formation parameters in binder and filler; ℓ - wavelength of the stability loss mode along the OX_3 axis. For laminated and fibrous materials of periodical structure, the following relations hold

$$\alpha = \pi \frac{h}{\ell}; \ \alpha^{(1)} = \pi \frac{h^{(1)}}{\ell}; \ \alpha = \pi \frac{R_{12}}{\ell}, \ \alpha^{(1)} = \pi \frac{R}{\ell} \tag{12}$$

In Chapter IV of the monograph [4] for laminated and fibrous materials of periodical structure, using the strict approach within the limits of the piecewise homogeneous medium for comparatively rigid materials (three-dimensional linearized theory of elastic stability under small pre-critical deformations), asymptotic analysis of characteristic determinants was made for the case

$$\alpha >> 1; \ \alpha^{(1)} << 1; \ S^{(1)} << S \tag{13}$$

The inequalities (13) have the following meaning: the wavelength of the stability loss mode is considerably less than the binder's thickness within one period of the structure; the wavelength of the stability loss mode is considerably larger than the binder thickness within one period of the structure; the composite material with low concentration of the filler is analyzed. For that case in [4], the first term of the asymptotic representation for critical loading is calculated; for fibrous material, the relation (IV.70), page 188 was derived in [4] and for laminated material, the expression (IV.102), page 198, was given. The following condition is assumed to be satisfied

$$\frac{G^{(1)}}{G} << \frac{S}{S^{(1)}} \tag{14}$$

This is the relation between the ratio of filler and binder concentration and the ratio of their rigid characteristics. The following interpretation of the condition (14) is proposed: the more rigid is the filler, the lower its concentration should be. When the conditions (13) and (14) are satisfied, from respective relations [4], for laminated and fibrous materials, it follows

$$\sigma_{np} \approx GS \tag{15}$$

Since the materials are studied with low filler concentration (third inequality (13)), the expressions (9) and (15) are nearly identical and have the form

$$\sigma_{np} \approx G \tag{16}$$

i.e., the ultimate strength is approximately equal to the shear modulus of the binder.

Consequently, the conditions (13) and (14) approximately specify the applicability limits of the results [1]. It should be noted that when the conditions (13) and (14) are satisfied, the results [1] are also identical to the results of the continuum approach, although the results [1] are obtained in a more cumbersome way.

In Chapter IV [4], a comparison is made between numerical results for various forms of stability loss (page 223). That comparison demonstrates that the continuum approach results are of the same order as the results obtained within the framework of the piecewise-homogeneous medium for various form of the stability loss. However, the results obtained by the use of the approximate approach [1] for particular forms of stability loss differ by one order from the results based on approach on the basis of the piecewise-homogeneous medium.

It is reasonable to conclude therefore that the continuum approach leads to more precise results when compared with approximate approach [1], although the latter is made within the limits of the piecewise-homogeneous medium with approximate evaluation of the character of interaction between the filler and the binder.

The examples cited demonstrate: (1) the need for further development of the effective approach within the limits of the piecewise-homogeneous medium utilizing the three-dimensional linearized stability theory; (2) the necessity of the clear-cut specification of the applicability limits of various approximate approaches based on the approximate presentation of the interaction between the binder and the filler. The latter conclusion is of particular importance since in a considerable number of published studies, the approximate procedures for the estimation of the filler-binder interaction were used.

NEW RESULTS

Some new results will be discussed that were obtained utilizing the approach within the limits of piecewise-homogeneous medium with the use of the three-dimensional linearized elasticity theory. It should be pointed out that in that approach to obtain numerical results at the final stage, numerical procedures should be used.

Two fibers interference at highly elastic deformations. From above considertions, it follows that utilizing the strict approach within the framework of piewise homogeneous medium for fibrous materials, the numerical results were obtair for a single fiber only. These numerical results are applicable to composite materials with rather low filler content only. The problem arises of quantitative estimation of the applicability of the results obtained to composite materials. A partial solution may be obtained by solution of the problem of stability of tw parallel fibers accounting for fiber interference in the analysis of the stabili loss. Studies along these lines were made for two fibers at small precritical c formations using the apparatus of the monographs [4,6] and under highly elastic deformations using the apparatus of the monographs [5,7]. The results presente below were obtained by the author in association with S. D. Akbarov and M. A. Cherevko. Following notations are introduced: (τ_q, θ_q, X_3) - circular cylindrica coordinate systems related to each fiber's axis; ℓ - wavelength of the loss of stability mode along the OX_3 axis, along the fibers; R and R_{12} - fiber radius and distance between centers of fibers cross-sections; ζ_i and $\zeta_i^{(1)}$ $(i = 1,2,3)$ - roots of respective equations [4,7], calculated for binder and filler; $I_n(x)$ an $K_n(x)$ - Bessel functions for purely imaginary argument and Macdonald function. Under the assumption that in the precritical state, a uniform state occurs caus by compression (shortening) along the OX_3 axis, for problem evaluation general s lution [4,7] were used. The potentials included in these general solutions are represented in the following form: for the binder

$$\psi = \gamma \sin\gamma X_3 \sum_{n=-\infty}^{+\infty} \sum_{q=1}^{2} A_n^q K_n(\zeta_1 \gamma \tau_q) \exp in(\theta_q - \beta_q); \; \gamma = \frac{\pi}{\ell};$$

$$\Psi = \cos\gamma X_3 \sum_{n=-\infty}^{+\infty} \sum_{q=1}^{2} [B_n^q K_n(\zeta_2 \gamma \tau_q) + C_n^q K_n(\zeta_3 \gamma \tau_q)] \exp in(\theta_q - \beta_q);$$

$$A_{-n}^q \equiv A_n^{-q}; \; B_{-n}^q \equiv B_n^{-q}; \; C_{-n}^q \equiv C_n^{-q}; \tag{17}$$

$$\tau m A_o^q \equiv \tau m B_o^q \equiv \tau m C_o^q = 0; \; q = 1,2$$

and for the filler

$$\psi^{(1)q} = \gamma \sin\gamma X_3 \sum_{n=-\infty}^{+\infty} A_n^{(1)q} I_n(\zeta_1^{(1)} \gamma \tau_q) \exp in(\theta_q - \beta_q); \; q = 1,2;$$

$$\Psi^{(1)q} = \cos\gamma X_3 \sum_{n=-\infty}^{+\infty} [B_n^{(1)q} I_n(\zeta_2^{(1)} \gamma \tau_q) + C_n^{(1)q} I_n(\zeta_3^{(1)} \gamma \tau_q)] \exp in(\theta_q - \beta_q); \tag{18}$$

$$A_{-n}^{(1)q} \equiv \bar{A}_{-n}^{(1)q}; \; B_{-n}^{(1)q} \equiv \bar{B}_n^{(1)q}; \; C_{-n}^{(1)q} \equiv \bar{C}_n^{(1)q}; \; \tau_m A_o^{(1)q}$$

$$\equiv \tau_m B_o^{(1)q} \equiv \tau_m C_o^{(1)q}$$

In (17) and (18), β_q denotes the angle from which in each cylindrical coordinate system the angle θ_q is calculated. Utilizing the apparatus [4-7] and potentials in the form (17) and (18) for the problem of two fibers instability taking into account their interference, the characteristic equation in the form of infinite determinant is derived; it was proved that the determinant obtained is a normal type determinant allowing to consider the finite order determinant in the approximate analysis. In the course of calculation of numerical results, the determinants of several orders were evaluated to ensure the necessary precision.

Numerical results were obtained for the following four mode of stability loss: 1) stability loss in the plane where fibers are located (bending mode of stability loss); 2) stability loss in the plane where fibers are located in anti-phase; 3) stability loss from the plane where fibers are located in one phase (bending mode of stability loss); 4) stability loss from the plane in which the fibers are located, in antiphase. As a result of the analysis of numerical results, it is shown that in real conditions the third mode occurs since for that mode, minimal shortening is obtained. Consequently, two fibers should lose their stability in the bending mode from the plane where fibers are located. This conclusion is analogous to intuitive notions.

As a result of the numerical data analysis, the conclusion was reached that the critical shortening of two fibers to a precision of 5% does not differ from the critical shortening for one fiber when the following condition is satisfied

$$\frac{R_{12}}{R} \geq 5 \tag{19}$$

It is thus concluded that when the condition (19) is satisfied, the interference of fibers under the stability loss may not be taken into account in the composite material. It should be noted that this conclusion was reached on the basis of solution of the problem for two fibers; further precision may be introduced for more complicated structures.

From the condition (19), the restrictions may be derived for that value of filler concentration, at which the interference of the fibers under the stability loss does not occur and consequently for composite fibrous material, the solution for a single fiber may be used. For example, for tetragonal (square) and hexagonal packing the following estimates were derived respectively

$$S^{(1)} \leq 0.06; \; S \leq 0.07 \tag{20}$$

As was already mentioned, the estimates (20) may be more precise, when more complicated structures are analyzed (row of fibers, biperiodical fiber system).

Fracture of fibrous materials with metallic matrix with elastic-plastic deformations. Consider a fibrous material with low filler concentration, for that reason, the interference of fibers under the stability loss will not be accounted for. The study was made using the three-dimensional linearized stability theory at small precritical deformations [4-7]. The filler is assumed to be a linear-elastic material; metallic matrix is considered as an isotropic incompressible elastic-plastic body, the deformational theory of plasticity will be used. In our analysis, the concept of continuing loading is employed, the occurrence of additional unloading zones under stability loss will not be taken account of. We note that as a result of loading, the matrix and the fiber undergo equal shortening along the OX_3 axis, its direction coincides with that of the fiber axis.

In that formulation, the author has obtained his results in collaboration with M. A. Cherevko. In the analysis, a power relation between the stress and deformation intensities was assumed in the form

$$\sigma_u = A\varepsilon_u^k \tag{21}$$

As a result of non-coincidence of transverse expansion coefficients for elastic fiber and elastic-plastic incompressible matrix, an inhomogeneous precritical state occurred in the matrix that depended on radial coordinate. In the analysis of the linearized stability problem after separation of variables, an eigenvalue problem was obtained for the system of ordinary equations with variable coefficients. For the solution of that eigenvalue problem, numerical procedure was used. The numerical results obtained prove the significant effect of plastic properties of the matrix on the magnitude of critical shortening.

As an example, the composite fibrous material [8] will be considered with a matrix (binder) of pure aluminum and a filler of stainless steel. In [8], an experimental study was made of that composite material with low filler concentration ($S^{(1)} \approx 0,04$). It follows that conditions (20) are satisfied. Consequently, under stability loss, the interference of fibers should not occur. Using the procedure mentioned, theoretical calculation was made; as a result, the critical value of the shortening $\varepsilon_{kp} \approx 5,9\%$ was obtained. Experimentally, for the material considered, the critical value of that shortening was determined [8], at which fracture of the material occurs $\varepsilon_{kp} \approx 6,0\%$. A satisfactory coincidence of theoretical and experimental results has been obtained.

CONCLUSIONS

From the above results, it follows that the use of three-dimensional linearized stability theory of deformed bodies allows with a sufficient degree of precision to determine the ultimate strength under compression of unidirectional composite materials and to predict the fracture characteristics. An urgent need exists for development of analogous theories for fibrous materials with higher filler concentration (taking account of the fiber interference under stability loss) in brittle fracture as well as for development of all aspects of the theory in non-brittle fracture (temperature strength, strength under prolonged loading, strength under plastic deformations, etc.).

REFERENCES

[1] Rozen, B., "Mechanics of composition strengthening", in: Fibrous Composite Materials, Moscow, "Mir" Publishers, pp. 54-96, 1967.

[2] Guz, A. N., "On stability theory construction for unidirectional fibrous materials", Prikladnaya Mekhanika, Vol. 5, No. 2, pp. 62-70, 1969.

[3] Guz, A. N., "On the theoretical ultimate strength determination under the reinforced materials compression", Dopovidi AN of Ukranian SSR, Series A, No. 3, pp. 236-238, 1969.

[4] Guz, A. N., "Stability of three-dimensional deformed bodies", Kiev: "Naukova dumka" Publishers, 276 p., 1971.

[5] Guz, A. N., "Stability of elastic bodies under the finite deformations", Kiev: "Naukova dumka" Publishers, 272 p., 1973.

[6] Guz, A. N., "Stability theory principles of mining workings", Kiev: "Naukova dumka" Publishers, 204 p., 1977.

[7] Guz, A. N., "Stability of elastic bodies under the universal compression", Kiev: "Naukova dumka" Publishers, 144 p., 1979.

[8] Pinnel, M. R. and Lawley, A., "Correlation of uniaxial yielding and substructure in aluminum-stainless steel composites", Metallurgical Transactions, Vol. 1, No. 5, pp. 1337-1348, 1970.

INFLUENCE OF STRESS INTERACTION ON THE BEHAVIOR OF OFF-AXIS UNIDIRECTIONAL COMPOSITES

M. J. Pindera and C. T. Herakovich

Virginia Polytechnic Institute and State University
Blacksburg, Virginia 24061

ABSTRACT

The influence of combined states of stress on the shear response along material principal directions in off-axis, unidirectional, composite coupons is examined for systems whose matrix material obeys the von Mises yield condition. Such analysis is motivated by trends in experimental data observed for at least two composite systems that indicate deviations from pure shear behavior along material directions for various off-axis configurations.

The yield function for plane stress of a transversely isotropic composite lamina consisting of stiff, linearly elastic fibers and a von Mises matrix material is formulated in terms of Hill's elastic stress concentration factors and a single plastic constraint parameter. The above are subsequently evaluated on the basis of observed average lamina and constituent response for the Avco 5505 boron-epoxy system. It is shown that inclusion of residual stresses in the yield function together with the incorporation of Dubey and Hillier's concept of generalized yield stress for anisotropic media in the constitutive equation correctly predicts the trends observed in experiments. The incorporation of the strong axial stress interaction necessary to predict the correct trends in the shear response is directly traced to the high residual axial stresses in the matrix induced during fabrication of the composite.

INTRODUCTION

The use of the off-axis tensile specimen for determination and/or verification of elastic properties and fracture strength of unidirectional composite materials is well documented, Sih [1]. As most composite materials exhibit varying degrees of nonlinear response prior to failure in an off-axis tension test, this specimen would also appear well suited for verification of the various assumptions in a number of nonlinear theories recently proposed in the literature, as well as some of their consequences. An indirect outgrowth of certain assumptions employed in one such theory, for instance, is a recently proposed method for determination of nonlinear lamina shear stress-strain response on the basis of the 10° off-axis tension test [2].

The nonlinear response of unidirectional composites is a complex phenomenon which can be caused by a number of diverse mechanisms such as nonlinear constituent behavior, damage accumulation through fiber or matrix cracking or debonding, fiber rotation or any combination of the above to mention the better known causes. The extension of results of the uniaxial test along principal material directions to constitutive models for multiaxial loading situations is not always straightforward due to the possibility of various stress interactions takin place at the microlevel. Foy [3], for example, demonstrated the influence of the transverse normal stress component on the shear strain response (and vice versa) along the material axes for composites with stiff, linearly elastic fibers and ductile matrices using a finite element, micromechanical analysis. He showed that for combined normal and shear loading in a fixed ratio yielding initiated sooner and the extent of inelastic behavior was more pronounced than for either normal or shear stress acting alone, Figure 1.

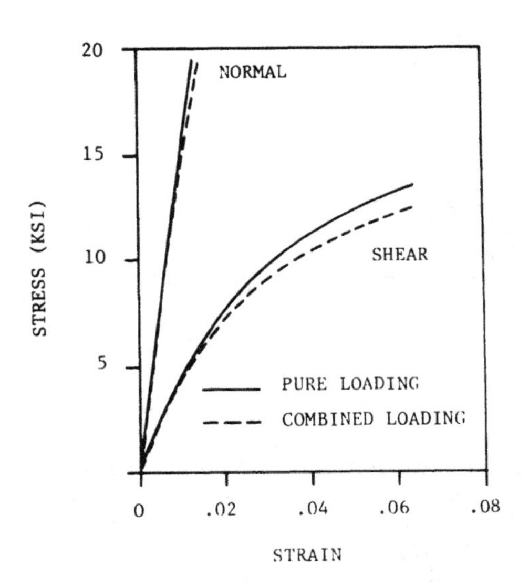

Fig. (1) - Stress-strain curves for unidirectional boron/epoxy (50% fiber volume fraction) under combined loads. Normal:shear load ratio is 8:3, [3]

A number of macromechanical constitutive models have been proposed in the literature [4-8] to describe the nonlinear response of lamina with stiff, elastic fibers and significantly more compliant and often ductile matrices. The early attempts [4,5] disregarded the possibility of stress interaction in the nonlinear range during combined loadings, while in the later formulations, this effect was taken into account, often in an heuristic manner. One such model du to Hashin et al [6] is reminiscent of the total deformation theory of plasticit in that the stress interaction is accounted for through a loading function, bas upon an invariant formulation, which is used to determine the different functio al forms of the various strain components. The loading function is limited to quadratic terms in transverse normal and shear stresses thus eliminating the po sibility of the axial normal stress influence on the transverse and shear strai behavior.

Data generated by Cole and Pipes [9] are employed in this paper to re-examine and study the extent of stress interaction in off-axis unidirectional boron-epoxy coupons with the aid of Hashin's model and a micromechanical approach utilizing a yield function concept based on Hill's method [10] for determining the elastic stress concentration factors. The results of the micromechanical analysis are subsequently employed in Dubey and Hillier's generalized yield stress formulation [11] to study the nonlinear shear stress-strain response along the principal material directions in off-axis tension coupons. Several fiber orientations are considered including the 10° coupon which has been proposed for the determination of pure shear response.

The influence of stress interaction on the shear response in the off-axis tensile tests of Cole and Pipes is shown in Figure 2 along with the pure shear

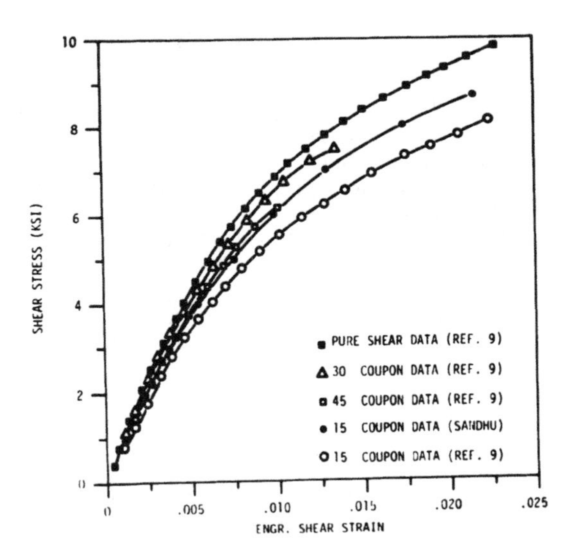

Fig. (2) - Experimental shear stress-strain response along material principal directions for various off-axis tension coupons. Avco 5505 boron-epoxy system

response obtained from unidirectional tubes. These experimental results show that the extent of inelastic behavior is not a monotonic function of fiber orientation. The inelastic shear strains (for a specified shear stress) decrease in magnitude as the fiber angle is increased from 15° to 30°, but as the fiber angle is increased further to 45° the inelastic strains increase. (Similar behavior has been observed by the authors for the graphite-polyimide off-axis coupons). One of the goals of this paper is to provide an explanation for such behavior.

In the first part of the paper, the model of Hashin et al is briefly outlined. It is then used to predict the shear stress-strain response along material axes using unidirectional data taken from [9]. An empirical observation is made that modification of Hashin's model by incorporation of an axial normal stress interaction term significantly improves correlation with the experimental data. Based on the above results, micromechanical formulation of the yield function for the composite in the presence of stiff fibers is employed along with Dubey's generalized yield stress concept to examine the effect of the σ_{11} stress component on yielding in the matrix phase and its consequences on the nonlinear shear response along material directions. It is shown that incorporation of the presence of residual curing stresses in the composite leads to the correct prediction of trends in the experimental shear stress-strain data of Cole and Pipes.

STRESS INTERACTION MODEL

Following classical concepts of deformation theory of plasticity, Hashin et al [6] assume that the inelastic part of strain can be expressed in the following manner.

$$\varepsilon_{ij}^{P} = \sigma_{ij} f_{ij}(L) \qquad \text{(no summation)} \tag{1}$$

where L is some general quadratic loading function of the stresses. Since composites which employ stiff fibers such as boron or graphite behave linearly (or nearly so) under loads applied in the fiber direction, the plastic response is assumed to be limited to transverse and shear strains. Furthermore, the effect of the stress in the fiber direction on the remaining strain components is neglected and thus equation (1) is expressible as follows:

$$\begin{aligned} \varepsilon_{12}^{P} &= \frac{\sigma_{12}}{2G_{12}} f_{12}\,(\alpha^2\sigma_{22} + \beta\sigma_{12}^2) \\ \varepsilon_{22}^{P} &= \frac{\sigma_{22}}{E_{22}} f_{22}\,(\alpha^2\sigma_{22} + \beta\sigma_{12}^2) \end{aligned} \tag{2}$$

where the interaction term $\sigma_{12}\sigma_{22}$ has been excluded on the basis of material symmetry arguments and only strains relevant to laminate analysis retained. The functional form of $f_{22}(L)$ and $f_{12}(L)$ is determined from uniaxial tension and torsion tests along material directions. It is noted that no explicit mention is made of the yield function and its relationship to the loading function in this formulation despite apparent separation of the total strain into elastic and inelastic or plastic portions. Combining plastic and elastic portions of strain, the relevant total strain components become:

$$\begin{aligned} \varepsilon_{11} &= \frac{\sigma_{11}}{E_{11}} - \frac{\nu_{12}}{E_{11}}\sigma_{22} \\ \varepsilon_{22} &= -\frac{\nu_{12}}{E_{11}}\sigma_{11} + \frac{\sigma_{22}}{E_{22}} + \frac{\sigma_{22}}{E_{22}}\left[\left(\frac{\sigma_{22}}{\sigma_y}\right)^2 + \left(\frac{\sigma_{12}}{\tau_y}\right)^2\right]^{\frac{M-1}{2}} \\ \varepsilon_{12} &= \frac{\sigma_{12}}{2G_{12}} + \frac{\sigma_{12}}{2G_{12}}\left[\left(\frac{\sigma_{22}}{\sigma_y}\right)^2 + \left(\frac{\sigma_{12}}{\tau_y}\right)^2\right]^{\frac{N-1}{2}} \end{aligned} \tag{3}$$

where the constants M, N, σ_y and τ_y are obtained from the Ramberg-Osgood approximation of uniaxial tests.

THEORETICAL-EXPERIMENTAL-CORRELATION

Data presented by Cole and Pipes on the Avco 5505 boron-epoxy system have been employed to determine the Ramberg-Osgood parameters in Hashin's model. These are given in the Appendix. Predicted shear response in the material coordinate system for the various off-axis fiber orientations is plotted in Figure 3. While Hashin's model predicts monotonically decreasing response with increasing off-axis angle, the experimental data points to a reversal around 30° off-axis angle, Figure 2.

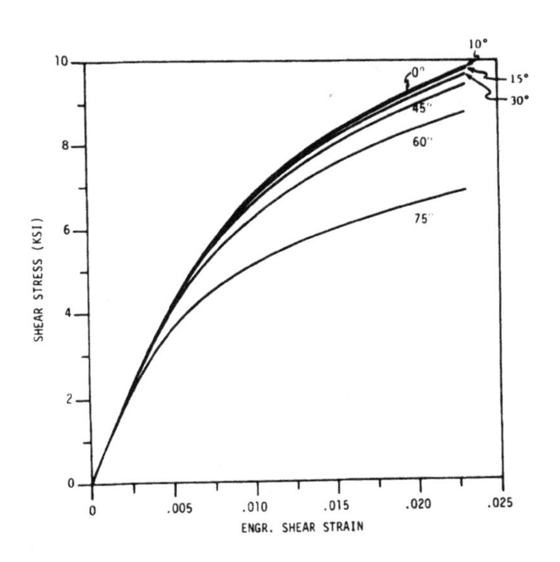

Fig. (3) - Predicted shear stress-strain response along material principal directions for various off-axis tension coupons. Hashin et al model. Avco 5505 boron-epoxy system

The concept of stress interaction in an elastoplastic material can be illustrated by considering the shear stress-strain response along axes inclined at some angle with respect to the loading direction of an isotropic tensile specimen. The hypothetical isotropic material is taken to have the same shear properties as the considered boron-epoxy system for illustration purposes. Using the J_2 isotropic theory (in this case the flow and deformation approaches are coincident) the shear response along the inclined axes is given by:

$$\varepsilon_{12} = \frac{\sigma_{12}}{2G_{12}} + \frac{\sigma_{12}}{2G_{12}} \left[\frac{2}{3} + \frac{1}{3} (\tan^2\theta + \cot^2\theta)\right]^{\frac{N-1}{2}} \left(\frac{\sigma_{12}}{\tau_y}\right)^{N-1} \quad (4)$$

with yield governed by:

$$\sigma_{12}^2 \left[\frac{2}{3} + \frac{1}{3} (\tan^2\theta + \cotan^2\theta)\right] = S^2, \; 0<\theta<90° \tag{5}$$

(or alternatively, $\sigma_{12} = \frac{\sqrt{3}}{2} S \sin 2\theta$) where S is the yield stress in pure shear.

Equation (4), plotted in Figure 4, predicts monotonically increasing shear response along rotated axes with increasing angle between 0° and 45° with reversal occurring at 45°. This follows from the symmetry of the yield function,

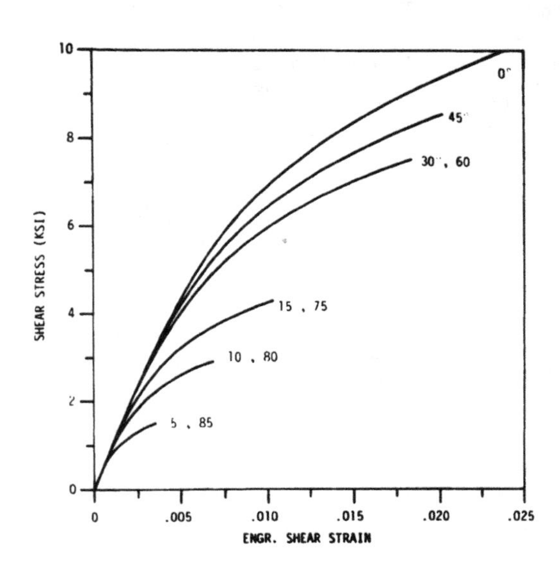

Fig. (4) - Predicted shear stress-strain response along rotated axes for an equivalent von Mises isotropic coupon in tension. Based on pure shear response of Avco 5505 Boron-epoxy system

equation (5), about the 45° angle. For comparison purposes, Hashin's theory, Figure 3, yields the following relation for the shear response along material axes inclined at the same angle as the reference axes in the hypothetical isotropic specimen

$$\varepsilon_{12} = \frac{\sigma_{12}}{2G_{12}} + \frac{\sigma_{12}}{2G_{12}} \left[1 + \left(\frac{\tau_y}{\sigma_y}\right)^2 \tan^2\theta\right]^{\frac{N-1}{2}} \left(\frac{\sigma_{12}}{\tau_y}\right)^{N-1} \tag{6}$$

The absence of the resolved shear stress reversal in the above model can be directly traced to the absence of the axial stress (σ_{11}) interaction term.

MICROMECHANICAL CONSIDERATIONS

Assuming that the matrix can be modelled as a von Mises solid that controls the nonlinear response of the composite, the yield function of the lamina can be determined from the knowledge of the stress concentration factors at yield. We take the view that these stress concentration factors are related to the average elastic concentration factors through a single plastic constraint parameter. This appears to be partially substantiated by various finite-element, micromechanical studies, Dvorak and Rao [12], which have revealed that although yielding usually starts at the fiber/matrix interface, the stress vector remains nearly radial under proportional loading. The yield zone on the other hand spreads rapidly throughout the entire matrix with increasing external tractions.

The elastic stress concentration factors can be determined from the knowledge of the average lamina as well as constituent response following Hill's outline [10]. The relationship between average matrix and external stresses at yield, including residual stresses, is then assumed to be expressible as follows:

$$\begin{bmatrix} \bar{\sigma}_{11m} \\ \bar{\sigma}_{22m} \\ \bar{\sigma}_{12m} \end{bmatrix} = K_p \begin{bmatrix} B_{11} & B_{12} & 0 & \bar{\sigma}_{11} & & \bar{\sigma}_{11mr} \\ B_{21} & B_{22} & 0 & \bar{\sigma}_{22} & + & \bar{\sigma}_{22mr} \\ 0 & 0 & B_{66} & \bar{\sigma}_{12} & & 0 \end{bmatrix} \tag{7}$$

where subscript m refers to matrix, r residual and no subscript external stresses. Also, $\bar{\sigma}_{33m} = \bar{\sigma}_{33mr}$ where $\bar{\sigma}_{22mr} = \bar{\sigma}_{33mr} = \frac{\bar{\sigma}_{11mr}}{A}$, and the elastic stress concentration factors $B_{11},\ldots,B_{66}$ are given in the Appendix. Here, K_p is the previously mentioned plastic constraint factor to be determined from experimental data, and the single scale factor A is a consequence of transverse isotropy of the composite.

Employing the above formulation, von Mises yield function for the transversely isotropic lamina in plane stress becomes:

$$\begin{aligned} &\frac{1}{3}(B_{11}^2 - B_{11}B_{21} + B_{21}^2)\bar{\sigma}_{11}^2K_p^2 + \frac{1}{3}(B_{12}^2 - B_{12}B_{22} + B_{22}^2)\bar{\sigma}_{22}^2K_p^2 \\ &+ \frac{1}{3}(2B_{11}B_{12} + 2B_{21}B_{22} - B_{11}B_{22} - B_{21}B_{12})\bar{\sigma}_{11}\bar{\sigma}_{22}K_p^2 \\ &+ \frac{1}{3}(1 - \frac{1}{A})\bar{\sigma}_{11mr}(2B_{11}-B_{21})\bar{\sigma}_{11}K_p + \frac{1}{3}(1 - \frac{1}{A})\bar{\sigma}_{11mr}(2B_{12}-B_{22})\bar{\sigma}_{22}K_p \\ &+ B_{66}^2\bar{\sigma}_{12}^2K_p^2 = K^2 - \frac{1}{3}[(1 - \frac{1}{A})\bar{\sigma}_{11mr}]^2 \end{aligned} \tag{8}$$

where K is the yield shear stress of the matrix material.

The importance of the residual stresses cannot be overemphasized. Employing a single inclusion composite cylinder model and temperature dependent material properties for this particular system, the average curing stresses have been estimated to be:

$$\begin{bmatrix} \bar{\sigma}_{11mr} \\ \bar{\sigma}_{22mr} \\ \bar{\sigma}_{33mr} \end{bmatrix} = 2786 \begin{bmatrix} 1 \\ 1/3 \\ 1/3 \end{bmatrix} \text{psi} \tag{9}$$

Although time dependent response has not been considered due to insufficient data, the reduction in residual stresses due to viscoelastic response will be partially off-set by the use of a single inclusion composite cylinder model which generally underestimates induced stresses.

The above yield function, equation (8), is plotted in Figure 5 in terms of the resolved shear stress along the material axes and the off-axis angle. Also plotted are the yield functions of an equivalent isotropic material and a transversely isotropic lamina with no residual stresses. It is clearly seen that the reversal of the shear stress-strain response evident in Cole's and Pipes' data can be explained on the basis of initial yielding in the matrix. To do this, we formulate a generalized yield stress function according to Dubey's and Hillier's suggestion as follows:

$$\bar{\sigma} = C_{ij}\sigma_{ij} + \sqrt{D_{ijk\ell}\sigma_{ij}\sigma_{k\ell}}$$

and assume that the plastic shear strain is a function of $\bar{\sigma}$. The coefficients C_{ij}, $D_{ijk\ell}$ are determined from the yield function given by equation (8) and the pure shear response. These are given in the Appendix. The resulting shear stress-strain curves along material axes for the various off-axis angles are presented in Figure 6 indicating the extent of stress interaction. It is seen that

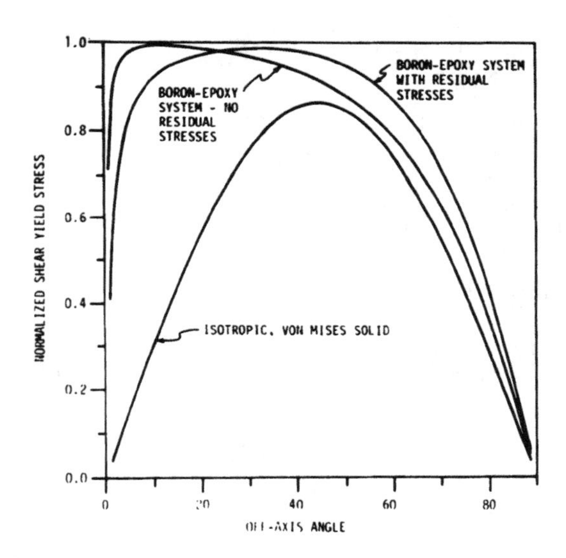

Fig. (5) - Yield shear stress of Avco 5505 boron-epoxy system along the material principal directions of an off-axis tension coupon as a function of the fiber orientation

the use of the 10° off-axis specimen for determination of pure shear response can lead to noticeable deviations for a material system that can be modelled by the assumed set of governing equations.

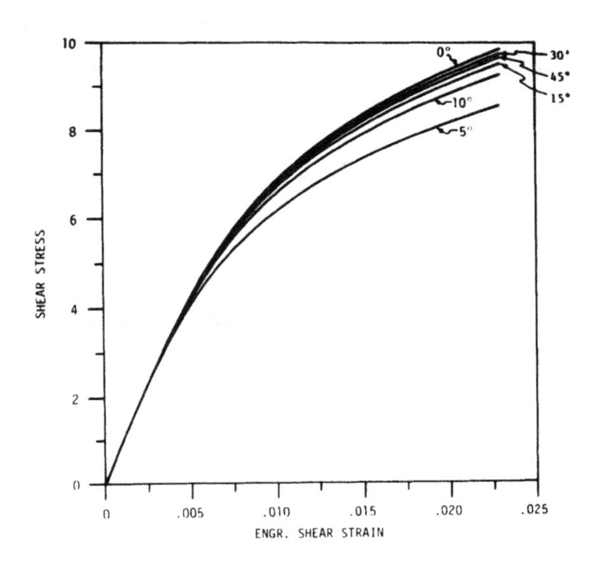

Fig. (6) - Predicted shear stress-strain response along material principal directions for various off-axis tension coupons based on Dubey and Hillier's generalized yield stress concept and micromechanical considerations including residual stresses. Avco 5505 boron-epoxy system

CONCLUSIONS

The deviations from pure shear response along material axes in an off-axes tension test for the Avco 5505 boron-epoxy system have been explained to a large extent on the basis of micromechanical formulation of a yield surface and a generalized yield stress concept. The model is applicable to systems with stiff, linearly elastic fibers and ductile matrices whose response can be approximated by that of a von Mises solid. The trends evident in the data of Cole and Pipes for the shear response along the material principal directions are traced to the strong axial stress ($\bar{\sigma}_{11}$) interaction which in turn follows from relatively significant residual stresses. The model predicts that in the absence of these stresses the influence of axial stress on nonlinear shear stress-strain response for low off-axis fiber orientations is only significant in the 0°-5° off-axis range. When the residual stresses are included, however, this influence is noticeable up to approximately 15°. This is interesting in view of the recently proposed test method for determination of pure nonlinear shear stress-strain response on basis of the 10° off-axis tension test.

ACKNOWLEDGEMENT

This work was supported by the NASA-Virginia Tech Composites Program NASA CA NCCI-15. This support is gratefully acknowledged.

APPENDIX

i) Ramberg-Osgood parameters for the Avco 5505 boron-epoxy system

M	σ_y(psi)	N	τ_y(psi)
3.434	26,155.2	4.430	9,593.9

ii) Elastic stress concentration factors $B_{11},\ldots,B_{66}$

$$B_{11} = \frac{1}{v_m}\left[\frac{E_m}{(E_f-E_m)^2-(\nu_f E_m-\nu_m E_f)^2}\right]\left[\frac{(E_f-E_m)(E_f-E_{11})}{E_{11}} - \frac{(\nu_f E_m-\nu_m E_f)(\nu_f E_{11}-\nu_{12}E_f)}{E_{11}}\right]$$

$$B_{12} = \frac{1}{v_m}\left[\frac{E_m}{(E_f-E_m)^2-(\nu_f E_m-\nu_m E_f)^2}\right]\left[\frac{(E_f-E_m)(\nu_f E_{11}-\nu_{12}E_f)}{E_{11}} - \frac{(\nu_f E_m-\nu_m E_f)(E_f-E_{22})}{E_{22}}\right]$$

$$B_{21} = \frac{1}{v_m}\left[\frac{E_m}{(E_f-E_m)^2-(\nu_f E_m-\nu_m E_f)^2}\right]\left[\frac{-(\nu_f E_m-\nu_m E_f)(E_f-E_{11})}{E_{11}} + \frac{(E_f-E_m)(\nu_f E_{11}-\nu_{12}E_f)}{E_{11}}\right]$$

$$B_{22} = \frac{1}{v_m}\left[\frac{E_m}{(E_f-E_m)^2-(\nu_f E_m-\nu_m E_f)^2}\right]\left[\frac{-(\nu_f E_m-\nu_m E_f)(\nu_f E_{11}-\nu_{12}E_f)}{E_{11}} + \frac{(E_f-E_m)(E_f-E_{22})}{E_{22}}\right]$$

$$B_{66} = \frac{1}{v_m}\left[\left(\frac{G_f-G_{12}}{G_f-G_m}\right)\left(\frac{G_m}{G_{12}}\right)\right]$$

where

$G_m = 0.19 \times 10^6$ psi $G_f = 24.0 \times 10^6$ psi $G_{12} = 0.88 \times 10^6$ psi

$E_m = 0.49 \times 10^6$ psi $E_f = 58.0 \times 10^6$ psi $E_{11} = 30.1 \times 10^6$ psi

$\nu_m = 0.31$ $\nu_f = 0.20$ $E_{22} = 2.87 \times 10^6$ psi

$v_m = 0.50$ $\nu_{12} = 0.225$

iii) Determination of generalized yield stress. From experiment

$\varepsilon_{12}^P = \frac{\sigma_{12}}{2G_{12}}\left(\frac{\sigma_{12}}{\tau_y}\right)^{N-1}$. Also, $\bar{\sigma} = \sqrt{2D_{66}}\,\sigma_{12}$ in contracted notation. Thus

$\varepsilon_{12}^P = \frac{\sigma_{12}}{2G_{12}}\left(\frac{\bar{\sigma}}{\tau_y\sqrt{2D_{66}}}\right)^{N-1}$. From longitudinal tension and compression, we obtain:

$C_1 = \xi\sqrt{D_{11}}$ in contracted notation, where

$$\xi = \frac{(X_c - X_t)}{(X_c + X_t)}$$

X_t, X_c being the magnitudes of yield stress in tension and compression, respectively. Similarly in transverse tension and compression

$$C_2 = \eta\sqrt{D_{22}}$$

where

$$\eta = \frac{(Y_c - Y_t)}{(Y_c + Y_t)}$$

From the invariance of $\bar{\sigma}$, we also have:

$$\sqrt{D_{11}} = \left(\frac{S}{X_t}\right) \frac{\sqrt{2D_{66}}}{(1+\xi)}$$

and

$$\sqrt{D_{22}} = \left(\frac{S}{Y_t}\right) \frac{\sqrt{2D_{66}}}{(1+\eta)}$$

where S is the yield stress in pure shear. Thus the function $\dfrac{\bar{\sigma}}{\tau_y\sqrt{2D_{66}}}$ becomes:

$$\frac{\bar{\sigma}}{\tau_y\sqrt{2D_{66}}} = \frac{1}{\tau_y} \left[\left(\frac{S}{X_t}\right) \frac{\xi}{(1+\xi)} \bar{\sigma}_{11} + \left(\frac{S}{Y_t}\right) \frac{\eta}{(1+\eta)} \bar{\sigma}_{22} + \sqrt{ \left(\frac{S}{X_t}\right)^2 \frac{\bar{\sigma}_{11}^2}{(1+\xi)^2} + \left(\frac{S}{Y_t}\right)^2 \frac{\bar{\sigma}_{22}^2}{(1+\eta)^2} + \bar{\sigma}_{12}^2 } \right]$$

The parameters S, X_t, Y_t, ξ and η have been evaluated from the yield function given by equation (8) and the experimental data. The plastic constraint factor K_p has been determined on the basis of yielding, i.e., deviation from linearity, in transverse tension.

REFERENCES

[1] Sih, G. C., "Fracture Mechanics of Composite Materials", Fracture of Composite Materials, Sih and Tamuzs, editors, Sijthoff and Noordhoff, p. 111, 1979.

[2] Chamis, C. C. and Sinclair, J. H., "10° Off-Axis Tensile Test for Shear Properties in Fiber Composites", Experimental Mechanics, Vol. 17, No. 9, p. 339, September 1977.

[3] Foye, R. L., "Theoretical Post-Yielding Behavior of Composite Laminates - Part I - Inelastic Micromechanics", J.C.M., Vol. 7, p. 178, April 1973.

[4] Petit, P. H. and Waddoups, M. E., "A Method for Predicting the Nonlinear Behavior of Laminated Composites", J.C.M., Vol. 3, p. 2, January 1969.

[5] Hahn, H. T. and Tsai, S. W., "Nonlinear Elastic Behavior of Unidirectional Composite Laminae", J.C.M., Vol. 7, p. 102, January 1973.

[6] Hashin, Z., Bagchi, D. and Rosen, B. W., "Nonlinear Behavior of Fiber Composite Laminates", NASA CR-2313, April 1974.

[7] Sandhu, R. S., "Nonlinear Behavior of Unidirectional and Angle-Ply Laminates", J. Aircraft, Vol. 13, No. 2, pp. 104-111, February 1976.

[8] Jones, R. M. and Morgan, M. S., "Analysis of Nonlinear Stress-Strain Behavior of Fiber-Reinforced Composite Materials", AIAA Journal, Vol. 15, No. 12, December 1977.

[9] Cole, B. and Pipes, R., "Filamentary Composite Laminates Subjected to Biaxial Stress Fields", AFFDL-TR-73-115, 1974.

[10] Hill, R., "Elastic Properties of Reinforced Solids: Some Theoretical Principles", J. Mech. Phys. Solids, Vol. 11, pp. 357-372, 1963.

[11] Dubey, R. and Hillier, M. J., "Yield Criteria and the Bauschinger Effect for a Plastic Solid", Transactions of the ASME, Paper No. 71-Met-P, Vol. 94, Series D, pp. 228-230, 1972.

[12] Dvorak, G. J. and Rao, M. S. M., "Axisymmetric Plasticity Theory of Fibrous Composites", Int. J. Engr. Sci., Vol. 14, pp. 361-373, 1976.

CHARACTERIZATION OF MATRIX/INTERFACE-CONTROLLED STRENGTH OF UNIDIRECTIONAL COMPOSITES

H. T. Hahn

Washington University
St. Louis, Missouri 63130

J. B. Erikson

National Defense Research Institute
Stockholm, Sweden

and

S. W. Tsai

Air Force Wright Aeronautical Laboratories
Wright-Patterson Air Force Base, Ohio 45433

ABSTRACT

This paper presents a method of characterizing the matrix/interface-controlled strength, including scatter, of unidirectional composites under combined loading. The failure envelope is represented by a second order polynomial and the scatter is described by the strength vector whose magnitude has a Weibull distribution. The method is based on the assumption of failure originating at inherent cracks parallel to the fibers and holds promise as a means of accounting for the size effect under combined loading. In addition, the energy release rate approach is discussed as a physical model and the appropriate data are presented.

INTRODUCTION

Many theories are available for representation of the strengths of unidirectional composites, as surveyed in [1] and [2]. As regards the matrix/interface-controlled strength, which is the topic of discussion in the present paper, these theories can be classified by the degree of interaction between the normal and shear stresses and also between tension and compression. For example, the modified von Mises-Hill criterion [1] takes into account the coupling between the transverse normal and shear stresses; however, it requires a different polynomial when the transverse stress is compressive. The maximum stress or maximum strain criterion incorporates no coupling at all; it is based on the independence of the failure modes. The tensor polynomial criterion [1], however, recognizes full coupling and employs only one polynomial.

All of the theories have their advantages and disadvantages. However, one or all of the following three factors can be cited as a reason for preferring one theory to the others: physical foundation, goodness of fit with the data and convenience.

In most cases, the goodness of fit tests have been performed on the off-axis strength without much success; the off-axis strength decreases rapidly with increasing off-axis angle, thus contributing to the possibility of visual deception in graphical comparisons. Consequently, it is difficult to distinguish one theory from another graphically.

Convenience depends on the type of application. For example, the tensor polynomial criterion is simpler in digital applications whereas the maximum strain criterion is more convenient in graphical applications.

As for the physical foundations behind the failure theories, one may observe that the von Mises-Hill criterion assumes failure to be independent of hydrostatic stress and that the maximum stress or maximum strain criterion is based on the noninteraction among the failure modes. However, no physical foundation has been provided for the tensor polynomial criterion.

The objective of the present paper is to discuss some of the physical evidences in support of the tensor polynomial criterion and propose a corresponding characterization procedure for the combined loading strength including scatter. The discussion is limited only to unidirectional polymer matrix composites subjected to transverse normal and longitudinal shear stresses so that composite failure is controlled by the matrix/interface properties. Thus the appropriate polynomial is of the form

$$F_2\sigma_2 + F_{22}\sigma_2^2 + F_{66}\sigma_6^2 = 1$$

The foregoing polynomial follows from the more general tensor polynomial in [1] in the absence of the longitudinal stress in the fiber direction, σ_1. However, recognizing the difference between the fiber-controlled failure and the matrix/interface-controlled failure, we propose that the reduced polynomial be used to describe the matrix/interface-controlled strength even when $\sigma_1 \neq 0$. It goes without saying that the fiber-controlled strength is then described by

$$F_1\sigma_1 + F_{11}\sigma_1^2 = 1$$

PHYSICAL BACKGROUND

A. Interrelationship Between Strength and Fracture Toughness

Just like homogeneous brittle materials, unidirectional composites exhibit, when the artificially introduced crack is parallel to the fibers [3], those fracture characteristics that are amenable to the linear elastic fracture mechanics predictions. In addition, the transverse or shear strength of unnotched composite is known to depend strongly on the inherent defects such as voids and interfacial debonds [4-6]. Thus, if these defects are regarded as typical cracks, the unnotched strength can be predicted from the fracture toughness and the size of the defect [7,8].

Suppose the unnotched strength is represented by the polynomial,

$$F_2\sigma_2 + F_{22}\sigma_2^2 + F_{66}\sigma_6^2 = 1 \quad (1)$$

where σ_2 is the transverse normal stress and σ_6 the longitudinal shear stress. The F's are the components of strength tensors.

Now, if the failure is assumed to initiate at an inherent crack of half length a_0 parallel to the fibers, Figure 1, equation (1) can be rewritten in terms of the nominal stress intensity factors k_2', k_6'

$$A_2k_2' + A_{22}k_2'^2 + A_{66}k_6'^2 = 1 \quad (2)$$

where

$$k_2' = \sigma_2\sqrt{a_0} \quad (3)$$

$$k_6' = \sigma_6\sqrt{a_0} \quad (4)$$

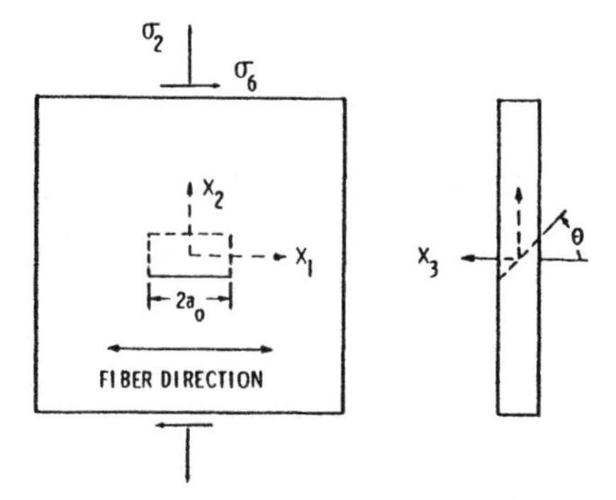

Fig. (1) - Composite lamina with a through-the-thickness crack

The parameters A's are related to the F's through

$$A_2 = F_2/\sqrt{a_0} \quad (5)$$

$$A_{22} = F_{22}/a_0 \quad (6)$$

$$A_{66} = F_{66}/a_0 \quad (7)$$

In the foregoing derivation, the surface of the inherent crack is not necessarily normal to σ_2, Figure 1. However, when σ_2 is tensile, the inherent crack can be assumed normal to σ_2 and the nominal stress intensity factors, k_2', k_6', reduce to the Mode I and Mode II stress intensity factors, k_1, k_2, respectively. Consequently, equation (2) can be considered in the first quadrant as representing a mixed-mode fracture toughness envelope under in-plane loadings. Moreover, through the use of the uniaxial strengths and corresponding fracture toughnesses, a_0 can be determined from

$$a_0 = (\frac{k_{1c}}{X_2})^2 \tag{8}$$

or

$$a_0 = (\frac{k_{2c}}{X_6})^2 \tag{9}$$

where X_2 and X_6 are the transverse tensile and shear strengths, respectively, and k_{1c} and k_{2c} are the Mode I and Mode II fracture toughnesses, respectively.

The foregoing hypothesis can be checked against the data for Scotchply 1002. The appropriate strength and fracture properties are listed in Table 1 [9]. The inherent crack half length is then

$$a_0 = 2.61 \text{ mm or } 3.53 \text{ mm} \tag{10}$$

depending on whether equation (8) or (9) is used.

TABLE 1 - STRENGTH AND FRACTURE PROPERTIES OF SCOTCHPLY 1002*

X_2(MPa)	X_6(MPa)	k_{1c}($MNm^{-3/2}$)	k_{2c}($MNm^{-3/2}$)
20.0	66.2	1.022	3.934
$F_2(GPa)^{-1}$	$F_{22}(GPa)^{-2}$	$F_{66}(GPa)^{-2}$	
42.79	360.76	228.24	

*Data taken from [9]

There are many factors that can contribute to the difference in the calculated values of a_0: test method, material variability, and the assumption of through-the-thickness crack. Although effects of the last two factors cannot be clearly defined at present, the first seems to point in the right direction. That is, k_{2c} was determined from a cantilever beam subjected to a concentrated load [3]. Since this test is closer to the short beam shear test than to the off-axis tension, the interlaminar strength should be used for X_6. Since the interlaminar shear strength is usually higher than the in-plane shear strength that was used in equation (9), the resulting a_0 will be smaller than 3.53 mm and hence closer to what is predicted by equation (8).

For the foregoing reasons and also since the overall comparison between the strength and fracture toughness under combined stress is of interest, we take the average value

$$a_0 = 3.07 \text{ mm} \tag{11}$$

and proceed to investigate the consequences. The fracture toughness tensor components follow upon substitution of the F's in Table 1 and a_0 into equations (5) and (7):

$$A_2 = 0.7735 \ (\text{MNm}^{-3/2})^{-1} \tag{12}$$

$$A_{22} = 0.1179 \ (\text{MNm}^{-3/2})^{-2} \tag{13}$$

$$A_{66} = 0.07459 \ (\text{MNm}^{-3/2})^{-2} \tag{14}$$

The resulting fracture toughness envelope is seen to agree well, in the first quadrant, with the experimental data in Figure 2. It should be noted that the data were obtained from the artificially introduced cracks which are normal to σ_2 [3].

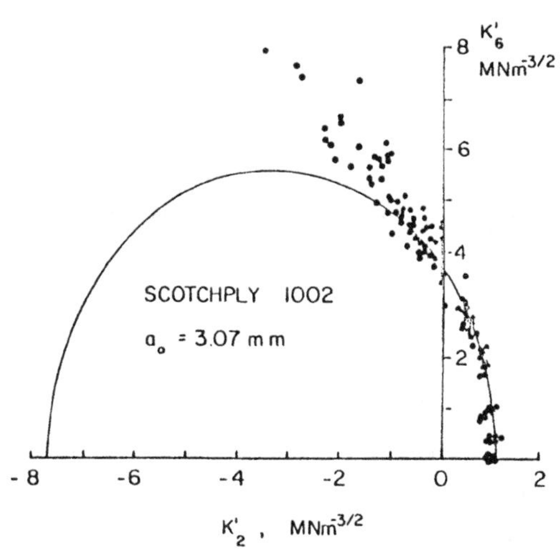

Fig. (2) - Interaction between k_2' and k_6'

The Mode II fracture toughness increases as transverse compression is applied because of the friction between the crack surfaces [9]. This increase is correctly, although lacking in quantitative accuracy, predicted by the polynomial strength criterion, but not by the other criteria. Also, the coupling between k_2' and k_6' in the first quadrant can be described neither by the maximum stress criterion nor by the maximum strain criterion.

Thus, it can be concluded that the polynomial strength criterion agrees, when the transverse stress is tensile, with the basic characteristics of the available fracture toughness envelope. As the transverse compression increases, however, the failure will, in general, start from an inherent crack whose surfaces are inclined to the applied stress. How this critical angle of inclination depends on the applied stresses should be known if the entire failure envelope is to be related to the fracture toughness. By way of illustrating this dependency, we discuss the energy release rate approach in the following subsection.

B. Application of Energy Release Rate Approach

Since the crack extension in unidirectional composites is at least macroscopically self-similar, the energy release rate can be calculated easily. That is, for a crack parallel to the fibers, the total energy release rate becomes [10]

$$G = G_1 + G_2 + G_3 \tag{15}$$

Here, the energy release rates G_1, G_2, G_3 resulting from k_1, k_2 and k_3, respectively, are given by

$$G_1 = \pi B_1 k_1^2, \qquad B_1 = \alpha \frac{1}{E_T} \tag{16}$$

$$G_2 = \pi B_2 k_2^2, \qquad B_2 = B_1 \left(\frac{E_T}{E_L}\right)^{1/2} \tag{17}$$

$$G_3 = \pi B_3 k_3^2, \qquad B_3 = \frac{1}{2(G_{TT} G_{LT})^{1/2}} \tag{18}$$

Note that k_3 is the Mode III stress intensity factor. The orthotropy correction factor α is defined by

$$\alpha = \frac{1}{\sqrt{2}} \left[\left(\frac{E_T}{E_L}\right)^{1/2} + \frac{E_T/G_{LT} - 2\nu_{TL}}{2}\right]^{1/2} \tag{19}$$

where E_L is the longitudinal modulus parallel to the fibers, E_T the transverse modulus, ν_{TL} the minor Poisson's ratio, G_{LT} the longitudinal shear modulus, and G_{TT} the transverse shear modulus.

We now consider a crack of length $2a_0$ which is not normal to σ_2 but inclined by an angle θ, Figure 1. In terms of the applied stresses σ_2 and σ_6, the stress intensity factors for this crack are expressed as follows:

$$k_1 = \sigma_2 \sqrt{a_0} \cos^2\theta, \quad \sigma_2 \geq 0 \tag{20}$$

$$k_1 = 0 \qquad , \quad \sigma_2 < 0 \tag{21}$$

$$k_2 = \sigma_6 \sqrt{a_0} \cos\theta \tag{22}$$

$$k_3 = \sigma_2 \sqrt{a_0} \sin\theta \cos\theta \tag{23}$$

Note that in the calculation of k_3 no friction has been taken into account.

When σ_2 is positive, G becomes stationary, i.e., $dG/d\theta = 0$, at the following angles:

$$\theta = 0 \text{ or } \pi/2 \tag{24}$$

$$\theta = \sin^{-1}\left[\frac{1}{2} - \frac{B_1+B_2(\sigma_6/\sigma_2)^2}{2(B_3-B_1)}\right]^{1/2} \tag{25}$$

On the other hand, if σ_2 is negative, then $G_1 = 0$ and the angle given by equation (25) is changed to

$$\theta = \sin^{-1}\left[\frac{1}{2} - \frac{B_2}{2B_3}\left(\frac{\sigma_6}{\sigma_2}\right)^2\right]^{1/2} \tag{26}$$

Elastic properties of commonly used composites are listed in Table 2 together with the calculated values of B's. For those composites, we have

$$2B_1 > B_3 \tag{27}$$

TABLE 2 - ELASTIC PROPERTIES

	Gl/Ep Scotchply 1002	Gr/Ep T300/5208	Gr/Ep AS/3501	B/Ep Narmco 5505
E_L(GPa)	34.47	181.33	137.69	207.53
E_T(GPa)	11.49	10.34	9.65	18.82
G_{LT}(GPa)	4.86	7.17	4.21	5.24
ν_{LT}	0.05	0.28	0.30	0.21
$G_{TT}^{(a)}$(GPa)	4.32	3.89	3.63	7.08
$B_1(\text{MPa})^{-1}$	81.24	66.43	86.35	54.16
$B_2(\text{MPa})^{-1}$	46.90	15.86	22.86	16.31
$B_3(\text{MPa})^{-1}$	109.12	94.69	127.88	82.11
Reference	[3]	[11]	[1]	[13]

(a) $G_{TT} = E_T/[2(1+\nu_{TT})]$, $\nu_{TT} = 0.33$

and therefore equation (25) cannot be satisfied. Thus, when σ_2 is tensile, the maximum energy release rate occurs at θ=0 independently of the shear stress σ_6 and

$$G_{max} = G\big|_{\theta=0} = \pi a_0(B_1\sigma_2^2 + B_2\sigma_6^2) \tag{28}$$

However, when σ_2 is compressive, the critical crack orientation depends on the stress ratio σ_6/σ_2. That is, if $|\sigma_6/\sigma_2| \geq (B_3/B_2)^{1/2}$, then

$$G_{max} = G|_{\theta=0} = \pi a_0 B_2 \sigma_6^2 \tag{29}$$

On the other hand, if $|\sigma_6/\sigma_2| < (B_3/B_2)^{1/2}$, then

$$G_{max} = G|_{\theta=\theta_0} = \pi a_0 B_2 \sigma_6^2 \left[\frac{1}{2} + \frac{1}{4}\frac{B_3}{B_2}\left(\frac{\sigma_2}{\sigma_6}\right)^2 + \frac{1}{4}\frac{B_2}{B_3}\left(\frac{\sigma_6}{\sigma_2}\right)^2\right] \tag{30}$$

where θ_0 satisfies equation (26). Note that, when $\sigma_6 = 0$, i.e., pure compressio[n] one obtains

$$\theta_0 = \pi/4 \tag{31}$$

and

$$G_{max} = \pi a_0 B_3 \sigma_2^2/4 \tag{32}$$

If the energy release rate is applicable, the uniaxial strengths are related to the critical energy release rate G_c by

$$X_6 = \left(\frac{G_c}{\pi a_0 B_2}\right)^{1/2} \tag{33}$$

$$\frac{X_2}{X_6} = \left(\frac{B_2}{B_1}\right)^{1/2} \tag{34}$$

$$\frac{X_2'}{X_6} = 2\left(\frac{B_2}{B_3}\right)^{1/2} \tag{35}$$

where X_2' is the compressive transverse strength. The combined-stress failure cr[i]teria then follows from equations (28)-(30). That is, when $\sigma_2 \geq 0$, equation (28) reduces to

$$\left(\frac{\sigma_2}{X_2}\right)^2 + \left(\frac{\sigma_6}{X_6}\right)^2 = 1 \tag{36}$$

When $\sigma_2 < 0$, on the other hand, two different criteria follow depending on the ratio, $|\sigma_6/\sigma_2|$. If $|\sigma_6/\sigma_2| > (B_3/B_2)^{1/2}$, then the shear strength is independent of σ_2, i.e.,

$$\sigma_6 = X_6 \tag{37}$$

However, if $|\sigma_6/\sigma_2| < (B_3/B_2)^{1/2}$, then the failure criterion is given by

$$\left(\frac{\sigma_2}{X_2'}\right)^2 + \left(\frac{\sigma_6}{X_6}\right)^2 \left[\frac{1}{2} + \frac{1}{16}\left(\frac{X_2'}{X_6}\right)^2\left(\frac{\sigma_6}{\sigma_2}\right)^2\right] = 1 \tag{38}$$

Table 3 lists the appropriate strength ratios for various composites, both experimental and predicted. Insofar as the ratio X_2/X_6 is concerned, the

TABLE 3 - STRENGTH RATIOS

Material	X_2/X_6		X_2'/X_6		X_6	Test Method for Shear
	Exp	Theory	Exp	Theory	(MPa)	
Gl/Ep Scotchply 1002	0.32	0.76	2.22	1.31	66.2	-
Gr/Ep T300/5208	0.60	0.49	3.71	0.82	67.6	[± 45] Laminate
Gr/Ep AS/3501-5	0.73	0.52	3.84	0.85	52.1	Rail Shear
B/Ep Narmco 5505	0.98	0.55	4.72	0.89	66.5	[± 45] Laminate

energy release rate approach seems to yield a fairly good correlation with the data for both graphite/epoxy composites. However, the correlation is very poor for the ratio X_2'/X_6 irrespectively of the type of material.

In any case, the energy release rate implies a strong interaction between σ_2 and σ_6. Aside from the applicability of the energy release rate approach, there are many assumptions, such as the noninteracting through-the-thickness cracks, the Mode III fracture in compression, etc., that have to be validated in order to account for the discrepancy between the theory and the data. Still, what is interesting is that the dependence on the inherent crack length of the strength under combined stresses simply follows from the failure envelope if σ_2 and σ_6 are replaced by $\sigma_2\sqrt{a_0}$ and $\sigma_6\sqrt{a_0}$, respectively. Thus, it is plausible to assume that equations (1) through (7) are valid in the entire σ_2-σ_6 space and that the A's are independent of a_0.

C. Derivation of Strength Distribution from Crack Length

Now that the relationship between the strength and the inherent crack length is known, we can derive a strength distribution under combined loading. To this end, we further note that the critical crack orientation is independent of the crack length, as shown in the preceding subsection. Thus, we can assume that the scatter in strength is solely due to the variation in the crack length.

Suppose the crack half length has the cumulative distribution

$$P(a) = \exp\left[-\left(\frac{\hat{a}}{a}\right)^{\alpha/2}\right] \tag{39}$$

where $\hat{a}$ and α are the characterization parameters. Since the A's in equations (5)-(7) can be considered as deterministic material constants, we obtain the following distributions of F's"

$$P(F_2) = \exp[-(\frac{\hat{F}_2}{F_2})^{\alpha}] \tag{40}$$

$$P(F_{22}) = \exp[-(\frac{\hat{F}_{22}}{F_{22}})^{\alpha/2}] \tag{41}$$

$$P(F_{66}) = \exp[-(\frac{\hat{F}_{66}}{F_{66}})^{\alpha/2}] \tag{42}$$

where $\hat{F}$'s correspond to $\hat{a}$ through equations (5)-(7).

We now introduce the concept of a strength vector [9]. Suppose failure occurs at the stresses σ_2 and σ_6. The pair (σ_2,σ_6) then defines, in the stress space, a vector which is called the strength vector. Since the failure envelope represents a set of pairs of failure stresses, any point on the failure envelope has a strength vector associated with it. From equations (1) and (5)-(7), the magnitude σ of this strength vector (σ_2,σ_6), called the combined strength hereafter, is shown to be proportional to $1/\sqrt{a}$:

$$\sigma = \frac{[(A_2^2+4A_{22})\cos^2\phi+4A_{66}\sin^2\phi]^{1/2}-A_2\cos\phi}{2(A_{22}\cos^2\phi+A_{66}\sin^2\phi)} \frac{1}{\sqrt{a}} \tag{43}$$

where

$$\phi = \tan^{-1}(\sigma_6/\sigma_2) \tag{44}$$

Then, introducing the characteristic combined strength,

$$\hat{\sigma} = \sigma\Big|_{a=\hat{a}} \tag{45}$$

we can rewrite equation (43) as

$$\frac{\sigma}{\hat{\sigma}} = (\frac{\hat{a}}{a})^{1/2} \tag{46}$$

Therefore, the probability of the combined strength being greater than σ, $R(\sigma)$, is expressed in the following simple form:

$$R(\sigma) = \exp[-(\frac{\sigma}{\hat{\sigma}})^{\alpha}] \tag{47}$$

Note that the characteristic combined strength $\hat{\sigma}$ varies with the angle ϕ. However, the distribution of the ratio $\sigma/\hat{\sigma}$ remains independent of ϕ. Also,

$\sigma/\hat{\sigma}$ has the same distribution as do the uniaxial strengths. In fact, equation (47) reduces to

$$R(X_2) = \exp[-(\frac{X_2}{\hat{X}_2})^\alpha] \tag{48}$$

$$R(X_2') = \exp[-(\frac{X_2'}{\hat{X}_2'})^\alpha] \tag{49}$$

for the tensile and compressive transverse strengths, respectively, and to

$$R(X_6) = \exp[-(\frac{X_6}{\hat{X}_6})^\alpha] \tag{50}$$

for the shear strength.

As is clear by now, the shape parameter α is independent of the mode of loading. Table 4 lists values of α for some composites available in the litera-

TABLE 4 - VALUES OF α

Material	Trans. Tension	Trans. Compression	Shear	No. of Specimens	Reference
GR/Ep AS/3501-5	21	26.8	12.6 (a)	10	[12]
AS/3501-5	16.9	-	20.9 (b)	5-6	[14]
AS/3501-5A	11.3	-	14.6 (c)	13-14	[15]
HTS/2256	9.4	5.5	7.4 (d)	9-15	[16]
B/Ep Narmco 5505	17.5	13.2	12.0 (d)	7-15	[13]

(a) Rail Shear
(b) 10-degree off-axis tension
(c) Short beam shear
(d) [±45] laminate tension

ture. No definitive relationship is seen between α and the mode of loading although some of the inconsistencies may be due to the difference in the test methods employed. Therefore, in the following section, we assume α is independent of the type of loading and adopt the procedure based on σ to characterize the combined strength data.

ANALYSIS OF COMBINED LOADING DATA

A. Characterization of Strength

Strength data were obtained by testing off-axis tubes in combined axial and torsional loadings. The experimental procedures are described in [17].

The strengths listed in Table 5 are fit by the equation

$$F_{02}\sigma_2 + F_{022}\sigma_2^2 + F_{066}\sigma_6^2 = 1 \tag{51}$$

where F_0's are determined by the least squares method. That is, equation (51) for each set of data $(\sigma_2^{(i)}, \sigma_6^{(i)})$, is rearranged as follows

$$\begin{bmatrix} \sigma_2^{(1)} & \sigma_2^{(1)2} & \sigma_6^{(1)2} \\ \sigma_2^{(2)} & \sigma_2^{(2)2} & \sigma_6^{(2)2} \\ - & - & - \\ \sigma_2^{(n)} & \sigma_2^{(n)2} & \sigma_6^{(n)2} \end{bmatrix} \begin{bmatrix} F_{02} \\ F_{022} \\ F_{066} \\ \\ \end{bmatrix} = \begin{bmatrix} 1 \\ 1 \\ - \\ 1 \end{bmatrix} \tag{52}$$

TABLE 5 - STRESS COMPONENTS AT FAILURE

Specimen Number	σ_1 MPa	σ_2 MPa	σ_6 MPa	Specimen Number	σ_1 MPa	σ_2 MPa	σ_3 MPa
0-1	375.4	0	44.38	-45-2	97.22	-97.22	0
0-2	0	0	48.04	-45-3	155.4	-85.81	34.78
0-3	-375.4	0	-39.02	-60-1	55.60	11.05	51.34
0-4	0	0	-48.60	60-2	86.56	-34.81	-64.91
15-1	-259.4	-18.62	59.50	-60-3	116.1	-78.02	78.03
-15-1	139.2	9.99	37.30	90-1	0	12.82	0
-15-2	135.1	- 8.03	3.12	90-2	0	23.67	0
-15-3	323.2	-58.68	-66.20	90-2'(a)	0	20.26	0
15-3	11.49	4.26	- 9.50	90-3	0	0	-74.95
-30.1	118.4	12.78	45.26	90-3'(a)	0	0	-79.36
-30-2	193.8	-85.15	-17.79	90-4	0	8.04	0
-30-3	- 11.64	11.61	6.69	90-5	0	8.94	0
-45-1	- 13.79	25.03	5.62	90-6	0	0	70.12

(a) Retested after failure.

Using an abridged notation for equation (52), i.e.,

$$[\sigma]\{F_0\} = \{1\} \tag{53}$$

we can determine F_0's from the following equation:

$$\{F_0\} = ([\sigma]^T[\sigma])^{-1}[\sigma]^T\{1\} \tag{54}$$

where the subscripts T and -1 stand for the transpose and inversion, respectively. The results are

$$F_{02} = 3.376 \times 10^{-2}\ (\text{MPa})^{-1} \tag{55}$$

$$F_{022} = 4.721 \times 10^{-4}\ (\text{MPa})^{-2} \tag{56}$$

$$F_{066} = 2.384 \times 10^{-4}\ (\text{MPa})^{-2} \tag{57}$$

Next, in order to determine $\hat{F}$'s, a nondimensional strength parameter s is defined by

$$s = \sigma/\sigma_0 \tag{58}$$

where σ is the actual combined strength and σ_0 is the combined strength predicted by the failure envelope, equation (51). That is, if (σ_2,σ_6) is a pair of stress components at failure, then

$$\sigma = (\sigma_2^2+\sigma_6^2)^{1/2} \tag{59}$$

$$\sigma_0 = (\sigma_{02}^2+\sigma_{06}^2)^{1/2} \tag{60}$$

where $(\sigma_{02},\sigma_{06})$ satisfies equation (51) and

$$\frac{\sigma_2}{\sigma_{02}} = \frac{\sigma_6}{\sigma_{06}} = s \tag{61}$$

Recalling that σ has a Weibull distribution, equation (47), we deduce

$$R(s) = \exp[-(\frac{s}{\hat{s}})^{\alpha}],\quad \hat{s} = \frac{\hat{\sigma}}{\sigma_0} \tag{62}$$

The experimental values of s are plotted in Figure 3 according to the linearized format of equation (62);

$$\ell n(-\ell nR) = \alpha \ell ns - \alpha \ell n\hat{s} \tag{63}$$

The median rank was used for R:

$$R = 1 - \frac{j-0.3}{N+0.4} \tag{64}$$

where j is the ordinal number of strength and N the total number of data.

The data represented by the closed symbols are seen to deviate much from the trend exhibited by the open symbols. Two of those four data were obtained from badly misaligned 15-degree off-axis tubes; therefore, these can be regarded as invalid data. The remaining two did not exhibit any apparent anomalies. How-

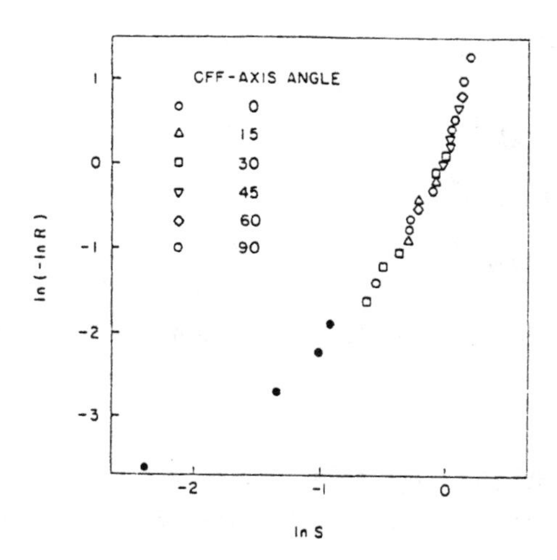

Fig. (3) - Linearized plot of s

ever, it is still possible that a slight misalignment may have caused premature failure since these two are 90-degree tubes.

Therefore, the aforementioned four points are discarded and new F_0's are calculated with the results

$$F_{02} = 3.126 \times 10^{-2} \ (\text{MPa})^{-1} \tag{65}$$

$$F_{022} = 4.428 \times 10^{-4} \ (\text{MPa})^{-2} \tag{66}$$

$$F_{066} = 2.391 \times 10^{-4} \ (\text{MPa})^{-2} \tag{67}$$

The distribution of s based on the newly determined F_0's is shown in Figure 4. The best-fit curve results from •

$$\alpha = 4.745, \ \hat{s} = 0.9695 \tag{68}$$

and the correlation coefficient is 0.9922. Furthermore, the corresponding $\hat{F}$'s are obtained from the following equations:

$$\frac{F_{02}}{\hat{F}_2} = \hat{s} \tag{69}$$

$$\frac{F_{022}}{\hat{F}_{22}} = \frac{F_{066}}{\hat{F}_{66}} = \hat{s}^2 \tag{70}$$

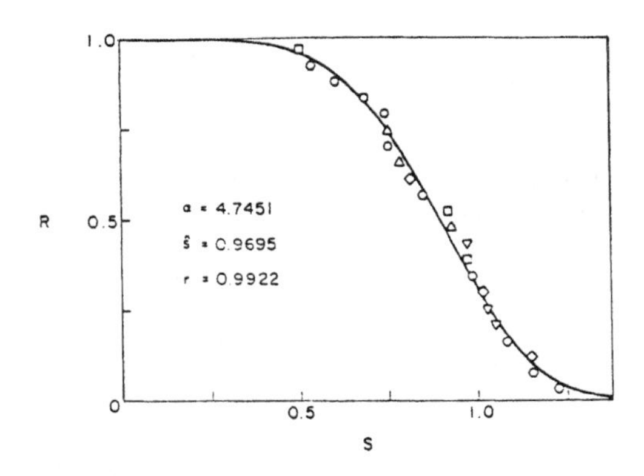

Fig. (4) - Distribution of s based on censored data

Thus, the matrix/interface-controlled strength of the composite studied is characterized by the distribution of σ, equation (47), and by the $\hat{F}$'s:

$$\hat{F}_2 = 3.224 \times 10^{-2} \ (\text{MPa})^{-1} \tag{71}$$

$$\hat{F}_{22} = 4.567 \times 10^{-4} \ (\text{MPa})^{-2} \tag{72}$$

$$\hat{F}_{66} = 2.466 \times 10^{-4} \ (\text{MPa})^{-2} \tag{73}$$

Now, given any applied stresses, the survival probability can be calculated.

As an example, suppose the following stresses are applied to the previously characterized composite:

$$\sigma_2 = -40 \text{ MPa}, \ \sigma_6 = 60 \text{ MPa} \tag{74}$$

The stress ratio $\sigma/\hat{\sigma}$ follows from equations (71)-(73) as

$$\frac{\sigma}{\hat{\sigma}} = 0.779 \tag{75}$$

Thus, the probability of survival, R, is given by

$$R = \exp[-\ (0.779)^{4.745}] = 0.737 \tag{76}$$

B. Size Effect

To see if the weakest link theory based on s can account for the size effect, some data were obtained from 90-degree tubes having gage length reduced to 3/8 the original gage length. Described in the following are the results, both analytical and experimental.

Following the same procedure as, e.g., in [18], but using s, we obtain the ratio of the characteristic strength parameters as

$$\frac{\hat{s}'}{\hat{s}} = \left(\frac{V}{V'}\right)^{1/\alpha} = \left(\frac{8}{3}\right)^{1/4.745} = 1.2296 \tag{77}$$

Here, the primed quantities are associated with the short tubes and V is the volume. The $\hat{F}$'s for the short tubes are then

$$\hat{F}_{2'}' = \hat{F}_2\hat{s}/\hat{s}' = 2.542 \times 10^{-2}\ (\text{MPa})^{-1} \tag{78}$$

$$\hat{F}_{22'} = \hat{F}_{22}\hat{s}/\hat{s}' = 2.939 \times 10^{-4}\ (\text{MPa})^{-2} \tag{79}$$

$$\hat{F}_{66'} = \hat{F}_{66}\hat{s}/\hat{s}' = 1.581 \times 10^{-4}\ (\text{MPa})^{-2} \tag{80}$$

The resulting failure envelope is compared with the experimental results in Figure 5. The broken curve is the best fit of the data. Although any definitive conclusions require more data, the comparison seems quite encouraging.

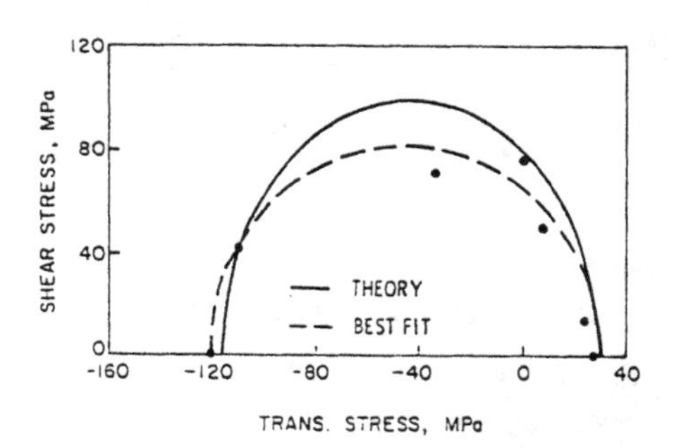

Fig. (5) - Predicted failure envelope and experimental data for short tubes

CONCLUSIONS

We have presented a method of characterizing the matrix/interface-controlled strength, including scatter, of unidirectional composites under combined transverse normal and shear loading. It was assumed that failure initiates at inherent cracks parallel to the fibers and that the scatter in strength is a manifestation of nonuniform crack length having a certain distribution. The energy release rate approach was discussed as a guide for a possible relationship between the combined strength and the inherent crack length.

The proposed failure criterion is a second-order polynomial. The scatter is described by the strength vector whose magnitude has a Weibull distribution. The method holds promise as a means of accounting for the size effect under combined loading.

In the energy release rate approach, actual defects were replaced by the noninteracting through-the-thickness cracks and the critical crack orientation was assumed to depend on the applied stresses in a deterministic manner. The relaxation of the first assumption calls for a three-dimensional stress analysis and the second assumption can be alleviated by employing a statistical approach, as was done for homogeneous materials in [19]. However, a major obstacle to the establishment of a strength-fracture toughness relationship seems to be the lack of an appropriate fracture criterion under combined state of stresses.

A similar approach was used in [20] to analyze the failure in the presence of stress gradients. The polynomial used included the longitudinal stress also, thus allowing a full interaction among the stress components. This approach will be applicable where the scatter in strength is independent of the type of loading.

ACKNOWLEDGEMENT

This paper is based on an Air Force Materials Laboratory technical report AFML-TR-78-85.

REFERENCES

[1] Wu, E. M., "Phenomenological Anisotropic Failure Criterion", Composite Materials, Vol. 2, G. P. Sendeckyj, Editor, Academic Press, New York, p. 353, 1974.

[2] Sandhu, R. S., "A Survey of Failure Theories of Isotropic and Anisotropic Materials", Air Force Flight Dynamics Laboratory, Wright-Patterson Air Force Base, Ohio 45433, AFFDL-TR-72-71, January 1972.

[3] Wu, E. M. and Reuter, R. C., Jr., "Crack Extension in Fiberglass Reinforced Plastics", Univ. of Illinois, T&AM Report No. 275, February 1965.

[4] Corten, H. T., "Influence of Fracture Toughness and Flaws on the Interlaminar Shear Strength of Fibrous Composites", Fundamental Aspects of Fiber Reinforced Plastic Composites, R. T. Schwartz and H. S. Schwartz, Editors, Interscience Pub., New York, p. 89, 1968.

[5] Greszczuk, L. B., "Micromechanics Failure Criteria for Composites", McDonnell Douglas Astronautics Co., Contract No. N00019-72-0221, May 1973.

[6] Vannucci, R. D., "Effect of Processing Parameters on Autoclaved PMR Polyimide Composites", Proceedings of the 9th National SAMPE Technical Conference, Vol. 9, SAMPE, p. 177, October 1977.

[7] Lauraitis, K., "Tensile Strength of Off-Axis Unidirectional Composites", Univ. of Illinois, T&AM Report No. 344, August 1971.

[8] Sih, G. C. and Chen, E. P., "Fracture Analysis of Unidirectional Composites", Journal of Composite Materials, Vol. 7, p. 230, 1973.

[9] Wu, E. M., "Strength and Fracture of Composites", Composite Materials, Vol. 5, L. J. Broutman, Editor, Academic Press, New York, p. 191, 1974.

[10] Sih, G. C. and Liebowitz, H., "Mathematical Theories of Brittle Fracture", Fracture: An Advanced Treatise, Vol. 2, H. Liebowitz, Editor, Academic Press, New York, p. 68, 1968.

[11] Hofer, K. E., Jr., Larsen, D. and Humphreys, V. E., "Development of Engineering Data on the Mechanical and Physical Properties of Advanced Composite Materials, Air Force Materials Laboratory, Wright-Patterson Air Force Base, Ohio 45433, AFML-TR-74-266, February 1975.

[12] Verette, R. M. and Labor, J. D., "Structural Criteria for Advanced Composites", Vol. II, Air Force Flight Dynamics Laboratory, Wright-Patterson Air Force Base, Ohio 45433, AFFDL-TR-76-142, Vol. II, March 1977.

[13] Shockey, P. D., Hofer, K. E. and Wright, D. W., "Structural Airframe Application of Advanced Composite Materials", Vol. IV, Air Force Materials Laboratory, Wright-Patterson Air Force Base, Ohio 45433, AFML-TR-69-101, Vol. IV, October 1969.

[14] Sendeckyj, G. P., presented at Mechanics of Composites Review, Dayton, Ohio, October 1977.

[15] Burroughs, B., Konishi, D. and Nadler, M., "Advanced Composites Serviceability Program", Progress Report No. NA-76-783-2, Rockwell International, Contract No. F33615-C-5344, April 1977.

[16] Grimes, G. C. et al, "A Study of the Stress-Strain Behavior of Graphite Fiber Composites to Assess the Stress Levels at which Significant Damage Occurs", Air Force Materials Laboratory, Wright-Patterson Air Force Base, Ohio 45433, AFML-TR-73-311, January 1974.

[17] Hahn, H. T. and Erikson, J., "Characterization of Composite Laminates Using Tubular Specimens", Air Force Materials Laboratory, Wright-Patterson Air Force Base, Ohio 45433, AFML-TR-77-144, August 1977.

[18] Knight, M. and Hahn, H. T., "Strength and Elastic Modulus of a Randomly Distributed Short Fiber Composite", Journal of Composite Materials, Vol. 9, p. 77, 1975.

[19] Batdorf, S. B. and Crose, J. G., "A Statistical Theory for the Fracture of Brittle Structures Subjected to Nonuniform Polyaxial Stresses", Journal of Applied Mechanics, Vol. 41, p. 459, 1974.

[20] Wu, E. M., "Failure Analysis of Composites with Stress Gradients", UCRL-80909, Lawrence Livermore Laboratory, August 1978.

DETERMINATION OF FRACTURE TOUGHNESS OF UNIDIRECTIONALLY FIBER-REINFORCED COMPOSITES

S. Parhizgar

University of Wisconsin
Platteville, Wisconsin 53818

L. W. Zachary

Iowa State University
Ames, Iowa 50010

and

C. T. Sun

University of Florida
Gainesville, Florida 32611

ABSTRACT

The fracture phenomena in orthotropic composite plates are examined and compared with the fracture in isotropic plates. The extent to which the principles of linear fracture mechanics are applicable to orthotropic plates is studied. It is shown that fracture toughness of unidirectional composites is independent of crack length but dependent on crack-fiber orientation. For experimental verification of the above principles, unidirectional glass epoxy material (Scotchply 1002) was used. The fracture toughness of Scotchply 1002 for different crack-fiber orientations is obtained by utilizing Solid Sap finite element program and compact tension specimens. An empirical formula relating the fracture toughness of the material for different crack-fiber orientation is found.

INTRODUCTION

The increasing use of orthotropic composite materials in aerospace technology requires a high degree of reliability and has created a great need for complete understanding of fracture phenomena in these materials. The proper way to begin a study of fracture in orthotropic materials is to compare their fracture with the fracture of isotropic materials. It should be determined whether or not "The Principles of Linear Fracture Mechanics" which are accepted universally as a reliable tool in the investigation of isotropic fractures are applicable to orthotropic materials. If these principles do not hold or only partially hold, a new or modified set of principles must be developed. The purpose of this paper is to fulfill this objective and to determine fracture toughness of unidirectionally fiber-reinforced composites.

REVIEW OF THE PRINCIPLES OF LINEAR FRACTURE MECHANICS

According to the principles of linear fracture mechanics, for an infinite plate with a central crack, Figure 1, the following can be stated:

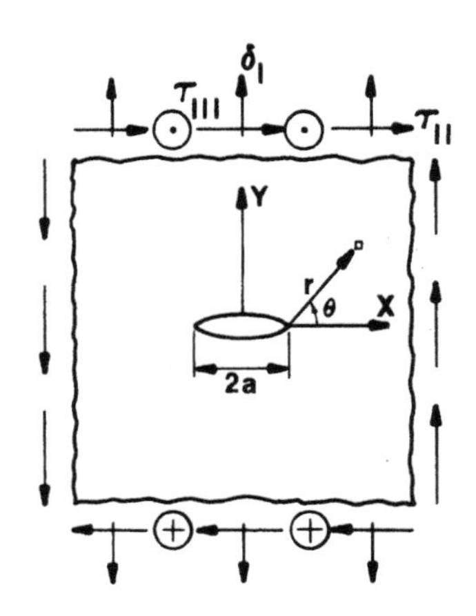

Fig. (1) - Plate containing central crack

a. The crack always advances along the original crack direction.

b. Based on loading conditions of Figure 1, crack tip displacement can be separated into three different modes: crack-opening mode or mode I; edge-slidin mode or mode II; and crack-tearing mode or mode III, Figure 2. Mode I correspon to σ_I, mode II corresponds to τ_{II} and mode III corresponds to τ_{III}. The superpo sition of these three modes gives the most general case of crack tip displacemen

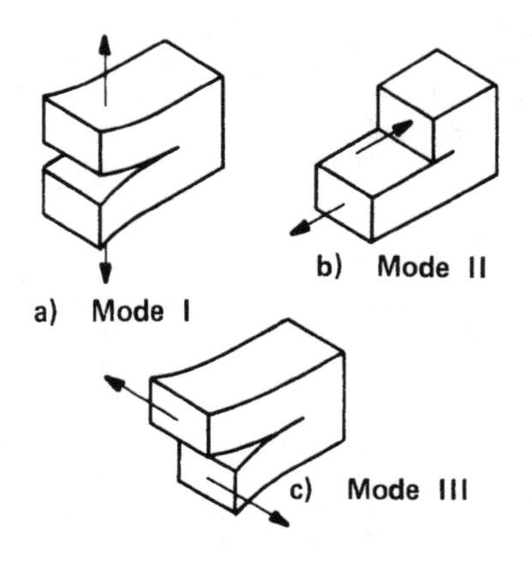

Fig. (2) - Crack tip failure modes

c. The crack tip stress and displacement equations for the above modes are given by the so-called Westergaard's equations [2]. The values of K_I, K_{II} and K_{III} in these equations are:

$$K_I = \sigma_I\sqrt{\pi a},\ K_{II} = \tau_{II}\sqrt{\pi a} \text{ and } K_{III} = \tau_{III}\sqrt{\pi a}$$

As can be seen from the Westergaard's equations, the crack tip stress distributions are independent of material orientations and material properties. Along any radial direction, see Figure 1, they are only functions of $1/\sqrt{r}$. K_I, K_{II} and K_{III} represent the intensity of such stress distributions and are called stress intensity factors. The critical values of stress intensity factors, the values corresponding to the start of crack growth, are called fracture toughness of mode I to mode III. They are shown as:

$$K_{IC} = \sigma_{Ic}\sqrt{\pi a},\ K_{IIC} = \tau_{IIc}\sqrt{\pi a},\ K_{IIIC} = \tau_{IIIc}\sqrt{\pi a}$$

where σ_{Ic}, τ_{IIc} and τ_{IIIc} are stresses at infinity corresponding to the start of crack growth and 2a is the crack length. Experimentally, it is proven that K_{IC}, K_{IIC} and K_{IIIC} are constant material properties. Once the fracture toughness of the material is known, the conditions at which the cracks start to grow can be predicted.

To understand the stress intensity factor and fracture toughness better, it must be noted that the stress intensity factor relates to fracture toughness in the same way that stress relates to strength. Strength is a material property whereas stress is not. In the same way, fracture toughness is a material property while stress intensity factor is not. Physically, fracture toughness indicates the ability of a material to resist crack propagation.

FRACTURE MECHANICS OF ORTHOTROPIC MATERIALS

Fracture phenomena in orthotropic materials are much more complex than fracture in isotropic materials. The crack tip stress and displacement equations are shown in Appendix 1. Careful study of these equations reveals the degree of complexity of orthotropic fracture and its significant difference from isotropic fracture. Because of these differences and because the principles of linear fracture mechanics are based on the fracture of isotropic materials, great care is needed in the application of these principles to orthotropic materials. When the principles are not applicable, new principles must be derived and checked with experimental data. To study the extent of application of linear fracture mechanics to orthotropic materials and the possible determination of constants K_{IC}, K_{IIC} and K_{IIIC} which characterize the fracture toughness, it is not only necessary to examine the stresses at the crack tip but also to consider the crack tip displacements, crack tip failure modes and the direction of crack growth as well. Only when all of these satisfy the principles of linear fracture mechanics can it be said that the principles are applicable to orthotropic materials. The major deviations of orthotropic fracture from the principles of linear fracture mechanics follow:

1. In orthotropic materials containing a crack, in general, the crack does not grow along the original crack orientation; i.e., in unidirectional composites

the crack always grows along the fiber orientation. This fact alters the principles of linear fracture mechanics, which assume in advance that the crack always grows in the direction of original crack.

2. After careful study of the crack tip displacements in orthotropic materials, (equations (2), (4) and (6), Appendix 1), it can be seen that these materials under the application of a single load mode, for example, symmetric loading of Figure 2a, in general, will have the crack tip displacements of mixed mode mode I and mode II (see Figure 2a and 2b). This also differs from the principles of linear fracture mechanics which are based on the separation of modes according to Figure 2.

3. Investigation of crack tip stresses for orthotropic materials, equations (1), (3) and (5), reveals that these stresses are functions of material complex parameters μ's (see Appendix 1) which, in turn, are functions of material properties and orientations. This also differs from the principles of linear fracture mechanics which were developed on the basis that crack tip stress distributions are independent of material properties and orientations.

Consideration of the above three facts indicates that, in general, the principles of linear fracture mechanics do not hold for orthotropic materials, and the possibility of finding three materials constants, K_{IC}, K_{IIC} and K_{IIIC}, to represent the fracture toughness of an orthotropic material does not exist. Thus each case of orthotropic materials must be considered separately, and the application of linear fracture mechanics to each individual case must be examined.

Obviously, the scope of this work does not permit a full investigation of different orthotropic cases, and therefore it has been limited to the case of unidirectionally fiber-reinforced composites. In this case, fibers are all along the same direction, Figure 3, and cracks always grow along the fiber orientations

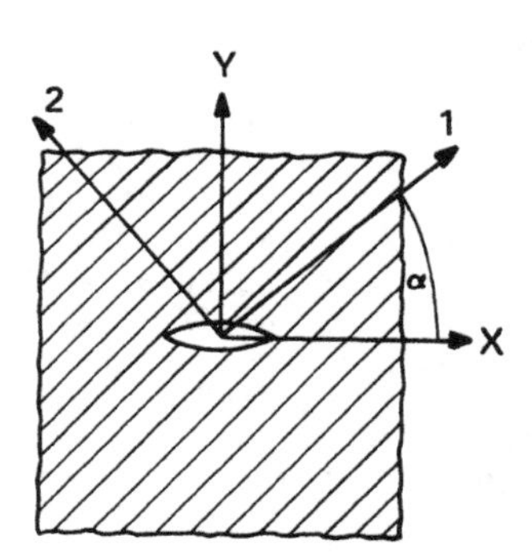

Fig. (3) - Composite plate with central crack

The two following different cases can be identified for unidirectional composite materials:

1. The special case of original crack along the fiber direction. (*In Figure* 3, the principal orthotropic axes, 1 and 2, coincident with the x and y axes, α=0). For this special case, all deviations from the principle of linear fracture mechanics explained above vanish, thus:

a. The crack grows along its original direction, which is the fiber direction. This is intuitively expected and also has been shown by our experiments.

b. Examination of crack tip displacement equations (2), (4) and (6) reveals that, for this case, the displacements are not mixed mode; i.e., for symmetric loading of Figure 2a, only the crack opening mode is present.

c. The material complex parameters μ's are constant for any fixed orientation of crack and fiber [6]. Therefore, the crack tip stress distribution equations (1), (3) and (5) are independent of material directional properties. Along any radial direction, see Figure 1, these stress distributions are only functions of $1/\sqrt{r}$. K_I, K_{II} and K_{III} represent the intensity of such stress distributions and are called stress intensity factors. The critical values of stress intensity factors, the values corresponding to the start of crack growth, are shown as K_{IC0}, K_{IIC0} and K_{IIIC0}. They are constant material properties. This fact has been verified experimentally and is shown in the experimental part of this work. The last indices in K_{IC0}, K_{IIC0} and K_{IIIC0} represent the crack fiber orientation $\alpha=0$.

Consideration of the above three facts indicates that the principles of linear fracture mechanics hold for the case of a crack along the fiber orientation. Most of the cracks and flaws created in the manufacturing processes are in the fiber direction. Furthermore, as is intuitively evident and has been shown by our experiments, this direction is the weakest and shows the least resistance against crack growth. The fracture toughness K_{IC0}, K_{IIC0} and K_{IIIC0} obtained for this special case has very significant value in design and is called the critical fracture toughness of materials.

2. The general case of fiber oriented at some angle α with respect to the crack, Figure 3. For this case we have:

a. The crack grows along the fiber orientation, which is not the original crack direction.

b. Equations (2), (4) and (6) show that crack tip displacements are mixed mode; i.e., for symmetric loading of Figure 2a, both crack opening and edge-sliding modes, Figure 2a and 2b are present.

c. As indicated before, for any fixed fiber orientation α, the material complex parameters μ's are constants and therefore crack tip stress distributions along any radial direction are only functions of $1/\sqrt{r}$. The intensities of these stress distributions at the start of crack growth are shown as $K_{IC\alpha}$, $K_{IIC\alpha}$ and $K_{IIIC\alpha}$ represent the crack fiber orientation α.

The crack tip stress distributions are the most important factors in fracture mechanics, and they are the major cause of crack advances. The crack tip displacements and the direction of crack growth are not equally important. Therefore, even though the cracks do not advance along their original direction and crack tip displacements cannot be separated, the intensity of the above stress distributions at the start of crack growth, $K_{IC\alpha}$, $K_{IIC\alpha}$ and $K_{IIIC\alpha}$, can still be considered as constant material properties fracture toughness for fixed fiber orientation α. $K_{IC\alpha}$, $K_{IIC\alpha}$ and $K_{IIIC\alpha}$ are functions of fiber orientation. In the experimental part of this work, it has been shown that for any fixed fiber orientation α, $K_{IC\alpha}$ is indeed a constant material property independent of crack length. $K_{IC\alpha}$ has been determined for several different angles α. These values then have been plotted against fiber orientation α. An empirical formula has been developed to relate $K_{IC\alpha}$, the fracture toughness when the crack and fiber are at an angle α, to K_{IC0}, the fracture toughness when the crack is along the fiber orientation. In the same way, $K_{IIC\alpha}$ and $K_{IIIC\alpha}$ can be related to K_{IIC0} and K_{IIIC0} respectively.

EXPERIMENTAL VERIFICATIONS

In order to establish full confidence in the principles discussed previously they should be verified experimentally. Obviously, the limited scope of this work does not permit the verification of all aspects of orthotropic fracture and different modes of failure. Therefore, only the experimental verification of the crack opening mode for unidirectionally fiber-reinforced composites will be established for fiber glass epoxy (Scotchply 1002, manufactured by 3M Company, Minneapolis, Minnesota). The remaining modes of failure can be verified by similar procedures.

To determine the fracture toughness of isotropic materials, the so-called "compliance calibration technique" is widely used. This method is based on energy release rates per unit of crack surface area A generated by the crack growth. According to this principle, the energy release rate is given by $G = P^2/2)(\partial\lambda/\partial A)$ where P is the applied load, λ is the compliance (that is, inverse spring constant), and $\partial\lambda/\partial A$ is the rate of compliance's change with respect to the crack surface area. To find $\partial\lambda/\partial A$ experimentally, numerous specimens with different crack lengths are needed and the procedure is very tedious and involved. Furthermore, the compliance calibration technique is not valid for the orthotropic materials when the crack does not advance along the original crack direction. This is because the compliance calibration technique is based on the principle of linear fracture mechanics, which assumes in advance that the crack grows along its original direction. Therefore, a method for determination of fracture toughness suitable to orthotropic materials must be developed.

As will be seen in this investigation, combinations of a single standard compact tension specimen and its finite element model can be used to determine the fracture toughness of materials. A brief outline of the procedures follows:

1. A compact tension specimen (or any other standard specimen) is made and the load corresponding to the start of the crack growth is determined.

2. A finite element model of the specimen is developed, and crack tip stress distributions corresponding to the load in step 1 are determined.

3. The above stress distributions are related to K_I, K_{II} and K_{III}, stress intensity factors, according to equations (1), (3) and (5). These equations are used to calculate K_I^r, K_{II}^r and K_{III}^r, stress intensity factors corresponding to the stress at a point distance r from the crack tip. These values are then plotted

versus r. The extrapolations of these curves to the K_I^r, K_{II}^r or K_{III}^r axes (r=0) are the fracture toughness of the material. Chan et al [4] first used this method to find the stress intensity factor of isotropic materials.

The advantages of this method are as follows:

a. Only one specimen is needed for determination of the fracture toughness.

b. This method can also be used for cases in which the crack advances at an angle of the original crack direction.

c. By avoiding the immediate crack tip area and relating stresses at a short distance from the crack tip to K, the effect of crack tip plastic zone on the results is minimized.

d. Since this method directly relates crack tip stresses to fracture toughness, it gives the most accurate results.

e. The results can be easily checked by repeating steps one to three above for a different crack length.

In this investigation, the above method has been used for the verification of the orthotropic fracture. Following is a discussion of the detailed procedures of this investigation:

1. Determination of the elastic constants of the material: The elastic constants of Scotchply 1002 were determined by attaching electrical resistance strain gauges to three standard tension test (dog bone) specimens with fiber orientations of 0, 45-45 and 90 degrees [5]. Two specimens were prepared for each fiber orientation. Plots of experimental values were made, and elastic constants E_1, E_2, ν_{12}, ν_{21} and G_{12} were obtained from the corresponding plot. The elastic constants were also calculated by the laws of mixture [5]. All of these values are shown in Table 1, including the values obtained from the manufacturer by direct contact. Since the results are very close to each other and in order to be able to compare this work with the work of other investigators, the elastic constants given by the manufacturer are used whenever possible. These values are given in the last column of Table 1.

TABLE 1 - ELASTIC CONSTANTS OF SCOTCHPLY 1002

Material's constants*	Experimental values	Laws of mixture values	Manufacturer's values	Values used in this research
E_1	40.7	34.5	40.0	40.0
E_2	8.1	11.0	8.3	8.3
G_{12}	4.1	3.2	-	4.1
ν_{12}	0.29	0.28	0.26	0.26
ν_{21}	0.07	0.06	-	0.05

*E and G in $(10)^9$ Pascal.

2. Preparation of specimens: Fiber glass-epoxy, Scotchply 1002 with stacking sequence of [0°/90°/13(0°)/90°/0°] was used. Unidirectional materials, 5 mm thick, were obtained by removing the first two and the last two layers of the 17 ply panel. Standard compact tension specimens with w = 50.8 mm, Figure 4, were

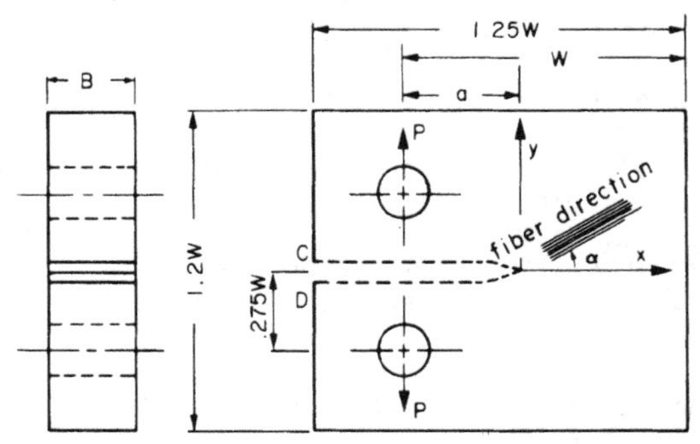

Fig. (4) - Compact tension specimen

made from unidirectional glass-epoxy material for different crack lengths (a/w = 0.3, 0.4, 0.5, 0.6, 0.7) and different fiber orientation (α = 0, 15, 30, 45, 60, 75 and 90 degrees). For each crack length and fiber orientation, at least two specimens were made. The crack tip was sharpened by using a jeweler's saw and razor blades. The specimens were then loaded by an electrohydraulic machine (MTS), and load versus crack-opening displacement, the displacement of point C with respect to point D in Figure 4, was recorded for all specimens. The maximu load, P_{max}, is considered as the load corresponding to the start of crack growth Figure 5 shows load versus crack-opening displacement for a specimen with a/w = 0.5 and α = 90 degrees. P_{max} for the other specimens are given in Tables 2-4.

3. Development of finite element models: Finite element models of all spec mens with different crack lengths had to be developed. When the cracks are alon the fiber orientation, α=0, the displacements of the specimens are symmetric wit respect to the x axis, see Figure 4. Therefore, the finite element model of a half specimen is suitable for this case. Figure 6 shows the computer plot for a half specimen model with quadrilateral elements and a/w = 0.5. The crack tip el ments are shown in Figure 7. Since the stress distributions along the x axis are required, the elements along this axis are well-refined. The area of the crack tip elements is only 0.0645 mm^2. This mesh has 544 elements and 602 nodes In the process of developing this mesh, the influences of shape and size of crac tip elements on the crack tip stress distributions were fully considered. Also, the mesh band widths were minimized to reduce the computer cost.

The finite element models for other crack lengths, a/w = 0.3, 0.4, 0.6 and 0.7, were obtained by developing two computer programs capable of generating nodes and elements of a new mesh with different crack lengths from the previous mesh [1]. The band width of the new mesh was still kept at a minimum.

When fiber orientation is at an angle with respect to the crack, the displac ment of the specimen is not symmetric, and thus a full specimen finite element

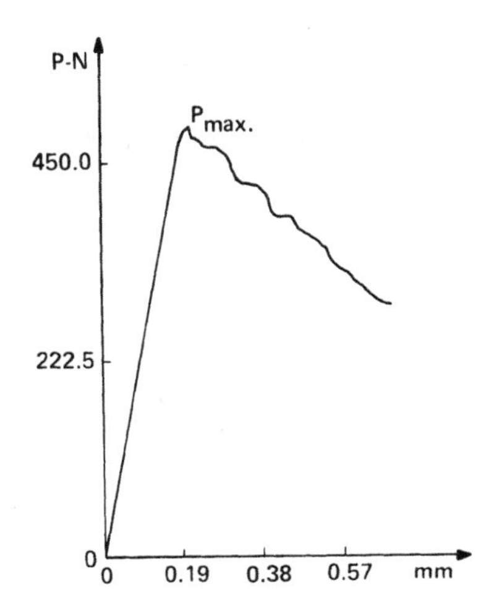

Fig. (5) - Load versus crack opening displacement for $\frac{a}{w}$ = 0.5 and α = 90°

TABLE 2 - FRACTURE TOUGHNESS FOR THE CASE OF CRACK ALONG THE FIBER DIRECTION

α	a/w	P_{max}(N)	$K_I B\sqrt{w}/P$	$K_{IC0} B\sqrt{w}$ (N)	K_{IC0} (MPa$\sqrt{m}$)	% Difference from mean
0	.3	351	5.0	1755	1.557	1
0	.5	182	9.1	1656	1.469	5
0	.7	85	21.4	1819	1.614	4

TABLE 3 - FRACTURE TOUGHNESS FOR THE CASE OF CRACK AT ANGLE 45° WITH RESPECT TO THE FIBER DIRECTION

α	a/w	P_{max}(N)	$K_I B\sqrt{w}/P$	$K_{IC45} B\sqrt{w}$ (N)	K_{IC45} (MPa$\sqrt{m}$)	% Difference from mean
45	.3	409	6.1	2495	2.214	2
45	.5	240	10.1	2400	2.130	2
45	.7	107	23.0	2461	2.184	0

TABLE 4 - FRACTURE TOUGHNESS FOR DIFFERENT CRACK-FIBER ORIENTATION

α	a/w	P_{max} (N)	$K_I B\sqrt{w}/P$	$K_{IC\alpha} B\sqrt{w}$ (N)	$K_{IC\alpha}$ (MPa$\sqrt{m}$)
0	.5	182	9.1	1656	1.469
30	.5	227	9.6	2179	1.933
45	.5	258	10.0	2580	2.289
60	.5	320	10.4	3328	2.953
90	.5	489	10.5	5135	4.557

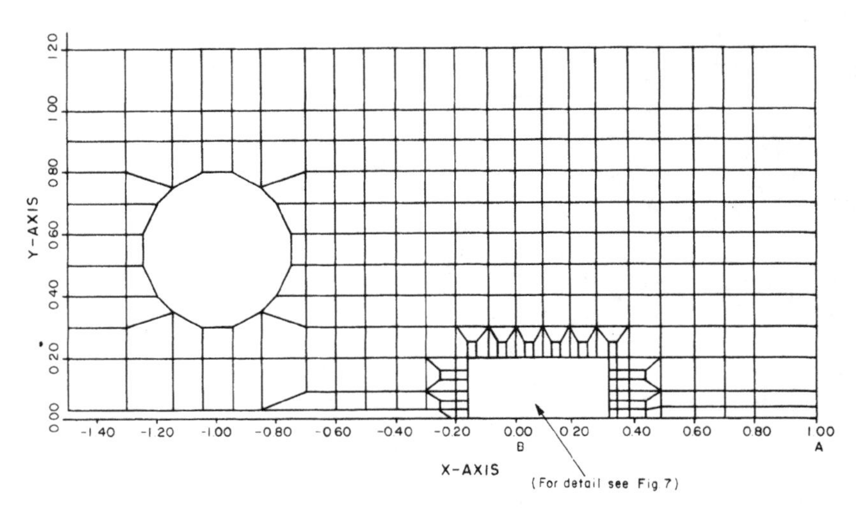

Fig. (6) - Half specimen mesh, $\frac{a}{w} = 0.5$

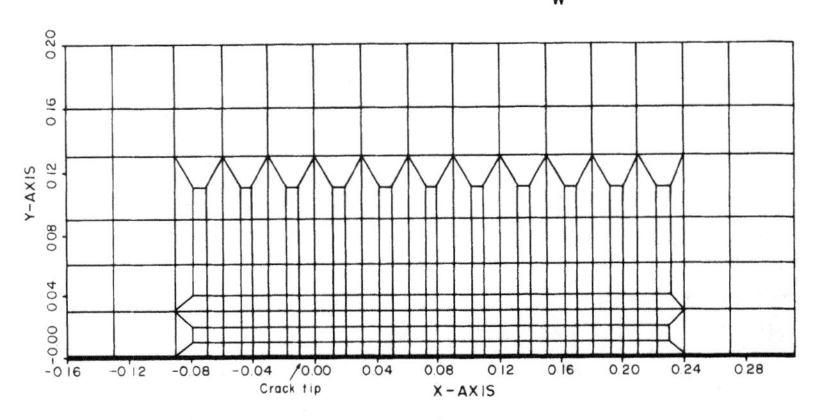

Fig. (7) - Crack tip elements

model is required. The full specimen meshes are also obtained by developing two computer programs capable of generating nodes and elements of a full specimen mes from a half specimen mesh. Again, the band width of the new mesh was kept at a minimum.

A Solid Sap finite element program, which is a general purpose program capable of handling orthotropic elements, was used. The linear elastic action was assumed and the Solid Sap program was run for all the models. A dummy load of P = 100 lbs. was used, see Figure 4. A computer program was developed to plot the normal crack tip stress σ_y versus the distance from the crack tip r. Figure 8 shows a typical crack tip stress distribution, σ_y along the x axis for the mode with a/w = 0.5 and α = 45 degrees.

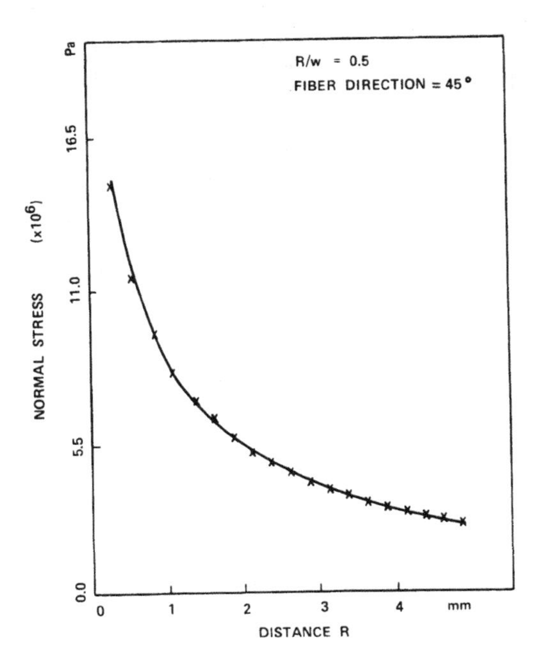

Fig. (8) - Crack tip stress distribution

4. Determination of stress intensity factors and fracture toughness: The above stresses are related to stress intensity factor K_I as indicated by equation (1). Using this equation for $\Theta=0$ and the above crack tip stress distribution, stress intensity factors K^r corresponding to distance r from the crack tip can be calculated and plotted against r. Extrapolation of these values to r=0 is the stress intensity factor for load P. The stress intensity factor corresponding to the start of crack growth ($P = P_{max}$, Figure 5) is the fracture toughness of the material. Figure 9 shows the computer plot of normalized K^r versus r/w, for a model with a/w = 0.5 and α = 45.

The fracture toughness for different crack lengths and fiber orientations was obtained by using the above method. The results are listed in Tables 2-4.

The crack tip displacement is an important factor in the study of fracture mechanics. Using crack tip deformations, the separation of modes, Figure 2, as occurs in isotropic materials, or the existence of mixed mode, as occurs in orthotropic materials, can be verified. Since the crack tip displacements are very small, they must be magnified for plotting purposes. Figure 10 shows typical crack tip displacement (magnified one hundred times) for the model with a/w = 0.5 and α = 45 degrees. It clearly indicates the existence of both crack opening and edge-sliding modes.

5. Results and Conclusions: Considering the results of Tables 2-4, the following conclusions can be drawn:

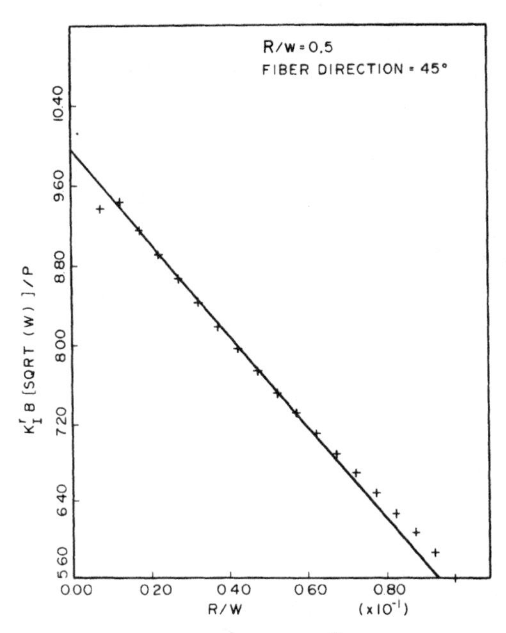

Fig. (9) - Normalized K^r versus r/w

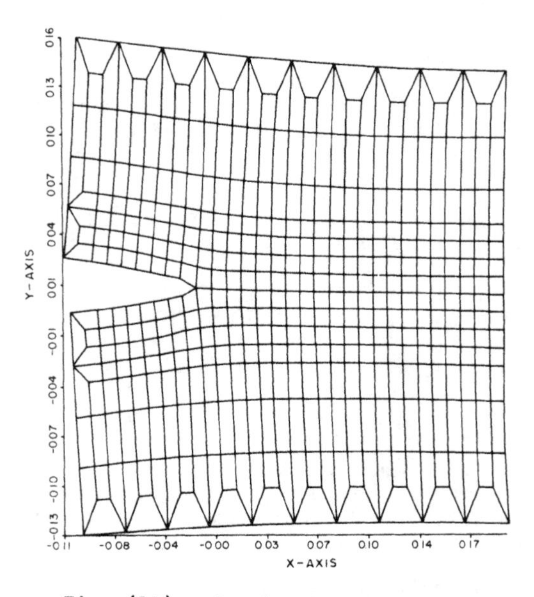

Fig. (10) - Crack tip displacement

a. For the case of cracks along the fiber direction, $\alpha=0$, the values of fracture toughness for different crack lengths are given in Table 2. The maximum percent difference between these values is 5%. Thus, for the case of cracks along the fiber orientation, the fact that the K_{ICO}, fracture toughness, is a constant material property independent of crack length was verified. After noting that for this case cracks always grow along their original directions and that the crack tip displacements can be separated into the three different failure modes, it can be concluded that the principles of linear fracture mechanics hold completely for the case of cracks along the fiber direction.

b. For the case of fiber oriented at some angle α (45°) with respect to the cracks, the values of fracture toughness are given in Table 3. The maximum percent difference between these values is 2%. Thus for each fixed angle α, i.e., $\alpha = 45°$, the fact that $K_{IC\alpha}$, fracture toughness, is a constant material property independent of crack length was verified. It must be remembered that for this case cracks do not grow along their original direction and the crack tip displacements cannot be separated into three different modes. These two facts violate the principles of linear fracture mechanics which are based on the growth of cracks along their original directions and the separation of crack tip displacements. But, since in the process of fracture, the direction of crack growth and the separation of crack tip failure modes are not as important as the intensity of crack tip stress distributions at the start of crack growth, it can be stated that the principles of linear fracture mechanics hold conditionally for the case of cracks at an angle α with respect to the fibers.

c. Table 4 gives the values of fracture toughness for a/w = 0.5 and different fiber orientation α. These values were plotted against fiber orientation α in Figure 11. Using this plot, the following empirical formula is found.

$$K_{IC\alpha} = (0.739\alpha^2 + 0.19\alpha + 1)K_{ICO}$$

This empirical formula relates $K_{IC\alpha}$, the fracture toughness when the crack and fiber are at an angle α (α in radian) with respect to each other, to K_{ICO}, the fracture toughness when the crack is along the fiber direction. In the same manner, a relationship between $K_{IIC\alpha}$ and K_{IICO} and between $K_{IIIC\alpha}$ and K_{IIICO} can be established.

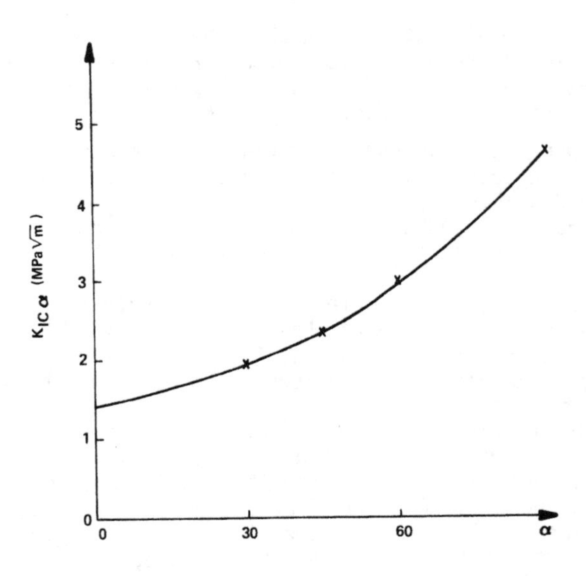

Fig. (11) - Fracture toughness versus angle α

REFERENCES

[1] Parhizgar, S., "Fracture Mechanics of Unidirectional Composite Materials", Iowa State University Dissertation, May 1979.

[2] Paris, P. C. and Sih, G. C., "Stress Analysis of Cracks", Symposium on Fracture Toughness Testing and Its Applications, ASTM 381, pp. 30-83, 1965.

[3] Sih, G. C., Paris, P. C. and Irwin, G. R., "On Cracks in Rectilinearly Anisotropic Bodies", Int. J. of Fracture Mechanics, 1, No. 3, pp. 189-203, 1965.

[4] Chan, S. K., Tuba, I. S. and Wilson, W. K., "On the Finite Element Method i Linear Fracture Mechanics", Eng. Fracture Mechanics, 2, pp. 1-17, 1970.

[5] Jones, R. M., Mechanics of Composite Materials, New York: McGraw-Hill, 197

[6] Lekhnitskii, S. G., Anisotropic Plates, 2nd Edition, Translated by T. Chero New York: Gordon and Breach, 1968.

APPENDIX 1

The crack tip stress and displacement equations for an infinite orthotropic plate with central crack subjected to the loading condition of Figure 1 follow [3]:

a. Crack opening mode, mode I ($\sigma_I \neq 0$, $\tau_{II} = \tau_{III} = 0$)

$$\sigma_x = K_I/\sqrt{2\pi r}\,\mathrm{Re}\,[\mu_1\mu_2/(\mu_1-\mu_2)(\mu_2/\sqrt{\cos\Theta+\mu_2\sin\Theta} - \mu_1/\sqrt{\cos\Theta+\mu_1\sin\Theta})]$$

$$\sigma_y = K_I/\sqrt{2\pi r}\,\mathrm{Re}\,[1/(\mu_1-\mu_2)(\mu_1/\sqrt{\cos\Theta+\mu_2\sin\Theta} - \mu_2/\sqrt{\cos\Theta+\mu_1\sin\Theta})] \qquad (1)$$

$$\tau_{xy} = K_I/\sqrt{2\pi r}\,\mathrm{Re}\,[\mu_1\mu_2/(\mu_1-\mu_2)(1/\sqrt{\cos\Theta+\mu_1\sin\Theta} - 1/\sqrt{\cos\Theta+\mu_2\sin\Theta})]$$

and

$$u = K_I\sqrt{2r/\pi}\,\mathrm{Re}\,[1/(\mu_1-\mu_2)(\mu_1 p_2\sqrt{\cos\Theta+\mu_2\sin\Theta} - \mu_2 p_1\sqrt{\cos\Theta+\mu_1\sin\Theta})]$$

$$(2)$$

$$v = K_I\sqrt{2r/\pi}\,\mathrm{Re}\,[1/(\mu_1-\mu_2)(\mu_1 q_2\sqrt{\cos\Theta+\mu_2\sin\Theta} - \mu_2 q_2\sqrt{\cos\Theta+\mu_1\sin\Theta})]$$

b. Edge-sliding mode, mode II ($\tau_{II} \neq 0$, $\sigma_I = \tau_{III} = 0$)

$$\sigma_x = K_{II}/\sqrt{2\pi r}\,\mathrm{Re}\,[1/(\mu_1-\mu_2)(\mu_2^2/\sqrt{\cos\Theta+\mu_2\sin\Theta} - \mu_1^2/\sqrt{\cos\Theta+\mu_1\sin\Theta})]$$

$$\sigma_y = K_{II}/\sqrt{2\pi r}\,\mathrm{Re}\,[1/(\mu_1-\mu_2)(1/\sqrt{\cos\Theta+\mu_2\sin\Theta} - 1/\sqrt{\cos\Theta+\mu_1\sin\Theta})] \qquad (3)$$

$$\tau_{xy} = K_{II}/\sqrt{2\pi r}\,\mathrm{Re}\,[1/(\mu_1-\mu_2)(\mu_1/\sqrt{\cos\Theta+\mu_1\sin\Theta} - \mu_2/\sqrt{\cos\Theta+\mu_2\sin\Theta})]$$

and

$$u = K_{II}\sqrt{2r/\pi}\,\mathrm{Re}\,[1/(\mu_1-\mu_2)(p_2\sqrt{\cos\Theta+\mu_2\sin\Theta} - p_1\sqrt{\cos\Theta+\mu_1\sin\Theta})]$$

$$(4)$$

$$v = K_{II}\sqrt{2r/\pi}\,\mathrm{Re}\,[1/(\mu_1-\mu_2)(q_2\sqrt{\cos\Theta+\mu_2\sin\Theta} - q_1\sqrt{\cos\Theta+\mu_1\sin\Theta})]$$

c. Crack tearing mode, mode III ($\tau_{III} \neq 0$, $\sigma_I = \tau_{II} = 0$)

$$\tau_{xz} = -K_{III}/\sqrt{2\pi r}\,\mathrm{Re}\,[\mu_3/\sqrt{\cos\Theta+\mu_3\sin\Theta}]$$
$$\tau_{yz} = K_{III}/\sqrt{2\pi r}\,\mathrm{Re}\,[1/\sqrt{\cos\Theta+\mu_3\sin\Theta}] \qquad (5)$$

$$w = K_{III}\sqrt{2r/\pi}\,\mathrm{Re}\,[\sqrt{\cos\Theta+\mu_3\sin\Theta}/(c_{45}+\mu_3 c_{44})] \qquad (6)$$

$$K_I = \sigma_I\sqrt{\pi a},\ K_{II} = \tau_{II}\sqrt{\pi a} \text{ and } K_{III} = \tau_{III}\sqrt{\pi a} \qquad (7)$$

where r, Θ and a are shown in Figure 1; u, v, w are displacements in x, y and z directions; μ_1, μ_2, μ_3 are called material complex parameters, they characterize the degree of anisotropy of the materials, and they depend on crack-fiber orienta tion α, Figure 3 [6]; K_I, K_{II}, K_{III} are given by equation (7); they are called stress intensity factors; p_1, p_2, q_1, q_2, c_{45}, c_{44} are related to material elastic constants and materials complex parameters as:

$$p_1 = a_{11}\mu_1^2 + a_{12} - a_{16}\mu_1 \qquad p_2 = a_{11}\mu_2^2 + a_{12} - a_{16}\mu_2$$
$$q_1 = a_{12}\mu_1 + a_{22}/\mu_1 - a_{26} \qquad q_2 = a_{12}\mu_2 + a_{22}/\mu_2 - a_{26} \qquad (8)$$
$$c_{44} = a_{55}/(a_{44}a_{55} - a_{45}^2) \qquad c_{45} = -a_{45}/(a_{44}a_{55} - a_{45}^2)$$

ON FAILURE MODES OF UNIDIRECTIONAL COMPOSITES UNDER COMPRESSIVE LOADING

L. B. Greszczuk

McDonnell Douglas Astronautics Company
Huntington Beach, California 92647

ABSTRACT

Experimental and theoretical studies are presented on failure modes of unidirectional composites subjected to compressive loading. Failure modes investigated include: fiber microbuckling, fiber-matrix debonding followed by microbuckling, interaction failure (transverse splitting) and fiber strength failure. Use of large scale, nearly perfect model composites is made in studying the various failure modes as well as the influence of fiber-matrix interface on the compressive strength. Approximate equations are presented for predicting the microbuckling failure of composites, compressive strength of composites with unbonded fibers as well as the interaction failure of unidirectional composites whereby the axial compressive strength is shown to be dependent on the transverse tensile strength of composites. A discussion is also presented on the various failure modes as applied to practical composites.

INTRODUCTION

In the past, a number of investigators have studied microbuckling as a failure mode for unidirectional composites subjected to compressive loading in the direction of fibers. Although microbuckling has been found to be a valid failure mode for composites made with very flexible resins ($E_r \lesssim 5x10^4$ psi), it does not appear to be the critical mode of failure in practical composites such as graphite/epoxy and boron/epoxy made with resins having typical Young's moduli of $E_r \approx 5x10^5$ psi. In the case of graphite/epoxy composites, the microbuckling theory predicts compressive strength values on the order of 600 ksi - 800 ksi, whereas the measured values range from ≈90 ksi for composites made with ultra high modulus graphite fibers to ≈250 ksi for composites made with high strength fibers. Thus, there appears to be a lower energy failure mode than microbuckling. The present paper presents some recent results on the failure modes of composites subjected to compressive loading, as influenced by properties of constituents, fiber-matrix interface and composite microstructure.

COMPRESSIVE FAILURE MODES

Several of the failure modes by which unidirectional composites can fail if subjected to compressive loading are illustrated in Figure 1 and discussed in

Sections that follow.

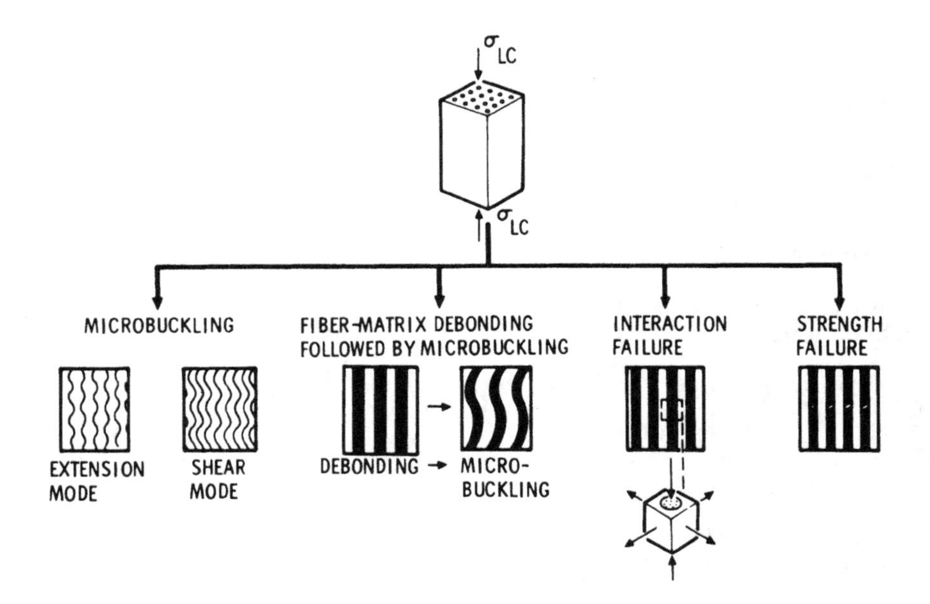

Fig. (1) - Failure modes for unidirectional composites subjected to compressive loading

Microbuckling Compression Failure. Fiber microbuckling as a failure mode for unidirectional composites subjected to compressive loading was apparently first suggested by Dow [1]. Idealizing the fibers and the matrix as laminae (two-dimensional model) and using column on elastic foundation approach as well as energy methods, a number of authors [2-6] derived pertinent equations for predicting the critical loads to cause microbuckling in the extensional and shear modes, both of which are illustrated in Figure 1. The three-dimensional solution for microbuckling of a single round fiber surrounded and bonded by matrix material of infinite extent has been obtained by Sadowsky, et al [7] and Herrmann et al [8].

Of the two microbuckling modes shown in Figure 1, microbuckling in the extension mode applies to composites with reinforcement content $k \lesssim 20\%$, whereas microbuckling in the shear mode applies to composites with $k \gtrsim 20\%$. Thus, only the latter failure mode is of practical interest. Using energy method, an approximate solution for microbuckling in the shear mode of a multifiber reinforce composite was obtained by Greszczuk as [9]:

$$\sigma_{CL} = G_{LT} + \lambda\pi^2 \; k \; E_f\left(\frac{R}{\ell}\right)^2 \qquad (1)$$

where σ_{CL} is the composite stress at failure, G_{LT} is the shear modulus of composite, λ is the fiber end fixity parameter (λ = 0.25 for simply supported ends and λ=1 for fixed ends), k is the fiber volume fraction, E_f is the Young's modulus of the fibers, R is the fiber radius and ℓ is the buckle wave length or spec men length. If the specimen length is large compared to the fiber radius, then

$$\sigma_{CL} = G_{LT} \tag{2}$$

To establish validity of equations (1) and (2), experiments were performed on carefully prepared, nearly perfect fiber-reinforced model composites [9,10]. Test variables in this investigation were: fiber properties, matrix properties, specimen size, reinforcement shape as well as size, array and end fixity of the reinforcing fibers. Figures 2 and 3 show several of the model composites. Both

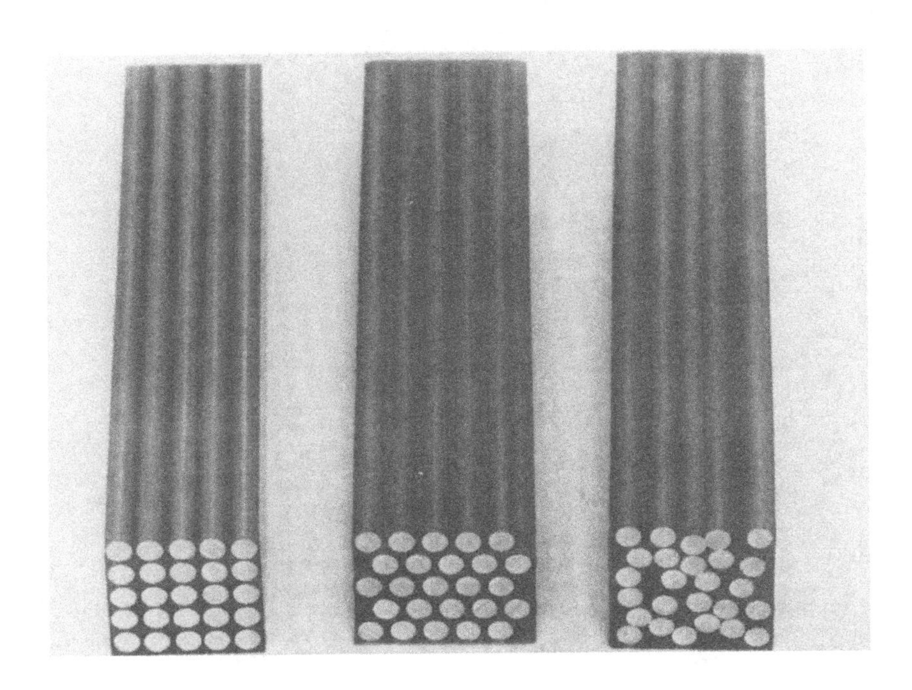

Fig. (2) - Composite test specimens with square, hexagonal and arbitrary fiber arrays

stainless steel and aluminum rods were used as the reinforcing fibers. The Young's moduli of resin materials used in the composite models ranged from 2.5 ksi to 457 ksi. Figure 4 shows typical load-deflection curves for composites made with aluminum rods and three different resins. All composites made with Resin A failed by elastic microbuckling. Composites made with Resin B exhibited some nonlinearity in the load-deflection curve near the point of failure, indicating initiation of yielding of the reinforcement. Composites made with Resin C failed by compression yielding. Similar results were obtained for composites reinforced with stainless steel rods. A comparison of test data with theoretical results predicted by equation (1) is shown in Figures 5 and 6. Figure 5 shows that in the case of composites made with low modulus resin, microbuckling theory appears to predict the failure load fairly accurately; however, as shown in Figure 6, if the resin shear modulus in increased $G_r \gtrsim 5$ ksi, the critical failure

Fig. (3) - Composite test specimens of various sizes

NOTE: REINFORCEMENT MATERIAL CONSISTED OF 0.078-IN.-DIA 6061-T6 ALUMINUM RODS ARRANGED IN SQUARE ARRAY. REINFORCEMENT CONTENT WAS k = 53.8% DEFLECTION OBTAINED FROM MACHINE HEAD TRAVEL, AND WAS CORRECTED FOR MACHINE COMPLIANCES

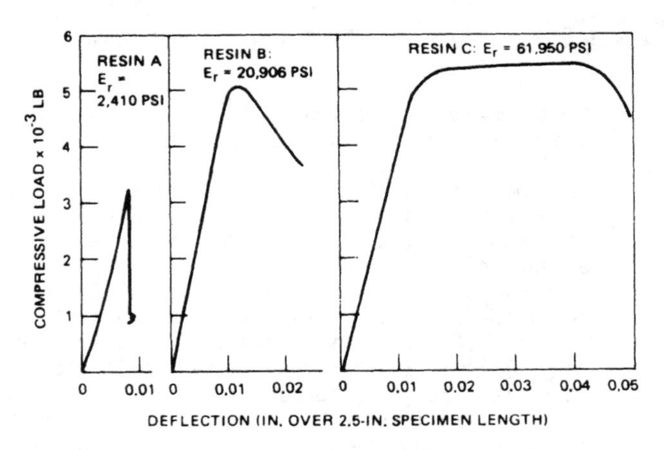

Fig. (4) - Typical compressive load-deflection curves for aluminum rod-reinforced composites made with resins having different moduli of elasticity

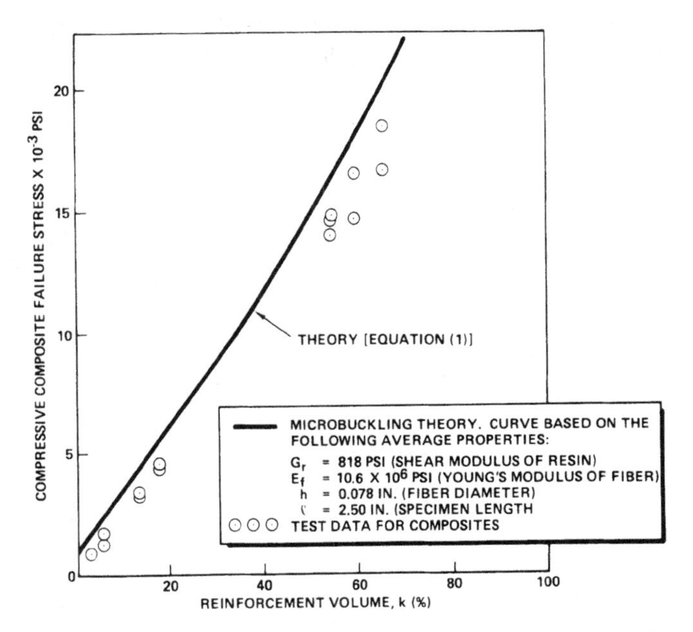

Fig. (5) - Test-theory comparison of compressive microbuckling strength of circular fiber-reinforced composites having various fiber volume fractions

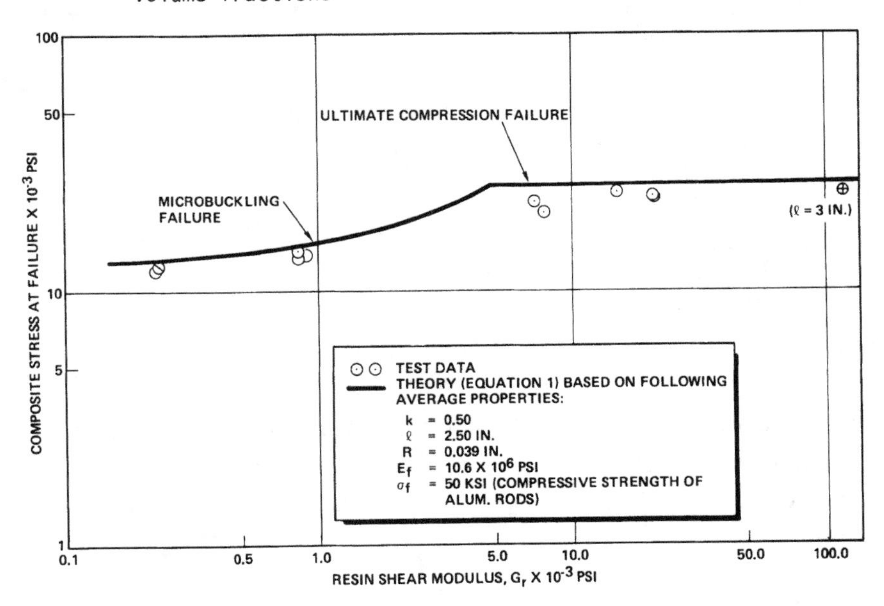

Fig. (6) - Test-theory comparison of compressive strength of composites made with aluminum rods as influenced by shear modulus of resin

mode is by ultimate compression failure of the reinforcement. Similar results for composites made with stainless steel rods are shown in Figure 7. The resin

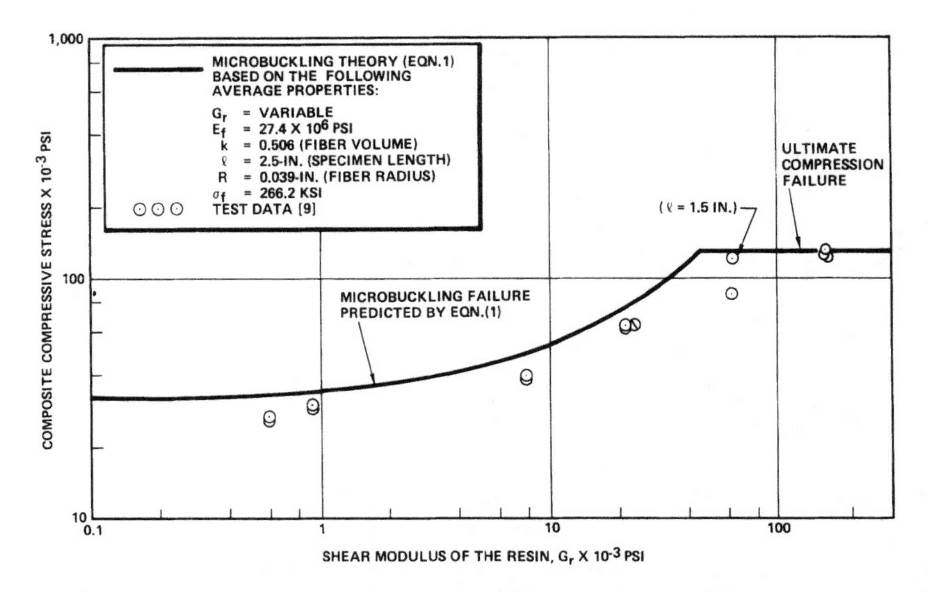

Fig. (7) - Test-theory comparison of compressive strength of composites consisting of stainless steel rods and resins with different shear moduli

shear modulus at which transition from microbuckling type failure to ultimate compression failure takes place can be estimated by setting σ_{CL} equal to the compressive strength of composite as predicted by the law of mixture equation and replacing G_{LT} by the following*

$$G_{LT} = \frac{G_r}{1+2\beta(\frac{G_r}{G_f} - 1)} + 1-\beta \tag{3}$$

$$\beta = \sqrt{k/\pi}$$

For the case when $G_r/G_f << 1$, the critical value for the resin shear modulus, G_r^*, at which transition from microbuckling to ultimate compression failure takes place becomes

*Equation (3) is an approximate expression for predicting shear modulus of composites from properties of constituents and was obtained from an approximate evaluation of integral given in [11] by equation (17).

$$G_r^* \leq \{F_f k - \pi^2 E_f k(\frac{R}{\ell})^2\}\{\frac{1-2\beta}{\beta+(1-\beta)(1-2\beta)}\} \quad (4)$$

where in addition to the previously defined terms F_f is the compressive strength of the fiber. In arriving at equation (4), contribution of matrix to ultimate strength of composite has been neglected. The values of G_r^* calculated from equation (4) for aluminum and steel fiber-reinforced composites are G_r^* = 4.8 ksi and G_r^* = 38.6 ksi, respectively.

The microbuckling test data for stainless steel rod reinforced composites appears to be somewhat lower than the predicted values of the corresponding data for aluminum rod-reinforced composites. Post-test examination of stainless steel rod reinforced composites shows presence of internal helical cracks in the matrix and debonding of the matrix at the interface. Figure 8 shows the fan-shaped internal helical cracks initiating of the fiber matrix interface whereas Figure 9 shows an oblique view of these cracks. The matrix appears to be "peeling" away from the reinforcement. The failure appears to be initiating at the fiber matrix interface, propagating along the interface and terminating in a helical crack These experimental observations of the modes of failure led to additional study effort on the role of fiber-matrix interface. Intuitively, one would expect that the fiber-matrix interface should have a significant effect on the strength as well as the failure modes of composite subjected to compressive loading.

Influence of Fiber-Matrix Interface on Compression Failure. To establish the influence of fiber-matrix interface on the compressive strength of unidirectional composites, model composites were made with stainless steel fibers which received different types of surface preparation to achieve different values of the interface strength. For any given surface preparation, the apparent interface shear strength was obtained from pullout tests on single rods imbedded 0.5 inch in the matrix. Composite models made of fibers with different types of surface preparation were tested in axial compression. Figure 10 shows the compressive strength of composites as a function of the apparent shear strength of the interface. It is obvious from these results that the fiber-matrix interface is one of the key parameters influencing the axial compressive strength of composites. Not only was the interface found to affect the compressive strength but also the failure modes. Whereas the specimens with low values of τ_A failed by microbuckling, those with high interface strength failed by compression yielding.

The compressive strength of a composite with unbonded or poorly bonded fibers can be estimated from the following equation:

$$\sigma_{LC}^* = \frac{G_r}{[\frac{2}{\sqrt{1-\alpha^2}} \tan^{-1}\sqrt{\frac{1+\alpha}{1-\alpha}} + 1 - \alpha - \frac{\pi}{2}]} + \pi^2 k\, E_f(\frac{R}{\ell})^2 \quad (5)$$

where in addition to the previously defined terms

$$\alpha = \sqrt{\frac{4k_{if}}{\pi}} \quad (6)$$

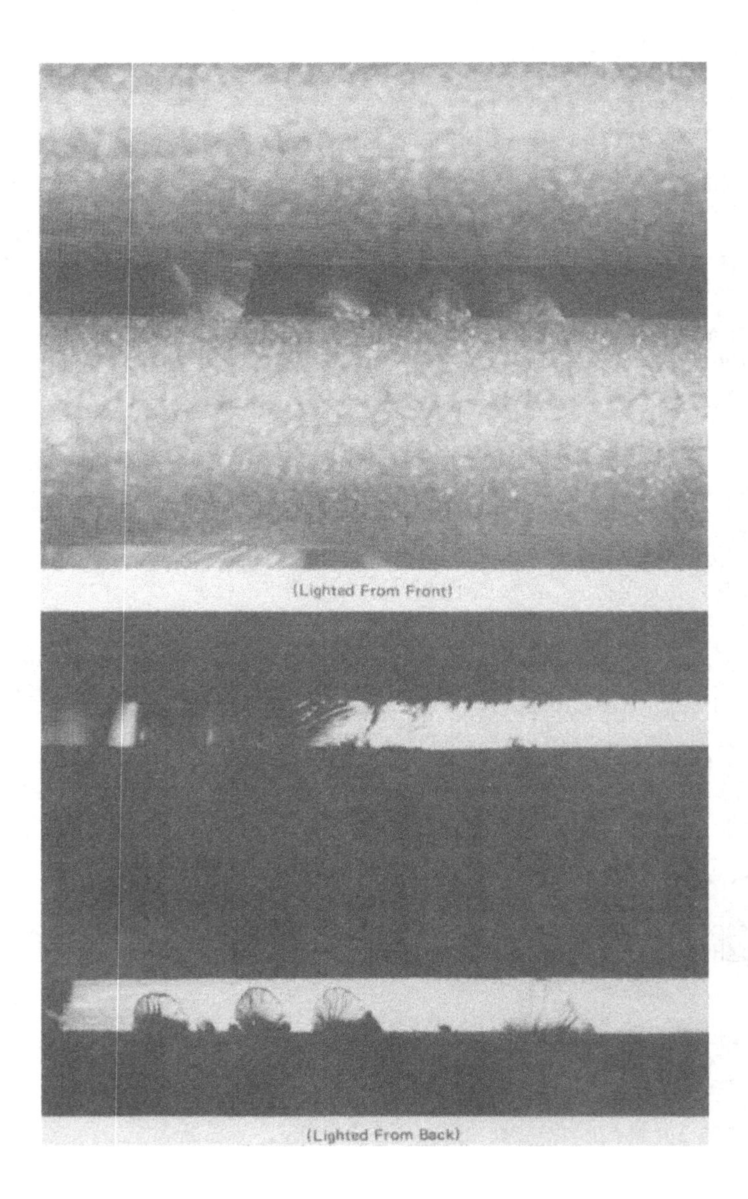

Fig. (8) - Internal fan-shaped subsurface helical cracks in stainless steel fiber-reinforced composite made with Resin C

Fig. (9) - Oblique view of helical subsurface cracks and resin "peeling" from the fiber surface (20x)

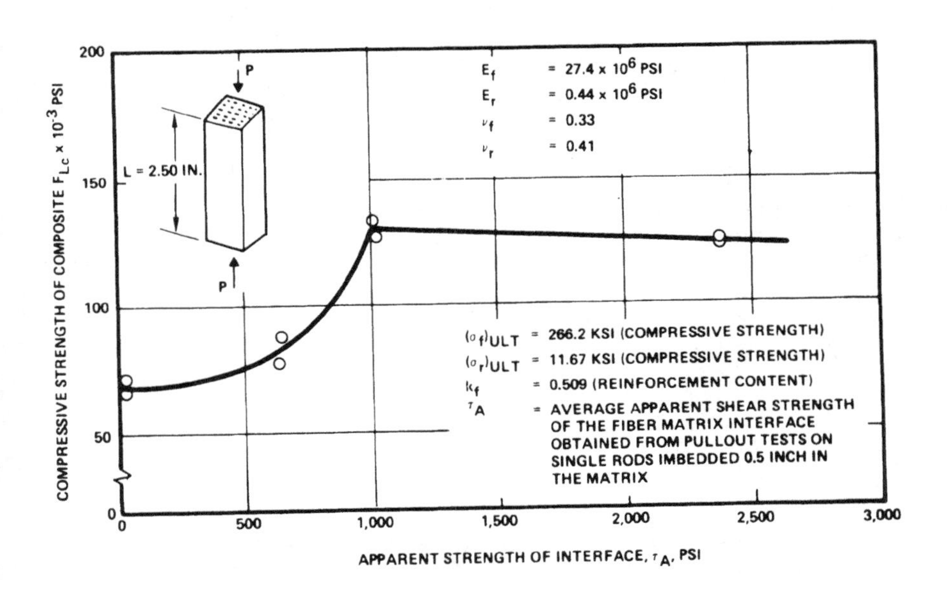

Fig. (10) - Influence of fiber-matrix interface on compressive strength of composites

and k_{if} is the volume fraction of unbonded fibers. Equation (5) was derived by assuming, as a first approximation, that the shear modulus of a composite containing unbonded fibers is the same as the shear modulus of a solid containing cylindrical voids in the amount equal to the amount of the unbonded fibers. Table 1 shows a comparison of results predicted from equation (5) with test data

TABLE 1 - COMPRESSIVE STRENGTH OF COMPOSITES CONTAINING UNBONDED FIBERS; TEST VERSUS THEORY*

Reinforcing Fiber	Fiber Modulus $E_f \times 10^{-6}$(psi)	Matrix Modulus $E_r \times 10^{-6}$(psi)	Matrix Poisson's Ratio ν_r	Fiber Content k (%)	Composite Stress at Failure x 10^{-3}psi		
					Test	Theory (Eqn 5)	Test/Theory
S. Steel	27.4	0.437	0.41	51.5	68.13	76.72	0.89
S. Steel	27.4	0.448	0.41	51.6	71.27	77.79	0.92
Aluminum	10.6	0.0028	0.47	53.8	11.80	13.98	0.84
Aluminum	10.6	0.0028	0.47	53.8	11.90	13.98	0.85

*Fiber diameter was 0.078 inch; Specimen length was 2.5 inches.

on composites made with unbonded fibers. The data for aluminum fiber reinforced composites is a poor indicator of the validity of equation (5), as the contribution to the overall composite strength is due primarily to the second term in equation (5). In the case of stainless steel fiber reinforced composites, the first term contributes significantly more to the composite strength than does th second term. As to the mechanism by which fiber matrix debonding can take place in actual composites the latter is discussed in the next section.

Interaction Failure under Compressive Loading. Because of the differences in Poisson's ratios between the fibers and the matrix, transverse stresses are induced when a composite is subjected to compressive loading in the fiber directic Figure 11. Since there are no externally applied transverse loads, the magnituc and distribution of internal stresses acting on the boundaries of a repeating el ment has to be such that

$$\int_{-\ell}^{\ell} \sigma_x dy = 0$$

$$\int_{-\ell}^{\ell} \sigma_y dx = 0 \qquad (7)$$

Thus, the stresses acting on the boundaries of a repeating element are self-equilibrating. For the case where $\nu_f < \nu_r$, the transverse stresses will be tensil midway between the fibers and compressive near the corners of the repeating element, as shown in Figure 11. Even though the induced transverse tensile stresse are low compared to the applied compressive stresses ($\sigma_x/\sigma_{LC} \simeq 0.01$ to 0.03), bu

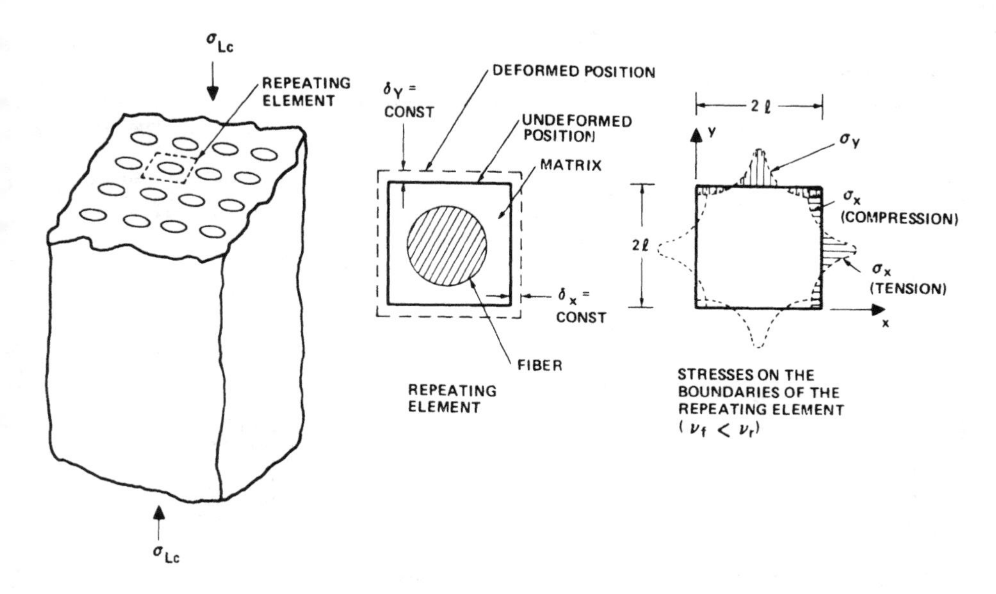

Fig. (11) - Internal stress state in composite subjected to compressive loading

so is the transverse tensile strength compared to axial compressive strength ($F_{Tt}/F_{LC} \simeq 0.03$ to 0.06). Consequently, it is quite possible for the induced transverse tensile stresses to cause fiber matrix debonding, or to cause interaction failure which is shown in Figure 11 as the third failure mode. An approximate interaction equation for predicting the compressive strength of composites was obtained by the author as

$$F^*_{LC} \simeq \frac{F_{LC}F_{Tt}}{F_{Tt}-KF_{LC}} \tag{8}$$

where

$$K \simeq -(k - \sqrt{\frac{4k}{\pi}})(\nu_f-\nu_r)\,\frac{E_T}{E_L} \tag{9}$$

$$F_{LC} = F_f k + \sigma^*_r(1-k) \tag{10}$$

where σ^*_r is the compressive stress in the matrix at a strain equal to the fiber failure strain, F_f is the compressive strength of the fibers, which as a first approximation can be assumed equal to the fiber tensile strength, ν_f and ν_r are the Poisson's ratios of the fibers and matrix respectively, E_L and E_T are Young's moduli of composite in the fiber and transverse directions, and F_{Tt} is the transverse tensile strength of a composite. Equation (8) was derived assuming that the inplane transverse tensile strength of composite, F_{Tt}, is the same as the transverse tensile strength through the thickness. Comparison of test data for graphite/epoxy composites with the predictions from equation (8) is shown in Figure 12. As transverse tensile strength of the composite increases, the interac-

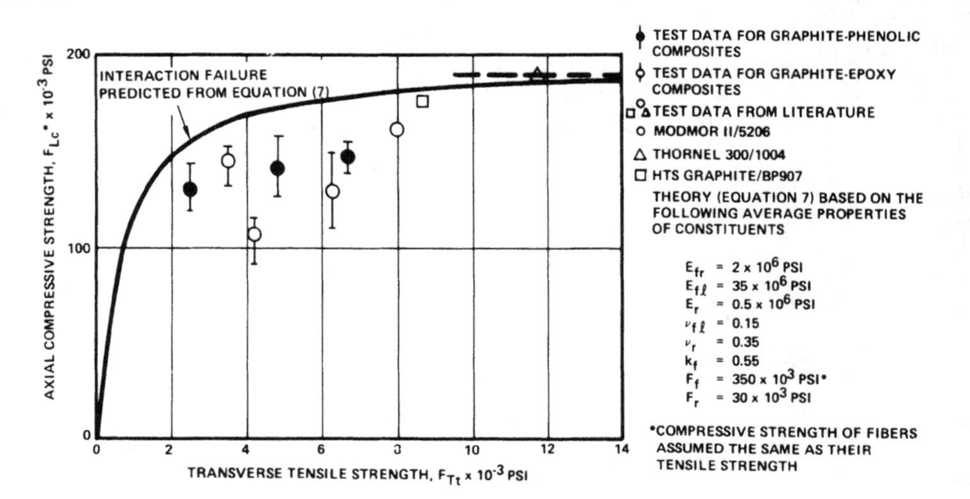

Fig. (12) - Test-theory comparison of compressive strength of graphite/epoxy and graphite/phenolic unidirectional composites

tion equation reduces to the law-of-mixtures relationship given by equation (10) and shown in Figure 12 by the dashed line. By taking into account differences in the inplane and through-the-thickness transverse tensile strength as well as the influence of voids and unbonded fibers, a better test-theory correlation is expected than shown in Figure 12. The work in this area is still in progress; however, from the results presented in Figure 12, it appears that the interactio failure appears to be a valid failure mode for composites subjected to compressi loading in the fiber direction.

An independent observation which gives support to the existence and influenc of transverse tensile stresses in unidirectional composites subjected to compres sive loading is that longitudinal cracks were observed in specimens tested in compression. Typical photographs of the failure zone in graphite/epoxy specimen tested in compression are shown in Figure 13. Most of the cracks appear to be parallel to the lamination planes, thus supporting the hypothesis that the trans verse tensile strength in the thickness direction may be lower than the inplane transverse tensile strength.

The final compression failure mode shown in Figure 1 is by strength failure of the fibers. The compressive strength based on this failure mode is given by equation (10), as was noted on the previous page.

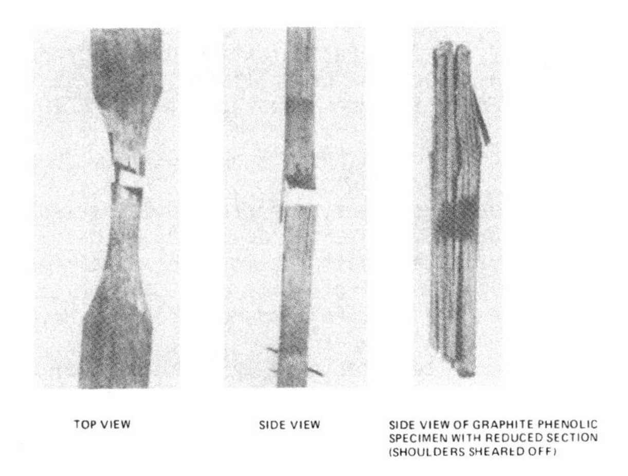

Fig. (13) - Compression failure in graphite/epoxy and graphite/phenolic unidirectional composites

DISCUSSION AND CONCLUSIONS

Experimental results presented on model composites show that for composites made with low Young's modulus matrix, microbuckling is the critical failure mode. The stress to cause microbuckling failure is shown to increase with the matrix shear modulus G_r up to a critical value of G_r^* when a transition to non-microbuckling failure mode takes place. Although the expression for G_r^* given by equation (4) was based on assumption of perfect fiber-matrix bond, it is apparent from the results presented in subsequent sections that the properties of the fiber-matrix interface and the transverse tensile strength, F_{tT}, of composite will influence G_r^*. To account for the effect of F_{tT} on G_r^* requires that one replace $F_f k$ in equation (4) by F_{LC}^* as given by equation (8). For composites with $G_r > G_r^*$, the interaction failure theory as given by equation (8) appears to predict the compressive failure stress of composite more accurately than does the law of mixtures equation [equation (10)]. Test data as well as experimentally observed failure modes in actual composites also support the interaction failure theory. Further evidence of the strong influence of the fiber-matrix interface strength on the axial compressive strength is obtained from test results on model composites made with fibers having different types of surface preparation so to achieve different values of the fiber-matrix interface strength. These tests show that as the apparent interface shear strength increases, so does the axial compressive strength. Since the transverse strength of composite and the fiber-matrix interface shear strength appear to be related, this too points out to some type of interaction failure. Other factors which have not been discussed here but are known to affect composite failure loads under compressive loading are the inelastic and nonlinear behavior of the constituents and initial defects in the composite such as bowed, twisted or misaligned fibers. Consideration of these is especially important when applying analytical results to actual composites.

REFERENCES

[1] Dow, N. F. and Gruntfest, I. J., "Determination of most needed potentially possible improvements in materials for ballistic and space vehicles", General Electric Company, Space Sciences Laboratory, TIS R60SD389, June 1980.

[2] Schuerch, H., AIAA Journal, Vol. 4, No. 1, pp. 102-106, January 1966.

[3] Hayashi, T., "On the shear instability of structures caused by compressive loads", AIAA Paper No. 65-770, presented at the Joint Meeting of the American Institute of Aeronautics and Astronautics, The Royal Aeronautical Society and the Japan Society for Aeronautics and Space Sciences, Aircraft Design and Technology Meeting, Los Angeles, California, November 1965.

[4] Rosen, B. W., "Fiber composite materials", ASM, Metals Park, Ohio, Chapter 3, 1965.

[5] Hayashi, T., "On the shear instability of structures caused by compressive loads", Proceedings, 16th Japan National Congress for Applied Mechanics, pp. 149-157, 1966.

[6] Chung, W.-Y. and Testa, R. B., Journal of Composite Materials, Vol. 3, pp. 58-80, January 1969.

[7] Sadowsky, M. A., Pu, S. L. and Hussain, M. A., Journal of Applied Mechanics Vol. 34, Series C, No. 4, pp. 1011-1016, December 1967.

[8] Herrmann, L. R., Mason, W. E. and Chan, S. T. K., Journal of Composite Materials, Vol. 3, pp. 212-226, 1967.

[9] Greszczuk, L. B., AIAA Journal, Vol. 13, No. 10, pp. 1311-1318, October 197

[10] Greszczuk, L. B., Composite Materials: Testing and Design (Third Conferenc ASTM STP 546, 1974.

[11] Greszczuk, L. B., "Interfiber stresses in filamentary composites", AIAA Journal, Vol. 4, No. 7, pp. 1274-1280, July 1971.

SECTION V

LAMINATES

DELAMINATION OF T300/5208 GRAPHITE/EPOXY LAMINATES

S. C. Chou

Army Materials and Mechanics Research Center
Watertown, Massachusetts 02172

INTRODUCTION

As graphite fiber-epoxy resin (Gr/Ep) composites are being used more and more in high performance structures, the need to predict fracture directions or to locate fracture origins becomes apparent. Stress and failure analyses have become primary considerations when corrective action decisions have to be made for repairing or redesigning failed components. This may occur even in the early stage of manufacturing. Observed failure modes in Gr/Ep composites include fiber breakage, matrix cracking, fiber-matrix debonding, delamination of laminates, etc. They are too numerous to be listed. This investigation will only concern with failure by the process of delamination in Gr/Ep laminates.

Composite delamination has been the subject of many past investigations. Several mathematical results have been obtained and compared to experiments on different Gr/Ep composites and specimens. The assumed failure criteria also differ from using the tensile stress to shear stress or the combination of both. Pagano and Pipes [1] considered delamination to occur in a laminate when the interlaminar tensile stress reaches a critical value. They employed an optimization procedure to design a Gr/Ep specimen, HTS/ERLA 2256, with a stacking sequence of $[(\pm 25)_2/90]_s$. The maximum interlaminar tensile stress was supposed to occur at the mid-plane of the laminate where delamination is assumed to take place. This was demonstrated experimentally. The same approach was employed by Harris and Orringer [2] using the Hercules Magnamite AS/3501-6 Gr/Ep with a stacking sequence of $[(\pm 26)_2/90]_s$. Delamination, however, occurred at the interface between the angle-ply and 90°-ply instead of the mid-plane as claimed in [1]. They attributed this difference to the combined effect of interlaminar tensile and shear stresses as being the cause of delamination rather than the sole action of interlaminar normal stress.

An attempt is made to investigate the effect of shear stresses on the delamination of Gr/Ep laminates with stacking sequences of $[(\pm\theta)_2/90]_s$ and $[\theta/-\theta_2/\theta/90]_s$ that were fabricated especially for this study. The stress analysis was made by using the stress hybrid finite element model under the condition of generalized plane strain. The failure threshold is determined from the tensor polynomial strength theory [3] and the theory of strain energy density function [4]. Experimental results revealed that transverse cracking always preceded delamination and hence they should be interpreted accordingly.

MATERIAL PROPERTIES AND CONFIGURATION OF TEST SPECIMEN

The T300/5208 Gr/Ep prepreg tape was used to fabricate the composite laminate test specimens. The material properties are given by

$$E_{11} = 22 \times 10^6 \text{ psi}; \; E_{22} = 1.54 \times 10^6 \text{ psi}$$
$$\nu_{23} = \nu_{12} = 0.28; \; G_{12} = G_{23} = 0.81 \times 10^6 \text{ psi} \quad (1)$$

These properties are approximately the same as those of HTS/ERLA 2256 used in [1] and AS/3501-6 used in [2]. The slight amount of difference in material properties should not have a serious influence on the failure mode of delamination. Therefore, the comparison should be valid.

Laminate specimens with dimensions 12 in. x 12 in. were fabricated from T300 /5208 Gr/Ep prepreg tapes in accordance with the supplier's recommendations for tooling, layup and cure cycle. Five of the laminates are made with a stacking sequence of $[(\pm\theta_2)/90]_s$ while the remaining specimens have a stacking sequence of $[\theta/-\theta_2/\theta/90]_s$. This is done on purpose to test the laminate response due to changes in the stacking sequence. The values of θ are 5°, 15°, 25°, 35° and 45°. The thickness of the ten-ply laminates average around 0.055 to 0.060 in. with a fiber volume fraction of approximately 55%. Test coupons of 1 in. x 8 in. were prepared from the full size laminate sheet. End tabs made of Glass/Ep with dimensions 1 in. x 2 in. x 0.125 in. are then bended to each test coupon leaving a gage section of 1 in. x 4 in. All specimens were instrumented with a FAE-12-35 PL strain gage and stored in a desiccator under vacuum prior to testing in the ambient environment.

TEST RESULTS

Each specimen was tested individually in an electro-hydraulic, servo-controlled closed-loop testing machine. The test was performed under the load control mode with a loading rate of one pound per second. Material damage is monitored constantly with two traveling microscopes (30X magnification) placed at opposing edge of the specimen while load and strain were also recorded to detect failure initiation. Tests were terminated as soon as delamination is detected. Since the loading rate was extremely slow, the measured quantities correspond closely to the threshold of failure.

Having established the monitoring technique, specimens were tested and found to crack in the transverse* direction in the 90° plies before delamination occurr at the interface between the angle-ply and 90°-ply. The number of transverse cracks increased with load until delamination marking the termination of a partic lar test. The strain and nominal stress at the first sight of transverse cracking

*Transverse cracks in the 90° plies were observed in planes oriented approximatel normal to the direction of applied tension. Note that delamination occurs in a plane parallel to the applied load.

and onset of delamination are summarized in Table 1. These results represent

TABLE 1 - TEST RESULTS FOR $[(\pm\theta)_2/90]_s$ AND $[\theta/-\theta_2/\theta/90]_s$ LAMINATES

θ		5°	15°	25°	35°	45°
First Sight of Transverse Crack	Strain (%)	0.493	0.326	0.301	0.351	0.532
	Stress (ksi)	69.7	39.6	29.1	20.5	15.9
Onset of Delamination	Strain (%)	0.697	0.383	0.336	0.406	0.62
	Stress (ksi)	99.3	46.7	30.8	23.9	18.2

the average values of the two types of specimens with stacking sequences of $[(\pm\theta)_2/90]_s$ and $[\theta/-\theta_2/\theta/90]_s$. Since the number of tests is not sufficient to establish a meaningful statistical interpretation on the variations of stacking sequence, this effect will no longer be pursued. The difference between the two strain levels for a fixed value of θ in Table 1 gives an indication of the relative number of transverse cracks created before delamination. For instance, the largest number of transverse cracks were created for $\theta = 5°$ while the smallest number corresponds to $\theta = 25°$ where only two or three transverse cracks were observed before delamination.

A series of photographs showing the delamination of five different types of T300/5208 Gr/Ep laminates distinguished by $[(\pm5)_2/90]_s$, $[(\pm15)_2/90]_s$, $[(\pm25)_2/90]_s$ $[(\pm30)_2/90]_s$ and $[(\pm45)_2/90]_s$ are shown in Figures 1 to 5. Each Figure contains a group of four photos. The left-hand side group shows the edge view of the $[\theta/-\theta_2/\theta/90]_s$ laminate while the right-side group shows the edge view of the $[(\pm\theta)_2/90]_s$ laminate. The top group is magnified 30 times and the bottom group is magnified approximately 200 times. Note that the meandering of the cracks in the 90° plies shown by the bottom group of the photographs in Figures 1 to 5 is the influence of transverse cracking on delamination.

The event of delamination is found to occur very quickly and cannot be accurately monitored by the traveling microscope. It is not possible to determine whether delamination initiated near the front of a transverse crack which is situated at the interface of the angle-ply and 90°-ply or at some other locations where the delamination threshold has been exceeded. This question cannot be adequately answered in the absence of a more refined stress analysis which includes the influence of transverse cracking and its interaction with the free edge of the laminate. These considerations, however, are beyond the scope of the present investigation. In what follows, delamination will be assumed to occur in a state of generalized plane strain without the influence of transverse cracking.

STRESS CALCULATIONS

The prerequisite of applying any failure criterion to predict composite failure requires a reliable stress solution. In this respect, the stress state near the free edge of the test coupon is of primary concern. Because of the complex nature of the stresses in the vicinity of free edges which may acquire high elevation, some preliminary discussions are in order. First, there is the possible

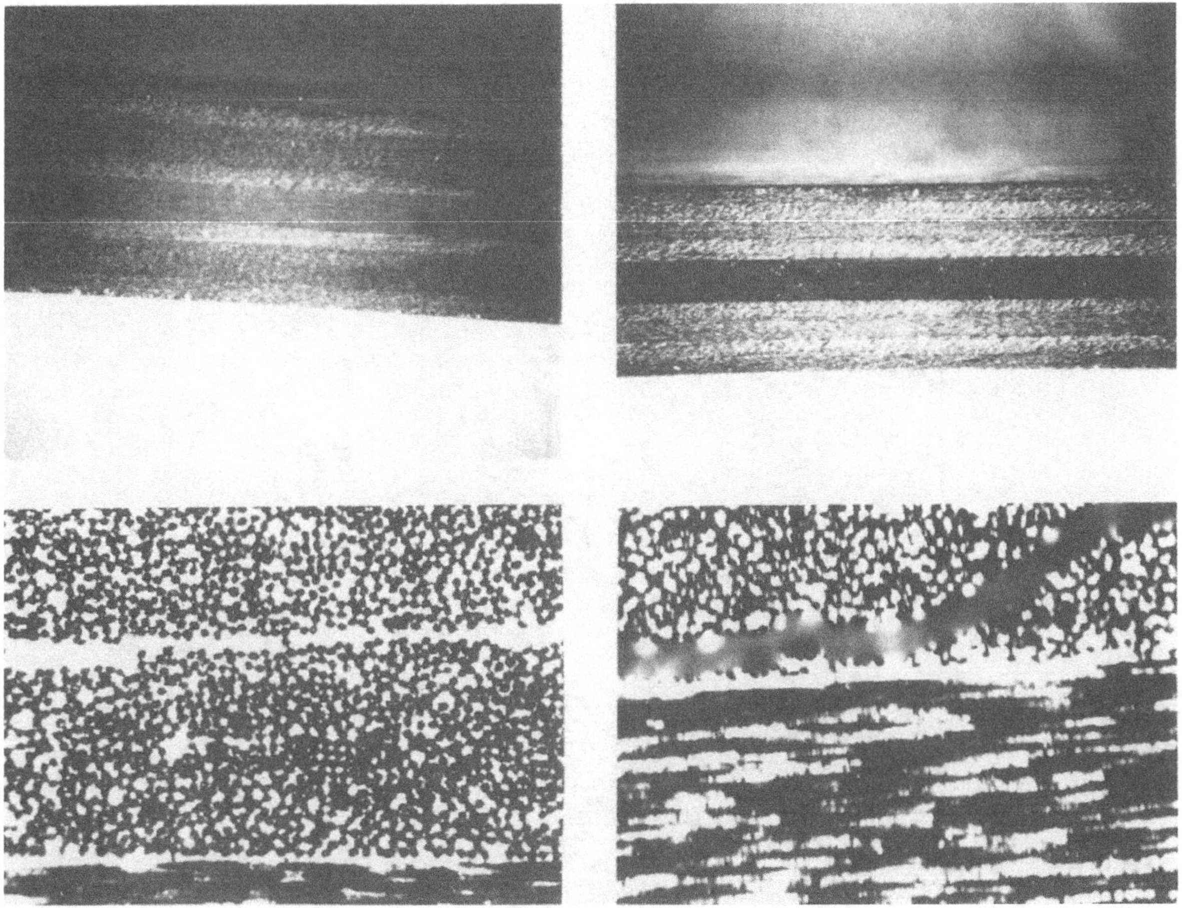

Fig. (1) - Typical delamination of T300/5208 graphite/epoxy $[(\pm 5)_2/90]_s$ laminates

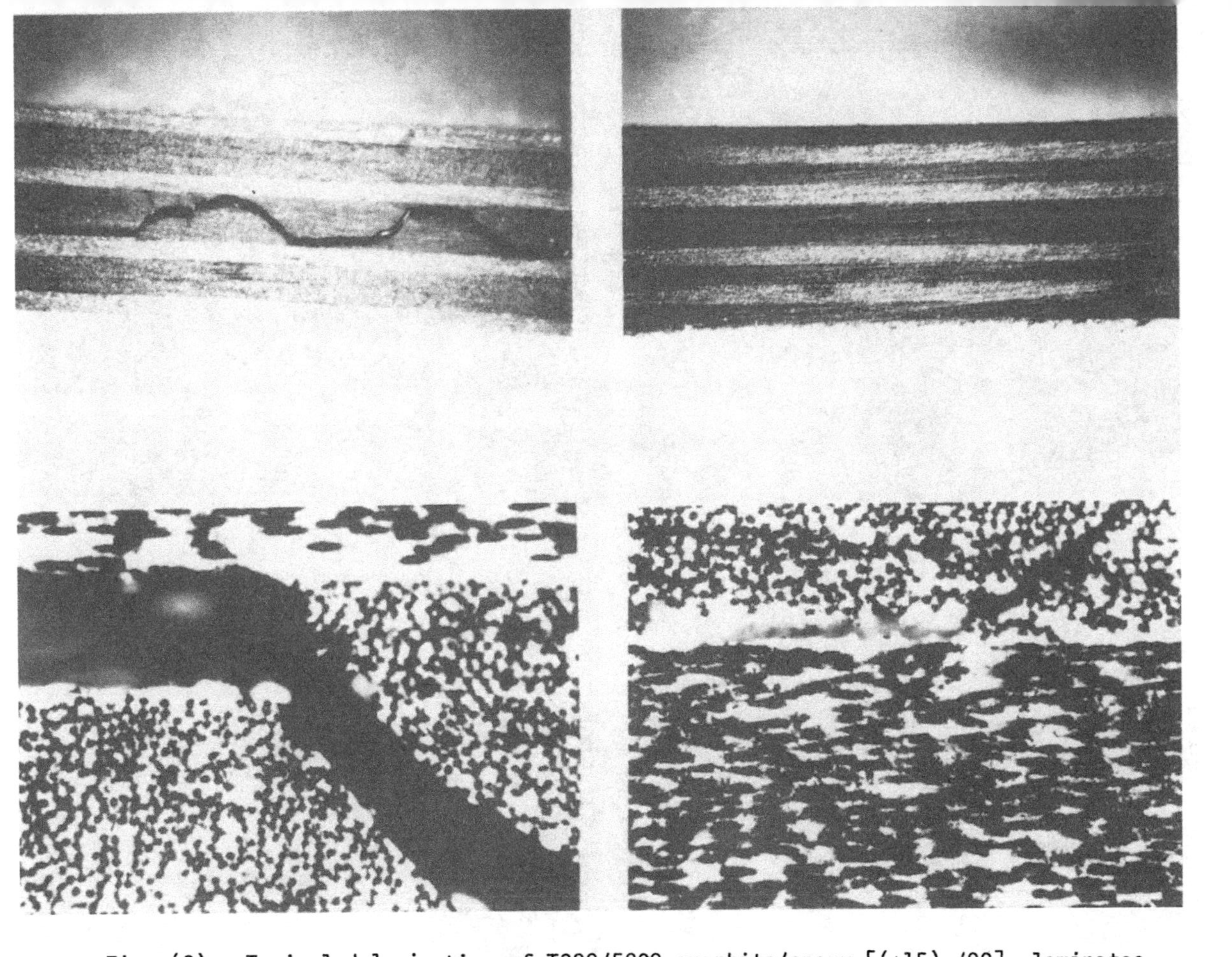

Fig. (2) - Typical delamination of T300/5208 graphite/epoxy $[(\pm15)_2/90]_s$ laminates

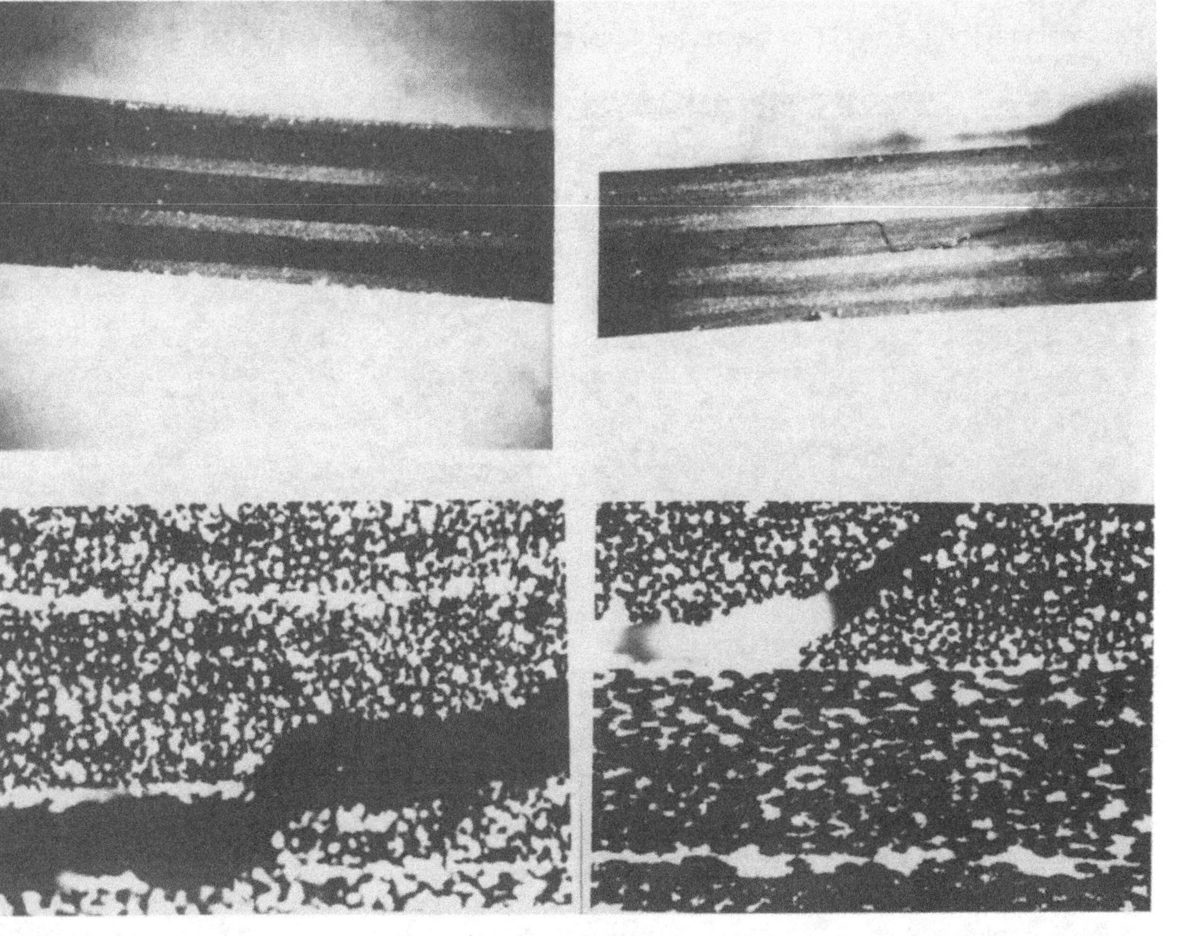

Fig. (3) - Typical delamination of T300/5208 graphite/epoxy $[(\pm 25)_2/90]_s$ laminates

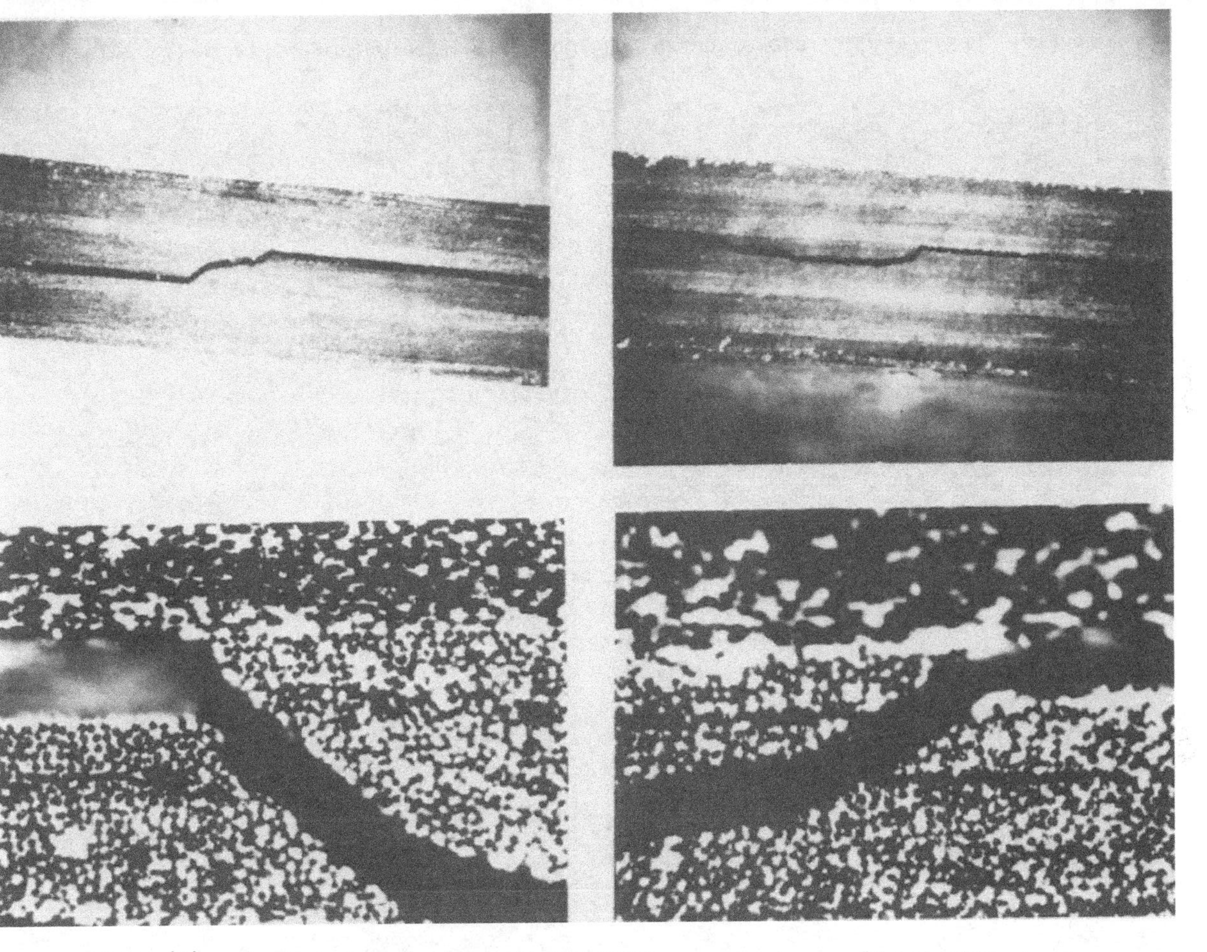

Fig. (4) - Typical delamination of T300/5208 graphite/epoxy $[(\pm 35)_2/90]_s$ laminates

Fig. (5) - Typical delamination of T300/5208 graphite/epoxy $[(\pm45)_2/90]_s$ laminates

high stress gradients near the interface and free edges of two adjoining anisotropic materials which may be simulated mathematically by the character of certain stress singularity. Next, the appropriate satisfaction of boundary conditions can also significantly influence the character of the solution.

Stress singularity. The problem of stress singularity has been studied by Ting and Chou [5,6] for a state of generalized plane strain. They considered the problem of edge singularity in an anisotropic composite which takes the form $r^{-\lambda}F(r,\theta)$, where λ depends on the elastic constants of the material. For the T300/5208 Gr/Ep laminate, the order of the stress singularity at the free edge of the interface between the $\theta°$ and 90° plies is governed by the exponent λ whose values for the five different types of laminates are given in Table 2. It is

TABLE 2 - ORDER OF STRESS SINGULARITY AT THE FREE EDGE OF THE INTERFACE BETWEEN (90/θ) PLIES, T300/5208

θ	5°	15°	25°	35°	45°
λ	0.054	0.0522	0.0476	0.0402	0.0304

seen that the order of stress singularity is much less than that at a crack tip with λ = 0.500. For r equal to 10^{-6} unit, the term $r^{-\lambda}$ has a value of approximately 2.0. This corresponds to microscopic scale size which is not of the concern of the present study being confined only to macroscopic scale analysis. The contribution of the singular term to the stress state will therefore be neglected.

Boundary conditions. The satisfaction of traction free edge boundary conditions is also important. Spilker and Chou [7] used the stress hybrid finite element method and showed that indeed the free-edge boundary conditions can be satisfied for the five types of laminates considered.

Figure 6(a) illustrates the element geometry, coordinate system and boundary conditions. Due to symmetry, only one quarter of the laminate needs to be analyzed. Figure 6(b) shows the degrees of freedom for the element in a typical layer. The region of analysis is separated into two. More elements smaller in

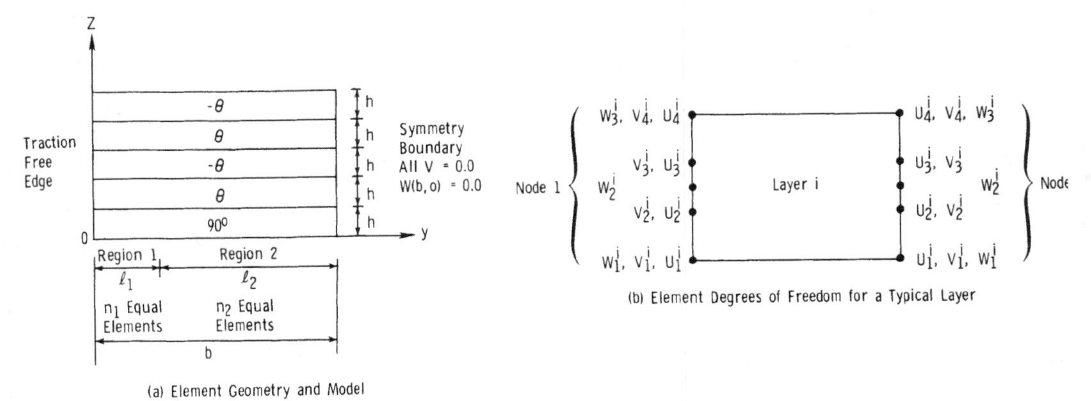

Fig. (6) - Element geometry and degrees of freedom

size are placed in regions marked 1 in Figure 6(a) near the traction free edge where high stress gradients are expected. In this region, $\ell_1 = 0.1b$ is subdivided into 5 equal elements. For region 2, the segment $\ell_2 = 0.9b$ is subdivided into 6 equal elements. No subdivisions are necessary in the z-direction, because the developed elements are already multi-layered. This grid pattern was selected on the basis of convergence studies. Additional refinements will have negligible influences on the stresses. Since the boundary conditions are satisfied exactly, there are only three non-zero stresses σ_x, σ_{xz} and σ_z at any point along the edge. Figure 7 shows the distribution of σ_x through the thickness of the laminates at

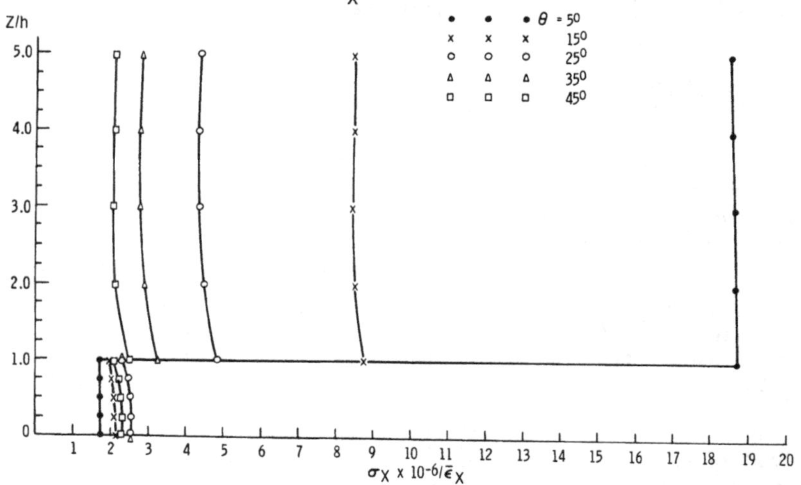

Fig. (7) - σ_x at the traction free edge for $[(\pm\theta)_2/90]_s$ laminate

the free edge for all the five cases. As it is expected that the difference of σ_x between the angle-plies and 90°-ply decreases as θ is increased. Figure 8 displays the distribution of the shear stress, σ_{xz}. The peak values occur at the interfaces of the $\pm\theta$-plies. The shear stress vanishes at the midplane $z/h = 0.0$ because of symmetry. It is interesting to note that maximum shear stresses occur at the interfaces for $\theta = 15°$ while minimum shear stresses corresponded to $\theta = 45°$. Figure 9 gives the variations of interlaminar normal stress σ_z with z/h. The maximum normal tensile stress occurred at the midplane of the laminate for all five cases. Approximately the same values of $(\sigma_z)_{max}$ were found for $\theta = 25°$ and 35° laminates. The lowest $(\sigma_z)_{max}$ corresponds to $\theta = 5°$. At the interface of angle-ply and 90°-ply $z/h = 1.0$, the stress σ_z is smaller than that at the midplane, but the shear stress σ_{xz} has a local maximum which may influence the failure mode.

TENSOR POLYNOMIAL STRENGTH THEORY

Using the rectangular Cartesian tensor notation, the quadratic tensor polynomial failure criterion [3] may be expressed as

$$F_i\sigma_i + F_{ij}\sigma_i\sigma_j \leq 1, \quad i,j = 1,2,\ldots,6 \tag{2}$$

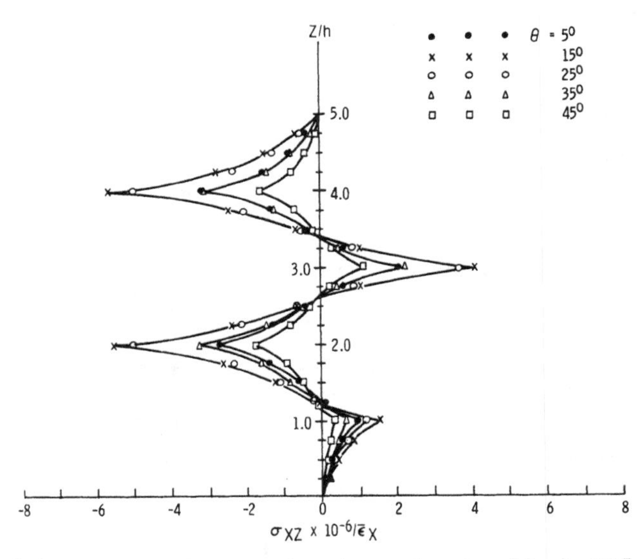

Fig. (8) - σ_{xz} at the traction free edge for $[(\pm\theta)_2/90]_s$ laminate

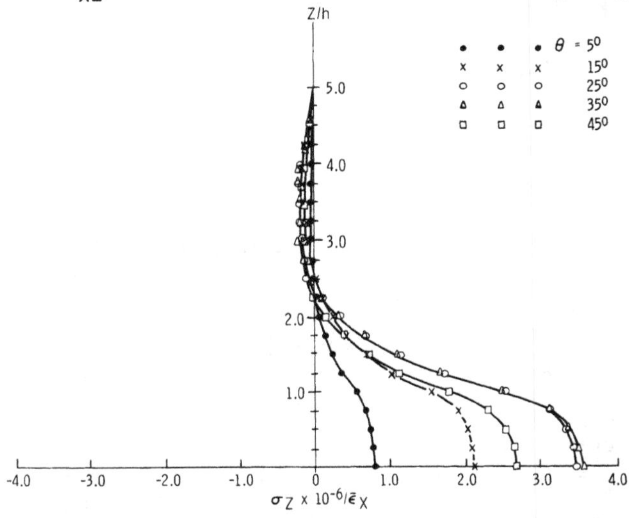

Fig. (9) - σ_z at the traction free edge for $[(\pm\theta)_2/90]_s$ laminate

where the failure stress components F_i and F_{ij} are expressed in terms of engineering strength measurements [3]. The measured strengths (in ksi) of T300/5208 Gr/Ep are given by [8]

$$
\begin{array}{lll}
\text{Longitudinal Tension:} & X_1 = 210 & \\
\text{Longitudinal Compression:} & X_1' = 200 & \\
\text{Transverse Tension:} & X_2 = 6.1 & \\
\text{Transverse Compression:} & X_2' = 21 & \\
\text{Shear in 1-2 Plane:} & X_6 = 13 & \\
\text{Biaxial Strength in} & \text{1-direction: } \tilde{\sigma}_1 = 230 & \\
 & \text{2-direction: } \tilde{\sigma}_2 = -15 &
\end{array}
\qquad (3)
$$

The application of the tensor polynomial strength theory, however, also requires a knowledge of the interlaminar strengths where it is not available for the T300 /5208 laminates. Therefore, it is necessary to assume $X_3 = X_2$, $X_3' = X_2'$, $X_5 = X_6$ and $X_4 = 0.5X_6$. These values may be substituted into equation (2) to yield

$$
\begin{aligned}
&F_1 = -0.000387;\ F_2 = F_3 = 0.116;\ F_4 = F_5 = F_6 = 0 \\
&F_{11} = 0.0231 \times 10^{-3};\ F_{22} = F_{33} = 7.81 \times 10^{-3} \\
&F_{44} = 23.67 \times 10^{-3};\ F_{55} = F_{66} = 5.92 \times 10^{-3} \\
&F_{23} = -7.81 \times 10^{-3};\ F_{12} = F_{13} = 0.0207 \times 10^{-3}
\end{aligned}
\qquad (4)
$$

in which F_i and F_{ij} have the units $(\text{ksi})^{-1}$ and $(\text{ksi})^{-2}$, respectively. Under the considerations, the strength vector for any complex stress state can be obtained and compared with the actual stress vector. Failure is assumed to occur when the magnitude of the actual stress vector exceeds that of the strength vector.

The failure criterion is applied to every point along the traction free edge for all five cases at the strain level in Table 1 when the first sight of transverse cracking was detected. It is found that the failure extends beyond the 90 plies into the angle-plies, which is unlike the experimental observation. Since both transverse cracking and delamination are considered to be closely associated with the transverse strengths, the disagreement between the theory and experiment could possibly be attributed to uncertainties in the transverse strengths. Bearing this in mind, a series of tests on 90° laminar coupon specimens with various widths was carried out and the results are shown in Figure 10. Note that the average strength of the specimens with 0.25 in. and 0.5 in. width is indeed approximately 6.1 ksi. However, the strength of those specimens with 1.0 in. width is higher. This is in contrast to the weak link theory of brittle materials. A possible explanation is that the specimens possess different surface to volume ratios. The transverse tension could possibly be greater than 6.1 ksi and as la as 10 ksi. Several values were used in the tensor polynomial strength theory to predict the failure region. For a transverse tension of 10 ksi, the failure zon is confined to the 90°-plies for all five cases which agrees with experiment. T ratios of actual stress to strength vector at points along the free edge for all

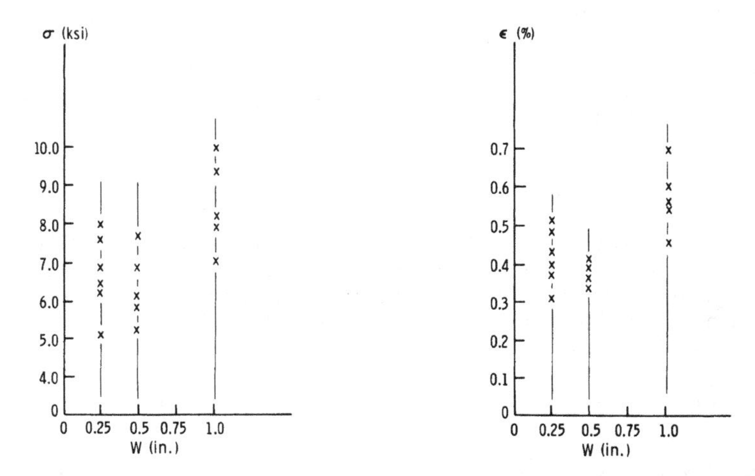

Fig. (10) - Test results for $(90)_8$ T300/5208 graphite/epoxy laminar with various widths

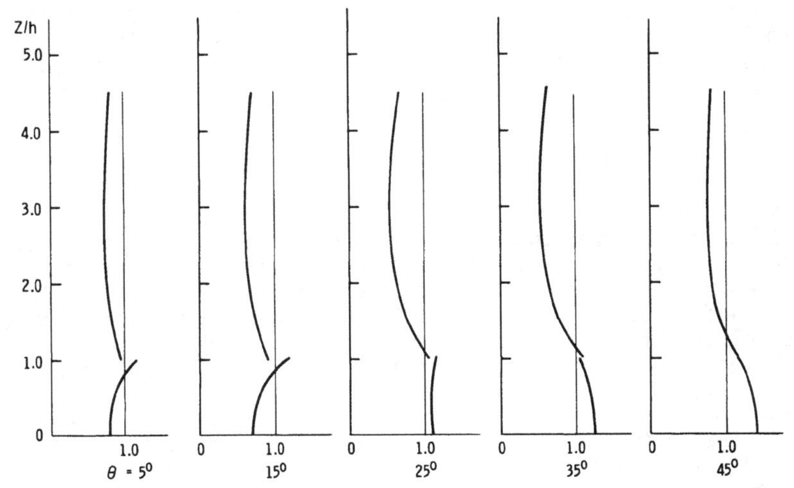

Fig. (11) - Ratio of stress vector to strength vector at traction free edge of $[(\pm\theta)_2/90]_s$ laminate

five cases with X_2 = 10 ksi are displayed in Figures 11. These points lie in the region where the actual stress to strength vector ratio is greater than unity represent failure. The adjustment made on the transverse tensile strength X_2 to allow agreement between theory and experiment requires justification. Other factors such as curing stress may also have to be considered. Those strength constants in equation (4) that must be altered for X_2 = 10 ksi are listed below:

$$F_{22} = F_{33} = 4.76 \times 10^{-3}$$

$$F_{23} = -4.76 \times 10^{-3}; \; F_{12} = F_{13} = 0.0604 \times 10^{-3} \tag{5}$$

STRAIN ENERGY DENSITY THEORY

The theory of strain energy density as proposed by Sih [4] assumes that failure occurs when the energy stored in an element of material reaches a critical value. The quantity dW/dV is the strain energy density function which can be easily computed from the stresses σ_i and strains ε_i:

$$\frac{dW}{dV} = \frac{1}{2}\sigma_i\varepsilon_i, \; i = 1,2,\ldots,6 \tag{6}$$

In the case of a linear, isotropic and elastic material, dW/dV can be divided into two components: one accounts for dilatation and the other for distortion. Hence, the combination of both effects is considered in the failure analysis. Their individual contributions are weighed mathematically* by minimizing and/or maximizing dW/dV with respect to the coordinate variables which are referenced from the possible failure sites in the material. To this end, a length parameter r can be introduced such that

$$\frac{dW}{dV} = \frac{S}{r} \tag{7}$$

in which S can be referred to as the strain energy density factor [4]. The form of equation (7) applies to all materials in continuum mechanics theories and is in no way restricted to linear elasticity as assumed in equation (6). When r is a fixed distance, then either dW/dV or S can be used to analyze failure. In the absence of plasticity, the onset of rapid fracture or failure corresponds to a critical value of S, i.e., S_c or critical value of dW/dV, i.e., $(dW/dV)_c$.

In order to determine the location of failure by fracture, S may be minimized** with reference to the coordinate variables y and z in Figure 6(a). Actual fracture is assumed to take place when $S_{min} = S_c$, i.e.,

$$S_c = r_o\left(\frac{dW}{dV}\right)_c \tag{8}$$

In equation (8), r_o may be interpreted as the radius of a core region outside of which the continuum mechanics analysis is valid.

---

* The dilatation component when neglected arbitrarily in dW/dV leads to the Von Mises' yield criterion in the classical theory of plasticity.

** The locations where S attains maximum correspond to failure by yielding.

The five different types of laminates mentioned earlier are analyzed by application of the strain energy density theory. The locations of the critical strain energy density factor are in the midplane of the 90°-plies, i.e., y = .02b and z=0 from the free edge. Typical plots of dW/dV versus y at z=0 and z at y = .02b are shown graphically in Figure 12. The values of $(dW/dV)_c$ at

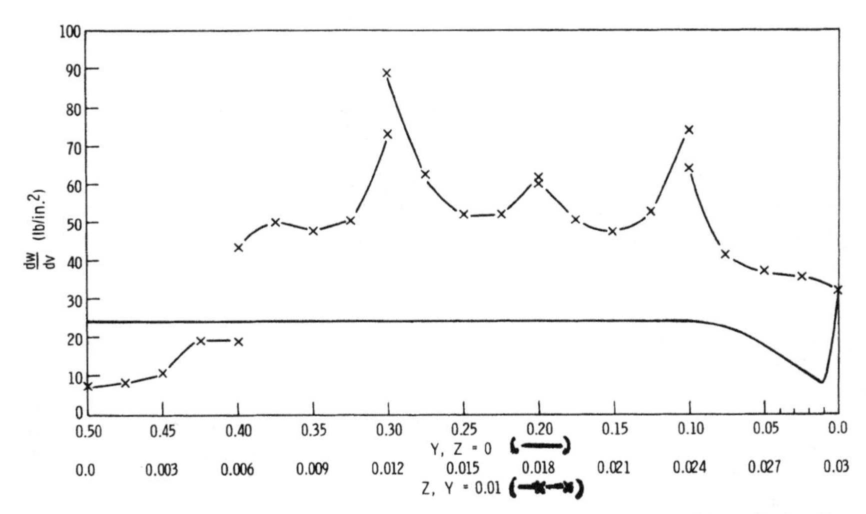

Fig. (12) - Strain energy density distribution for $[(\pm25)_2/90]_s$ laminate at ε = 0.3%

the strain level in Table 1 when transverse cracking first occurs are calculated and given in Table 3. These critical values also fall within the scatter band

TABLE 3 - CRITICAL VALUES OF STRAIN ENERGY DENSITY FUNCTION FOR T300/5208 Gr/Ep

θ	5°	15°	25°	35°	45°
$(dW/dV)_c$(lb/in³)	19.1	9.1	7.9	10.5	23.1

of the strain energy density functions calculated from the results shown in Figure 10.

Assuming that T300/5208 Gr/Ep has a critical strain energy density factor of S_c = 0.15 in-lb/in², then the size of the core region r_0 can be estimated from equation (8) with the aid of Table 3. Such a region determines the limiting scale length of the analysis such that minute surface defects or imperfections on the specimen free edges are excluded. Refer to Table 4 for values of r_0 as θ is varied from 5°, 15°,---, 45°. Since transverse cracks are observed to initiate inside the specimen at approximately y = .02b and z=0 and propagate toward the free edge, it is possible to estimate the imperfection size:

$$\text{Imperfection size} = y_{at(dW/dV)_c} - r_0 \tag{9}$$

whose numerical values are given in Table 4.

TABLE 4 - SIZES OF CORE REGION AND EDGE IMPERFECTION

θ	5°	15°	25°	35°	45°
r_o (in.) for S_c = 0.15	0.0079	0.0165	0.0189	0.0143	0.0065
Imperfection size (in.)	0.0021	0	0	0	0.0035

The definition and determination of S_c for the T300/5208 Gr/Ep require further study. Nevertheless, it is encouraging to note that the theoretical and experimental results agree well and show that delamination failure can be predicted with a reasonable degree of accuracy.

SUMMARY AND DISCUSSION

Laminates with stacking sequences of $[(\pm\theta)_2/90]_s$ and $[\theta/-\theta_2/\theta/90]_s$ for five different θ values were fabricated and tested. Transverse cracking has been observed to occur prior to delamination. This additional failure mode complicates the stress analysis, because it introduces stress singularities into the mathematical analysis in addition to those created by delamination. The specimen may have to be modified in order to isolate the mode of delamination failure from the others. This is left for future research.

Despite the shortcomings mentioned earlier, the generalized plane strain stress solution coupled with the tensor polynomial strength theory and the theory of strain energy density factor gave predictions that are close to the damages observed experimentally. Transverse cracking may or may not influence the results reported in this investigation. Certainly for the case θ = 25°, Table 1 shows that its effect is negligible. It is felt that the present models when used with care can yield useful information on the delamination of Gr/Ep laminates.

ACKNOWLEDGEMENT

The author wishes to express his appreciation to Dr. E. M. Wu of Lawrence Livermore National Laboratory for his valuable technical discussions on the subject and providing some of specimen materials and unpublished test results on T300/5208 Gr/Ep, and to Dr. B. M. Halpin of AMMRC who also provided specimen materials.

REFERENCES

[1] Pagano, N. J. and Pipes, R. B., "Some observations on the interlaminar strength of composite laminates", Int. J. Mech. Sci., Vol. 15, pp. 679-688, 1973.

[2] Harris, A. and Orringer, O., "Investigation of angle-ply delamination specimen for interlaminar strength test", J. Comp. Materials, Vol. 12, pp. 285-299, 1978.

[3] Wu, E. M., "Phenomenological anisotropic failure criterion", Composite Materials, Vol. 2, G. P. Sendeckyj, ed., 1974.

[4] Sih, G. C., "A special theory of crack propagation", Methods of Analysis and Solution of Crack Problems, Vol. I, G. C. Sih, ed., Noordhoff International Publishing, Leyden, pp. XXI-XLV, 1973.

[5] Ting, T. C. T. and Chou, S. C., "Edge singularities in anisotropic composites", to appear in International Journal of Solids and Structures.

[6] Ting, T. C. T. and Chou, S. C., "Stress singularities in laminated composites", presented at 2nd USA-USSR Symposium on Fracture of Composite Materials, Lehigh University, Bethlehem, Pa., March 1981.

[7] Spilker, R. L. and Chou, S. C., "Edge effects in symmetric composite laminates: importance of satisfying the traction free edge condition", J. Comp. Materials, Vol. 14, pp. 2-20, January 1980.

[8] Wu, E. M., Private communication and unpublished data, Lawrence Livermore National Laboratory.

TRESS SINGULARITIES IN LAMINATED COMPOSITES

T. C. T. Ting

University of Illinois at Chicago Circle
Chicago, Illinois 60680

S. C. Chou

Army Materials and Mechanics Research Center
Watertown, Massachusetts 02172

ABSTRACT

The singular nature of the stress distribution near the free edge of an interface in a laminated composite is analyzed by using the method originally due to Stroh for solving dislocation problems. With this approach, the degenerate case of uncoupling between the displacement components u_3 and u_1, u_2 presents no difficulties in finding the eigenvalues and eigenvectors of the elasticity constants. Numerical examples are presented for the order of singularity at the interface of an angle-ply graphite/epoxy laminated composite in which each layer is assumed to be orthotropic material with the direction of the fibers inclined at various angles θ with the x_3-axis.

INTRODUCTION

The stress singularity at the free edge of an interface in a laminated composite is one of the factors responsible for the delamination when the composite is subject to an external load. An analytical solution which is valid for the whole composite is next to impossible to obtain. Several approximate numerical solutions are available which show a good agreement between them for points away from the free edge [1-3]. For points near the free edge, numerical solutions are not capable of predicting an infinite stress and it is where the discrepancies between various approximate solutions occur. Wang and Choi [4] used an eigenfunction expansion technique to determine the stresses in the interface. However, the completeness of the eigenfunction expansion is an open question [5]. Moreover, Wang and Choi used the method originally due to Lekhnitskii [6] which, as we will show in the paper, breaks down under certain special situations.

Although the order of singularity does not depend on the stacking sequence of the lamina in the composite, the coefficients of the singular terms which are related to the intensity factor do. This suggests that one might use a special finite element at the free edge points with regular finite elements elsewhere so that the exact behavior of the singularity is prescribed in the special finite element while the intensity factors at the free edge points are determined by solving the complete boundary value problem. With this in mind we present here a means to determine the nature of the singularities at a free edge point. We use the approach which was originally due to Stroh [7] and was further developed by Barnett and his co-workers [8] for solving dislocation problems in anisotropic elastic solids.

BASIC EQUATIONS

Let σ_{ij}, ε_{ij} and u_i be the stress, strain and displacement, respectively, in a fixed rectangular coordinate system x_i. The strain-displacement, stress-strain and equilibrium equations can be written as

$$\varepsilon_{ij} = \tfrac{1}{2}(\partial u_i/\partial x_j + \partial u_j/\partial x_i) \tag{1}$$

$$\sigma_{ij} = c_{ijk\ell}\varepsilon_{k\ell} \tag{2}$$

$$\partial\sigma_{ij}/\partial x_j = 0 \tag{3}$$

where repeated indices imply summation and

$$c_{ijk\ell} = c_{k\ell ij} = c_{jik\ell} \tag{4}$$

are the elasticity constants.

We assume that ε_{ij} and hence σ_{ij} are independent of x_3. Let

$$u_i = \upsilon_i f(Z) + U_i \tag{5}$$

$$\sigma_{ij} = \tau_{ij}\, df(Z)/dZ + T_{ij} \tag{6}$$

$$Z = x_1 + px_2 \tag{7}$$

where p is the eigenvalue, υ_i and τ_{ij} are the associated eigenvectors. T_{ij} is the stress associated with U_i while U_i is obtained by integrating Eq. (1), keeping in mind that ε_{ij} are independent of x_3. Thus we have, [4],

$$\left.\begin{aligned} U_1 &= u_1^o + \omega_2 x_3 - \omega_3 x_2 - \left(\frac{a_1}{2}\, x_3 + a_4 x_2\right)x_3 \\ U_2 &= u_2^o + \omega_3 x_1 - \omega_1 x_3 - \left(\frac{a_2}{2}\, x_3 - a_4 x_1\right)x_3 \end{aligned}\right\} \tag{8}$$

$$U_3 = u_3^o + \omega_1 x_2 - \omega_2 x_1 + (a_1 x_1 + a_2 x_2 + a_3) x_3 \tag{8 (Cont'd)}$$

where u_i^o, ω_i $(i = 1,2,3)$ and a_j $(j = 1 \text{ to } 4)$ are constants. It is not difficult to see that u_i^o and ω_i represent rigid body translation and rotation, respectively. a_1 and a_2 represent a bending in the (x_1,x_3) and (x_2,x_3) planes, respectively. a_3 is the uniform extension in the x_3-direction and a_4 is the twisting of the body about the x_3-axis.

To determine the eigenvalue p and the eigenvectors υ_i and τ_{ij}, we substitute Eqs. (5) and (6) into (1-3) ignoring the terms associated with U_i and T_{ij} since they can be treated separately. We have

$$\tau_{ij} = (c_{ijk1} + pc_{ijk2})\upsilon_k \tag{9}$$

$$D_{ik}\upsilon_k = 0 \tag{10}$$

where

$$D_{ik} = Q_{ik} + p(R_{ik} + R_{ki}) + p^2 T_{ik} \tag{11}$$

$$\left.\begin{aligned} Q_{ik} &= c_{i1k1} \\ R_{ik} &= c_{i1k2} \\ T_{ik} &= c_{i2k2} \end{aligned}\right\} \tag{12}$$

The vanishing of the determinant D_{ik} yields a sextic equation for p. Since the eigenvalues are all non-real [6,7], we have three pairs of complex conjugates for p. The associated eigenvectors υ_i and τ_{ij} are obtained from Eqs. (9) and (10).

We see from Eqs. (11) and (12) that p is determined from $c_{ijk\ell}$. υ_i and τ_{ij} are then determined from Eqs. (9) and (10). In [6] whose approaches were used by Wang and Choi [5], Lekhnitskii determined p from the stiffness constants $s_{ijk\ell}$. Two stress functions which satisfy the equations of equilibrium are then used to determine τ_{ij} and υ_i. The problem with that approach is that $\tau_{22} = 1$ regardless of the value of p. For materials which are symmetric with respect to the (x_1,x_2) plane, u_3 is uncoupled from u_1 and u_2 and $\tau_{22} = 0$ for one of the p's [9]. Therefore, Lekhnitskii's approach is not applicable for problems in which $\tau_{22} = 0$.

Although the material properties of a composite are often given in terms of the stiffness constants $s_{ijk\ell}$, one can transform them to $c_{ijk\ell}$ before analyzing the stress singularities.

ELASTICITY CONSTANTS VERSUS STIFFNESS CONSTANTS

In view of the symmetry property of $c_{ijk\ell}$, Eq. (4), we write Eqs. (2) and (4) as

$$\sigma_q = c_{qt}\varepsilon_t , \qquad c_{qt} = c_{tq} , \tag{13}$$

where

$$\left.\begin{array}{lll} \sigma_1 = \sigma_{11} , & \sigma_2 = \sigma_{22} , & \sigma_3 = \sigma_{33} , \\ \sigma_4 = \sigma_{23} , & \sigma_5 = \sigma_{13} , & \sigma_6 = \sigma_{12} , \end{array}\right\} \tag{14}$$

$$\left.\begin{array}{lll} \varepsilon_1 = \varepsilon_{11} , & \varepsilon_2 = \varepsilon_{22} , & \varepsilon_3 = \varepsilon_{33} , \\ \varepsilon_4 = 2\varepsilon_{23} , & \varepsilon_5 = 2\varepsilon_{13} , & \varepsilon_6 = 2\varepsilon_{12} . \end{array}\right\} \tag{15}$$

Notice that the transformations between σ_{ij} and σ_i are not exactly the same as between ε_{ij} and ε_i. The transformations between $c_{ijk\ell}$ and c_{qt} are as follows

$$ij \text{ or } k\ell = \left\{\begin{array}{l} 11 \\ 22 \\ 33 \\ 23 \\ 31 \\ 12 \end{array}\right. , \qquad q \text{ or } t = \left\{\begin{array}{l} 1 \\ 2 \\ 3 \\ 4 \\ 5 \\ 6 \end{array}\right. \tag{16}$$

Hence, for instance, we have

$$c_{12} = c_{1122} , \qquad c_{14} = c_{1123} , \qquad c_{64} = c_{1223} \tag{17}$$

In composites, Eq. (2) is sometimes written as

$$\varepsilon_{ij} = s_{ijk\ell}\sigma_{k\ell} \tag{18}$$

where $s_{ijk\ell}$ are the stiffness constants. They are related to the elasticity constants by

$$c_{ijk\ell}s_{k\ell pq} = \delta_{ip}\delta_{jq} \tag{19}$$

where δ_{ip} is the Kronecker delta. Since $s_{ijk\ell}$ possess the same symmetry property as $c_{ijk\ell}$, we may write Eq. (18) as

$$\varepsilon_q = s_{qt}\sigma_t , \qquad s_{qt} = s_{tq} \tag{20}$$

While the transformations between σ_{ij}, ε_{ij} and σ_t, ε_t are given by Eqs. (14) and (15), the transformations between the subscripts of $s_{ijk\ell}$ and s_{qt} follow Eq. (16) with the following modifications: if either q or t is larger than 3, $s_{qt} = 2s_{ijk\ell}$. If both q and t are larger than 3, $s_{qt} = 4s_{ijk\ell}$. Thus, for instance, we have

$$s_{12} = s_{1122}\,, \qquad s_{14} = 2s_{1123}\,, \qquad s_{64} = 4s_{1223} \tag{21}$$

STRESS SINGULARITIES AT A FREE EDGE

Consider a laminated composite whose cross section is shown in Fig. 1. To study the stress singularities at a free edge of an interface, we place the origin of the x_1,x_2 axes at one of the free edge points with the x_1-axis along the interface and the x_2-axis along the free edge. We choose $f(Z)$ of Eq. (5) in the form

$$f(Z) = Z^{1+\delta}/(1+\delta) \tag{22}$$

where δ is a constant. Since there are three pairs of complex conjugates for p, we write Eqs. (5) and (6) as

$$u_i = \sum_{L=1}^{3}\left\{A_L \upsilon_{i,L} Z_L^{1+\delta} + B_L \bar{\upsilon}_{i,L} \bar{Z}_L^{1+\delta}\right\}\Big/(1+\delta) + U_i \tag{23}$$

$$\sigma_{ij} = \sum_{L=1}^{3}\left\{A_L \tau_{ij,L} Z_L^{\delta} + B_L \bar{\tau}_{ij,L} \bar{Z}_L^{1+\delta}\right\} \tag{24}$$

where an overbar denotes the complex conjugate, A_L and B_L are constants and the subscript L identifies the three pairs of the eigenvalues. Using the polar coordinates (r,ϕ), we may write Z as [9]

$$Z = x_1 + px_2 = r\zeta \tag{25}$$

where

$$\zeta = \cos\phi + p\sin\phi \tag{26}$$

Equations (23) and (24) can then be written as

$$u_i = \frac{r^{1+\delta}}{1+\delta}\sum_{L=1}^{3}\left\{A_L \upsilon_{i,L} \zeta_L^{1+\delta} + B_L \bar{\upsilon}_{i,L} \bar{\zeta}_L^{1+\delta}\right\} + U_i \tag{27}$$

$$\sigma_{ij} = r^{\delta}\sum_{L=1}^{3}\left\{A_L \tau_{ij,L} \zeta_L^{\delta} + B_L \bar{\tau}_{ij,L} \bar{\zeta}_L^{\delta}\right\} \tag{28}$$

Equations (27) and (28) apply to the material with the elasticity constants

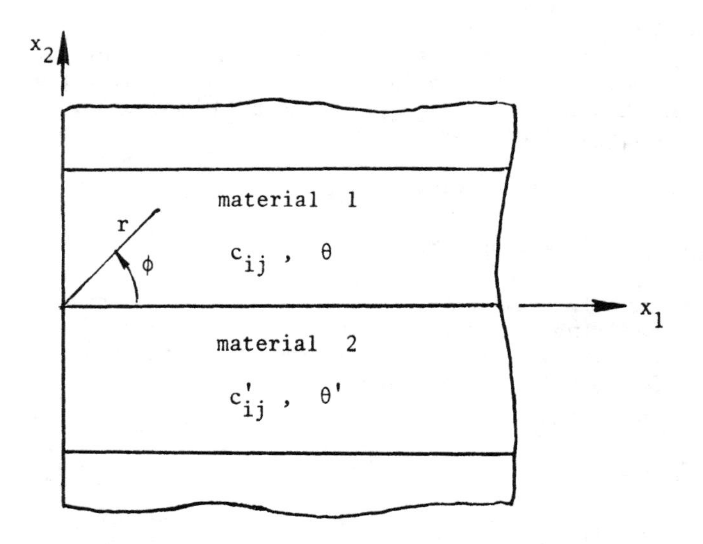

Fig. (1) - Two layers (θ/θ') of an angle-ply laminated composite

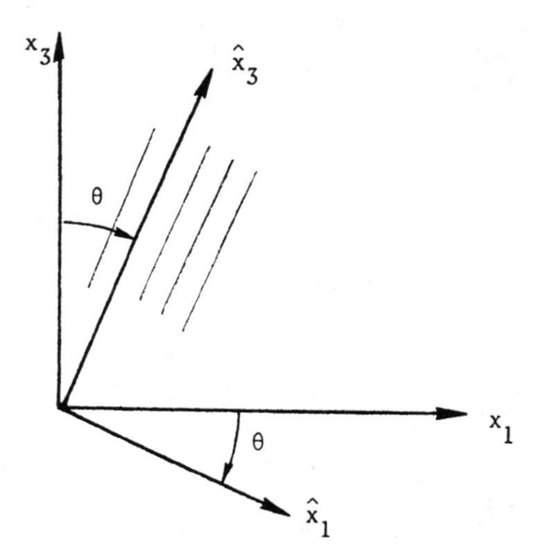

Fig. (2) - Principal directions of an angle-ply laminate

c_{ij}, Fig. 1. For the material with the elasticity constants c'_{ij}, similar equations may be written with a prime added to all quantities except r and δ. By applying the interface continuity condition along $\phi = 0$ and the stress free condition on $\phi = \pm\pi/2$, one obtains 12 linear homogeneous equations for A_L, B_L, A'_L and B'_L which can be written as

$$\underset{\sim}{K}\underset{\sim}{q} = \underset{\sim}{0} \tag{29}$$

where $\underset{\sim}{K}$ is a square matrix and $\underset{\sim}{q}$ is a column matrix whose elements are $A_L, B_L, \ldots$ A nontrivial solution for $A_L, B_L, \ldots$ exists if the determinant of $\underset{\sim}{K}$ vanishes. This provides a root for δ. If the real part of δ is negative, we see from Eq. (28) that σ_{ij} is singular at $r = 0$. In applications, only the real or imaginary part of the right-hand sides of Eqs. (27) and (28) should be used.

If δ is real, one may choose without loss of generality $B_L = \bar{A}_L$. Equations (27) and (28) can then be written as

$$u_i = \frac{r}{1+\delta} \sum_{L=1}^{3} \left\{ M_L \mathrm{Re}\left[\upsilon_{i,L} \zeta_L^{1+\delta}\right] + N_L \mathrm{Im}\left[\upsilon_{i,L} \zeta_L^{1+\delta}\right] \right\} + U_i \tag{30}$$

$$\sigma_{ij} = r^{\delta} \sum_{L=1}^{3} \left\{ M_L \mathrm{Re}\left[\tau_{ij,L} \zeta_L^{\delta}\right] + N_L \mathrm{Im}\left[\tau_{ij,L} \zeta_L^{\delta}\right] \right\} \tag{31}$$

where Re and Im stand for real and imaginary, respectively, and M_L, N_L are real constants.

ORTHOTROPIC MATERIALS

For orthotropic materials in which the axes of symmetry are the (x_1, x_2, x_3) axes, s_{ij} vanish except [10,11]

$$s_{11} = 1/E_1, \quad s_{22} = 1/E_2, \quad s_{33} = 1/E_3$$

$$\left.\begin{aligned} & s_{44} = 1/G_{23}, \quad s_{55} = 1/G_{31}, \quad s_{66} = 1/G_{12} \\ & s_{12} = s_{21} = -\nu_{21}/E_2 \\ & s_{13} = s_{31} = -\nu_{31}/E_3 \end{aligned}\right\} \tag{32}$$

$$s_{23} = s_{32} = -\nu_{32}/E_3$$

where E_1, E_2, E_3, G_{23}, G_{31}, G_{12}, ν_{21}, ν_{31} and ν_{32} are the engineering constants. In terms of c_{ij}, we have [10]

$$
\begin{aligned}
c_{11} &= (s_{22}s_{33} - s_{23}^2)/s_o \\
c_{22} &= (s_{33}s_{11} - s_{13}^2)/s_o \\
c_{33} &= (s_{11}s_{22} - s_{12}^2)/s_o \\
c_{12} &= c_{21} = (s_{13}s_{23} - s_{12}s_{33})/s_o \\
c_{13} &= c_{31} = (s_{12}s_{23} - s_{13}s_{22})/s_o \\
c_{23} &= c_{32} = (s_{12}s_{13} - s_{23}s_{11})/s_o \\
c_{44} &= G_{23}\,, \quad c_{55} = G_{31}\,, \quad c_{66} = G_{21}
\end{aligned}
\tag{33}
$$

where

$$
s_o = 2s_{12}s_{23}s_{31} + s_{11}s_{22}s_{33} - s_{11}s_{23}^2 - s_{22}s_{13}^2 - s_{33}s_{12}^2 \tag{34}
$$

In an angle-ply graphite/epoxy laminated composite, the material is assumed to be orthotropic with respect to the $(\hat{x}_1, x_2, \hat{x}_3)$ coordinate system in which the $\hat{x}_3$-direction is the direction of the fibers which make an angle θ with the x_3-axis, Fig. 2. In the rest of the paper, we will assume that materials 1 and 2 in Fig. 1 are made of the same orthotropic material except that the angles θ and θ' are different. Thus $\hat{c}_{ij}$ are the same for both materials. The relations between c_{ij} and $\hat{c}_{ij}$ can be shown to be

$$
\begin{aligned}
c_{11} &= c^4\hat{c}_{11} + 2c^2s^2(\hat{c}_{13} + 2\hat{c}_{55}) + s^4\hat{c}_{33} \\
c_{12} &= c^2\hat{c}_{12} + s^2\hat{c}_{32} \\
c_{13} &= (c^4 + s^4)\hat{c}_{31} + c^2s^2(\hat{c}_{11} + \hat{c}_{33} - 4\hat{c}_{55}) \\
c_{15} &= cs\left\{c^2(\hat{c}_{13} + 2\hat{c}_{55} - \hat{c}_{11}) - s^2(\hat{c}_{13} + 2\hat{c}_{55} - \hat{c}_{33})\right\} \\
c_{22} &= \hat{c}_{22} \\
c_{23} &= c^2\hat{c}_{32} + s^2\hat{c}_{12} \\
c_{25} &= cs(\hat{c}_{32} - \hat{c}_{12}) \\
c_{33} &= c^4\hat{c}_{33} + 2c^2s^2(\hat{c}_{13} + 2\hat{c}_{55}) + s^4\hat{c}_{11} \\
c_{35} &= cs\left\{-c^2(\hat{c}_{31} + 2\hat{c}_{55} - \hat{c}_{33}) + s^2(\hat{c}_{13} + 2\hat{c}_{55} - \hat{c}_{11})\right\}
\end{aligned}
\tag{35}
$$

$$\left.\begin{aligned}
c_{44} &= c^2\hat{c}_{44} + s^2\hat{c}_{66} \\
c_{46} &= cs(\hat{c}_{44} - \hat{c}_{66}) \\
c_{55} &= (c^2 - s^2)^2\hat{c}_{55} + c^2s^2(\hat{c}_{11} - 2\hat{c}_{13} + \hat{c}_{33}) \\
c_{66} &= c^2\hat{c}_{66} + s^2\hat{c}_{44}
\end{aligned}\right\} \quad \begin{matrix}(35)\\ (\text{Cont'd})\end{matrix}$$

where, for simplicity, we used the notations

$$c = \cos\theta\ , \qquad s = \sin\theta \tag{36}$$

Notice that

$$\begin{aligned}
c_{11} + 2c_{13} + c_{33} &= \hat{c}_{11} + 2\hat{c}_{13} + \hat{c}_{33} \\
c_{55} - c_{13} &= \hat{c}_{55} - \hat{c}_{13} \\
c_{44} + c_{66} &= \hat{c}_{44} + \hat{c}_{66} \\
c_{12} + c_{23} &= \hat{c}_{12} + \hat{c}_{23} \\
c_{44}c_{66} - c_{46}^2 &= \hat{c}_{44}\hat{c}_{66} - \hat{c}_{46}^2
\end{aligned} \tag{37}$$

Therefore, these are invariants under rotation of the (x_1, x_3) axes.

NUMERICAL EXAMPLES

We use the following engineering constants [12] for each layer which is an orthotropic material referred to the $(\hat{x}_1, x_2, \hat{x}_3)$ coordinate system:

$$\left.\begin{aligned}
E_1 &= E_2 = 2.1 \times 10^6 \text{ psi} \\
E_3 &= 20 \times 10^6 \text{ psi} \\
G_{12} &= G_{23} = G_{31} = 0.85 \times 10^6 \text{ psi} \\
\nu_{21} &= \nu_{31} = \nu_{32} = 0.21
\end{aligned}\right\} \tag{38}$$

With Eq. (38), $\hat{s}_{ij}$ and $\hat{c}_{ij}$ are obtained from Eqs. (32) and (33) and c_{ij} and c'_{ij} associated with θ and θ' in the two layers are determined from Eq. (35). Equations (9-12) then provide the eigenvalues p_L $(L = 1,2,3)$ and the associated eigenvectors $\upsilon_{i,L}$ and $\tau_{ij,L}$. By substituting Eqs. (27,28) or (30,31) in the stress-free boundary conditions at $\phi = \pm\pi/2$ and the continuity conditions at $\phi = 0$, one obtains 12 homogeneous linear equations for the constants A_L, B_L, A_L', B_L' $(L = 1,2,3)$ or M_L, N_L, M_L', N_L' $(L = 1,2,3)$. This is Eq. (29). For a non-trivial solution of these constants, the determinant of the coefficient matrix $\underset{\sim}{K}$ must vanish. δ in Eqs. (27) and (28) is a root of this determinant. When the real part of δ is negative, we have a stress singularity at the free edge.

We will first present the result on the eigenvalues p_L. The dependence of the eigenvalues on the orientation of the fibers θ is shown in Fig. 3a. For the material given by Eq. (38), all p's are purely imaginary and can be written as

$$p_L = \pm i\beta_L \qquad (L = 1,2,3) \tag{39}$$

If the material were isotropic, $\beta_L = 1$ for all L. In Fig. 3a we see that all β_L are nearly equal to one for small θ. At first we thought that the curves for p_1 and p_2 intersect at $\theta \cong 6.5$ degrees. This would imply that p is a double root and the analyses presented here need a modification (see [9]) for $\theta = 6.5$ degrees. However, further calculations for p_L for θ between 5 and 8 degrees show that they do not intersect, Fig. 3b. Therefore, Eqs. (27) and (28) are valid for the material given by Eq. (38) with an arbitrary orientation θ.

We next present the order of singularity δ for various combinations of the angle-ply laminates. This is shown in Table 1. Three combinations of θ and θ' are given. In the first group $\theta' = 0$ and θ varies from 15 to 90 degrees. The largest singularity occurs at (90/0) angle-ply composite where $\delta = -3.33888 \times 10^{-2}$. In the second group $\theta' = 90$ degrees and θ varies from 0 to 75 degrees. The largest δ occurs at (0/90) angle-ply composite and has the same δ value as the (90/0) angle-ply composite as it should be. The last group is for $(\theta/-\theta)$ angle-ply composites in which θ varies from 15 to 75 degrees. This group has been studied by Wang and Choi [4] and our results agree with their calculations.

DISCUSSION AND CONCLUDING REMARKS

For $\theta = 0$ and 90 degrees, $\tau_{22} = 0$ and the method proposed in [6] and used in [4] is not applicable. We present the case of $\theta = 90$ degrees. For this case, the eigenvalues are:

$$\left.\begin{aligned} p_1 &= 4.76716\, i \\ p_2 &= 0.63439\, i \\ p_3 &= 1.00000\, i \end{aligned}\right\} \tag{40}$$

The eigenvectors υ_i and τ_{ij} associated with p_3 are

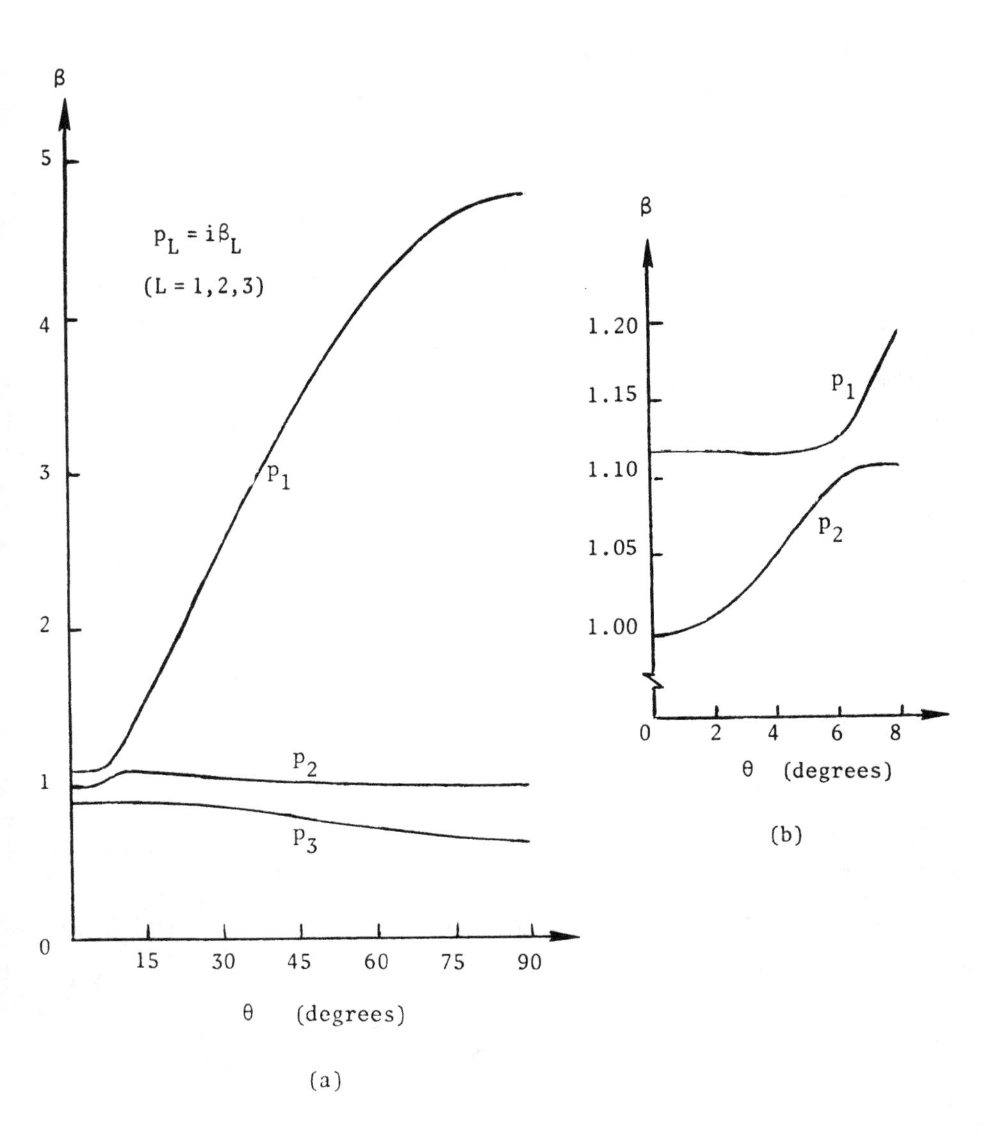

Fig. (3) - Dependence of the eigenvalues p_L on the angle θ of the angle-ply laminate

TABLE 1 - STRESS SINGULARITY δ AT THE FREE EDGE OF THE INTERFACE BETWEEN TWO ANGLE-PLY (θ/θ') GRAPHITE/EPOXY LAMINATES

θ	δ		
	$\theta' = 0$	$\theta' = 90^\circ$	$\theta' = -\theta$
0	------	-3.3388×10^{-2}	------
15	-1.3528×10^{-4}	-3.2814×10^{-2}	-6.4322×10^{-4}
30	-2.6286×10^{-3}	-2.8682×10^{-2}	-1.1658×10^{-2}
45	-9.6461×10^{-3}	-2.0575×10^{-2}	-2.5575×10^{-2}
60	-1.9866×10^{-2}	-1.0519×10^{-2}	-2.3346×10^{-2}
75	-2.9388×10^{-2}	-2.6785×10^{-3}	-8.9444×10^{-3}
90	-3.3388×10^{-2}	------	------

$$\left.\begin{array}{ll} \upsilon_1 = \upsilon_2 = 0 \,, & \upsilon_3 = 1.0 \\ \tau_{11} = \tau_{22} = \tau_{33} = \tau_{12} = 0 \,, & \\ \tau_{23} = 0.85\,i \,, & \tau_{13} = 0.85 \end{array}\right\} \quad (41)$$

In [6,4], τ_{22} is assumed to be unity which contradicts the actual solution.

After finding a δ from the determinant of $\underset{\sim}{K}$ in Eq. (29), one obtains $\underset{\sim}{q}$ whose elements are A_L, B_L... Therefore, A_L, B_L,... are related and since $\underset{\sim}{q}$ is non-unique to an arbitrary multiplicative constant there remains only one arbitrary constant for the singular terms in Eqs. (27) and (28). This arbitrary constant can be determined only if one considers the complete boundary value problem. If a finite element scheme is employed to solve the complete boundary value problem, one may incorporate in the finite element scheme a special element at each free edge point for which the singular nature of the stresses is determined by the present analyses, Eqs. (27) and (28).

In using the results of the analyses presented here for the interpretation of physical phenomena such as the delamination, one should not just compare the magnitudes of the singularity δ at each interface. While it is true that the larger the magnitude of δ the stronger the singularity is, the stress singularity as given by Eq. (28) also depends on the constants A_L, B_L. If these constants are vanishingly small, then the stress singularity may be considered as very weak.

ACKNOWLEDGEMENTS

The numerical calculation presented in Fig. 3b was performed by Miss Renata Zwierski. The work presented here is supported by the U.S. Army Materials and Mechanics Research Center, Watertown, Massachusetts, through contract DAAG 46-80-C-0081.

REFERENCES

[1] Herakovich, C. T., "On Thermal Edge Effects in Composite Laminate," Int. J. Mech. Sci., Vol. 18, No. 3, 1976, 129-134.

[2] Wang, A. S. D. and Cross, F. W., "Edge Effects on Thermally Induced Stresses in Composite Laminates," J. Composite Materials, Vol. 11, 1977, 300-312.

[3] Pipes, R. B., Visson, J. R. and Chou, T. W., "On the Hygrothermal Responses of Laminate Composite Systems," J. Composite Materials, Vol. 10, 1976, 129-148.

[4] Wang, S. S. and Choi, I., "Boundary Layer Thermal Stresses in Angle-Ply Composite Laminates," Modern Developments in Composite Materials and Structures, ed. by J. R. Vinston, ASME, 1979, 315-341.

[5] Dempsey, J. P. and Sinclair, G. B., "On the Stress Singularities in the Plane Elasticity of the Composite Wedge," J. of Elasticity, Vol. 9, 1979, 373-391.

[6] Lekhnitskii, S. G., "Theory of Elasticity on an Anisotropic Elastic Body," (translated by P. Fern), Holden-Day, Inc., San Francisco, 1963.

[7] Stroh, A. N., "Steady State Problems in Anisotropic Elasticity," J. Math. Phys., Vol. 41, 1962, 77-103.

[8] Bacon, D. J., Barnett, D. M. and Scattergood, R. O., "Anisotropic Continuum Theory of Lattice Defects," Progress in Materials Science, ed. by B. Chalmers, J. W. Christian and T. B. Massalski, Vol. 23, No. 2-4, Pergamon Press, 1978, 51-262.

[9] Ting, T. C. T. and Chou, S. C., "Edge Singularities in Anisotropic Composites," to appear in Int. J. Solids Structures.

[10] Jones, R. M., "Mechanics of Composite Materials," McGraw-Hill, 1975.

[11] Christensen, R. M., "Mechanics of Composite Materials," John Wiley & Sons, 1979.

[12] Pipes, R. B. and Pagano, N. S., "Interlaminar Stresses in Composite Laminates under Uniform Axial Extension," J. Composite Materials, Vol. 4, 1970, 538-548.

THE RELATIONSHIP OF STIFFNESS CHANGES IN COMPOSITE LAMINATES TO FRACTURE-RELATED DAMAGE MECHANISMS

K. L. Reifsnider and A. Highsmith

Virginia Polytechnic Institute and State University
Blacksburg, Virginia 24061

ABSTRACT

The monotonic or cyclic loading of continuous fiber laminated composite materials produces various types of micro-damage prior to fracture. Each time one of these micro-events occurs, the local and global stiffness of the material is altered, causing several types of variations in the tensor modulus components of the laminae and laminate. This paper attempts to establish specific relationships between these variations in stiffness and the factors which control the laminate fracture event.

INTRODUCTION

One of the most distinctive features of fibrous composite materials is the manner in which the stiffness of those materials changes during quasi-static and cyclic loading. These changes can be large enough to significantly influence the response of the materials to loading, and are directly related to various damage mechanisms which affect strength and life in addition to the stiffness characteristics [1-7]. For oriented fibrous composites, there are generally four independent (tensor) stiffness components which represent the in-plane behavior of composite laminae and two additional ones if three dimensional response is represented. Hence, a four (or six) parameter system is available which can be used to quantitatively identify, monitor, and characterize damage mechanisms. Changes in the stiffness components can also be used directly in the calculation of materials response in the presence of the damage so characterized. Moreover, stiffness changes are equally well suited to the characterization of notched or unnotched specimens or components.

Generally speaking, the damage events that cause changes in the tensor stiffness components eventually lead to fracture of the laminate if the loading is sufficiently severe. However, our understanding of the fracture event is not well developed, especially from the standpoint of mechanics. It is difficult, for example, to determine exactly where and how fracture initiates and exactly what local and global events control the situation which produced the fracture. More specifically, it is difficult to identify and describe the state of stress and state of strength of composite laminates after damage has occurred, so that the nature and likelihood of fracture can be anticipated.

It is the premise of this paper that changes in tensor stiffness components can be used to experimentally monitor and analytically describe the micro-events which define the damage state of composite laminates, and that these changes can be directly linked to certain of those events thought to be important in the precipitation of the final fracture event.

RELATIONSHIPS BETWEEN STIFFNESS CHANGES AND FRACTURE

The scope of discussion in this paper will be limited to the relationships outlined in Figure 1. That figure is a flow chart (from bottom to top) which id

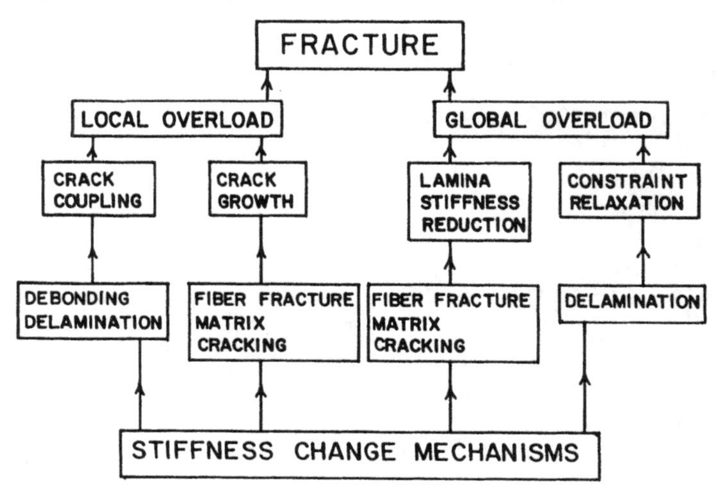

Fig. (1) - Flow chart relating stiffness change mechanisms to fracture

tifies several damage mechanisms which cause stiffness changes and couples their influence to fracture through two rather general categories, "local overload" an "global overload". Three generic damage mechanisms are addressed, matrix cracking, delamination and fiber fracture. However, these mechanisms contribute to fracture-related situations in different ways and in different combinations.

The specific nature of the relationships outlined in Figure 1 will be discussed in detail below under the two general headings of local and global overload. Since the exact nature of fracture in composite laminates has not been es tablished, a more precise connection is not available. It should also be mentio that the discussion will apply to both notched and unnotched specimens in most cases although most of the data will betaken from experience with unnotched coup specimens.

GLOBAL OVERLOAD

A. Lamina Stiffness Reduction

If damage develops fairly uniformly over the region where fracture eventually occurs, the engineering consequence may be global in the sense that the damage may influence fracture by effecting a change in the stiffness of the damaged laminae which, in turn, causes a redistribution of stress that overloads th remaining plies. Referring to Figure 1, there are two ways in which that can happen, by matrix cracking or by fiber fracture. We consider matrix cracking first.

Matrix cracking occurs in various ways, but the most important cracking mode in the present context is the formation of cracks through the thickness of off axis plies, generally along fiber directions or perpendicular to the principal tensile load axis. Such cracks have the appearance shown in Figure 2 when

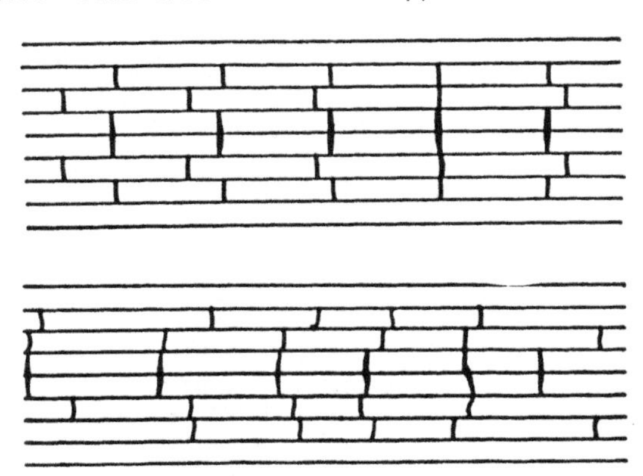

Fig. (2) - Observed (top) and predicted (bottom) characteristic damage states for matrix cracking in a $[0,\pm45,90]_s$ graphite epoxy laminate

viewed from the edge of such a laminate. In that figure, a $[0,90,\pm45]_s$ graphite epoxy laminate crack pattern is shown along with a pattern for that laminate predicted by Reifsnider et al [8-10]. The predicted pattern is based on the observation that such matrix cracks form in regularly spaced arrays (within the variability caused by specimen manufacturing, etc.). The spacing of the cracks is a laminate property, determined only by lamina properties, and the orientations and stacking of the laminae. The patterns so formed are called "characteristic damage states" (CDS) and are found to be stable in the sense that once the CDS forms, essentially no more transverse matrix cracks form until crack coupling or some other damage mode is introduced. Numerous such CDS's have been predicted and observed [8-10].

This type of matrix cracking can be represented as a lamina stiffness change by introducing appropriate reductions in the tensor stiffness components for the specific CDS that is appropriate for a given laminate. The details of that process have been discussed in an earlier publication [11]. Table 1 shows some data that illustrates the results of such lamina stiffness reductions for two laminate types. The stresses (in the global system with "x" in the direction of loading and "y" the in-plane transverse axis) for each ply before and after matrix cracks form are also shown. The laminates were picked to illustrate the effects under discussion. (A $[90_3,0]_s$ laminate is not likely to be a popular engineering choice). However, several aspects of the results are generic. For the $[90_3,0]_s$ laminate, the 0 degree plies are under transverse tension for tensile laminate loading. When cracks form in the 90 degree plies, the tensile stress in the 0 degree plies increases by 174 percent. For the $[0,\pm45]_s$ laminate, the

TABLE 1

Laminate/ Condition	Ply (degrees)	Stress Component (MPa) (percent change)			Predicted/Observed Stiffnes Component Decreases (percen			
		σ_x	σ_y	τ_{xy}	ΔE_x	$\Delta \nu_{xy}$	ΔG_{xy}	ΔD_{yy}
Glass/Epoxy	90	68.4	- 4.41	0				
$[90_3,0]_s$					-	-	-	-
undamaged	0	225.4	13.3	0				
Glass/Epoxy	90	0	-12.13	0	47.3	74.3	0	2.0
$[90_3,0]_s$		(100)	(175)		/	/	/	/
matrix cracks in 90° plies	0	430.6	36.4	0	38.8	71.0	0	0
Glass/Epoxy	-45	81.9	11.3	-21.4				
$[0,\pm45]_s$	+45	81.9	11.3	21.4	-	-	-	-
undamaged	0	267.0	-22.6	0				
Glass/Epoxy	-45	74.21 (10)	6.3 (44)	-40.2 (88)	4.3	20.6	11.8	11.0
$[0,\pm45]_s$	+45	74.21 (10)	6.3 (44)	40.2 (88)	/ 9.4	/ 8.1	/ 28.4	/ 5.3
matrix cracks in 45° plies	0	282.2 (6)	-12.5 (45)	0				

compressive stress in the 0 degree plies decreases (by 45 percent) when cracks form in the matrix of the 45 degree plies along the fiber directions. The shea stress in the 45 degree plies increases when damage is introduced. The stiffne changes in the laminate that accompany these stress redistributions (along with predicted changes) are also shown in Table 1. Again, there are some generic di ferences. Only the axial Young's modulus, E_x, is altered significantly for the $[90_3,0]_s$ laminate, while all four stiffnesses (D_{yy} is the transverse bending stiffness) change for the $[0,\pm45]_s$ case. It was nearly impossible to isolate m trix cracking alone for the $[\pm45,0]_s$ laminate. The predicted stiffness changes are not in good agreement with measured ones for that situation.

While the magnitude of stiffness changes and consequent stress redistri tions may be smaller for laminates more commonly used for engineering purposes, they are not generally insignificant. Graphite epoxy quasi-isotropic laminates can be expected to show of the order of 10 percent E_x changes, for example, und fatigue loading to failure. Even a few percent can be of critical importance t vibration or stability considerations. These changes are directly related, thr global stress redistributions, to the states of stress and strength prior to fracture.

Fiber fracture may also contribute to lamina stiffness reductions and, therefore, to global overloads. The details of such a process are incompletely developed at this time, but an argument based on elasticity solutions for a prolate ellipsoid imbedded in an infinite body provided by Eshelby and further developed by Christensen and Russel can be used to determine some related information [12-14]. Figure 3 is a schematic diagram of a "fiber" which we idealize

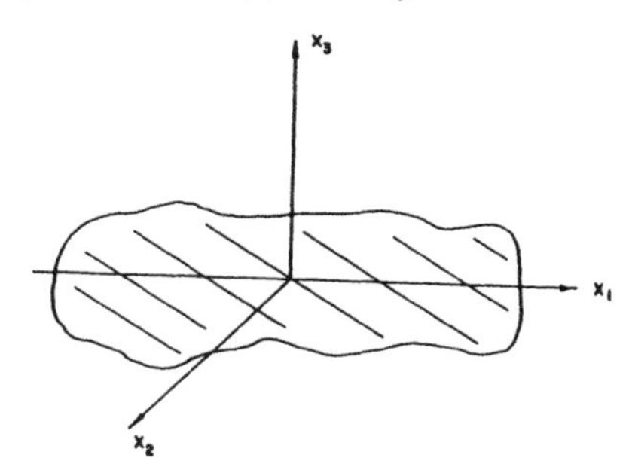

Fig. (3) - Coordinate system for fiber fracture model

as a prolate ellipsoid with major axis along the unit coordinate axis as shown. If the ratio of the semi-monor axis, b, to the semi-major axis, a, is given by K = b/a, then for dilute concentrations (essentially no interaction between fiber stress fields) the longitudinal Young's modulus can be written as

$$E_L = E + E\rho \frac{\frac{\Delta\mu}{2(1-\nu)}(3\Delta\lambda+2\Delta\mu) + \frac{E(1-2\nu)}{2(1+\nu)}\Delta\lambda + \frac{E(1+2\nu^2)}{(1+\nu)(1-2\nu)}\Delta\mu}{\Delta\mu(3\Delta\lambda+2\Delta\mu)K^2\,\frac{1+\nu}{1-\nu}\ln\frac{2}{K} - \frac{5-4\nu}{4(1-\nu)} + \frac{E}{2(1-\nu)}(\Delta\lambda+\Delta\mu)+\mu(3\lambda+2\mu)} \qquad (1)$$

where

$\Delta\lambda = \lambda^*$ (fiber) - λ_m (matrix)

$\Delta\mu \quad \mu^*$ (fiber) - μ_m (matrix)

and E and ν are the matrix Young's modulus and Poisson's ratio respectively. A similar expression can be written for the composite Poisson's ratio as a function of K. The relationship of equation (1) to our fiber fracture problem is established by observing that as continuous fibers break they change their aspect ratio K which is reflected in the changes observed and predicted in the modulus E_L. In addition to the limitations of the model mentioned above, the validity of the analysis scheme is further weakened by the fact that matrix damage at the fiber fracture positions is not included in any way.

Figure 4 shows a plot of equation (1), using the material constants given below for a graphite epoxy composite material.

Graphite: E = 206.7 GPa, G = 27.56 GPa, ν = 0.2

Epoxy: E = 3.45 GPa, G = 1.27 GPa, ν = 0.35

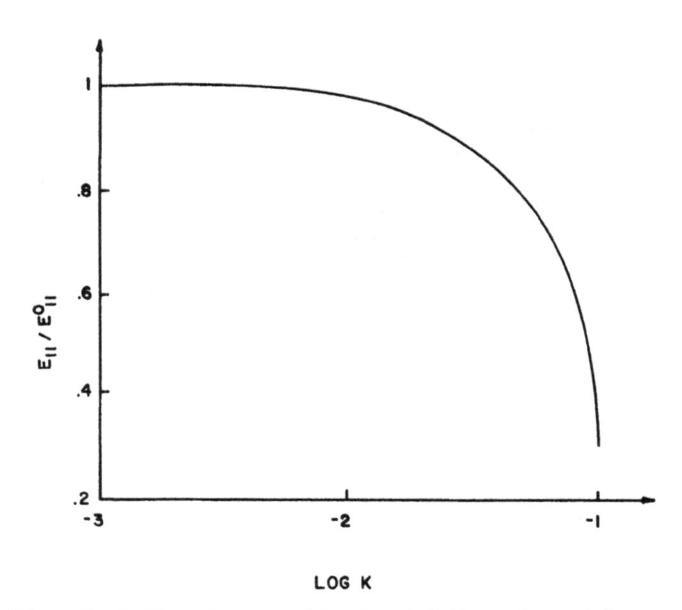

Fig. (4) - Variation in normalized axial Young's modulus as a function of broken fiber aspect ratio

(The results for ν as a function of K are similar). Perhaps the most distinctive feature of Figure 4 is the prediction that K must reach 0.001 or so before any significant reduction in the fiber direction stiffness occurs. If this prediction is accurate, the effect of fiber fracture on the longitudinal stiffness and Poisson's ratio of a lamina is not likely to be observed until fracture is imminent, if at all. Corresponding results for inplane shear stiffness are not available.

B. Constraint Relaxation

As indicated in Figure 1, global overload can be caused by lamina stiffness reduction, or by constraint relaxation due to delamination. In the undamaged state, adjacent plies in a laminate are bonded together, so that, when the laminate is subjected to loads, the inplane strain components in adjacent plies must be equal at the ply interfaces. For the special case of a symmetric laminate subjected to inplane loads only, the inplane strain components are essentially uniform throughout the laminate (the stress components on the other hand vary through the thickness as they depend on the orientation of the lamina in question). If an axial load is applied to a laminate which has a free edge, equ librium requires that near the free edge out of plane stress components σ_z and τ_{xz} be nonzero [15]. Depending on the stacking sequence and the applied load, these interlaminar stresses may be sufficiently large to fail the matrix materia that bonds adjacent plies together and cause a delamination. When the material delaminates, the plies on either side of the delamination are no longer constrai to have all inplane strain components equal. The relaxation of this constraint causes a redistribution of stress throughout the laminate and results in a decrease in laminate stiffness [16].

Consider for example a $[0,\pm45,90]_s$ T300/5208 graphite epoxy laminate subjected to a stress resultant Nx. Stress components determined by laminate analysis for such a laminate in the undamaged state are shown in Table 2. An

TABLE 2

Damage	Ply	σ_x (KPa)	σ_y (KPa)	τ_{xy} (KPa)
Undamaged	0	628	0.0	0.0
	+45	158	89	102
	-45	158	89	-102
	90	42	-177	0.0
Delaminated	0	710	- 18	0.0
	+45	99	8.9	45
	-45	99	8.9	- 45
	90	45	0.0	0.0

analysis of moment equilibrium indicates that large tensile values of σ_z exist at the interface between the 45° and 90° plies, and experiment shows that, in fact, this material delaminates at those interfaces. The analysis actually predicts that the largest tensile σ_z occurs at the midplane, but this interface is stronger than the interface between the 45° and 90° plies because it is not as resin-rich. The results of another laminate analysis for which the 90° plies are completely delaminated from the rest of the laminate and the laminate assembly is subjected to the same load are also shown in Table 2. The stress in the 90° plies in the laminate loading direction is slightly larger in the delaminated case, but because of constraint relaxation the stress carried by these plies transverse to the direction of loading is reduced from -177 KPa in the undamaged state to zero in the delaminated state. In the +45° and 45° plies the stress in the load direction is reduced by 37 percent. Because of the change in the axial load carried by the offaxis plies which results from delamination, the stress in the fiber direction of the 0° plies is increased by 13 percent. Since the 0° plies tend to control laminate behavior in the case of axial loading, the stiffness of the laminate is reduced approximately in proportion to the increase in stress in the 0° plies. In this case, the laminate stiffness was reduced by 13 percent.

LOCAL OVERLOAD

A. Crack Growth

In Figure 1 we identified local overloading as the other generic situation which precipitates fracture. Crack growth can occur by matrix cracking whereby cracks in off-axis plies continue to grow through the laminate thickness into other neighboring off-axis plies. Such growth is likely only if neighboring plies have very similar fiber orientations. In the plane of a ply, crack growth can also occur by fiber fracture. In that situation, the change in stiffness is dependent upon the mode of growth. Two limiting cases are illustrated in Figure 5.

For purposes of discussion, we assume that the fiber fractures are concentrated in a slot with length a in a unidirectional specimen having width b and

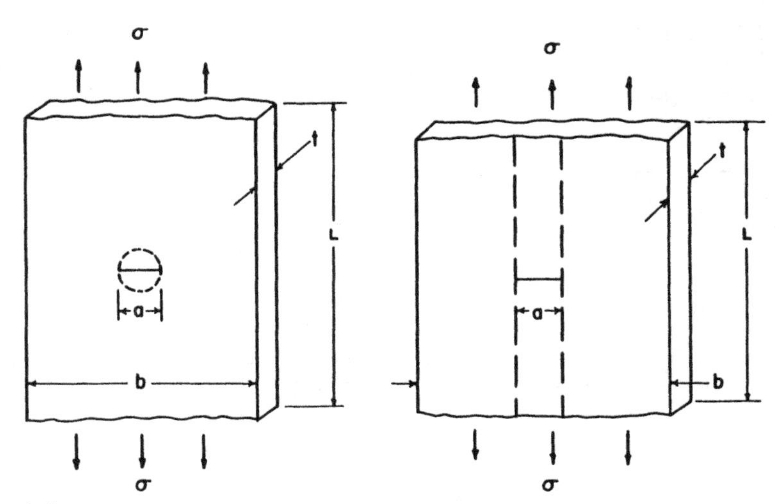

Fig. (5) - Schematic diagrams of localized (a) and splitting (b) models for strain energy release rate calculations

thickness t. If we make the assumption (similar to the Griffith assumption) that strain energy is released in the neighborhood of the broken fibers over a region with diameter "a" (as shown in Figure 5a), and invoke the usual relationship between the change in compliance of the total specimen (under a constant applied strain) and the strain energy release rate due to the fiber fracture, one can express the increased compliance of the specimen in terms of the original compliance, c_0, as shown in equation (2).

$$c = c_0 \left[1 + \frac{\pi a^2}{4b\ell}\right] \tag{2}$$

If one assumes that the fiber fractures over the length "a" are accompanied by complete splitting of those fractured fibers over the total length ℓ of the specimen, then the total volume of material above and below the fracture line releases its strain energy as suggested by Figure 5b. In that instance, the relationship of the increased compliance to the original compliance of the specimen is given by equation (3).

$$c = c_0 \left[1 + \frac{2a}{b}\right] \tag{3}$$

The differences between these two limiting expressions is instructive. When the strain energy release is highly localized, the dependence on the length of fracture caused by the fiber failures is quadratic, and the specimen length enters the expression as shown by equation (2). In the instance when extensive longitudinal splitting occurs, the dependence on the local fracture path length is linear, and there is no dependence on the length of the specimen, as shown in equation (3). These differences can be identified in the laboratory, and the correlation between these expressions and experimental data can serve as an indication of the degree of longitudinal splitting that occurs. Also, if significant longitudinal splitting occurs, additional cracks that form may release less energy than earlier ones. These expressions and our experience with them deserve more discussion than we can include in the present discussion. We will have to limit our discussion to the example data shown in Figure 5. In that figure, the predictions of equations (2) and (3) are shown (in normalized form) for a range of different effective crack lengths as measured by a compliance gage with a gage length of 50 mm. The data were taken from unidirectional glass epoxy specimens with "slots" introduced by breaking glass fibers prior to specimen manufacture. The top curve and bottom curve in Figure 6 are the predicted relationships from

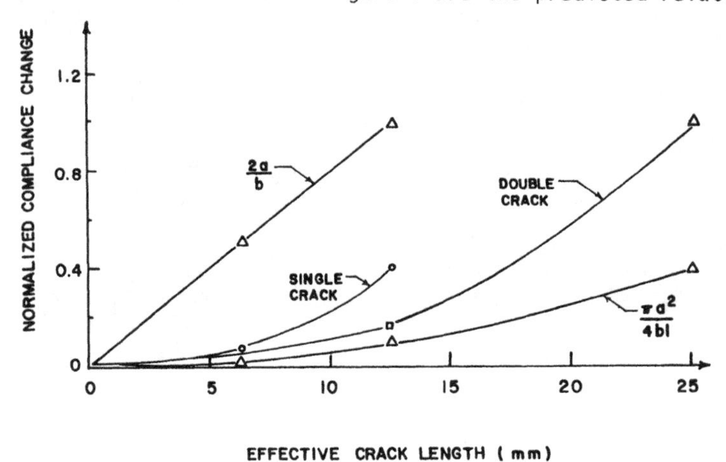

Fig. (6) - Normalized compliance change, predicted and observed

equations (3) and (2), respectively. The upper curve of the two between the predictions was generated from data points taken from uncracked, 6.2 mm cracked and 12.4 mm cracked specimens (25 mm wide and 0.83 mm thick). (The curve was "eye-balled" through the data for ease in associating the points). The lower data curve was sketched from data taken from specimens having two 6.2 mm cracks and two 12.4 mm cracks, respectively, one directly above the other in the 50 mm gage section. The data suggest that the damage due to the broken fibers is neither as local as that modelled by Figure 5a or as global as that modelled by Figure 5b. A more complete discussion of this subject will be given in subsequent papers. It is clear, however, that stiffness changes due to crack growth by fiber fracture as represented here can be identified, represented and measured, and that laminate fracture as a result of such growth could occur.

B. Crack Coupling

As noted in Figure 1, crack coupling is another mode of damage that can cause local overloading and stiffness changes. While it is true that sufficient magnification would reveal many types of cracks, we restrict our attention to two types, the matrix cracks that form through the total thickness of one or more plies (as shown in Figure 2) and cracks that form at fiber fracture positions. Coupling of the matrix cracks in neighboring plies occurs by delamination; one crack or the other grows along a ply interface until a crack in the neighboring ply is encountered whereupon the cracks couple together. Since many laminated composite materials are very insensitive to crack tip stresses for such matrix cracks, this has nearly the same effect as a crack growing directly through the

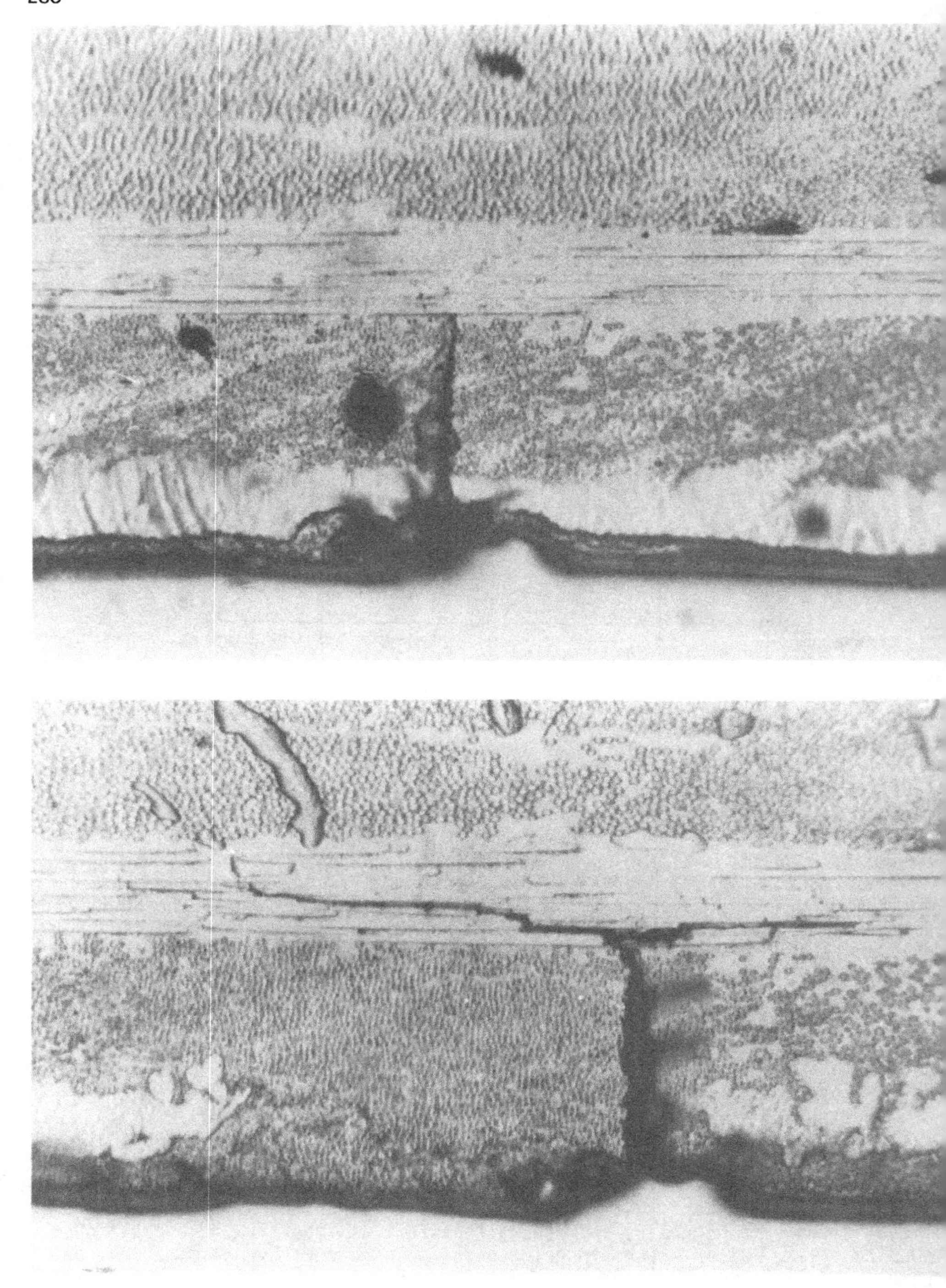

Fig. (7) - Growth of a transverse crack in the 90° plies of a $[90_2,0,\pm45]_s$ laminate (top) into the 0° adjacent ply (bottom)

thickness of the two plies.

A different coupling situation is shown in Figure 7. In that figure, a crack in the 90° plies of a $[90_2,0,\pm45]_s$ graphite epoxy laminate couples up with cracks in fiber bundles by a debonding process as shown in the sequence Figure 7a, 7b. In this way, cracks in off axis plies can grow into a 0° ply and contribute to the fracture process. It should also be mentioned that we have observed a higher density of fiber bundle fractures, for the situation used an an example above, in the neighborhood of matrix cracks in the double 90° plies.

As mentioned in earlier publications, stiffness changes due to crack coupling have been detected, but a thorough investigation of those changes has not yet been conducted [8].

CLOSURE

Matrix cracking, delamination and fiber fracture damage mechanisms combine in various ways to produce damage modes such as crack coupling, crack growth, lamina stiffness reduction and constraint relaxation which can cause global or local overloads leading to fracture. This process is reflected in laminate tensor stiffness changes which are quantitative indicators of the precise nature and extent of the damage development. The stiffness changes also serve the essential purpose of interpretation by providing the material property changes needed in a stress analysis to determine the state of stress and state of strength of the laminate in the damaged condition.

ACKNOWLEDGEMENTS

The authors wish to gratefully acknowledge the support of this research by the National Science Foundation under Grant No. CME-7680213, monitored by Dr. C. Astill. They also express their appreciation to Barbara Wengert for typing the manuscript and to G. K. McCauley for assistance in preparing the figures.

REFERENCES

[1] Salkind, M. J., "Fatigue of Composite Materials", Composite Materials: Testing and Design (Second Conference), ASTM STP 497, Philadelphia, 1972.

[2] Nevadunsky, J. J., Lucas, J. J. and Salkind, M. J., "Early Fatigue Damage Detection in Composite Materials", J. of Composite Materials, 9:394, 1975.

[3] Reifsnider, K. L., Stinchcomb, W. W. and O'Brien, T. K., "Frequency Effects on a Stiffness-Based Fatigue Failure Criterion in Flawed Composite Specimens", in "Fatigue of Filamentary Composite Materials", K. L. Reifsnider and K. N. Lauraitis, eds., ASTM STP 636, Philadelphia, 1977.

[4] Johnson, W. S., "Characterization of Fatigue Damage Mechanism in Continuous Fiber Reinforced Metal Matrix Composites", Ph.D. Dissertation, Duke University, Durham, 1979.

[5] Stinchcomb, W. W., Reifsnider, K. L. and Williams, R. S., "Critical Factors for Frequency Dependent Fatigue Processes in Composite Materials", Experimental Mech., 16(9):343, 1976.

[6] Hahn, H. T. and Kim, R. Y., "Fatigue Behavior of Composite Laminate", J. of Comp. Materials, 10:156, 1976.

[7] O'Brien, T. K., "An Evaluation of Stiffness Reduction as a Damage Parameter and Criterion for Fatigue Failure in Composite Materials", Ph.D. Dissertation, Virginia Polytechnic Institute and State University, Blacksburg, 1978.

[8] Reifsnider, K. L., Stinchcomb, W. W. and Henneke, E. G., "Defect-Property Relationships in Composite Materials", AFML-Technical Report 76-81; Part 1, April 1976, Part II, June 1977, Part III, April 1979.

[9] Reifsnider, K. L. and Talug, A., "Analysis of Fatigue Damage in Composite Laminates", Int. J. Fatigue, pp. 3-11, 1980.

[10] Reifsnider, K. L., "Some Fundamental Aspects of the Fatigue and Fracture Response of Composite Materials", Proc. 14th Annual Meeting of Society of Engineering Science, November 14-16, 1977, Lehigh University.

[11] Highsmith, A. and Reifsnider, K. L., "Stiffness Reduction Mechanisms in Composite Laminates", Proc. ASTM Conf. on Damage in Composite Laminates, November 12-14, 1980, Bal Harbour, Florida.

[12] Eshelby, J. D., "The Determination of the Elastic Field of an Ellipsoidal Inclusion, and Related Problems", Proc. Roy. Soc. London, Vol. A241, p. 376, 1957.

[13] Russel, W. B., "On the Effective Moduli of Composite Materials, Effect of Fiber Length and Geometry at Dilute Concentrations", Z. Angew. Math. Phys., Vol. 24, p. 581, 1973.

[14] Christensen, R. M., Mechanics of Composite Materials, John Wiley and Sons, New York, p. 90.

[15] Pagano, N. J. and Pipes, R. B., "Some Observations on the Interlaminar Strength of Composite Laminates", Int. Journal of Mechanical Science, Vol. 15, pp. 679-688, 1973.

[16] O'Brien, T. K., "Characterization of Delamination Onset and Growth in a Composite Laminate", NASA Technical Meoorandum 81940, January 1981.

THREE DIMENSIONAL FINITE ELEMENT ANALYSIS OF DAMAGE ACCUMULATION IN COMPOSITE LAMINATE

J. D. Lee

The George Washington University
Washington, D.C. 20052

ABSTRACT

A three dimensional finite-element computer program has been developed to analyze layered fiber-reinforced composite laminate. This program is capable of: (1) calculating the detailed stress distribution, (2) identifying the damage zone and mode of failure, (3) analyzing the damage accumulation, and (4) determining the ultimate strength of the composite laminate.

FUNDAMENTALS

In this paper, attention is focused on a biaxially loaded composite laminate consisting of several fiber-reinforced composite layers each with a specified fiber orientation.

A basic assumption has been made in this analysis: Each layer is modeled as a linear elastic continuum with hexagonal anisotropy about fiber orientation; individual fibers and matrix are not considered.

The general stress-strain relation for a linear anisotropic elastic solid can be written as

$$\varepsilon_{ij} = S_{ijmn}\sigma_{mn} \tag{1}$$

where

$$S_{ijmn} = S_{mnij} = S_{jimn} = S_{ijnm} \tag{2}$$

The number of independent elastic constants is 21. However, for each layer, the unidirectional fiber-reinforced composite has axis symmetry about the fiber orientation, the number of independent elastic constants is reduced to five and, in engineering notation, equation (1) can be expressed as:

$$\begin{bmatrix} \varepsilon_L \\ \varepsilon_T \\ \varepsilon_z \\ \gamma_{Tz} \\ \gamma_{zL} \\ \gamma_{LT} \end{bmatrix} = \begin{bmatrix} E_L^{-1} & -\nu_{LT}E_L^{-1} & -\nu_{LT}E_L^{-1} & 0 & 0 & 0 \\ -\nu_{LT}E_L^{-1} & E_T^{-1} & -\nu_{Tz}E_T^{-1} & 0 & 0 & 0 \\ -\nu_{LT}E_L^{-1} & -\nu_{Tz}E_T^{-1} & E_T^{-1} & 0 & 0 & 0 \\ 0 & 0 & 0 & 2(1+\nu_{Tz})E_T^{-1} & 0 & 0 \\ 0 & 0 & 0 & 0 & G_{LT}^{-1} & 0 \\ 0 & 0 & 0 & 0 & 0 & G_{LT}^{-1} \end{bmatrix} \begin{bmatrix} \sigma_L \\ \sigma_T \\ \sigma_z \\ \sigma_{Tz} \\ \sigma_{zL} \\ \sigma_{LT} \end{bmatrix} \quad (3)$$

where the L-axis denotes the fiber orientation. Then it is straightforward to obtain

$$\begin{bmatrix} \sigma_L \\ \sigma_T \\ \sigma_z \\ \sigma_{Tz} \\ \sigma_{zL} \\ \sigma_{LT} \end{bmatrix} = \begin{bmatrix} d_{11} & d_{12} & d_{13} & 0 & 0 & 0 \\ d_{12} & d_{22} & d_{23} & 0 & 0 & 0 \\ d_{13} & d_{23} & d_{33} & 0 & 0 & 0 \\ 0 & 0 & 0 & d_{44} & 0 & 0 \\ 0 & 0 & 0 & 0 & d_{55} & 0 \\ 0 & 0 & 0 & 0 & 0 & d_{66} \end{bmatrix} \begin{bmatrix} \varepsilon_L \\ \varepsilon_T \\ \varepsilon_z \\ \gamma_{Tz} \\ \gamma_{zL} \\ \gamma_{LT} \end{bmatrix} \quad (4)$$

where

$$\begin{aligned} d_{11} &= E_L(1-\nu_{Tz}^2)/\delta \\ d_{22} &= d_{33} = E_T(1-\nu_{LT}^2 E_T/E_L)/\delta \\ d_{12} &= d_{13} = E_T(1+\nu_{Tz}^2)\nu_{LT}/\delta \\ d_{23} &= E_T(\nu_{Tz}+\nu_{LT}^2 E_T/E_L)/\delta \\ d_{44} &= 0.5\ E_T(1+\nu_{Tz}) \\ d_{55} &= d_{66} = G_{LT} \\ \delta &= 1-\nu_{TZ}^2 - 2\nu_{LT}^2(1+\nu_{Tz})E_T/E_L \end{aligned} \quad (5)$$

BIAXIALLY LOADED COMPOSITE LAMINATE

Let the specimen under consideration be a rectangular plate made of layered composite material, with a centered circular hole, subjected to in-plane biaxial loading, Figure 1. The direction of the fiber is characterized by the angle ϕ

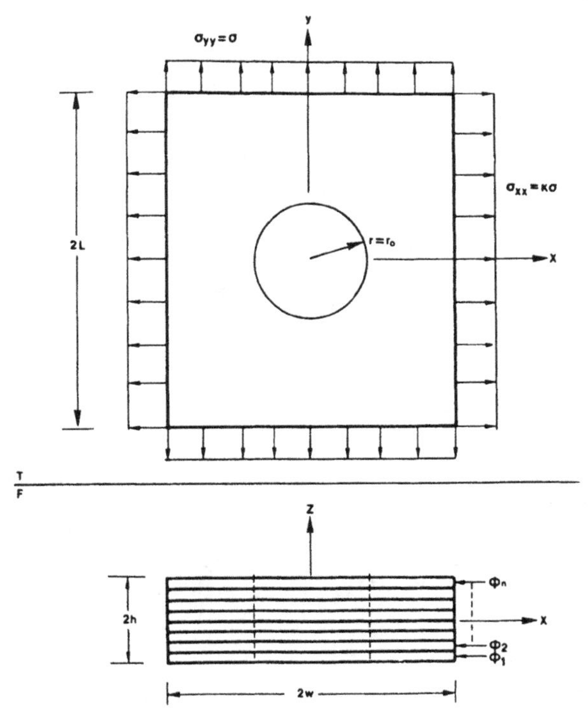

Fig. (1) - The top view and the front view of a composite laminate

between the fiber and the x-axis. Thus, the arrangement of n layers for the entire laminate can be represented by $[\phi_1,\phi_2,\ldots\phi_n]$. In this paper, let the attention be focused on a symmetric ply configuration which may be represented as

$$[\phi_1,\phi_2,\ldots\phi_i]_s \equiv [\phi_i,\ldots\phi_2,\phi_1,\phi_1,\phi_2,\ldots\phi_i] \tag{6}$$

The top four plies of a symmetric ply configuration $[0,45,-45,90]_s$ is shown in Figure 2. It is noticed that this configuration has a mirror symmetry with respect to x-y plane at z=0 and a periodicity of π about z-axis [1]. If the loading condition also has the same characteristics, then it is obvious that

$$\begin{aligned} u_i(-x,-y,z) &= -\,u_i(x,y,z), \qquad i,j = 1,2 \\ \sigma_{ij}(-x,-y,z) &= \sigma_{ij}(x,y,z) \\ u_z(-x,-y,z) &= u_z(x,y,z) \end{aligned}$$

$$\sigma_{zi}(-x,-y,z) = -\sigma_{zi}(x,y,z),\ i = 1,2$$

$$\sigma_{zz}(-x,-y,z) = \sigma_{zz}(x,y,z) \tag{7}$$

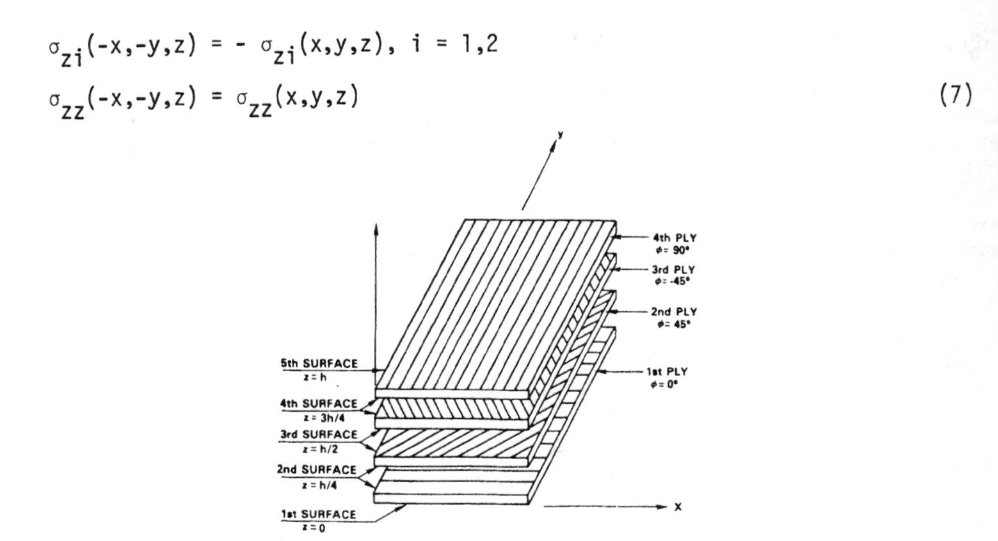

Fig. (2) - The top four plies of a symmetric ply configuration $[0,45,-45,90]_s$

Taking the mirror symmetry and the periodicity into consideration, one may set up a finite-element mesh that covers only a quarter of the entire laminate. An how, it is straightforward to solve the problem by a three dimensional finite-e ment computer program. Let the finite-element mesh consist of a family of 8-no brick type element, Figure 3. After the displacements of each nodal point bein

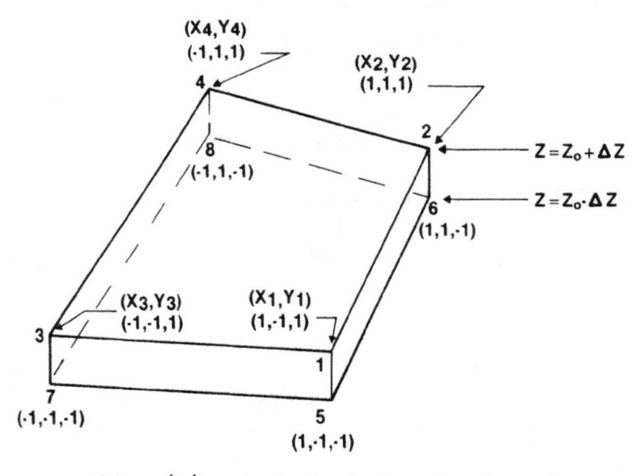

Fig. (3) - A typical 8-node element

obtained, then the six components of stresses, σ_{xx}, σ_{yy}, σ_{zz}, σ_{yz}, σ_{zx}, σ_{xy} at any point within any element can be found in terms of the nodal point displacements of that element [2]. Then the stress components in (L,T,z) coordinate sy tem, Figure 4, can be written as

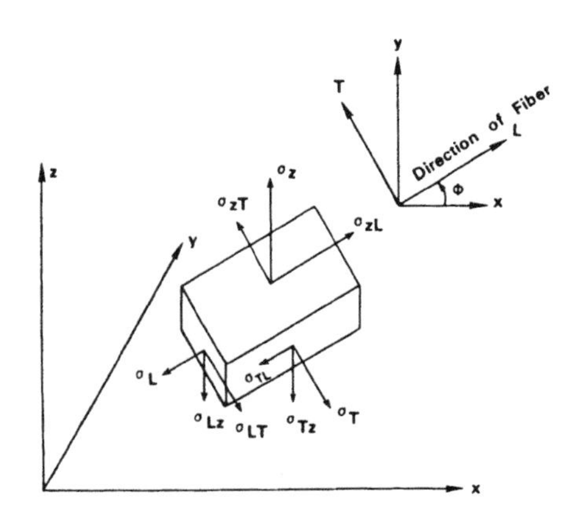

Fig. (4) - The stress components in (L,T,z) coordinate system

$$\begin{aligned}
\sigma_L &= \sigma_{xx}\cos^2\phi + \sigma_{yy}\sin^2\phi + \sigma_{xy}\sin2\phi \\
\sigma_T &= \sigma_{xx}\sin^2\phi + \sigma_{yy}\cos^2\phi - \sigma_{xy}\sin2\phi \\
\sigma_{LT} &= (\sigma_{yy}-\sigma_{xx})\sin\phi\cos\phi + \sigma_{xy}\cos2\phi \qquad (8)\\
\sigma_{Lz} &= \sigma_{xz}\cos\phi + \sigma_{yz}\sin\phi \\
\sigma_{Tz} &= - \sigma_{xz}\sin\phi + \sigma_{yz}\cos\phi
\end{aligned}$$

FAILURE CRITERION

For any given failure criterion, based on the stress and displacement field, one may identify the damage zone. In this work, consider only the breakage of fibers and failure of matrix. The failure criterion of fiber breakage and matrix failure adopted in this paper are briefly discussed as follows:

(1) Fiber Breakage - It is assumed that the fibers of a certain element break if the stresses at the center of that element satisfy one of the following inequalities

$$\begin{aligned}
&\sigma_L > \sigma_{FN} \\
&(\sigma_{LT}^2+\sigma_{Lz}^2)^{1/2} > \sigma_{FS}
\end{aligned} \qquad (9)$$

In such a case, the stiffness matrix of that element, referring to equation (4), will be reduced to zero.

(2) Matrix Failure - It is assumed that the matrix of a certain element fail if the stresses at the center of that element satisfy one of the following inequalities

$$
\begin{aligned}
&\sigma_T > \sigma_{MN} \\
&(\sigma_{TL}^2+\sigma_{Tz}^2)^{1/2} > \sigma_{MS}
\end{aligned}
\tag{10}
$$

In such a case, the stiffness matrix of that element will be reduced to

$$
[d] = \begin{bmatrix} d_{11} & 0 & d_{13} & 0 & 0 & 0 \\ 0 & 0 & 0 & 0 & 0 & 0 \\ d_{13} & 0 & d_{33}' & 0 & 0 & 0 \\ 0 & 0 & 0 & 0 & 0 & 0 \\ 0 & 0 & 0 & 0 & d_{55} & 0 \\ 0 & 0 & 0 & 0 & 0 & 0 \end{bmatrix}
\tag{11}
$$

The failure strength, σ_{FN}, σ_{FS}, σ_{MN} and σ_{MS} are determined experimentally and taken as the input data for the computer program. This failure criterion permits the identification of the damage zone and its mode of failure for a specified amount of applied loading.

NUMERICAL RESULTS

For a specified amount of applied loading σ and a fixed biaxial load factor k, based on the failure criterion, it is possible to identify the damage zone an the mode of failure. After the stiffness matrix of the damage zone being modified, solve the boundary value problem again and see whether the damage zone is enlarged or not. If the damage zone is enlarged, then keep the applied loading unchanged and solve the problem again. If the damage zone is not enlarged, then increase the applied loading until new failure occurs. By doing so, it is possi ble to find the ultimate strength of the composite laminate.

For illustrative purpose, the input data and the numerical solutions of a particular problem are given as follows:

A. Input Data

1. Material Constants

$$
\begin{aligned}
E_L &= 19.52 \times 10^3 \text{ ksi} \\
E_T &= 1.404 \times 10^3 \text{ ksi} \\
G_{LT} &= 0.65 \times 10^3 \text{ ksi} \\
\nu_{LT} &= 0.264 \\
\nu_{Tz} &= 0.435
\end{aligned}
\tag{12}
$$

2. Failure Strength

$$\begin{aligned} \sigma_{FN} &= 180 \text{ ksi} \\ \sigma_{FS} &= 12 \text{ ksi} \\ \sigma_{MN} &= 8 \text{ ksi} \\ \sigma_{MS} &= 12 \text{ ksi} \end{aligned} \tag{13}$$

3. Ply Configuration

$$[0,45,-45,90]_S \tag{14}$$

4. Geometrical Parameters

$$\begin{aligned} 2L &= 6 \text{ in.} \\ 2W &= 6 \text{ in.} \\ 2h &= 0.0423 \text{ in.} \\ r_o &= 0.5 \text{ in.} \end{aligned} \tag{15}$$

B. Numerical Solutions

1. Uniaxial Loading, k=0

The maximum tensile stresses, Max.σ_L and Max.σ_T, among the top four layers are plotted as functions of θ at r = 0.514 in. in Figure 5 and Figure 6, respectively. The damage zones at various stages are shown in Figures 7-11. It

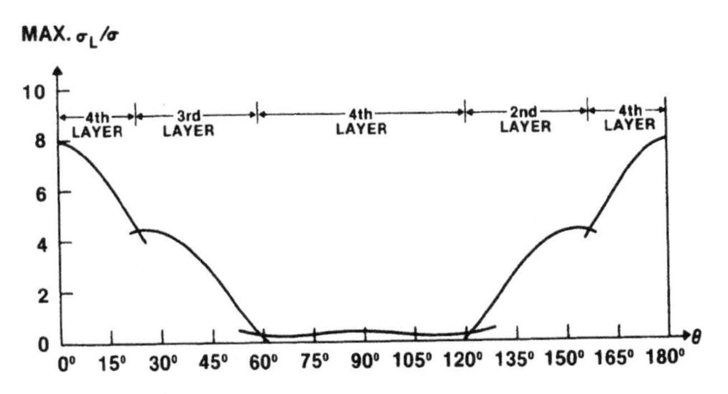

Fig. (5) - Max.σ_L versus θ at r = 0.514 in., k=0

is noticed that the maximum stress concentration factor is 7.92; matrix failure occurs at σ = 14.3 ksi; and the ultimate strength of the laminate is 22.6 ksi.

2. Biaxial Loading, k = 0.5

Similarly, Max.σ_L and Max.σ_T are plotted in Figure 12 and Figure 13,

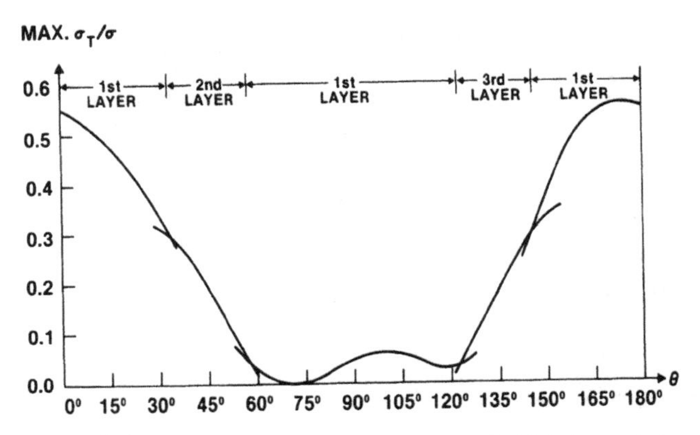

Fig. (6) - Max.σ_T versus θ at r = 0.514 in., k=0

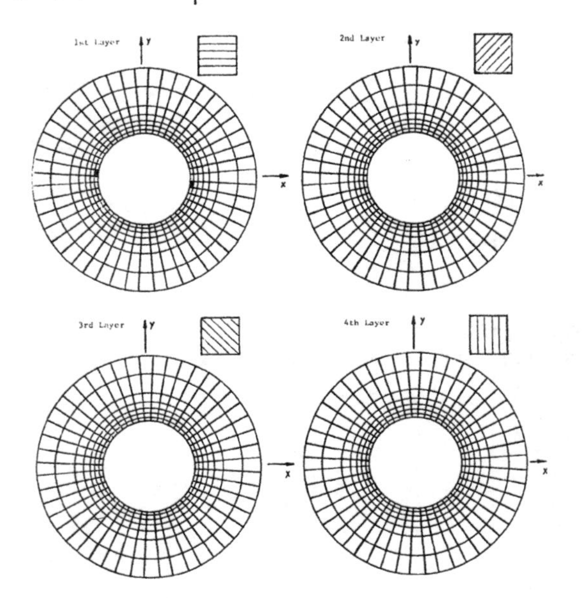

Fig. (7) - Damage zone at σ = 14.3 ksi, k=0
matrix failure ⊠ fiber breakage ■

respectively. The damage zones at various stages are shown in Figures 14-17. It is noticed that the maximum stress concentration factor is reduced to 6.57; matrix failure occurs at σ = 17.2 ksi; and the ultimate strength is increased to 27.4 ksi.

3. Equal Biaxial Loading, k=1

Similarly, Max.σ_L and Max.σ_T are plotted in Figure 18 and Figure 19, respectively. The damage zones at various stages are shown in Figures 20-24. I is noticed that the maximum stress concentration factor is further reduced to 5.

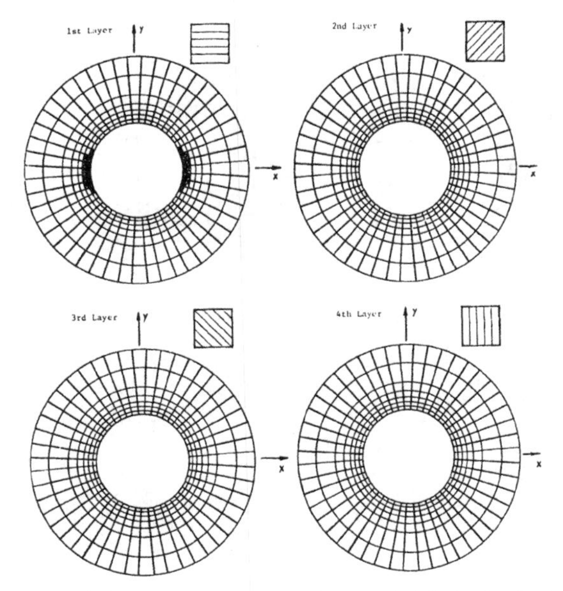

Fig. (8) - Damage zone at σ = 20.0 ksi, k=0
matrix failure ☒ fiber breakage ■

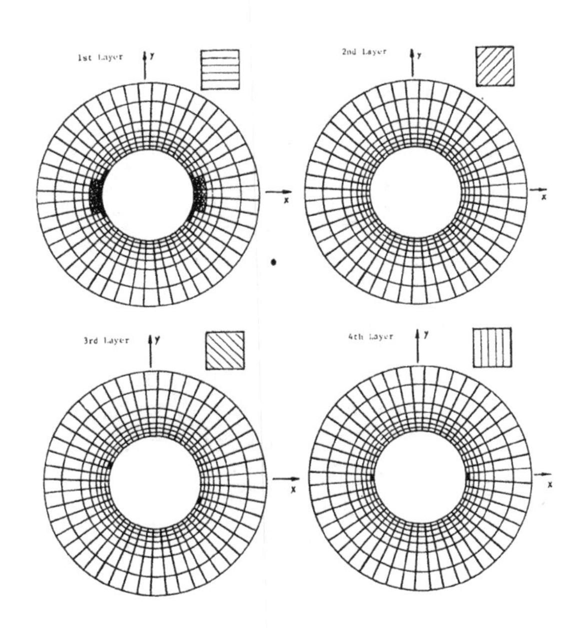

Fig. (9) - Damage zone at σ = 22.6 ksi, k=0
matrix failure ☒ fiber breakage ■

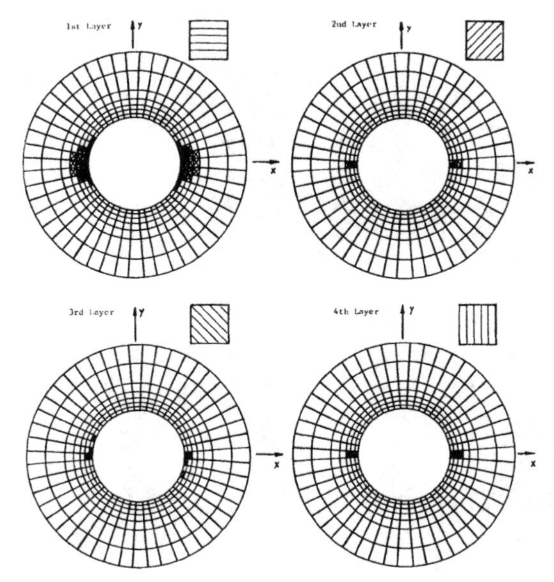

Fig. (10) - Damage zone at σ = 22.6 ksi, k=0
matrix failure ⊠ fiber breakage ■

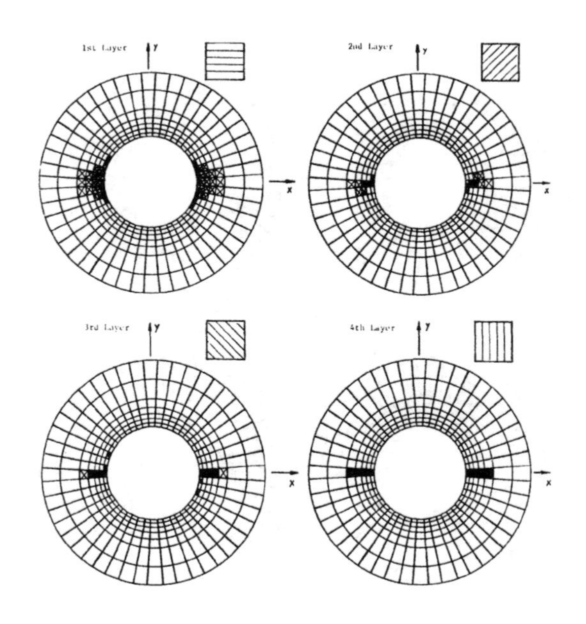

Fig. (11) - Damage zone at σ = 22.6 ksi, k=0
matrix failure ⊠ fiber breakage ■

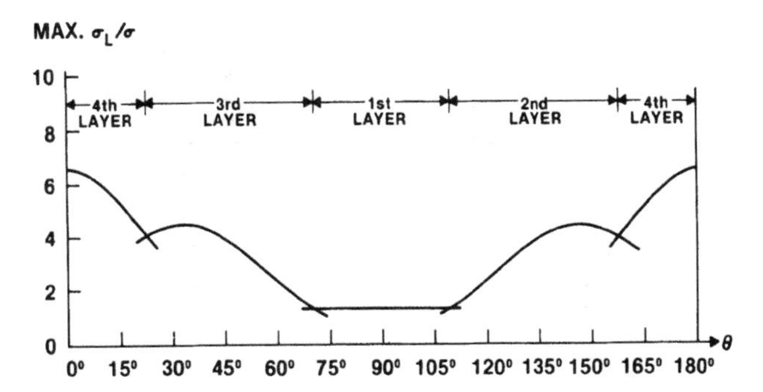

Fig. (12) - Max.σ_L versus θ at r = 0.514 in., k=0.5

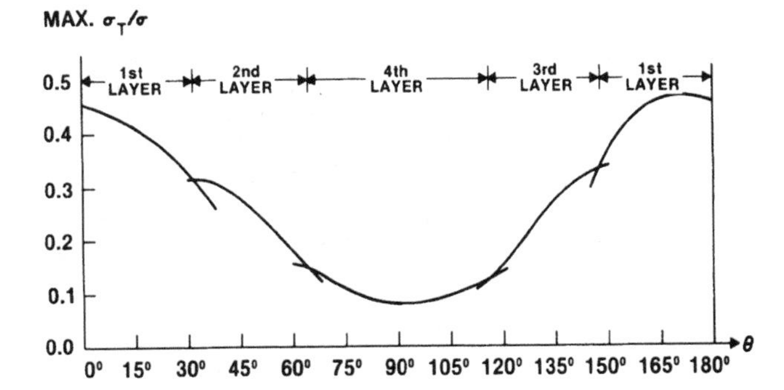

Fig. (13) - Max.σ_T versus θ at r = 0.514 in., k=0.5

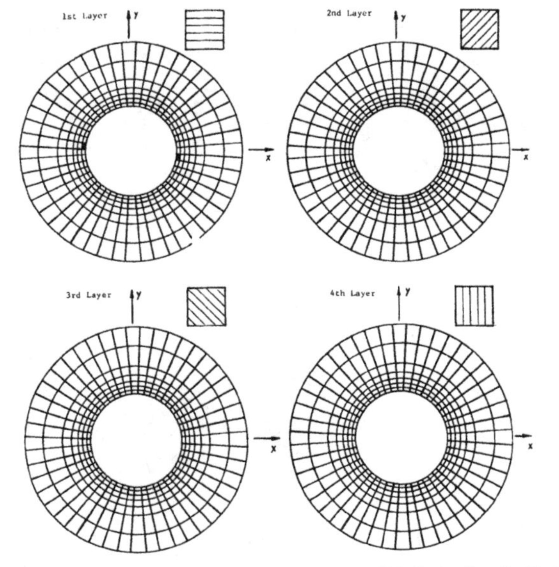

Fig. (14) - Damage zone at σ = 17.2 ksi, k=0.5
matrix failure ☒ fiber breakage ■

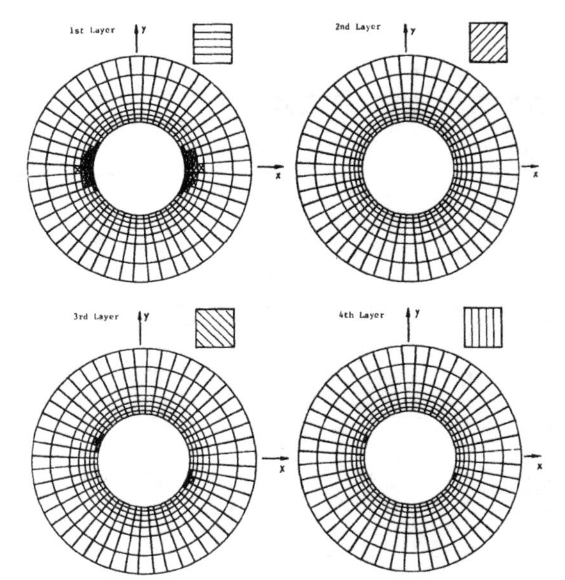

Fig. (15) - Damage zone at σ = 25.4 ksi, k=0.5
matrix failure ☒ fiber breakage ■

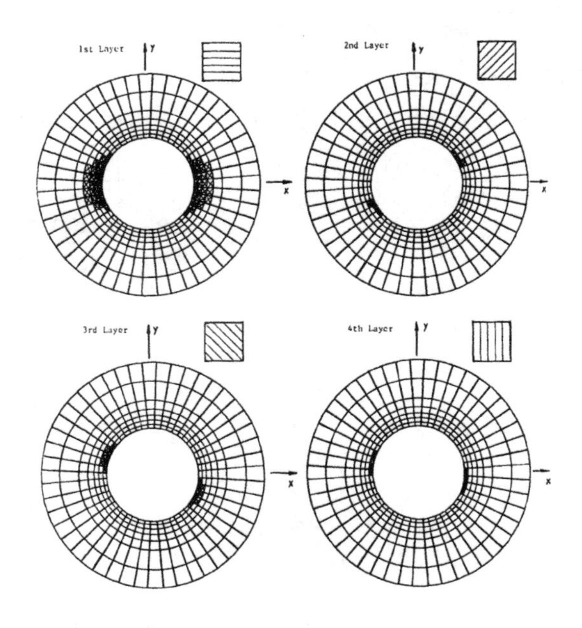

Fig. (16) - Damage zone at σ = 27.4 ksi, k=0.5
matrix failure ☒ fiber breakage ■

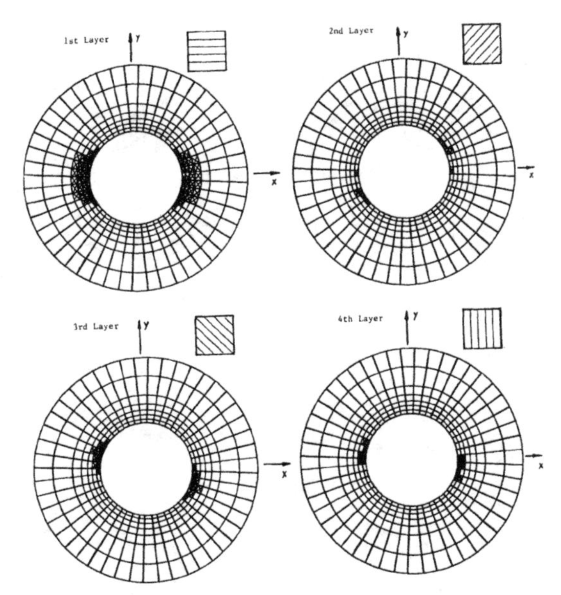

Fig. (17) - Damage zone at σ = 27.4 ksi, k=0.5
matrix failure ☒ fiber breakage ▮

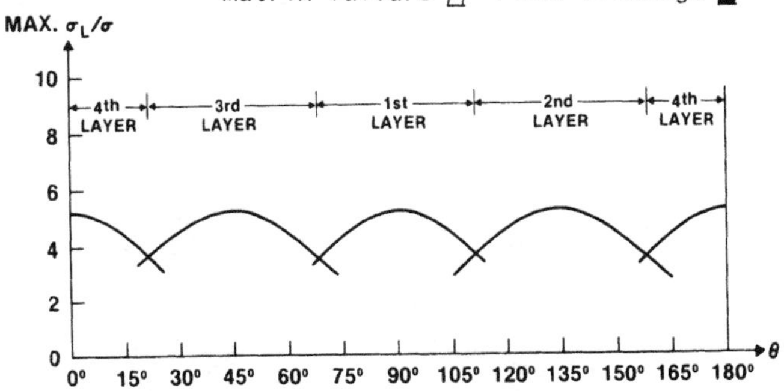

Fig. (18) - Max.σ_L versus θ at r = 0.514 in., k=1

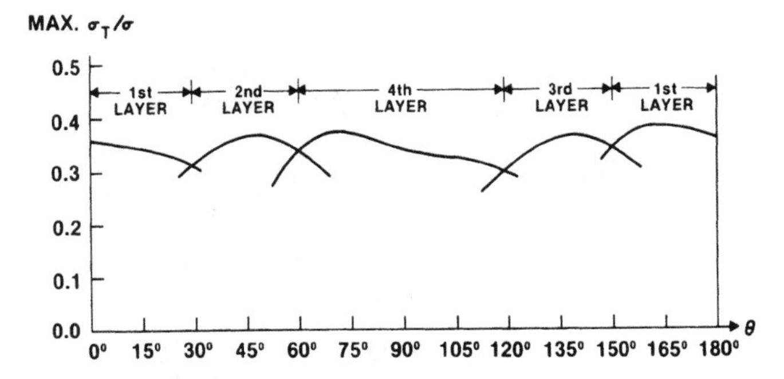

Fig. (19) - Max.σ_T versus θ at r = 0.514 in., k=1

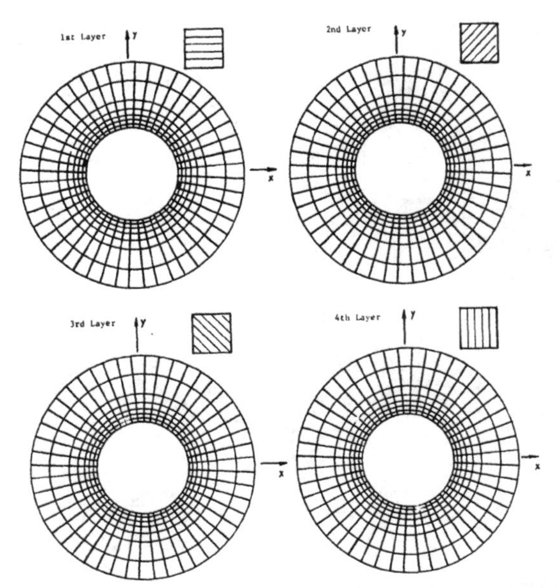

Fig. (20) - Damage zone at σ = 20.9 ksi, k=1
matrix failure ☒ fiber breakage ■

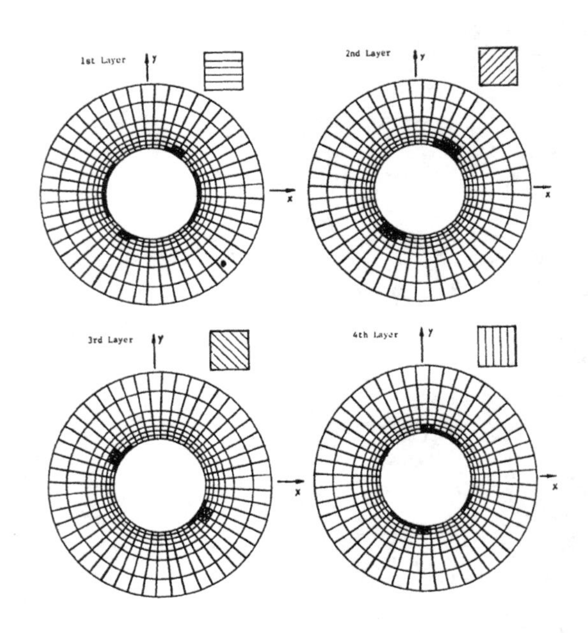

Fig. (21) - Damage zone at σ = 24.9 ksi, k=1
matrix failure ☒ fiber breakage ■

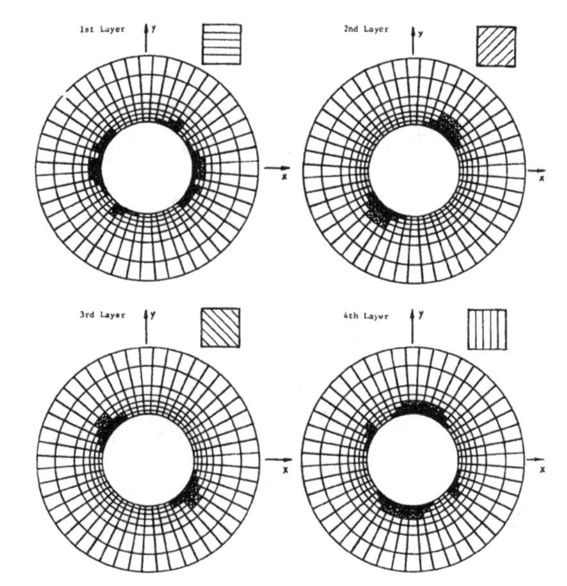

Fig. (22) - Damage zone at σ = 27.9 ksi, k=1
matrix failure ☒ fiber breakage ■

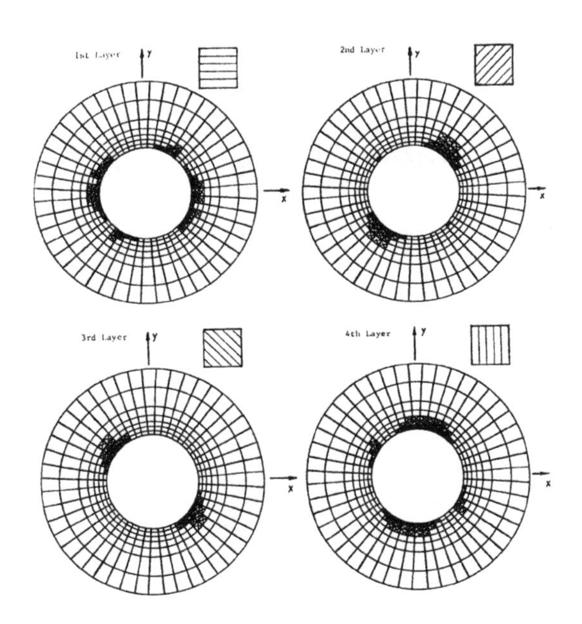

Fig. (23) - Damage zone at σ = 28.0 ksi, k=1
matrix failure ☒ fiber breakage ■

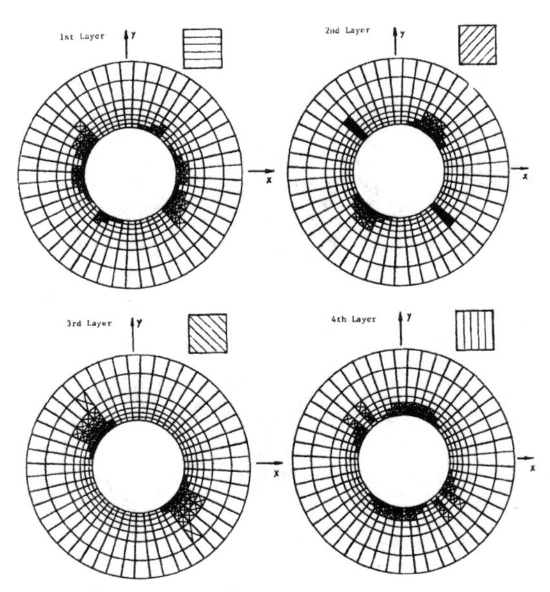

Fig. (24) - Damage zone at σ = 28.0 ksi, k=1
matrix failure ☒ fiber breakage ■

matrix failure occurs at σ = 20.9 ksi, and the ultimate strength is increased to 28.0 ksi; moreover, the final broken path of the specimen is along the line θ = 135°-315° instead of the line θ = 0°-180° as in the first two cases.

ACKNOWLEDGEMENT

The author wishes to acknowledge the financial support for this work from NASA-Langley Research Center (Grant NSG-1289).

REFERENCES

[1] Lee, J. D., Computers and Structures, 12, p. 319, 1980.

[2] Legerlind, L. J., Applied Finite Element Analysis, Wiley, New York, 1976.

POST-CRAZING ANALYSIS OF GLASS-EPOXY LAMINATES

D. G. Smith and J.-C. Huang

Tennessee Technological University
Cookeville, Tennessee 38501

ABSTRACT

A glass-epoxy material known as Scotchply SP-250 was characterized. Lamination theory together with the material's nonlinear ply shear curve was used to predict the post-crazing response of several laminates. These predictions were compared with tests on laminate coupons. A finite element program was developed for obtaining stress and failure analysis of laminated structures with stress concentrations loaded transversely or inplane. The program uses the nonlinear material model together with a doubly-curved, isoparametric, thick-plate or thick-shell element.

INTRODUCTION

Many glass-epoxies exhibit a considerable amount of nonlinear deformation prior to gross fracture. Under increasing load, certain plys within the laminate begin to fail by matrix cracking and splitting between fibers. The onset of matrix cracking gives the laminate a hazy, milkly, light-colored appearance, sometimes referred to as crazing. Beyond the onset of crazing the laminate compliance increases with increasing load; the crazing area grows and may extend to plys of other angles before the ultimate load is reached. The onset of crazing may occur at loads which are rather low compared to the laminate's ultimate load. For many structural applications the laminate's reserve strength beyond crazing may safely be utilized. This paper describes the beginning of an effort to obtain information on laminate behavior in the post-crazing region as well as the development of a general stress-failure analysis technique which incorporates the nonlinear material response.

POST-CRAZING CHARACTERIZATION OF SP-250 GLASS-EPOXY

The material used in this work is known as Scotchply SP-250 manufactured by the 3M Company. It is a high-strength, moldable, epoxy glass prepreg having a low cure temperature. The nominal ply thickness is 0.09 inch and the fiber volumes ratio is about 50 percent.

A. Material Ply Properties

Since mechanical characterization of this material had apparently not been reported it was necessary to determine the ply stiffness and strength properties. Tension and compression tests were run transverse and parallel to the fibers. For the compression tests a modified IITRI compression fixture [1] was used. The ply stiffness properties for tension (E_{11}^T, E_{22}^T, etc) and compression (E_{11}^c, E_{22}^c, etc) as well as the material's ultimate strengths and strains are tabulated below.

$E_{11}^T = 5.64 \times 10^6$ psi	$e_1^T = 24{,}000\ \mu\varepsilon$
$E_{22}^T = 1.74 \times 10^6$ psi	$X_1^C = 112$ ksi
$\nu_{12}^T = 0.299$	$X_2^T = 7.55$ ksi
$G_{12} = 0.680 \times 10^6$ psi	$e_2^T = 4{,}760\ \mu\varepsilon$
$E_{11}^C = 5.87 \times 10^6$ psi	$X_2^C = 25.0$ ksi
$E_{22}^C = 2.12 \times 10^6$	$e_2^C = 18{,}600\ \mu\varepsilon$
$\nu_{12}^C = 0.317$	$S_{12} = 7.23$ ksi
$X_1^T = 134$ ksi	$e_{12} = 19{,}700\ \mu\varepsilon$

The material's nonlinear shear behavior was of particular concern. For the shear tests the three-rail fixture was used. The ply shear curve is shown in Figure 1.

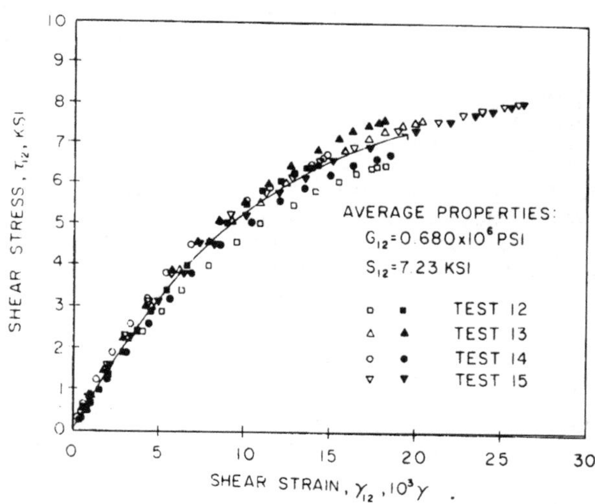

Fig. (1) -The Shear Stress-Strain Response for the Unidirectional $[0]_s$ SP-250 Glass-Epoxy.

B. Model of Laminate Nonlinear Response

An analytical model of the laminate nonlinear behavior was constructed. The numerical scheme is similar to the well-known procedure described by Rowlands [2] except that here the ply's actual nonlinear shear curve is used. Numerically the average laminate stresses are applied in increments. For each increment of stress the incremental laminate strain components are calculated using the laminate compliance matrix from the previous stress increment. These strain increments are then used to calculate increments of ply stresses for each ply using the stiffness matrices from the previous load increment. Following Sandhu [3] the nonlinear ply shear curve is fitted with a cubic spline interpolation to define the ply shear modulus at a given strain. The current tangent modulus is used in calculating the stiffness matrix. Thus the ply nonlinear shear behavior is incorporated into the predicted laminate strain response.

As the incremental loading continues failure eventually occurs in some ply(s). Once ply failure, as predicted by a failure formula (the Tsai-Wu formula [4] was used in all the following), is reached the ply is investigated to determine if the predicted failure is in the fiber or matrix. If the ply stress in the fiber direction $(\sigma_1)_k$ exceeds neither the ultimate tensile or compressive strength in the fiber direction, it is assumed that the failure is in the ply matrix. Once a ply fails, unloading of that ply occurs, and some of the failed ply's stresses are distributed to the remaining unfailed plys. Information on in situ unloading of failed plys is scare and various unloading behaviors have been proposed (see Hahn and Tsai [5] or Chou et al [6]). In the following it was assumed that, following failure by matrix cracking, the failed ply accepted no further transverse stresses. Thus under incremental loading the constants E_{22}, G_{12} were set approximately equal to zero and E_{11} retained its original value. Once the ply stress $(\sigma_1)_k$ exceeded the corresponding tensile or compressive strength it was assumed that the failure was in the fiber and that under further incremental loading all stiffness of the ply was lost. Thus E_{11}, E_{22} and G_{12} were all set approximately equal to zero. This procedure results in a predicted nonlinear laminate stress-strain curve which exhibits "knees" at the failures of the various plys. This idea of modifying ply stiffnesses as damage increases was recently given a formal mathematical basis by Nuismer [7].

The load increments are continued until by some definition total failure of the laminate is predicted. For the $[0/\pm45/90]_S$ laminate, failure is assumed to occur once fiber failure (as distinct from matrix failure) has occurred in two or more plys. For the angle plys, laminate failure is assumed to occur once the modified laminate stiffness becomes singular (i.e. physically large strains are predicted). A uniform definition of laminate failure applicable to a number of layups, is lacking.

C. Experimental and Predicted Laminate Response

Three angle ply laminates and one quasi-isotropic laminate of Scotchply SP-250 were tested to failure in uniaxial tension. The laminate layups were $[\pm30]_S$, $[\pm45]_S$, $[\pm60]_S$, and $[0/\pm45/90]_S$. Three tests were run for each layup. All laminates were 8 plys thick.

Figure 2 shows the nonlinear stress-strain response of the $[\pm 30]_s$ laminate. The predicted longitudinal strain ε_x is about 10 percent greater than the measured strain. The computed curves in Figure 2 exhibit correctly the decreasing stiffness with increasing load. The initial stiffness of the laminate is about 3.34 x 10^6 psi, and the final predicted stiffness is about 2.02 x 10^6 psi. The predicted ultimate stress is low, about 42 ksi as compared to an actual failure stress of about 60 ksi. Generally, strength predicitions using lamination theory fall below the acutal strength for angle plys. Chamis and Sullivan [8] have indicated that this may be due to the difference in the in situ ply strength and the ply strength measured in unidirectional coupons.

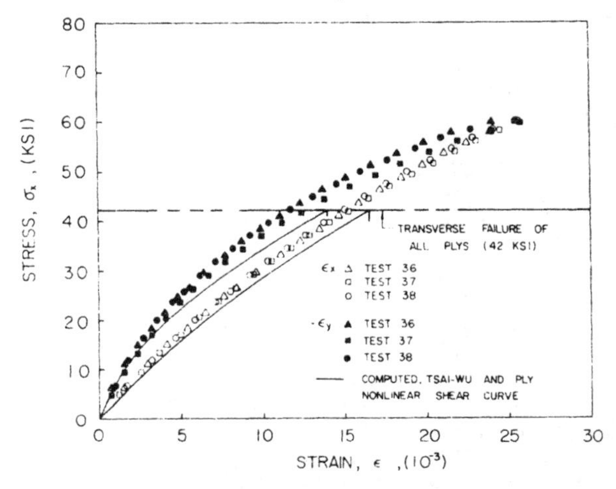

Fig. (2) - Stress-Strain Response of the $[\pm 30]_s$, SP-250 Glass-Epoxy Laminate.

Figure 3 shows the initial portion of the stress-strain curve for the $[\pm 45]_s$ laminate; the full curve is omitted. It can be seen that the laminate yields at a stress of about 17 ksi. Final separation finally occurs at a strain of near 100,000 $\mu\varepsilon$--a 10 percent elongation. As noted by Rotem and Hashin [9] the $[\pm 45]_s$ laminates exhibit a singular amount of large deformation prior to ultimate failure. The transverse and longitudinal strains both agree well with the test values up to the predicted failure load of 12 ksi.

The predicted and test response of the $[\pm 60]_s$ laminate is shown in Figure 4. Overall the predicted stiffness of the laminate is greater than the test stiffness. Predicted failure occurs at a stress of about 9 ksi; the actual failure stress was about 11 ksi.

The strain response of the $[0/\pm 45/90]_s$ laminate is shown in Figure 5. Transverse failure of the 90-degree plys is predicted at a stress of 14 ksi

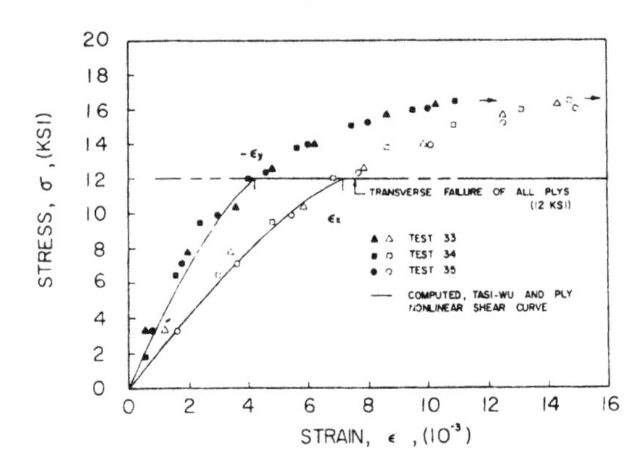

Fig. (3) - Stress-Strain Response of the $[\pm 45]_s$, SP-250 Glass-Epoxy Laminate--Initial Portion.

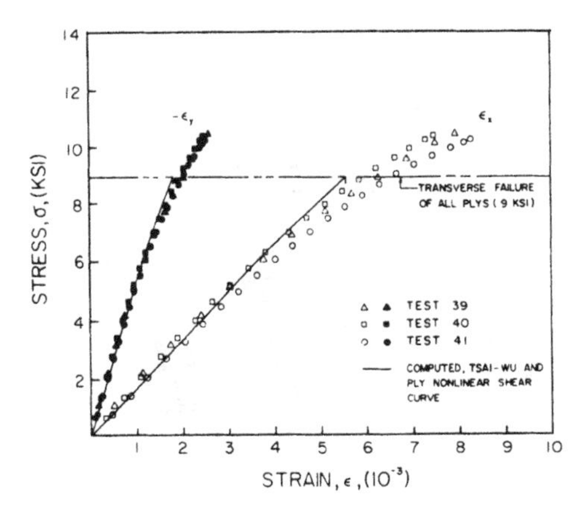

Fig. (4) - Stress-Strain Response of the $[\pm 60]_s$, SP-250 Glass-Epoxy Laminate.

followed by a transverse failure of the ±45-degree plys at a stress of 18 ksi. Predicted and test failure stresses are 53 ksi and 41 ksi, respectively. In constrast to the case of the angle plys, for the $[0/\pm 45/90]_s$ laminate, the prediction method over estimates the strength. The stiffness of the laminate is predicted very well, however. The predicted longitudinal stiffness decreases from an initial predicted value of 3.01 x 10^6 psi to a final value of 1.88 x 10^6 psi.

Figure 6 shows the compression behavior of the $[0/\pm 45/90]_s$ laminate. The prediction agrees very well with the test results. Predicted failure load was 51 ksi, the same as the mean test value.

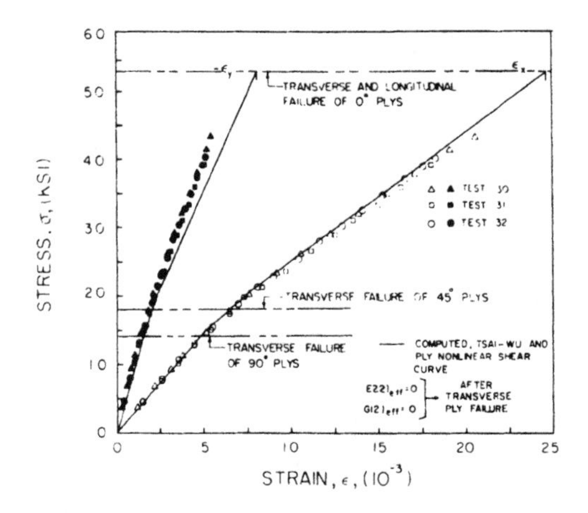

Fig. (5) - Stress-Strain Response of the $[0/\pm45/90]_s$, SP-250 Glass-Epoxy Laminate.

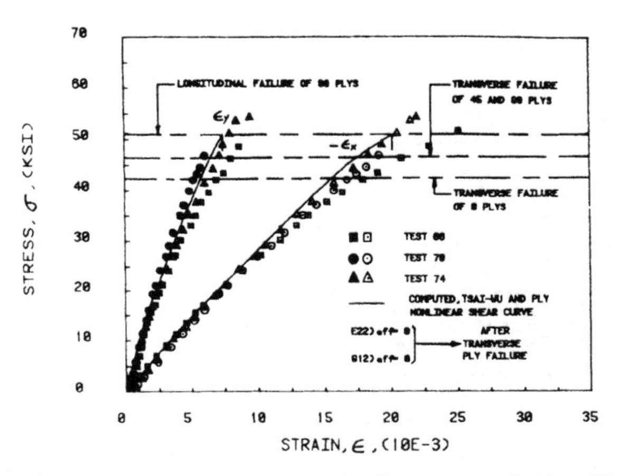

Fig. (6) - Compressive Stress-Strain Response of the $[0/\pm45/90]_s$, SP-250 Glass-Epoxy Laminate.

FAILURE ANALYSIS METHOD

A finite element program which incorporates the above nonlinear stress-strain model was developed. While laminate characterization tests were confined to laminates loaded with inplane forces only, the finite element program has the capability to analyze plates and shells containing bending as well as inplane forces.

A. Description of the Finite Element

The element used is a doubly-curved, isoparametric, quadratic, 8-

node, thick-shell element. The element is derived from the 16-node solid element by specializing the element so that strain energy of the stresses normal to the midsurface is ignored and by constraining lines initially normal to the midsurface to remain straight. The resulting element has 40 degrees of freedom--three displacements and two rotations for each of the eight nodes. Though the midsurface normals are to remain straight during deformation, these lines need not remain normal to the deformed mid-surface. Therefore, the ability to model transverse shear deformation is retained. Transverse shear is thought to be significant for laminate plates and shells.

Geometric as well as material nonlinearity were included. The incremental procedure is employed, using load increments of equal magnitude. After the application of each load increment the coordinates of the node are updated and the adjusted coordinates used in the computation of the stiffness for the next increment. The nonlinear shear stress-strain curve as well as the post failure unloading of the plies is modeled as previously explained. Details of the element are explained more fully in Reference [10].

B. Verification of the Computer Model

Two problems are included to illustrate the capabilities and limitations of the finite element computer program aside from the nonlinear material aspects. Examples included have known solutions, and thus provide good test cases.

1. Thin hyperbolic paraboloid shell. The boundary of this shell is assumed to be rigidly held against both displacements and rotations and subjected to a uniform load. The geometry is shown in Figure 7. The entire shell was modelled using only 4 elements. The results compared with those of Minich and Chamis [11] and Choi and Schnobrich [12] in Figure 8 show good agreement.

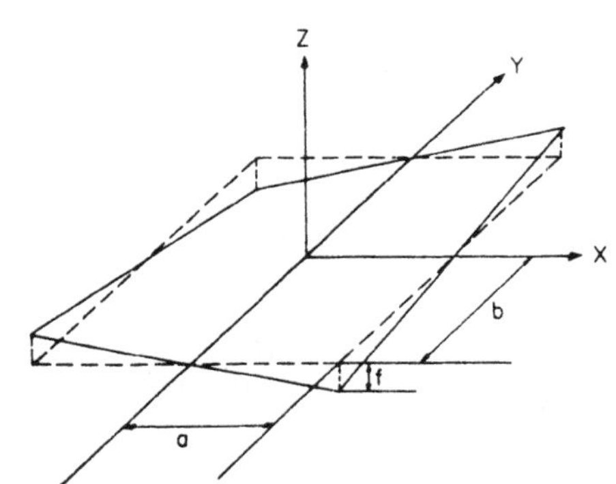

Fig. (7) - The Geometry of a Hyperbolic Paraboloid.

2. Cylindrical shell roof. In this test case bending action is severe, due to supports restraining deflection at the ends. The shell is supported on diaphragms as shown in Figure 9. These allow no displacement in

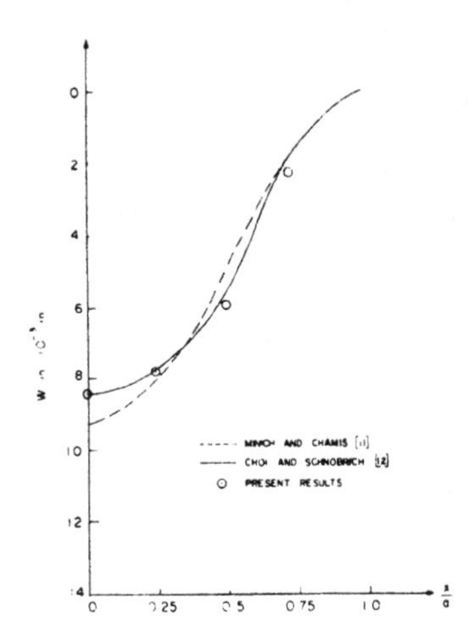

Fig. (8) - Vertical Deflection Across the Midspan of a Clamped Hyperbolic Paraboloid Under Uniform Load.

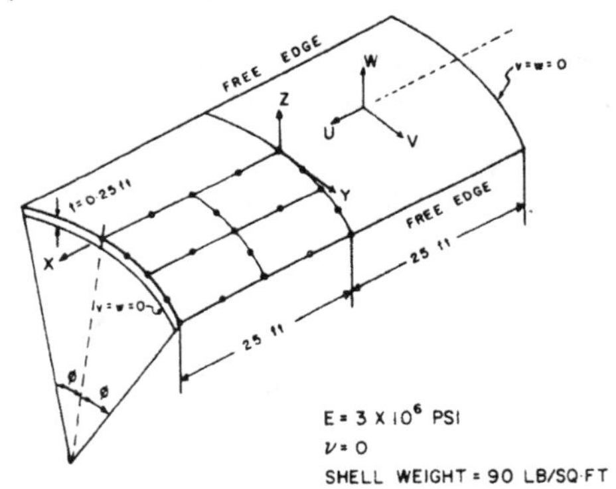

Fig. (9) - A Cylindrical Shell Roof.

their own plane. Due to symmetry only a quarter of the shell was actually analyzed. Vertical displacements of the shell at the mid-span section are compared with those of Pawsley [13] in Figure 10. The shell is well modelled by only one element.

RESPONSE OF GLASS-EPOXY LAMINATE WITH A HOLE

An example like that of Chow et al [14] was chosen. Three tensile coupons of SP-250 containing a hole were tested. The layup was $[0/\pm45/90]_s$, eight plys thick. During the load application the strain was monitered near

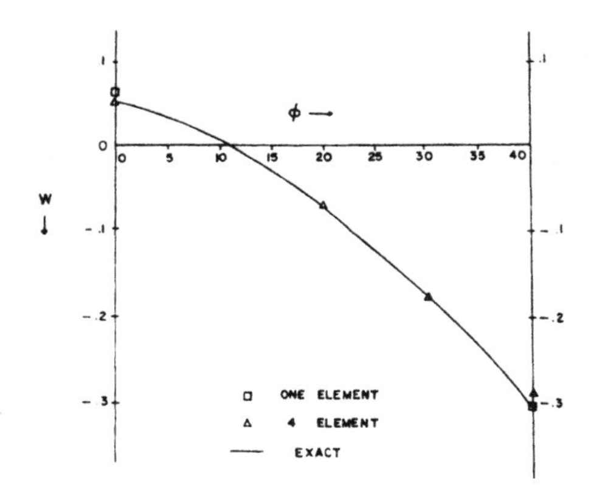

Fig. (10) - Midspan Vertical Displacement for the Cylindrical Shell Roof.

the hole by a 1/16-inch strain gage. The coupon dimensions and gage location are shown in Figure 11.

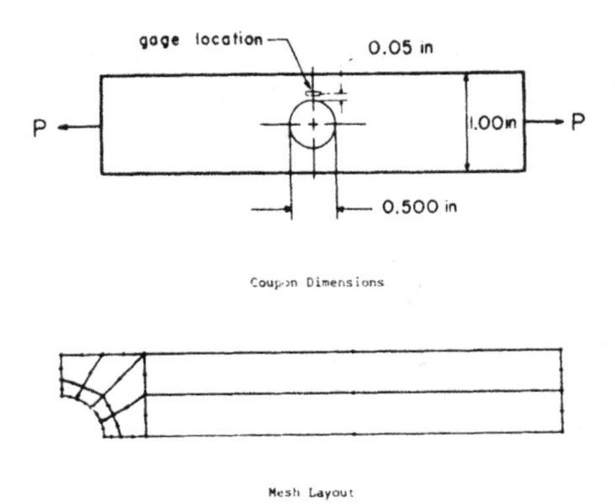

Fig. (11) - Dimensions and Mesh Layout for the $[0/\pm45/90]_s$ Coupon with a Hole.

The mesh layout for the computer simulation is shown in Figure 11. The number of nodes was the same as Chow's [14]. The computed response is compared with the three test responses in Figure 12. The two agree fairly well although the computed load is slightly higher than the test load. One would expect this, recalling that for an uniaxial test specimen the nonlinear material model and failure definition resulted in a failure load too high. The indicated computed failure was taken to be when two plys failed by fiber

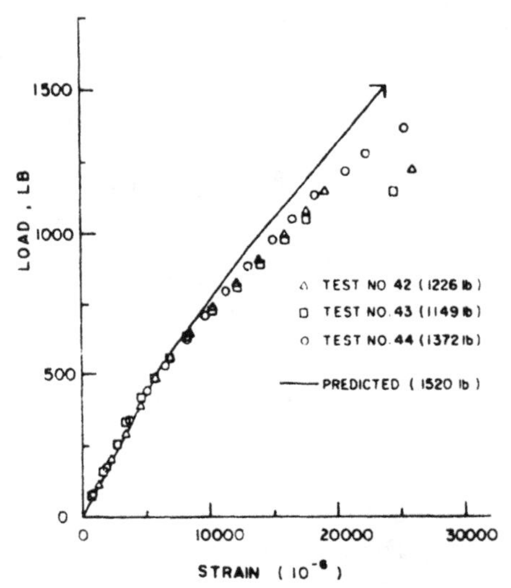

Fig. (12) - Comparison of the Computed and Test Strain Near a Hole in a $[0/\pm45/90]_s$, SP-250 Glass-Epoxy Laminate.

fracture. In this problem these failures occurred first, of course, in the elements on the hole edge. Figure 12 illustrates the power of the developed nonlinear finite element program. Although encouraging, the results are imperfect, due to an imperfect model of laminate strain behavior, and work continues to refine the model.

ACKNOWLEDGMENTS

The authors wish to acknowledge P. H. Huang, former M.S. student, who conducted most of the tests, and V. Nopratvarakorn and C. C. Chao, doctoral students, who carried out a great deal of the computer programming. This research was supported by the Ground Equipment and Missile Structures Directorate, U. S. Army Missile Command, Redstone Arsenal, Alabama, under Contract No. DAAK40-78-C-0165.

REFERENCES

[1] Hofer, K. E. Jr., and Rao, P. N., "A New Static Compression Fixture for Advanced Composite Materials," Journal of Testing and Evaluation, Vol. 5, No. 4, July 1977, pp. 278-283.

[2] Rowlands, R. E., "Flow and Failure of Biaxially Loaded Composites," Inelastic Behavior and Composite Materials, Ed. Carl T. Herakovich, ASME, 1975, pp. 97-125.

[3] Sandhu, R. S., "Ultimate Strength Analysis of Symmetric Laminates," Air Force Flight Dynamics Laboratory, AFFDL-TR-73-137, February 1974.

[4] Tsai, S. W. and Wu, E. M., "A General Theory of Strength for Anisotropic Materials," Journal of Composite Materials, Vol. 5, January 1971, pp. 58-80.

[5] Hahn, H. T. and Tsai, S. W., "On the Behavior of Composite Laminates After Initial Failures," Journal of Composite Materials, Vol. 8, July 1974, pp. 288-305.

[6] Chou, S. C., Orringer, O., and Rainey, J. H., "Post-Failure Behavior of Laminates: I - No Stress Concentration," Journal of Composite Materials, Vol. 10, October 1976, pp. 371-381.

[7] Nuismer, R. J., "Continuum Modeling of Damage Accumulation and Ultimate Failure in Fiber Reinforced Laminated Composite Materials," Research Workshop on Mechanics of Composite Materials, Duke University, October 1978.

[8] Chamis, C. C. and Sullivan, T. L., "In Situ Ply Strength: An Initial Assessment," NASA TM-73771, 1978.

[9] Rotem, A. and Hashin, Z., "Failure Modes of Angle Ply Laminates," Journal of Composite Materials, Vol. 9, April 1975, pp. 191-206.

[10] Smith, D. G. and Huang, Ju-chin, "Post-Crazing Stress Analysis of Glass-Epoxy Laminates," Tennessee Technological University, TTU-ESM-79-1, May 1979.

[11] Minich, M. C. and Chamis, C. C., "Doubly-Curved Variable-Thickness Isoparametric Heterogeneous Finite Element," Computers and Structures, Vol. 7, 1975, pp. 295-301.

[12] Choi, C. K. and Schnobrich, "Use of Non-conforming Modes in Finite Element Analysis of Plates and Shells," University of Illinois, UILU-ENGR-73-2019, 1073.

[13] Pawsley, S. F., The Analysis of Moderately Thick to Thin Shells by the Finite Element Method, Report No. UCSESM 70-12, Dept. of Civil Engineering, University of California, 1970.

[14] Chou, S. C., Orringer, O., and Rainey, J. H., "Post-Failure Behavior of Laminates: II - Stress Concentration," Journal of Composite Materials, Vol. 11, January 1977, pp. 71-78.

SECTION VI

STRESS AND STRENGTH ANALYSIS

GENERALIZED STRUCTURAL STRENGTH CRITERIA OF REINFORCED PLASTICS UNDER PLANE STRESS

A. M. Skudra and F. Ya. Bulavs

Riga Polytechnic Institute
226355 Riga, USSR

ABSTRACT

Strength criteria for reinforced plastics have been presented considering the possibility of fiber, matrix or bond failure.

INTRODUCTION

One of the principal trends in the development of modern technology is to use the materials available to the best advantage. This requires an accurate qualitative and quantitative estimation of the peculiar performance of the given materials under various load conditions. In the first place, this refers to the various types of reinforced plastics which have a wide range of engineering applications.

Multidirectionally reinforced plastics have a laminated structure. In practice, they are usually subjected to plane state of stress. When load is applied, the unidirectionally reinforced plies oriented at various angles do not fail simultaneously. The failure of the individual plies oriented in critical directions does not always coincide with the failure of the whole laminate. In many cases, the failure can be described as a two-stage process: at first, the material loses its continuity due to the matrix or bond failure in the most unfavorable oriented plies, and later, as the load increases, a complete failure results from fiber rupture.

In order to calculate the point at which continuity is lost and when the material fails completely, it is necessary to consider the strength properties of the individual unidirectionally reinforced plies oriented in various directions.

To simplify the calculation formulas, it will be assumed that the structure of the material is symmetrical to the midplane.

The calculation scheme for a multidirectionally reinforced laminate plastic under plane state of stress is seen in Figure 1.

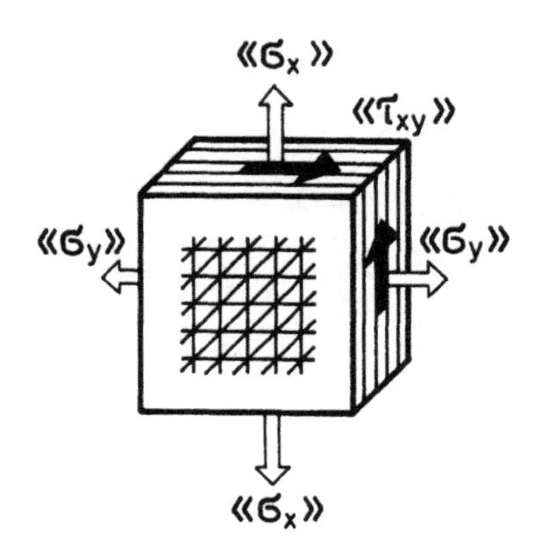

Fig. (1) - Calculation scheme

FIBER FAILURE

To determine the strength of a single unidirectionally reinforced ply involving fiber failure, the strength criterion according to which fibers undergo failure if the strain reaches its ultimate value will be used. Assuming that there is a stiff bond between the fibers and the matrix (i.e., there is no slip) the fiber strain is equal to the strain in the reinforced plastic in the reinforcement direction $<\varepsilon_{11}>$.

The individual plies of a laminate are reinforced in various directions given by angle β. For instance, Figure 2 shows an arbitrary ply "k", the strength criterion for which is the following:

$$<\varepsilon_{11}>_k = \varepsilon_{fR} \tag{1}$$

where ε_{fR} is the ultimate fiber strain under tension or compression.

Substituting the expression for $<\varepsilon_{11}>_k$ in (1), we obtain

$$<<\varepsilon_1>>\cos^2\beta_k + <<\varepsilon_2>>\sin^2\beta_k + <<\gamma_{12}>>\sin\beta_k\cos\beta_k = \varepsilon_{fR} \tag{2}$$

Strains $<<\varepsilon_1>>$, $<<\varepsilon_2>>$ and $<<\gamma_{12}>>$ are determined by the methods of lamination theory provided there is no slip between the plies. Considering the strain relationship, criterion (2) is given as

$$\begin{aligned}&<<\sigma_1>>(a_1\cos^2\beta_k + a_2\sin^2\beta_k + a_{12}\cos\beta_k\sin\beta_k) + <<\sigma_2>>(b_1\cos^2\beta_k \\ &\quad + b_2\sin^2\beta_k + b_{12}\cos\beta_k\sin\beta_k) + <<\tau_{12}>>(c_1\cos^2\beta_k + c_2\sin^2\beta_k \\ &\quad + c_{12}\cos\beta_k\sin\beta_k) = \varepsilon_{fR}\end{aligned} \tag{3}$$

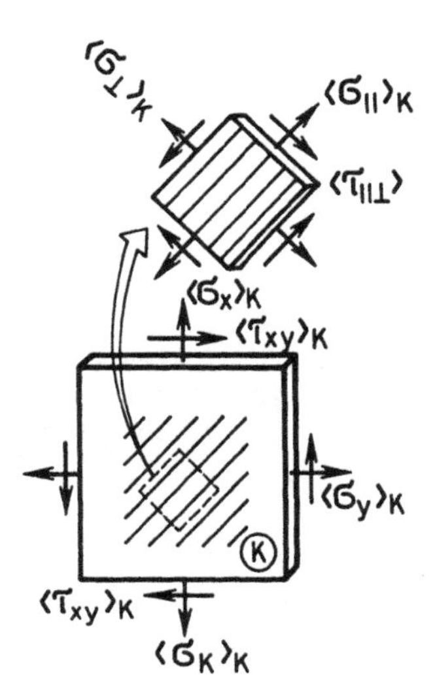

Fig. (2) - The stress state of an arbitrary ply

Criterion (3) shows the point of fiber failure in ply "k". This criterion should be applied repeatedly to all the plies having different fiber orientations. This many-fold calculation permits us to establish which of the plies will fail first, i.e., we obtain the critical angle β_k. Knowing the critical value of angle β_k and the structure of the laminate, criterion (3) determines the strength of the laminate as a whole. It can be explained in view of the fact that the failure of most loaded fibers causes the abrupt redistribution of stresses, usually connected with an avalanche failure of the whole material.

Thus, criterion (3) is the generalized structural strength criterion for reinforced plastics when the fibers are to fail first. Ratios a, b and c included in it depend upon the structure of the material and are calculated from the following expressions:

$$a_1 = \frac{\bar{A}_{22}\bar{A}_{66}-\bar{A}_{26}}{\bar{\Delta}}; \quad b_1 = \frac{\bar{A}_{16}\bar{A}_{26}-\bar{A}_{12}\bar{A}_{66}}{\bar{\Delta}};$$

$$c_1 = \frac{\bar{A}_{12}\bar{A}_{26}-\bar{A}_{22}\bar{A}_{16}}{\bar{\Delta}}; \quad a_2 = \frac{\bar{A}_{16}\bar{A}_{26}-\bar{A}_{12}\bar{A}_{66}}{\bar{\Delta}};$$

$$b_2 = \frac{\bar{A}_{11}\bar{A}_{66}-\bar{A}_{12}}{\bar{\Delta}}; \quad c_2 = \frac{\bar{A}_{12}\bar{A}_{16}-\bar{A}_{11}\bar{A}_{26}}{\bar{\Delta}};$$

$$a_{12} = \frac{\bar{A}_{12}\bar{A}_{26}-\bar{A}_{16}\bar{A}_{22}}{\bar{\Delta}}; \quad b_{12} = \frac{\bar{A}_{12}\bar{A}_{16}-\bar{A}_{26}\bar{A}_{11}}{\bar{\Delta}};$$

$$c_{12} = \frac{\bar{A}_{11}\bar{A}_{22}-\bar{A}_{12}^2}{\bar{\Delta}};$$

$$\bar{\Delta} = \bar{A}_{11}(\bar{A}_{22}\bar{A}_{66}-\bar{A}_{26}^2) - \bar{A}_{12}(\bar{A}_{12}\bar{A}_{66}-\bar{A}_{16}\bar{A}_{26}) + \bar{A}_{16}(\bar{A}_{12}\bar{A}_{26}-\bar{A}_{22}\bar{A}_{16}).$$

STRESSES IN A SINGLE PLY

In the general case of a plane state of stress when the arbitrary axis x, y coincide with the laminate symmetry axis 1 and 2, the stresses the directions of the elastic symmetry of the plies are calculated as follows (Skudra, 1980):

$$<\sigma_{||}>_k = \frac{1}{\bar{\Delta}}\,(<<\sigma_1>>a_1 + <<\sigma_2>>b_1 + <<\tau_{12}>>c_1) \tag{4}$$

$$<\sigma_{\perp}>_k = \frac{1}{\bar{\Delta}}\,(<<\sigma_1>>a_2 + <<\sigma_2>>b_2 + <<\tau_{12}>>c_2) \tag{5}$$

$$<\tau_{||\perp}>_k = \frac{1}{\bar{\Delta}}\,(<<\sigma_1>>a_3 + <<\sigma_2>>b_3 + <<\tau_{12}>>c_3) \tag{6}$$

The symbols introduced in the formulas are:

$$a_1 = \cos^2\beta_k\bar{A}_{66}(\bar{Q}_{11}\bar{A}_{22}-\bar{Q}_{12}\bar{A}_{12}) + \sin^2\beta_k\bar{A}_{66}(\bar{Q}_{12}\bar{A}_{22}-\bar{Q}_{22}\bar{A}_{12})$$
$$+ 2\sin\beta_k\cos\beta_k\bar{A}_{66}(\bar{Q}_{16}\bar{A}_{22}-\bar{Q}_{26}\bar{A}_{12});$$

$$b_1 = \cos^2\beta_k\bar{A}_{66}(\bar{Q}_{12}\bar{A}_{11}-\bar{Q}_{11}\bar{A}_{12}) + \sin^2\beta_k\bar{A}_{66}(\bar{Q}_{22}\bar{A}_{11}-\bar{Q}_{12}\bar{A}_{12})$$
$$+ 2\sin\beta_k\cos\beta_k\bar{A}_{66}(\bar{Q}_{26}\bar{A}_{11}-\bar{Q}_{16}\bar{A}_{12});$$

$$c_1 = \bar{A}_{11}\bar{A}_{22}(\bar{Q}_{16}\cos^2\beta_k + \bar{Q}_{26}\sin^2\beta_k + 2\bar{Q}_{66}\sin\beta_k\cos\beta_k);$$

$$a_2 = \sin^2\beta_k\bar{A}_{66}(\bar{Q}_{11}\bar{A}_{22}-\bar{Q}_{12}\bar{A}_{12}) + \cos^2\beta_k\bar{A}_{66}(\bar{Q}_{12}\bar{A}_{22}-\bar{Q}_{22}\bar{A}_{12})$$
$$- 2\sin\beta_k\cos\beta_k\bar{A}_{66}(\bar{Q}_{16}\bar{A}_{22}-\bar{Q}_{26}\bar{A}_{12});$$

$$b_2 = \sin^2\beta_k\bar{A}_{66}(\bar{Q}_{12}\bar{A}_{11}-\bar{Q}_{11}\bar{A}_{12}) + \cos^2\beta_k\bar{A}_{66}(\bar{Q}_{22}\bar{A}_{11}-\bar{Q}_{12}\bar{A}_{12})$$
$$- 2\sin\beta_k\cos\beta_k\bar{A}_{66}(\bar{Q}_{26}\bar{A}_{11}-\bar{Q}_{16}\bar{A}_{12});$$

$$c_2 = \bar{A}_{11}\bar{A}_{22}(\bar{Q}_{16}\sin^2\beta_k + \bar{Q}_{26}\cos^2\beta_k - 2\bar{Q}_{66}\sin\beta_k\cos\beta_k);$$

$$a_3 = \sin\beta_k\cos\beta_k\bar{A}_{66}[\bar{A}_{22}(\bar{Q}_{12}-\bar{Q}_{11}) + \bar{A}_{12}(\bar{Q}_{12}-\bar{Q}_{22})]$$
$$+ \bar{A}_{66}(\cos^2\beta_k-\sin^2\beta_k)(\bar{Q}_{16}\bar{A}_{22}-\bar{Q}_{26}\bar{A}_{12});$$

$$b_3 = \sin\beta_k\cos\beta_k\bar{A}_{66}[\bar{A}_{11}(\bar{Q}_{22}-\bar{Q}_{12}) + \bar{A}_{12}(\bar{Q}_{11}-\bar{Q}_{12})]$$
$$+ \bar{A}_{66}(\cos^2\beta_k-\sin^2\beta_k)(\bar{Q}_{26}\bar{A}_{11}-\bar{Q}_{16}\bar{A}_{12});$$

$$c_3 = \bar{A}_{11}\bar{A}_{22}[\sin\beta_k\cos\beta_k(\bar{Q}_{26}-\bar{Q}_{16}) + (\cos^2\beta_k-\sin^2\beta_k)\bar{Q}_{66}]$$

Ratios $\bar{A}_{ij}$ are calculated from the formula:

$$\bar{A}_{ij} = \frac{1}{\delta}\sum_{k=1}^{n} (\bar{Q}_{ij})_k(h_k-h_{k-1}),$$

where δ is the laminate thickness and $\bar{Q}_{ij}$ denotes the elastic properties of plies.

From (4)-(6), it follows that a unidirectionally reinforced single ply is in a plane state of stress even in the case of simple loading, as in $\langle\langle\sigma_2\rangle\rangle = \langle\langle\tau_{12}\rangle\rangle = 0$.

FAILURE OF THE POLYMER MATRIX

If the failure of the reinforced plastic was caused by the polymer matrix failure or the fiber-matrix bond failure, the generalized strength criterion is:

$$F(\langle\sigma_\perp\rangle_k;\ \langle\tau_{||\perp}\rangle_k) = 1 \qquad (7)$$

In this criterion $\langle\sigma_\perp\rangle_k$ and $\langle\tau_{||\perp}\rangle_k$ denote mean stresses in ply "k" which in our case of loading will fail first. The right-hand expression of criterion (7) depends upon the adopted criteria for the polymer matrix and the bond failure and on the ratio between $\langle\sigma_\perp\rangle_k$ and $\langle\tau_{||\perp}\rangle_k$.

The peculiarities of the unidirectionally reinforced plastic failure under stresses $\langle\sigma_\perp\rangle_k$ and $\langle\tau_{||\perp}\rangle_k$ have been discussed by Skudra and Bulavs [2]. If the polymer matrix bond fails under combined tensile and shear stresses, function F is expressed as follows:

$$F(\langle\sigma_\perp\rangle_k;\ \langle\tau_{||\perp}\rangle_k) = (1-\nu_m^2)\bar{\sigma}_r^2\,\frac{\langle\sigma_\perp^2\rangle_k}{(R_m^+)^2} + \frac{\langle\tau_{||\perp}^2\rangle_k}{T_m^2} \qquad (8)$$

Formula (8) is based on the assumption that the polymer matrix strength under combined tension and shear is determined according to an energy criterion.

If the failure of a unidirectionally reinforced ply oriented at angle β_k is typical for longitudinal shear, then

$$F(\langle\sigma_\perp\rangle_k;\ \langle\tau_{||\perp}\rangle_k) = \frac{1}{2}\left(\frac{\langle\sigma_\perp\rangle_k}{R_m^+}\right)^2 + (1+\nu_m)\left(\frac{\langle\tau^2_{||\perp}\rangle_k}{R_m^+}\right)^2$$

$$+ \frac{\langle\sigma_\perp\rangle_k}{2(R_m^+)^2}\sqrt{\langle\sigma_\perp^2\rangle_k + 4\langle\tau^2_{||\perp}\rangle_k} \tag{9}$$

If compressive and shear stresses act simultaneously

$$F(\langle\sigma_\perp\rangle_k;\ \langle\tau_{||\perp}\rangle_k) = \frac{1}{2}\left(\frac{\langle\sigma_\perp\rangle_k}{R_\perp^-}\right)^2 + (1+\nu_{\perp||})\left(\frac{\langle\tau_{||\perp}\rangle_k}{R_\perp^-}\right)^2$$

$$- \frac{\langle\sigma_\perp\rangle_k}{2(R_\perp^-)_k^2}\sqrt{\langle\sigma_\perp^2\rangle_k + 4\langle\tau^2_{||\perp}\rangle_k} \tag{10}$$

In formulas (8)-(10), the notation is the following: R_m^+; T_m - tensile and shear strengths of the polymer matrix; ν_m; $\nu_{\perp||}$ - Poisson's ratios of the polymer matrix and a unidirectionally reinforced plastic; $(R_\perp^-)_k$ - transverse compressive strength of a unidirectionally reinforced ply "k". Strength $R_\perp^-$ is calculated by means of the following formula [2]:

$$R_\perp = \frac{3{,}5\ R_m^+}{(1+\nu_{\perp||})\bar{\sigma}_r\sqrt{1-\nu_m^2}}$$

BOND FAILURE

To obtain an analytical expression for function F in the case of fiber-matrix bond failure the tensor form of strength criterion for an anisotropic material is used:

$$\pi_{\alpha\beta}\sigma_{\alpha\beta} + \pi_{\alpha\beta\gamma\delta}\sigma_{\alpha\beta}\sigma_{\gamma\delta} = 1 \tag{11}$$

Making use of interface strength symmetry and assuming that the bonds which make the mechanical interaction between the fibers and the matrix possible can only fail due to elongation, criterion (11) yields the formula for F [3]

$$F(\langle\sigma_\perp\rangle_k;\ \langle\tau_{||\perp}\rangle_k) = \frac{\langle\sigma_\perp\rangle_k\bar{\sigma}_r}{R_b} + \left(\frac{\langle\tau_{||\perp}\rangle_k\bar{\sigma}_{rz}}{T_b}\right)^2 \tag{12}$$

The value and the sign of stresses $\langle\sigma_\perp\rangle_k$ and $\langle\tau_{||\perp}\rangle_k$ for every particular case of loading is given by (5) and (6). If stress $\langle\sigma_\perp\rangle_k$ is compressive, it is introduced in (9)-(12) as a negative quantity.

Parameter $\bar{\sigma}_r$ and $\bar{\sigma}_{rz}$ in formulas (8) and (12) depend upon the structures of the material, fiber volume content, fiber arrangement and other characteristics of the fibers and the matrix. The values of these parameters are obtained by the methods of the theory of elasticity. Figures 3 and 4 show the dependence of these parameters on the ratio between fiber and matrix elastic moduli for various fiber volume contents.

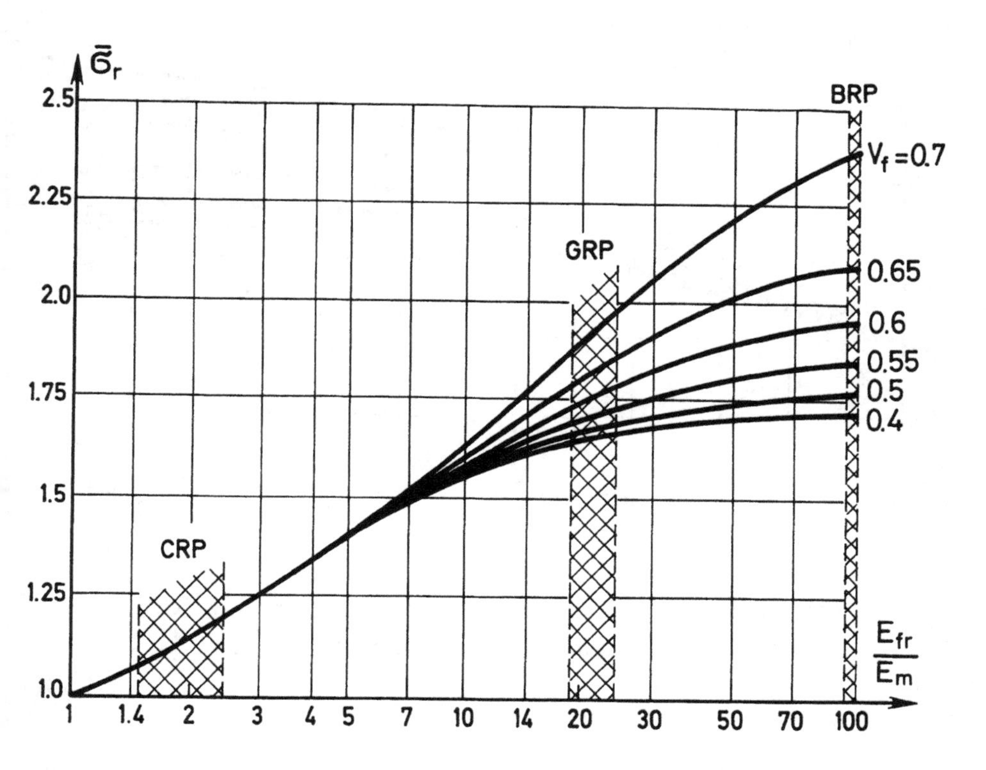

Fig. (3) - Dependence of $\bar{\sigma}_r$ on the fiber volume content (V_f) and on the ratio of elastic moduli of the fibers (E_{fr}) and the matrix (E_m)

R_b and T_b in formula (12) denote bond strength in tension and in shear. Experimentally, they are determined from the following relationships [3]:

$$R_b = R_\perp^+ \cdot \bar{\sigma}_r \tag{13}$$

$$T_b = T_{||\perp} \cdot \bar{\sigma}_{rz} \tag{14}$$

In the above formulas, $R_\perp^+$ and $T_{||\perp}$ are unidirectionally reinforced plastic strengths obtained in tests for transverse tension and longitudinal shear; parameters $\bar{\sigma}_r$ and $\bar{\sigma}_{rz}$ are obtained from curves of Figures 3 and 4.

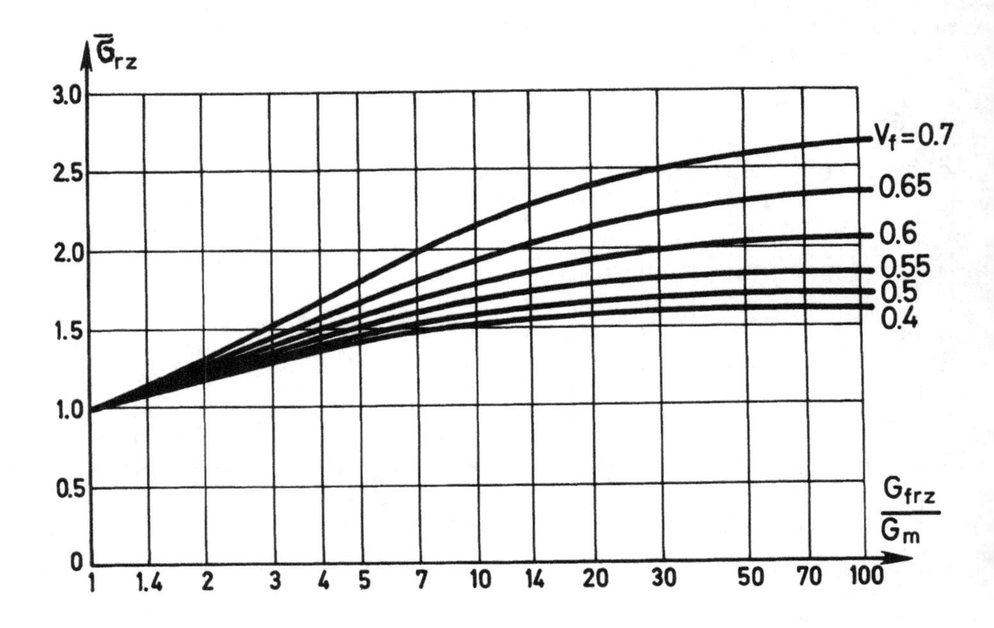

Fig. (4) - Dependence of $\bar{\sigma}_{rz}$ on the fiber volume content (V_f) and on the ratio of shear moduli of the fibers (G_{frz}) and the matrix (G_m)

BIAXIAL LOADING OF CROSS-PLIED LAMINATES

Ultimate strength and continuity loss curves for an epoxy-glass plastic unde biaxial loading are shown in Figure 5. Strength curves for a cross-plied plasti involving fiber failure have been plotted by continuous lines whereas dotted lin show continuity loss curves corresponding to the matrix failure. Criterion (3) has been used to plot ultimate strength curves, and criterion (7) for continuity loss curves as derived from formulas (8) and (9). Calculation formulas followin from those criteria are:

$$\langle\langle\sigma_1\rangle\rangle\bar{A}_{22} - \langle\langle\sigma_2\rangle\rangle\bar{A}_{12} = \varepsilon^{+}_{fR}(\bar{A}_{11}\bar{A}_{22}-\bar{A}^2_{12}) \qquad (15)$$

$$-\langle\langle\sigma_1\rangle\rangle\bar{A}_{12} + \langle\langle\sigma_2\rangle\rangle\bar{A}_{11} = \varepsilon^{+}_{fR}(\bar{A}_{11}\bar{A}_{22}-\bar{A}^2_{12}) \qquad (16)$$

$$\langle\langle\sigma_1\rangle\rangle\bar{A}_{22} - \langle\langle\sigma_2\rangle\rangle\bar{A}_{12} = \varepsilon^{-}_{fR}(\bar{A}_{11}\bar{A}_{22}-\bar{A}^2_{12}) \qquad (17)$$

$$-\langle\langle\sigma_1\rangle\rangle\bar{A}_{12} + \langle\langle\sigma_2\rangle\rangle\bar{A}_{11} = \varepsilon^{-}_{fR}(\bar{A}_{11}\bar{A}_{22}-\bar{A}^2_{12}) \qquad (18)$$

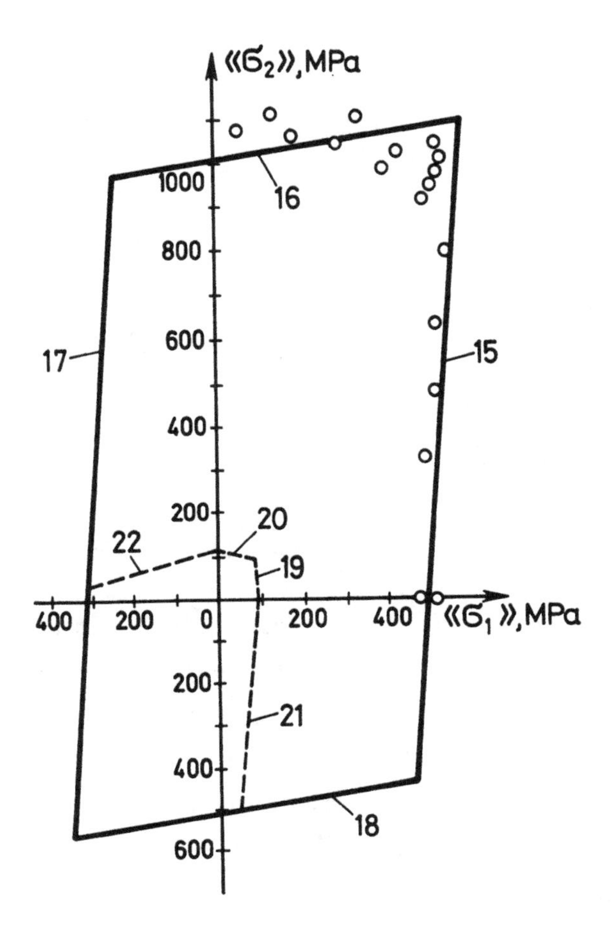

Fig. (5) - The strength envelope for a cross-ply laminate under biaxial loading

$$<<\sigma_1>>q_1 + <<\sigma_2>>q_2 = R_m^+ \tag{19}$$

$$<<\sigma_1>>q_3 + <<\sigma_2>>q_4 = R_m^+ \tag{20}$$

$$<<\sigma_1>>q_3 - <<\sigma_2>>q_4 = R_m^+ \tag{21}$$

$$- <<\sigma_1>>q_1 + <<\sigma_2>>q_2 = R_m^+ \tag{22}$$

The notations in the above are:

$$q_1 = \frac{Q_{12}\bar{A}_{22}-Q_{22}\bar{A}_{12}}{\bar{A}_{11}\bar{A}_{22}-\bar{A}_{12}^2}\,\bar{\sigma}_r\sqrt{1-\nu_m^2}$$

$$q_2 = \frac{Q_{22}\bar{A}_{11}-Q_{12}\bar{A}_{12}}{\bar{A}_{11}\bar{A}_{22}-\bar{A}_{12}^2}\,\bar{\sigma}_r\sqrt{1-\nu^2}$$

$$q_3 = \frac{Q_{22}\bar{A}_{22}-Q_{12}\bar{A}_{12}}{\bar{A}_{11}\bar{A}_{22}-\bar{A}_{12}^2}\,\bar{\sigma}_r\sqrt{1-\nu_m^2}$$

$$q_4 = \frac{Q_{12}\bar{A}_{11}-Q_{22}\bar{A}_{12}}{\bar{A}_{11}\bar{A}_{22}-\bar{A}_{12}^2}\,\bar{\sigma}_r\sqrt{1-\nu_m^2}$$

The numbers of formulas used have been indicated near the respective curves in Figure 5.

Test data given in the work by Jones [4] have also been used in Figure 5.

COMBINED AXIAL LOADING AND SHEAR IN A CROSS-PLIED REINFORCED PLASTIC

The case of a cross-plied plastic subjected to a load acting parallel to the direction of reinforcement is discussed in this section. Under such loading, some of the plies are in more unfavorable conditions, namely, those which are subjected to shear stresses $<\tau_{||\perp}>$ and tensile stresses $<\sigma_\perp>$ perpendicular to reinforcement. The failure mechanism of such plies depends on the ratio between the applied stresses $<<\tau_{12}>>$ and $<<\sigma_1>>$ and also on the ratio between the polymer matrix and bond strengths.

The case when the polymer matrix strength is lower than the bond strength is addressed here.

If the ratio $\frac{<<\tau_{12}>>}{<<\sigma_1>>}$ between the applied stresses exceed a certain value, the failure of the ply is due to the shear failure of the matrix. Thus, considering (9), from (7), we obtain:

$$<<\tau_{12}>> = \sqrt{u_1-u_2<<\sigma_1^2>> - <<\sigma_1>>\sqrt{u_2^2<<\sigma_1^2>> + \frac{2}{\nu_m}u_1u_2}} \qquad (23)$$

where

$$u_1 = \frac{(R_m^+\bar{A}_{66})^2}{(1+\nu_m)G_{||\perp}^2}$$

$$u_2 = \left(\frac{q_5}{q}\right)^2 \frac{\nu_m \bar{A}_{66}^2}{2G_{\|\perp}^2(1+\nu_m)}$$

$$q_5 = Q_{22}\bar{A}_{22} - Q_{12}\bar{A}_{12}$$

$$q_6 = \bar{A}_{11}\bar{A}_{22} - \bar{A}_{12}^2$$

Figures 6 give a schematic representation of the failure of the material according to criterion (23). It is evident that (23) is the criterion for deter-

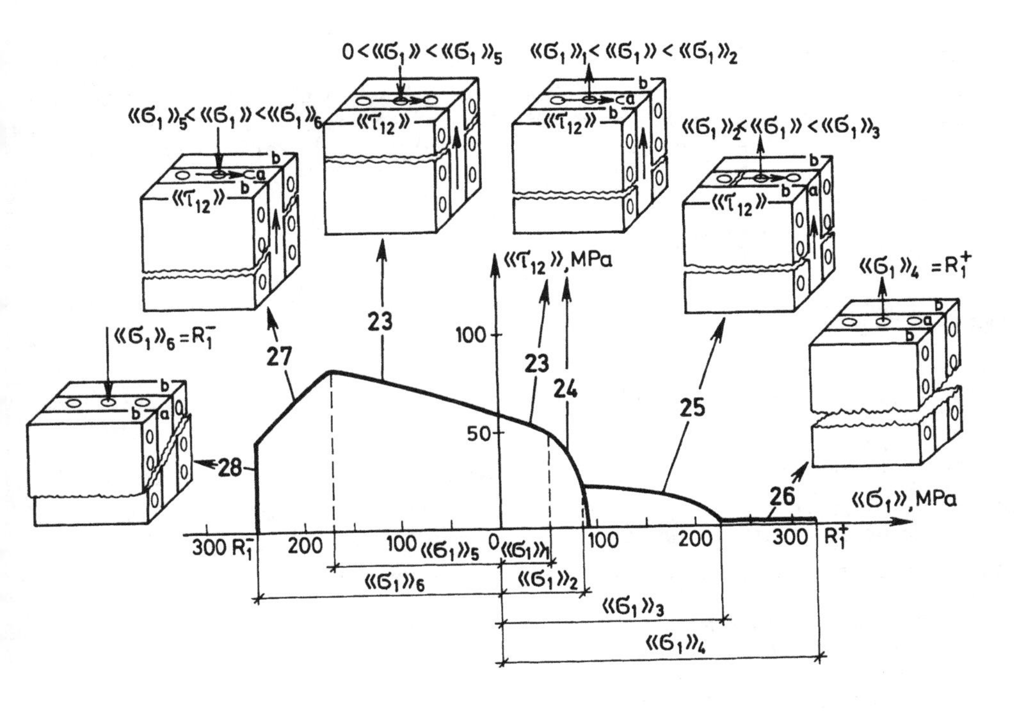

Fig. (6) - The strength envelope for a cross-ply laminate under combined axial and shear loadings

mining the shear failure of the polymer matrix in ply 'b'. After the failure of ply b, tensile stresses in ply 'a' increase abruptly. Usually when stresses increase in such an abrupt manner, ply 'a' fails, and therefore the failure of ply 'b' actually marks the failure of the whole laminate.

The reduction to a certain limit of the ratio between stresses $<<\tau_{12}>>/<<\sigma_1>>$ will change the mechanism of the matrix failure in ply 'b', i.e., bond failure will take place within the matrix. Thus, taking into consideration (8) from criterion (7):

$$<<\tau_{12}>> = \sqrt{u_3 - u_3 u_4 <<\sigma_1^2>>} \tag{24}$$

where

$$u_3 = \left(\frac{T_m \bar{A}_{66}}{G_{||\perp}}\right)^2$$

$$u_4 = \left(\frac{q_5}{q}\right)^2 \left(\frac{\bar{\sigma}_r}{R_m^+}\right)^2 (1-\nu_m^2)$$

If the stress ratio $<<\tau_{12}>>/<<\sigma_1>>$ does not exceed a certain limit, the failure of ply 'b' does not cause a simultaneous failure in ply 'a'. After the failure of ply 'b', a considerable redistribution of stresses takes place in the laminate. Considering the actual stress state of ply 'a', criterion (7) is used to obtain:

$$<<\tau_{12}>> = \sqrt{m_1^2 T_m^2 - u_5 <<\sigma_1^2>>} \tag{25}$$

where

$$u_5 = (1-\nu_m^2) \left[\frac{\bar{\sigma}_r (1-m_1) \nu_{\perp||} E_{||} T_m}{E_2 R_m^+}\right]^2$$

The notation in the above is the following: m_1 - ratio between the total volume of plies reinforced in direction 1 and the total volume of the material; E_2 - modulus of elasticity of the laminate in direction 2; $\nu_{\perp||}$; $E_{||}$ - Poisson's ratio and elastic modulus of a unidirectionally reinforced ply.

With a further increase of stress $<<\sigma_1>>$, fiber rupture takes place as shown schematically in Figure 6.

Criterion (3) in this case yields:

$$<<\sigma_1>> = m_1 [\psi E_{BZ} + (1-\psi) E_m] \varepsilon_{fR}^- \tag{26}$$

where ψ is the relative volume content of fibers; E_{BZ} and E_m are elastic moduli of the fibers and the matrix.

It should be observed that the shape of the final ultimate curve depends largely on the fiber arrangement ratio. Thus, for instance, if

$$m_1 = \frac{1-A}{1-A(1-\frac{E_\perp}{E_{||}})}$$

where

$$A = \frac{R_m^+}{E_{||}\nu_{\perp||}\bar{\sigma}_r\varepsilon_{fR}\sqrt{1-\nu_m^2}}$$

fiber rupture takes place simultaneously with the polymer matrix failure in ply 'a'.

The case of combined compressive and shear loading is discussed here. Varying the compressive stress $<<\sigma_1>>$ within the range from zero to a certain value, criterion (23) is valid provided that the stress $<<\sigma_1>>$ is introduced as a negative quantity.

A further increase of the compressive stress will change the mechanism of ply failure. It will not be typical for longitudinal shear but will be characteristic for transverse compression. Resorting to (10), and using criterion (7):

$$<<\tau_{12}>> = \sqrt{u_6-u_7<<\sigma_1^2>> - <<\sigma_1>>\sqrt{u_7^2<<\sigma_1^2>> + \frac{2}{\nu_{\perp||}}u_6u_7}} \tag{27}$$

where

$$u_6 = \frac{(R_\perp^-\bar{A}_{66})^2}{(1+\nu_{\perp||})G_{||\perp}^2}.$$

$$u_7 = \left(\frac{q_5}{q}\right)^2 \frac{\nu_{\perp||}\bar{A}_{66}^2}{2G_{||\perp}^2(1+\nu_{\perp||}^2)}$$

Criterion (27) is valid for a varying compressive stress $<<\sigma_1>>$ up to ultimate strength R_1^- for a cross-plied reinforced plastic under compressive load which is applied in reinforcement direction 1:

$$R_1^- = E_1\varepsilon_{fR}^- \tag{28}$$

where E_1 is the elastic modulus of a cross-plied plastic in direction 1.

Ultimate strength cruves based on formulas (23)-(28) for 1:2 cross-plied glass plastic are shown in Figure 6. The numbers near the curves indicate the formulas according to which the curves have been plotted.

In conclusion, it should be pointed out that the approach discussed in the present paper for determining the strength of reinforced plastics under combined load with given fiber and matrix properties and a known structure of the material. The approach also permits a reverse calculation, i.e., the choice of a suitable structure of the material to achieve the design strength.

REFERENCES

[1] Skudra, A. A., "The Strength of Helical-Wound Shells with Additional Reinforcement in Tangential Direction", in: Mechanics of Composite Materials, Vol. 3, Riga Polytechnic Institute, Riga, p. 75, 1980.

[2] Skudra, A. M. and Bulavs, F. Ya., "Structural Theory of Reinforced Plastics Riga, "Zinatne", 1978.

[3] Skudra, A. M., Kirulis, B. A. and Zaharov, A. V., "Bond Strength in Reinforced Plastics", in: Mechanics of Composite Materials, Vol. 1, Riga Polytechnic Institute, Riga, pp. 30-37, 1977.

[4] Jones, E. R., "Strength of Glass Filament Reinforced Plastics in Biaxial Loading", SPEJ, Vol. 25, pp. 50-53, March 1969.

ON-LINEAR PHENOMENOLOGICAL MODELS OF FIBRE-REINFORCED COMPOSITES

I. F. Obraztsov and V. V. Vasil'ev

Moscow Institute of Aviation Technology
USSR

Filamentary composites are usually considered as linearly elastic materials rior to failure. However, the deviation of fibre angles, resin crazing, or lasticity of the matrix may cause substantial nonlinearity of the material. In his paper, phenomenological models which seek to describe some aspects of non-inear behaviour of filamentary composites are presented. It is assumed that he material consists of a number of orthotropic unidirectional plies and that he properties of each macroscopically homogeneous ply are determined experimen-ally.

Consider an element of the ply before and after deformation, Figure 1. In-roducing an arbitrary direction α, the strain in this direction and the angle α' fter deformation are

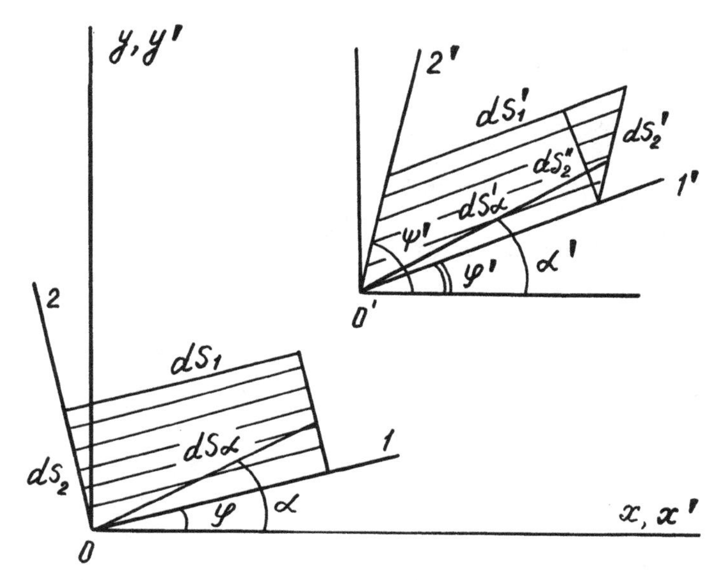

Fig. (1) - An element of the unidirectional ply before and after deformation

$$\varepsilon_\alpha = \frac{dS'_\alpha}{dS_\alpha} - 1, \quad \sin\alpha' = \frac{dy'}{dS'_\alpha} \tag{1}$$

For values of $\alpha = \phi$ and $\alpha = \pi/2+\phi$, equations (1) determine the strains ε_1 and ε_2 in the fibre and transverse directions, respectively, and the corresponding angles ϕ' or ψ' after deformation, Figure 1. Let γ_{12} and ω_{12} denote shear strai and angle of rotation in plane (1,2), namely

$$\gamma_{12} = \frac{\pi}{2} - (\psi'-\phi'), \quad \omega_{12} = \frac{1}{2}\,[(\phi'-\phi) - (\frac{\pi}{2} + \phi-\psi')] \tag{2}$$

For $\alpha = 0$, $\phi = 0$ and $\alpha = \pi/2$, $\phi = 0$, equations (1) and (2) determine the strain ε_x, ε_y, γ_{xy} and the angle of rotation ω_{xy} in plane (x,y). Using the relations:

$$dS_\alpha^2 = dx^2 + dy^2, \quad (dS'_\alpha)^2 = (dx')^2 + (dy')^2, \quad x' = x+u, \quad y' = y+v$$

where u and v denote the displacements, it follows that

$$\begin{aligned}
e_x^2 &= 1 + 2\,\frac{\partial u}{\partial x} + (\frac{\partial u}{\partial x})^2 + (\frac{\partial v}{\partial x})^2 \\
e_y^2 &= 1 + 2\,\frac{\partial v}{\partial y} + (\frac{\partial u}{\partial y})^2 + (\frac{\partial v}{\partial y})^2 \\
e_{xy} &= \frac{1}{e_x e_y}\,(\frac{\partial u}{\partial y} + \frac{\partial v}{\partial x} + \frac{\partial u}{\partial x}\frac{\partial u}{\partial y} + \frac{\partial v}{\partial x}\frac{\partial v}{\partial y}) \\
\sin 2\omega_{xy} &= \frac{1}{e_x e_y}\,[\frac{\partial v}{\partial x}\,(1 + \frac{\partial v}{\partial y}) - \frac{\partial u}{\partial y}\,(1 + \frac{\partial u}{\partial x})]
\end{aligned} \tag{3}$$

where $e_x = 1 + \varepsilon_x$, $e_y = 1 + \varepsilon_y$, $e_{xy} = \sin\gamma_{xy}$. Solving equations (3) for the de rivatives of the displacements gives

$$\begin{aligned}
\frac{\partial u}{\partial x} &= e_x\,\cos(\frac{\gamma_{xy}}{2} + \omega_{xy}) - 1 \\
\frac{\partial u}{\partial y} &= e_y\,\sin(\frac{\gamma_{xy}}{2} - \omega_{xy}) \\
\frac{\partial v}{\partial x} &= e_x\,\sin(\frac{\gamma_{xy}}{2} + \omega_{xy}) \\
\frac{\partial v}{\partial y} &= e_y\,\cos(\frac{\gamma_{xy}}{2} - \omega_{xy}) - 1
\end{aligned} \tag{4}$$

For values of $\alpha = \phi$ and $\alpha = \pi/2$ in equation (1), one can obtain from equations (1) and (2) the relations for ε_1, ε_2, γ_{12} similar to equations (3). Substituting equations (4) into these relations gives

$$e_1^2 = e_x^2\cos^2\phi + e_x e_y e_{xy}\sin2\phi + e_y^2\sin^2\phi \tag{5}$$

$$e_2^2 = e_x^2\sin^2\phi + e_x e_y e_{xy}\sin2\phi + e_y^2\cos^2\phi \tag{6}$$

$$e_{12} = \frac{1}{e_1 e_2}\left[(e_y^2 - e_x^2)\sin\phi\cos\phi + e_x e_y e_{xy}\cos2\phi\right] \tag{7}$$

$$\sin\phi' = \frac{1}{e_1}\left[e_x\sin\left(\frac{\gamma_{xy}}{2} + \omega_{xy}\right)\cos\phi + e_y\cos\left(\frac{\gamma_{xy}}{2} - \omega_{xy}\right)\sin\phi\right] \tag{8}$$

where $e_1 = 1 + \varepsilon_1$, $e_2 = 1 + \varepsilon_2$, $e_{12} = \sin\gamma_{12}$. Finally, consider the strain ε_2^o in the direction which remains orthogonal to the fibers direction, Figure 1

$$\varepsilon_2^o = \frac{dS_2''}{dS_2} - 1 \tag{9}$$

Noting that

$$dS_2'' = dS_2'\sin\left(\frac{\pi}{2} - \gamma_{12}\right),\quad dS_2' = dS_2(1+\varepsilon_2)$$

Equation (9) can be rewritten as

$$1 + \varepsilon_2^o = (1+\varepsilon_2)\cos\gamma_{12}$$

Using equations (6) and (7) gives

$$1 + \varepsilon_2^o = \frac{e_x e_y}{e_1}\sqrt{1 - e_{xy}^2} \tag{10}$$

Equations (5) - (8) and (10) represent the basic geometrical relations of the treated problem.

The equilibrium equations for a k-layered plate after deformation can be written in the following form:

$$t_x' = \sum_{i=1}^{k}\left(t_{1i}'\cos^2\phi_2' + t_{2i}'\sin^2\phi_i' - t_{12i}'\sin2\phi_i'\right)$$

$$t'_y = \sum_{i=1}^{k} (t'_{1i}\sin^2\phi'_i + t'_{2i}\cos^2\phi'_2 + t'_{12i}\sin 2\phi'_i)$$

$$t'_{xy} = \sum_{i=1}^{k} [(t'_{1i}-t'_{2i})\sin\phi'_i\cos\phi_i + t'_{12i}\cos 2\phi'_i] \tag{11}$$

where t'_x, t'_y, t'_{xy} are the average normal and shearing stress resultants in coordinate system (x',y'); t'_{1i}, t'_{2i} are the normal stress resultants parallel and orthogonal to the fiber direction in the ith ply; t'_{12i} is the in-plane shearing stress resultant.

First consider the composite with rigid fibers and matrix undergoing finite strains. The following hypotheses are assumed: (1) the longitudinal strain ε_1 is small and the fibers are linearly elastic; (2) the transverse normal strain ε_2^0 changes the spaces between the fibers and thus influences longitudinal and shearing properties of the ply. As a result, the constitutive relations for the unidirectional ply can be written in the form

$$t'_1 = \frac{\beta_1\varepsilon_1}{1+\varepsilon_2^0};\ t'_2 = \beta_2\varepsilon_2^0,\ t'_{12} = \frac{\beta_{12}\sin\gamma_{12}\cos\gamma_{12}}{1+\varepsilon_2^0} \tag{12}$$

The constant β_1 and the functions $\beta_2\ (\varepsilon_2^0)$, $\beta_{12}\ (\sin\gamma_{12})$ are available from tension and shear tests, such that the experimental curves provide the relations $t_1^0 = \beta_1\varepsilon_1$, $t_2^0 = \beta_0\varepsilon_2^0$, $t_{12}^0 = \beta_{12}\sin\gamma_{12}$, where stress resultants t_1^0, t_2^0, t_{12}^0 correspond to the initial dimensions of the specimens.

As an example, consider the case of uniaxial loading of an angle-ply laminat with fiber orientation $\pm\phi$ to the load axis x. In this case, $\gamma_{xy} = \omega_{xy} = 0$ and the problem reduces to solving the following system of equations:

$$\beta_1\varepsilon_1\frac{e_x}{e_y}\cos^2\phi + \beta_2(\varepsilon_2^0)\varepsilon_2^0 e_y^2\sin^2\phi - \beta_{12}(\sin\gamma_{12})\sin\gamma_{12}\cos\gamma_{12}\sin 2\phi = t'_x$$

$$\beta_1\varepsilon_1\frac{e_y}{e_x}\sin^2\phi + \beta_2(\varepsilon_2^0)\varepsilon_2^0 e_y^2\cos^2\phi + \beta_{12}(\sin\gamma_{12})\sin\gamma_{12}\cos\gamma_{12}\sin 2\phi = 0$$

$$(1+\varepsilon_1)^2 = e_x^2\cos^2\phi + e_y^2\sin^2\phi \tag{13}$$

$$1 + \varepsilon_2^0 = e_x e_y$$

$$\sin\gamma_{12} = \frac{(e_y^2-e_x^2)\sin\phi\cos\phi}{\sqrt{e_x^2\sin^2\phi+e_y^2\cos^2\phi}}$$

Equations (13) include 5 unknown strains e_x, e_y, ε_1, ε_2^0, γ_{12}.

The experimental curves for 0°, ±45° and 90° laminates, shown in Figure 2, were used for determination of the stiffness values β_1, β_2, β_{12}. The comparison

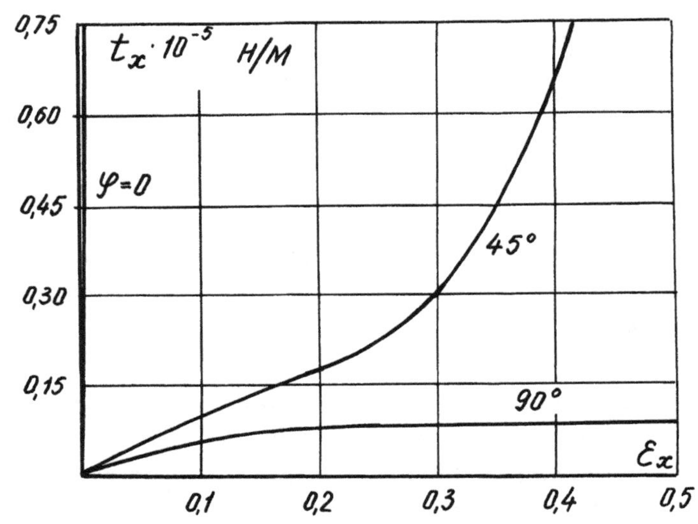

Fig. (2) - Experimental stress resultant-strain curves for 0°; ±45°; 90° laminates

of theoretical results (plotted points and solid lines respectively) for ±15°, ±30°, ±60°, ±75° laminates is shown in Figure 3.

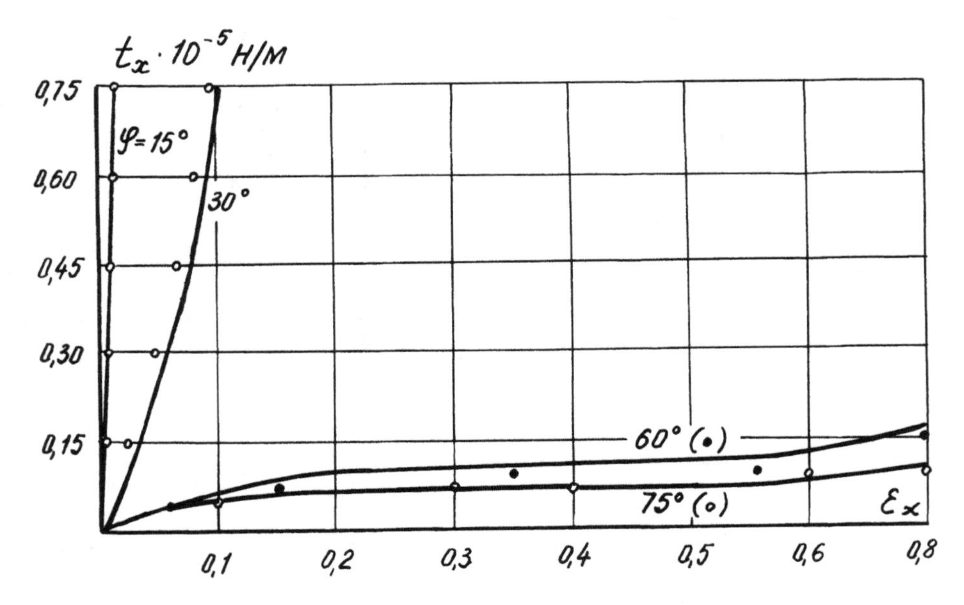

Fig. (3) - Experimental (solid lines) and theoretical (points) results for ±15°, ±30°, ±60°, ±75° laminates

The obtained relations may be used for prediction of equilibrium forms of a system of perfectly flexible fibers. Assuming $\beta_2 = \beta_{12} = 0$, then (for an arbitrary strain ε_1)

$$\frac{\beta_1\varepsilon_1 e_x \cos^2\phi}{e_y(1+\varepsilon_1)} = t'_x \quad \frac{\beta_1\varepsilon_1 e_y \sin^2\phi}{e_x(1+\varepsilon_1)} = t'_y$$
$$(1+\varepsilon_1)^2 = e_x^2\cos^2\phi + e_y^2\sin^2\phi \tag{14}$$

For the inextensible fibers ($\varepsilon_1 = 0$), equations (14) simplify to the following

$$\frac{e_x^2}{e_y^2} = \frac{t'_x}{t'_y}\,\mathrm{tg}^2\phi, \quad e_x^2\cos^2\phi + e_y^2\sin^2\phi = 1 \tag{15}$$

As an example, consider a filament-wound cylindrical pressure vessel of initial radius R and a $\pm\phi$ orientation of fibers. In this case $t'_x = PRe_y/2$, $t'_y = PRe_y$, where $e_y = I + \omega/R$, P is internal pressure, and ω is radial displacement. Figure 4 shows circumferential strain ε_y and axial strain ε_x as functions of the initial angle ϕ. Solid lines illustrate the solution of equations (15) and dotted lines - the solution of equations (14) ($PR/\beta_1 = 0.0133$).

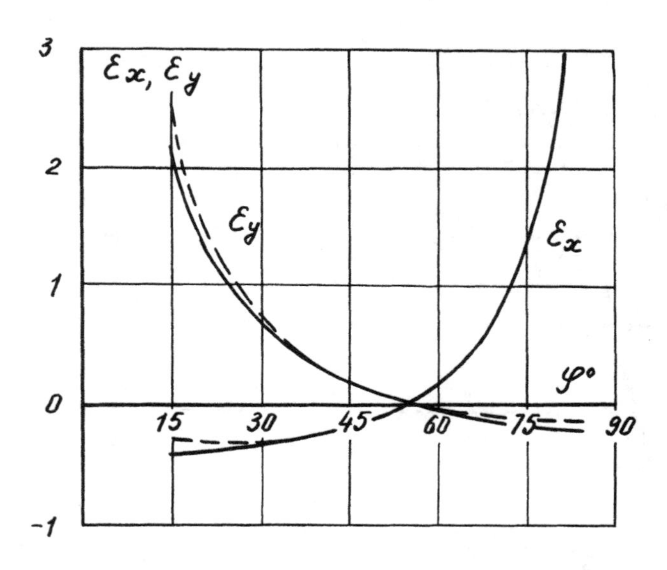

Fig. (4) - Axial (ε_x) and circumferential (ε_y) strains of a cylindrical shell as functions of the initial angle of winding

As previously mentioned, the non-linear behavior of composite materials may lso be caused by resin crazing. The transverse loading of a unidirectional ply sually causes the brittle resin to fracture between the fibers. The crazing ocurs when a certain combination of stresses σ_2 and τ_{12}, corresponding to the asumed failure criterion, achieves some ultimate level, i.e.,

$$\left(\frac{\overset{*}{\sigma}_2}{\bar{\sigma}_2}\right)^2 + \left(\frac{\overset{*}{\tau}_{12}}{\bar{\tau}_{12}}\right)^2 = 1$$

here $\bar{\sigma}_2$, $\bar{\tau}_{12}$ are the ultimate normal and shearing stresses. Assume that the ailed ply maintains the failure level stresses $\overset{*}{\sigma}_2$, $\overset{*}{\tau}_{12}$ (Figure 5), while the

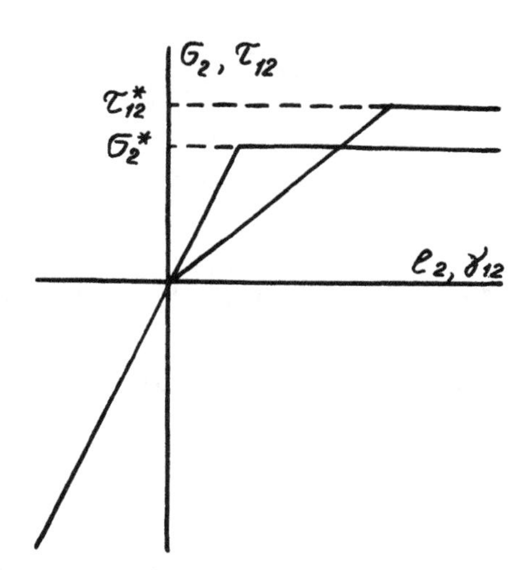

Fig. (5) - Stress-strain curves for the ply with failed matrix

ongitudinal stress σ_1 can continue to increase [1]. For these assumptions, Figre 6 illustrates the solution of equations (13) for ±45° laminate. It shows easonable agreement with experimental data of Rothem and Hashin [2]. (The echanical properties of a unidirectional ply were taken from the paper [2]). he stiffness of the laminate after resin crazing increases due to the change f orientation angle which is shown in Figure 6.

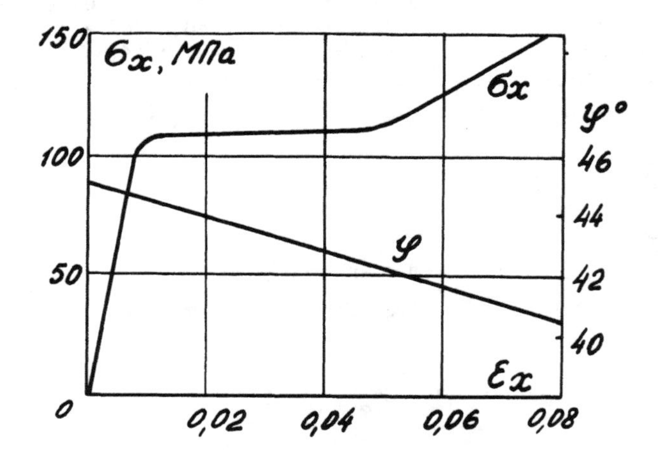

Fig. (6) - Stress-strain curve and ϕ as a function of ε_x for ±45° laminate

REFERENCES

[1] Vasil'ev, V. V. and Elpat'evskii, A. N., Polym. Mech., Vol. 5, pp. 915-920, 1967.

[2] Rothem, A. and Hashin, Z., J. Compos. Materials, Vol. 9, pp. 191-206, April 1975.

THE EFFECT OF INTERFACE STRUCTURE ON THE STRENGTH OF FIBROUS COMPOSITE MATERIALS

M. Kh. Shorshorov, L. M. Ustinov, O. V. Gusev, L. V. Vinogradov,
L. E. Gukasjan and A. G. Penkin

Baikov Institute of Metallurgy
Moscow, USSR

ABSTRACT

A model system of aluminum/boron fiber composite material which has a nonhomogeneous structure at the interface has been analyzed. The model system is loaded in the direction of the reinforcement. It has been shown that this material fractures by two mechanisms simultaneously: one of them is illustrated by the model of a bundle of fibers (cumulative mechanism) but the other one by the model of brittle fracture (noncumulative mechanism). Simultaneous activity of these mechanisms results in unmonotonic change of the strength of the aluminum/boron composite versus change of interface strength. A maximum of the strength of the aluminum/boron composite can be achieved at the interface strength which is less of maximum level, i.e., strength of matrix. The effect of brittle layers on the ultimate tensile strength of brittle fiber composite materials is determined by the methods of fracture mechanics. Two models of the fracture of brittle fiber/brittle layer systems have been demonstrated. The fracture criterion for each of them is based on comparing the critical stress intensity factor of the fiber with the stress intensity factor at the tip of the crack in the layer. The effect of the thickness of the layer on the fracture stress of the fiber has been derived.

INTRODUCTION

The ultimate tensile strength (UTS) of a fiber composite material (FCM) with metal matrix is estimated by well-known formulae or their modifications, based on low fiber volume fraction. Experience shows that these formulae give rather approximate values of UTS of FCM. They describe the effect of the geometrical structure of FCM, i.e., macrostructure which includes length and diameter of fibers, spaces between them, type and angle of reinforcing and so on. But these formulae do not consider microstructural peculiarities of FCM, especially microstructure of the interface. Therefore, it is not possible to use them in practice for selecting technological parameters of fabrication of FCM. These parameters strongly affect the interface structure which affects UTS and fracture mechanism of FCM. Small changes in the technological parameters of fabrication of FCM can dramatically change interface microstructure and affect UTS of composite materials. This is an important reason for the rather extensive scatter of UTS values of FCM which have similar macrostructures and properties of components. In the context of this paper, interfacial microstructure should be considered as the linking element between technological parameters of fabrication and properties of FCM.

Bonding zones of the components and intermetallic layers are elements of interface microstructure. They are products of physico-chemical reactions of the components of the interface. Characteristic size of the bonding zones is defined by their number and the relative density of bonding zones; the interface strength increases in proportion to this characteristic. The second microstructural element of the interface is defined by the depth of intrusion of the layer into the fiber if the layer has discrete character or by the thickness of the layer, if it has continuous character. In general, the characteristic size of the interfacial microstructure is ten or more times less than any macrostructural parameter of FCM.

In this paper, the effect of the density of bonding zones and the thickness of the brittle layer on UTS of some FCM are discussed.

ULTIMATE TENSILE STRENGTH OF ALUMINUM/BORON COMPOSITES WITH NONHOMOGENEOUS STRUCTURE OF THE INTERFACE

Experimental data show [1,2] that increasing the density of the bonding zones or the interface strength of aluminum/boron FCM in particular does not always result in increasing its UTS. This effect contradicts the well-known theor [3,4]. Attempts to understand this contradiction by postulated mechanisms of fracture have not given good results [5]. Analysis of conditional models of aluminum/boron with nonhomogeneous structure at the interface is more fruitful [6,7]. "Nonhomogeneous" means that the interfacial microstructure has discrete character - areas with strong bonding of the components (these are bonding zones or discrete intermetallic layers) are combined with areas which have zero strength of the interface. Homogeneous structure of the interface means that it has no areas with strong bonding anywhere or it has a continuous interface layer with constant thickness. In this paper, the analysis of UTS of aluminum/boron with nonhomogeneous structure of the interface is presented.

The composite material is treated as a system which consists of two parts. In the first part of the model, the interfacial strength is maximum, i.e., equal to the strength of the matrix, τ_m. In the second part, the interfacial strength is zero. The fracture of the first part is described by the noncumulative mechanism (NC-fracture), which is characterized by extremely localized fracture. This localization is the result of the high strength of the interface so that the first fracture of a fiber induces successive fractures of neighbouring fiber at approximately the same cross section [8,9]. The fracture of the second part of the model is described by the cumulative mechanism (C-fracture) which is characterized by an extremely non-localized process of fracture of fibers in the whole volume of FCM. The reason for this non-localized fracture process is the zero strength of the interface so that the first fracture of a fiber does not directly affect the fracture of neighbouring fibers at the same cross section [10]. Therefore, the failure of the first part of the model is similar to brittle fracture and that of the second part to the fracture of bundle of unbonded brittle fibers.

For the present, ignore the effect of the matrix in UTS of FCM. Then the model can be considered as two bundles of brittle fibers. Bundles I and II experience NC- and C-fracture, respectively. Both bundles sustain nominal load P. Fracture loads of the bundles I and II are P_1 and P_2. They are calculated by equations [3]:

$$P_1 = \sigma_{nc}[1 - F(\sigma_{nc})]S_1 \quad (1)$$

$$P_2 = \sigma_{max}[1 - F(\sigma_{max})]S_2 \quad (2)$$

where σ_{nc} is NC-fracture stress of the fibers, $F(\sigma_{nc})$ is distribution function for the strength of the fibers at σ_{nc}; S_1 and S_2 are cross sectional areas of of all fibers of the bundles I and II; σ_{max} is the maximum stress in the fibers at the moment of the C-fracture [10]; $F(\sigma_{max})$ is the distribution function for the strength of the fibers at σ_{max}. Since $S_1 + S_2 = S$ (S is the total cross sectional area of all fibers of both bundles), the symbol $\theta = S_1/S$ $(1 - \theta = S_2/S)$ is inserted in equations (1) and (2).

Further variations of fracture process of the model are considered. First, bundle I may fracture at a nominal load which is more than P_2. Then, the system fractures at the load P_{nc}. Second, bundle I may fracture, but at a nominal load which is less than P_2. Then, the system fractures at the load $P_m = P_2$, where P_2 is calculated from equation (2). Third, bundle II may fracture and induce immediate fracture of the system at the stress σ_{max} in all fibers and at nominal load P_b.

In the first variant of the model, only NC-fracture is defined; in the third variant, only C-fracture; but in the second one, both C-fracture and NC-fracture. Therefore, in this case, fracture has a mixed character (M-fracture). In terms of the fracture stress, the equations for the strength of the system become:

$$\sigma_{fnc} = \sigma_{nc}[1 - F(\sigma_{nc})] \quad (3)$$

$$\sigma_{fm} = \sigma_{max}[1 - F(\sigma_{max})](1-\theta) \quad (4)$$

$$\sigma_b = \sigma_{max}[1 - F(\sigma_{max})] \quad (5)$$

where σ_{fnc}, σ_{fm}, σ_b are the strengths of the system for NC-fracture (the first variant), M-fracture (the second variant), and C-fracture (the third variant), respectively.

Suppose that the distribution of fiber strength has the Weibull character. Then $(\ell\alpha\beta e)^{-1/\beta}$ is inserted in equations (4) and (5) instead of the term $\sigma_{max}[1 - F(\sigma_{max})]$ [3]. Here, ℓ is a characteristic length of the model system. It is supposed that $\ell = 0.25\ \ell_{cr}$ (ℓ_{cr} is the critical length of the fiber calculated from a well-known result [11]); α and β are parameters of the Weibull distribution function. The stresses σ_{fnc}, σ_{fm} and σ_b can be obtained after a series of operations using the equations, $F(\sigma) = 1 - \exp(-\ell\alpha\sigma^{\beta})$ [12] and $\bar{\sigma}_f = (\alpha\ell\beta e)^{1-\beta} \times \Gamma(1 + 1/\beta)$ [3]:

$$\sigma_{fnc} = \sigma_{nc} \exp[-(\frac{\sigma_{nc}}{m})^{\beta} (e\beta)^{-1} \tau_m^{-\beta/(1+\beta)} \theta^{-\beta/(1+\beta)} \tag{6}$$

$$\sigma_{fm} = m\tau_m^{1/(1+\beta)} \theta^{1/(1+\beta)} (1-\theta) \tag{7}$$

$$\sigma_b = m\tau_m^{1/(1+\beta)} \theta^{1/(1+\beta)} \tag{8}$$

where

$$m = [0,5\alpha d_f \Gamma(1+1/\beta)]^{-1/(1+\beta)} (0,25e\beta)^{-1/\beta} \tag{9}$$

where d_f is the fiber diameter and $\Gamma(1+1/\beta)$ is the Γ-function. These equations contain the parameter θ which also characterizes the strength of the interface o the system. Therefore, this parameter is represented as $\theta = \tau/\tau_m$ or $\theta = S_1/S$. The characteristic length of the model system is affected by the critical length of the fiber and the latter by the strength of the interface, i.e., τ or θ.

The value of σ_{nc} can be calculated from [9]

$$\sigma_{nc} = (1-V_f)(E_f \sigma_{y\ell} \varepsilon_m)/(fn)^{1/2} \tag{10}$$

where V_f is the volume fraction of the fibers, E_f is Young's modulus of the fibers, $\sigma_{y\ell}$ and ε_m are yield stress elongation to fracture of an unreinforced matrix. Here, it is considered that n=1. Equations (6)-(8) are shown in Figure 1 as functions of the parameter θ which characterizes the interface strength. The maximum efficiency of reinforcing is realized for C-fracture (curve 1, σ_b versus θ), and lower efficiency of the reinforcing corresponds to NC-fracture (curve 3, σ_{fnc} versus θ), or for M-fracture (curve 2, σ_{fm} versus θ). The mutual positions of these curves show that increasing θ results in transition from C-fracture in the first zone ($0 \leq \theta \leq \theta^{**}$) to NC-fracture in the second zone ($\theta^{**} \leq \theta \leq \theta_2^*$ to M-fracture in the third zone ($\theta_2^* \leq \theta \leq \theta_1^*$) and, finally, to NC-fracture again in the fourth zone ($\theta_1^* \leq \theta \leq 1$). It will be shown that upper curve is not realized in practice for aluminum/boron FCM. A typical curve for this kind of FCM has all four zones (solid line in Figure 1). Apparently, the maximum strength of aluminum/boron composites is reached in the third zone (M-fracture) at a (less than maximum) strength of the interface,

$$(\theta)_{max} = 1/(2+\beta) \tag{11}$$

It can be seen that $(\theta)_{max}$ is affected by β only.

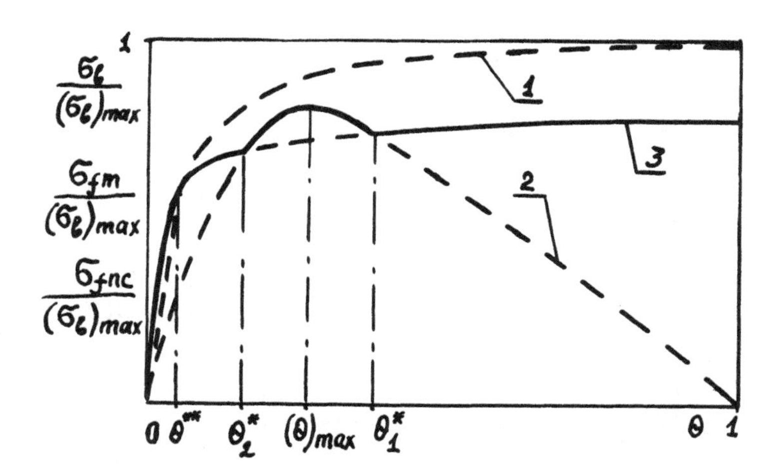

Fig. (1) - The effect of the interface strength on the UTS of aluminum/boron fiber composites presented in parametrical form (neglecting the effect of the matrix): 1. $\sigma_b/(\sigma_b)_{max}$; 2. $\sigma_{fm}/(\sigma_b)_{max}$; 3. $\sigma_{fnc}/(\sigma_b)_{max}$.
$(\sigma_b)_{max} = m\tau_m^{1/(1+\beta)}$

Experiments have been done to test theoretical results. Sheet samples of luminum/boron composite (thickness 1-2 mm) fabricated by hot rolling a packet f plasma sprayed monotypes of aluminum ADI/boron fibers had V_f = 0.38 or 0.55. ne interface strength was reduced by thermocycling between 20°C and 400°C; the ime of heating varied from 1 to 10s; the number of cycles from 1 to 150. The trength of the fibers is not changed by the thermocycling. Thermocycled samples ere tested under tensile or shear loading. The latter was necessary to estimate ne interface strength, τ. UTS of aluminum/boron was theoretically calculated y adding the effect of the matrix to equations (6)-(9). In general, good agreeent of experimental data with theoretical results has been found, Figure 2. But ne theory gives larger results than the experiment for the case of aluminum/boron ith V_f = 0,55 and low values of θ. This discrepancy is caused by breakage of the ibers during fabrication of the composite material. Better agreement between exerimental and theoretical results is obtained by accounting for this factor (doted curve in Figure 2).

The theoretical results can provide a guidance for the selection of matrix aterial. Typical calculated curves which show the effect of θ on the UTS of luminum/boron with different aluminum matrix alloys are presented in Figure 3. n this case, calculations are done by using the parameter, $\bar{\sigma}_f$ = 2800 MPa, β = 7, nd V_f = 0.5. Data for the matrix alloys are presented in Table 1.

Plasma sprayed aluminum ADI gives the worst results. It provides less staility of UTS values; a slight deviation of the parameters of fabrication of the omposite drastically changes UTS to the due steepness of the third part of the urve, Figure 3. Nevertheless, there is not much change in the interfacial microtructure and strength, i.e., θ. From this point of view, the aluminum alloy DI6

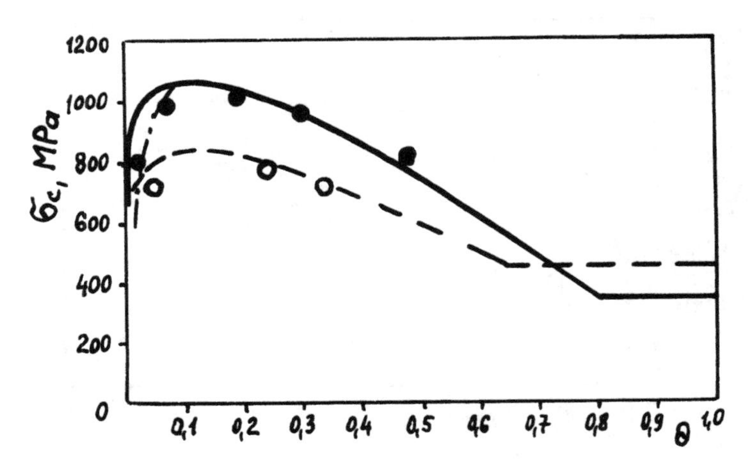

Fig. (2) - The effect of the interface strength on the UTS of aluminum/boron: • are theoretical and experimental data for V_f = 0.55; o are theoretical and experimental data for V_f = 0.38; -o- is theoretical data for V_f = 0.55 accounting for the fracture of boron fibers during the rolling of the composite

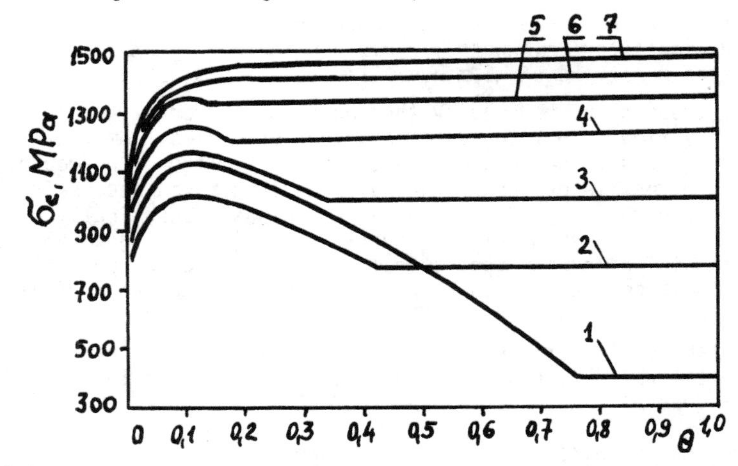

Fig. (3) - Theoretically calculated curves showing the effect of the interface strength and the properties of the matrix on the UTS of an aluminum/boron composite. 1. ADI, plasma sprayed; 2. ADI, annealed foil; 3. AMg2; 4. AMg6; 5. AD33 and M40; 6. Al +30%Be; 7. DI6

(quenched and aged) is the best matrix for aluminum/boron composites. It provides maximum strength σ_c and high stability of this characteristic for a wide range of θ values.

TABLE 1 - SOME MECHANICAL PROPERTIES OF ALUMINUM ALLOYS

Alloy	Condition	Yield stress $\sigma_{y\ell}$,MPa	UTS m', MPa	Fracture elongation ε_m,%
ADI	plasma sprayed	100	150	4
ADI	annealed foil	30	80	35
AMg2	annealed	80	190	23
AMg6	annealed	170	340	20
AD33	quenched and aged	270	300	15
M40	cladded	250	390	18
Al + 30%Be	annealed	300	400	18
D16	quenched and aged	290	440	19

ULTIMATE TENSILE STRENGTH OF FIBER COMPOSITE MATERIALS WITH BRITTLE INTERFACIAL LAYERS

It has been discovered that if the thickness of the brittle interfacial layer in FCM or the brittle deposit on the brittle fibers is higher than critical one then the strength of FCM or the fibers will decrease [13-15]. This effect is of great practical interest. Therefore, some work has appeared in recent years which attempts were done to explain the effect [13,14].

Recently, [15], it has been theoretically shown that the strength of the brittle fibers which have the brittle deposit starts to decrease when the thickness of the deposit becomes higher than some critical value t^*. The latter is calculated from

$$t^* = \frac{df}{2}\left[\sqrt{1 + \left(\frac{E_f\bar{\sigma}^n_\ell}{E_\ell\bar{\sigma}_f}\right)^{\beta_\ell}} - 1\right] \qquad (12)$$

where E_ℓ is Young's modulus of the deposit, $\bar{\sigma}_f$ is average strength of the fibers, β_ℓ is Weibull's factor which accounts for the variation in layer strength, and $\bar{\sigma}^n_\ell$ is the normalized strength of the layer (in this case, it is the strength of the layer with cross sectional area equal to the cross sectional area of the fiber). $\bar{\sigma}^n_\ell$ can be calculated from a modified Weibull's formula [12]:

$$\bar{\sigma}_{\ell 1}/\bar{\sigma}_{\ell 2} = (F_{\ell 2}/F_{\ell 1})^{1/\beta_\ell} \qquad (13)$$

where $\bar{\sigma}_{\ell 1}$ and $\bar{\sigma}_{\ell 2}$ are average values of the strengths of layers with cross sectional areas $F_{\ell 1}$ and $F_{\ell 2}$, respectively. To derive formula (13), it was assumed that if the layer thickness was less than t^*, then the fiber would fracture first and that the layer would fracture immediately too. If the layer thickness is higher than t^*, then it was assumed that the layer will fracture first inducing premature fracture of the fiber.

The authors of [16,17] have investigated some peculiarities of fracture of brittle layers in aluminum/steel wire composites tested in tension along the fibers.

The kinetics of fracture of the layer was observed through acoustic emission during tensile loading of the composite material. It was found that the brittle intermetallic layers markedly affect the deformation and fracture process in FCM The layer starts to fracture at ε = 1.2% elongation resulting from segments of similar length. A drastic decrease in the strength and fracture elongation of FCM aluminum/steel with an intermetallic layer ($FeAl_3$)(Fe_2Al_5) of thickness $\approx$ 20μm was found. In this case, the steel wire fractured by cleavage (brittle). If the layer was absent, the wire fractured by necking (ductile). Breaking the layer induces increased acoustic emission intensity $\dot{N}$ (region I in Figure 4). Cracks arising in the layer facilitate brittle fracture (cleavage) of the wires. This is the reason for the serrated curve $\dot{N}$-ε in region II of Figure 4.

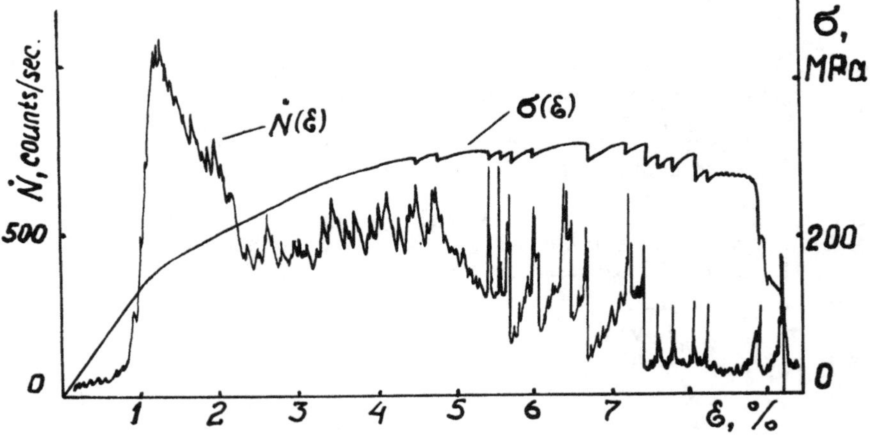

Fig. (4) - Intensity of acoustic emission $\dot{N}$ and flow stress σ as functions of elongation ε of the aluminum/steel wire composite material with V_f = 0.15 and intermetallic layer thickness of approximately 20μm

These experimental results make it possible to construct a model of the propagation of the cracks throughout the layer. In accordance to this model, Figure 5, cracks appear at the layer-matrix interface (e.g., point S). The crac propagates through the layer towards the layer/fiber interface and eventually reaches the fiber (point C). If the stress intensity at point C is not enough to cause fracture of the fiber, the crack will bypass the fiber completely (II). Otherwise, the crack will enter the fiber and will propagate through the layer and the fiber simultaneously (I, Figure 5).

This model is used for a brittle fiber/brittle layer system assuming that the third component (matrix) in the matrix/brittle fiber/brittle layer system

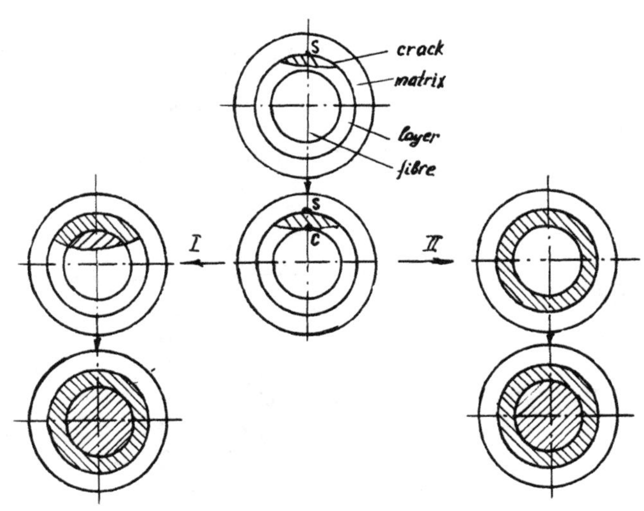

Fig. (5) - Two possible paths of transverse crack propagation in the fiber/layer system

has relatively low Young's modulus and does not noticeably affect the intensity stress factor at the tip of the crack (point C in Figure 5). Therefore, the following discussion may be applied to both brittle layer/brittle fiber and matrix/brittle layer/brittle fiber systems.

It is necessary to discuss two different types of fracture, of the brittle layer/brittle fiber system from the position of linear elastic fracture mechanics. The first type (point C, I, Figure 5) is analogous to the fracture of a semi-infinite plate with an edge crack which is normal to the interface. The tip of the crack is situated at the interface, Figure 6. The second type of

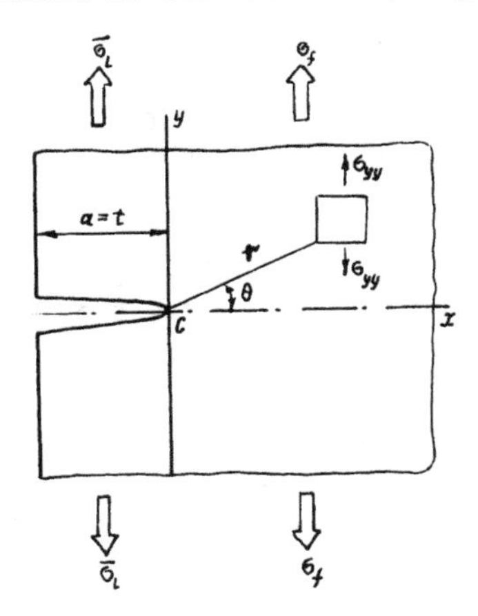

Fig. (6) - The first model of the fracture of the fiber/layer system

fracture (II, Figure 5) of the system can be discussed using a compound rod of infinite length which has a circumferential crack. The crack is normal to the interface of the components and the crack front is situated at the interface, Figure 7. In both models, the cracks completely divide one component, i.e., the layer (deposit). It is assumed that Poisson's factors of the components are equal. Nominal tension loads are applied to each component at infinite distance from the cracks. The values of the nominal loads are proportional to the Young'

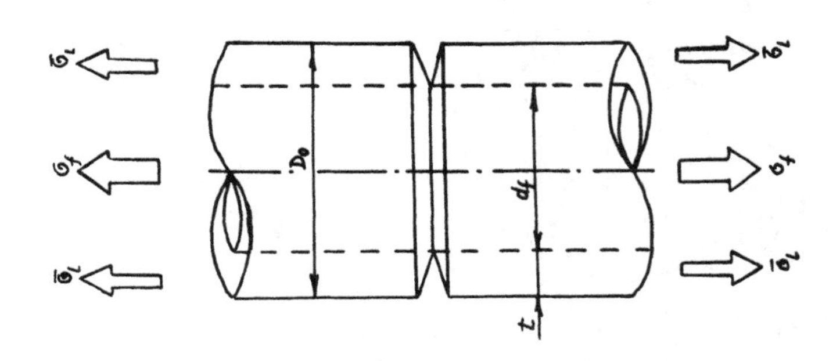

Fig. (7) - The second model of the fracture of the fiber/layer system

moduli of the components. The models are in plane strain. Delamination between the components is excluded. It would be simple to discuss these models by linear fracture mechanics if they were homogeneous. In this case, they are heterogeneous. In the general case, the components have different Young's moduli.

Some scientists have shown that the order of the stress singularity m, is affected in general by the ratio of the Young's moduli of the components [18-20] The larger the difference between the Young's moduli of the components, the larg the difference between m and the well-known value 1/2 which is characteristic of the singularity for homogeneous models. In practice, typical ratios of Young's moduli is not much different from (0.2 to 5 for FCM with metal matrix). Therefore, it is possible for this case to assume m equal to 1/2. Although this assumption results in rougher calculations, nevertheless, it gives a stress intensity factor like that for homogeneous materials, i.e., $kg/mm^{3/2}$. Therefore, it is possible to compare values calculated from the two different models of K_1 wi the critical stress intensity factor of the fiber.

The authors have shown [21-23] that the cleavage stress field at the tip of the crack changes drastically when it crosses the interface of components which have different Young's moduli. It has been theoretically shown by the method of this section [24] that K_1 for two-component models similar to that in Figure 6 i

$$K_1 = \sigma_2(2\pi\ell\ E_2/E_1)^{1/2} \tag{14}$$

where σ_2 is the nominal tensile stress in the second component, E_1 and E_2 are oung's moduli of the first and second components, respectively, and ℓ is the ength of the crack. If the tip of the crack is at least 0.25ℓ beyond the interace, then equation (14) will be transformed to the well-known result for a singleomponent model [25]:

$$K_1 = \sigma_2(2\pi\ell)^{1/2} \tag{15}$$

Comparing equations (14) and (15), it is rather easy to get κ which will be alled the heterogeneity factor

$$\kappa = (E_2/E_1)^{1/2} \tag{16}$$

It has been shown [21,22] that κ is not affected by the geometry of the sample. herefore, it can be assumed that κ is similar for the two models. It is known 25] that for the single-component edge crack model similar to Figure 6, K_1 is

$$K_1^{(I)} = 1.12\sigma(\pi a)^{1/2} \tag{17}$$

here σ is the nominal stress, a is length of the crack which is equal to the hickness of the layer t. The heterogeneity of the model is accounted for the actor κ by putting in equation (17):

$$K_1^{(I)} = 1.12\kappa\sigma_f(\pi t)^{1/2} \tag{18}$$

here σ_f is the nominal stress in the fiber when at fracture elongation of the ayer is larger than critical. σ_f can be defined from the condition of equality f strains at the interface.

$$\sigma_f = \bar{\sigma}_\ell \frac{E_f}{E_\ell} \tag{19}$$

here $\bar{\sigma}_\ell$ is fracture stress of the layer from equation (13). Substituting into quation (19)

$$\sigma_f = \bar{\sigma}_{\ell o} (E_f/E_\ell)(t_o/t)^{1/\beta_\ell} \tag{20}$$

fter putting equation (20) into equation (18)

$$K_1^{(I)} = 1.12\kappa\pi^{1/2}\,\bar{\sigma}_{\ell o}\, t_o^{\frac{1}{\beta_\ell}} \frac{E_f}{E_\ell}\, t^{\frac{1}{2}-\frac{1}{\beta_\ell}} \tag{21}$$

This equation shows the effect of the layer thickness on $K_1^{(I)}$: the latter increases with increasing thickness of the layer, Figure 8.

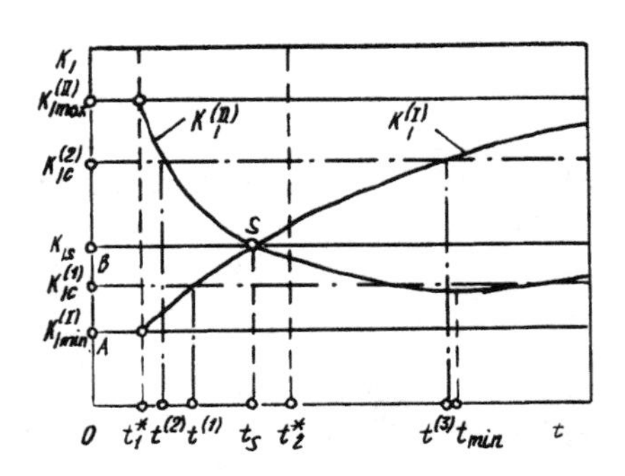

Fig. (8) - The effect of t on $K_1^{(I)}$ and $K_1^{(II)}$

While deriving the equation for the second model, $K_1^{(II)}$ is written which is appropriate for a homogeneous material [26]. After applying the heterogeneity factor and equation (19) instead of nominal stress, the result for the second heterogeneous model is:

$$K_1^{(II)} = 0.25\pi\kappa\bar{\sigma}_{\ell o} \frac{E_f}{E_\ell} \left(\frac{t_o}{t}\right)^{\frac{1}{\beta_\ell}} (d_f+2t)^{1/2}\left(0.45 + 0.9\,\frac{t}{d_f}\right) \tag{22}$$

It is important to note that the second model of the fracture of the fiber/layer system is realized only when the first model is not realized. This peculiarity is characterized by the specific manner of propagation of the crack in the fiber/layer system, Figure 5.

Analysis of equation (22) has shown that with increasing t, $K_1^{(II)}$ decreases to a minimum at $t = t_{min}$,

$$t_{min} = \frac{0.45 d_f}{3.88(\beta_\ell - 1)} \tag{23}$$

and then starts to increase, Figure 8. Calculated values of t_{min} for typical values of β_ℓ are shown in Table 2.

TABLE 2 - VALUES OF t_{min} FOR DIFFERENT VALUES OF β_ℓ

β_ℓ	3	4	5	6
t_{min}	0.06 d_f	0.04 d_f	0.03 d_f	0.023 d_f

Curves in Figure 8 of equations (21) and (22) cross at two points at which

$$t_s \simeq 0.01\ d_f \tag{24}$$

$$t_w \simeq 2.89\ d_f \tag{25}$$

The second intersection does not have practical meaning because t_w is 2-3 diameters of the fiber.

Equations (21) and (22) give a diagram, Figure 8, from which it can be seen which model (of the two models considered) must govern fracture in a concrete fiber/layer system. Consider first that t^* is less than t_s. Then the whole diagram can be divided into four general areas. Analysis of these areas is based on comparing the critical stress intensity factor of the fiber with $K_1^{(I)}$ and $K_1^{(II)}$.

<u>Area OA</u>. Here $K_{1c} \leq K_{1min}^{(I)}$, where $K_{1min}^{(I)}$ is the value of $K_1^{(I)}$ calculated by putting t^* in equation (21). When $t \geq t^*$, then the first fracture model is realized. The fracture stress of the fiber $\bar{\sigma}_f$ for this area is calculated from

$$\bar{\sigma}_f = \bar{\sigma}_{fi} \text{ if } 0 \leq t \leq t^* \tag{26}$$

$$\bar{\sigma}_f = \bar{\sigma}_{\ell o} \frac{E_f}{E_\ell} \left(\frac{t_o}{t}\right)^{\frac{1}{\beta_\ell}} \quad \text{if } t \geq t^* \tag{27}$$

Here, $\bar{\sigma}_{fi}$ is the primary strength of the fiber (without the layer). These equations are depicted by a graph which shows the effect of t on $\bar{\sigma}_f$, Figure 9a.

<u>Area AB</u>. Here, $K_{1min}^{(I)} < K_{Ic} \leq K_{Is}$. In this case, the fracture stress in the fiber is defined by equation (26) (for $0 \leq t \leq t^*$) and (27) (for $t \geq t^*$). Consider a concrete example. Assume that $K_{1c} = K_{1c}^{(1)}$, (Figure 8). If $0 \leq t \leq t^*$, then the fracture of the system starts with the fracture of a fiber at a fracture stress defined by equation (26). If $t^* \leq t < t^{(1)}$, then $K_1^{(I)} < K_{Ic} < K_1^{(II)}$. The system must fracture by the second model. If $t \geq t^{(1)}$, then we have $K_{1c} \leq K_1^{(I)}$ and the fracture will be described by the first model. The effect of t on $\bar{\sigma}_f$ for this area is also defined by equations (26) and (27), Figure 9a.

<u>Area BC</u>. Here $K_{1S} < K_{1C} \leq K_{1max}^{(II)}$. $K_{1max}^{(II)}$ is calculated by putting t^* into equation (22). If $0 \leq t \leq t^*$, then $\bar{\sigma}_f$ is defined by equation (26). If $t > t^*$ and $K_{1c} < K_1^{(II)}$ or $K_{1c} \leq K_1^{(I)}$, then $\bar{\sigma}_f$ is calculated by equation (27). But if $K_1^{(II)} < K_{1c} > K_1^{(I)}$, then the fracture of the layer does not induce immediate fracture of the fiber. This fracture can be realized only after increasing the nominal load up to the value at which K_1 reaches K_{1c} of the fiber. In this case, the stress at fracture is calculated by:

$$\bar{\sigma}_f = \frac{K_{1c}}{0.25\pi\kappa(d_f+2t)^{1/2}(0.45+0.9\,\frac{t}{d_f})} \tag{28}$$

Equation (28) is derived from equation (22) by the condition that $K_1^{(II)} = K_{1c}$ an the nominal stress in the fiber reaches the primary (without layer) value of the fiber's strength. This equation is applicable for the interval of values of t which can be defined by putting K_{1c} in equations (21) and (22). A typical graph showing the effect of t on $\bar{\sigma}_f$ for this area is presented in Figure 9b. Consider a concrete example. Assume that $K_{1c} = K_{1c}^{(II)}$, Figure 8. For the interval $t^* \leq t < t(2)$, $K_1^{(I)} < K_{1c}^{(2)} < K_1^{(II)}$. Therefore, the fracture of the system is described by the second model. For the interval of values $t^{(2)}<t<t^{(3)}$, $K_1^{(II)} < K_{1c}^{(2)} > K_1^{(I}$ In this case, the system is fractured by the second model, too. For values $t \geq t^{(3)}$, $K_1^{(I)} \geq K_{1c}^{(2)}$ and therefore the system is fractured according to the first model.

Area above point C. For this area, $K_{1c} > K_{1max}^{(II)}$. If $0 \leq t \leq t^*$, then the fracture stress of the fiber is defined by equation (26), but if $t>t^*$, then by equation (28) because in this case $K_{1c} > K_{1max}^{(II)} > K_1^{(I)}$. A typical graph showing the effect of t on $\bar{\sigma}_f$ for this area is presented in Figure 9a.

Assume now that $t^*>t_s$. Then the diagram can be divided into three areas: the area which is below the curve $K_1^{(I)}$ - t, the area which is above this curve, and the area in which this curve is situated. For the first area, the fracture stress of the fiber is defined by equation (26) if $t \leq t^*$ or by equation (27) if $t \geq t^*$. The effect of t on $\bar{\sigma}_f$ is presented in Figure 9a. For the second area, the fracture stress of the fiber is defined by equation (26) if $t \leq t^*$. For the third area, the fracture stress of the fiber is defined by equation (26) if $t \leq t^*$ and by equation (28) if $K_{1c} > K_1^{(I)}$ and by equation (27) if $K_{1c} < K_1^{(I)}$. A typical graph showing the effect of t on $\bar{\sigma}_f$ for this area is presented in Figure 9c.

Analysis of the diagram has shown that fracture of the layer does not always induce immediate fracture of the fiber. This situation occurs at condition $K_1^{(I)} < K_{1c} > K_1$ which corresponds to values $t^{(2)}<t<t^{(3)}$. The value of $t^{(3)}$ is

$$t^{(3)} = \left(\frac{E_\ell K_{1c}}{3.47\kappa E_f \bar{\sigma}_{\ell o} t_o^{1/\beta_\ell}}\right)^{(2\beta_\ell/\beta_\ell-2)} \tag{29}$$

It is necessary to note that secondary effects which can influence the fracture stress of the fiber/layer system are not considered in this analysis. It is known that in some cases [27] when $t<t^*$ the layer promotes a slight increase in fiber's strength. This effect is probably connected with healing of fiber surface defects by the layer. But this cannot significantly affect the general results of the analysis.

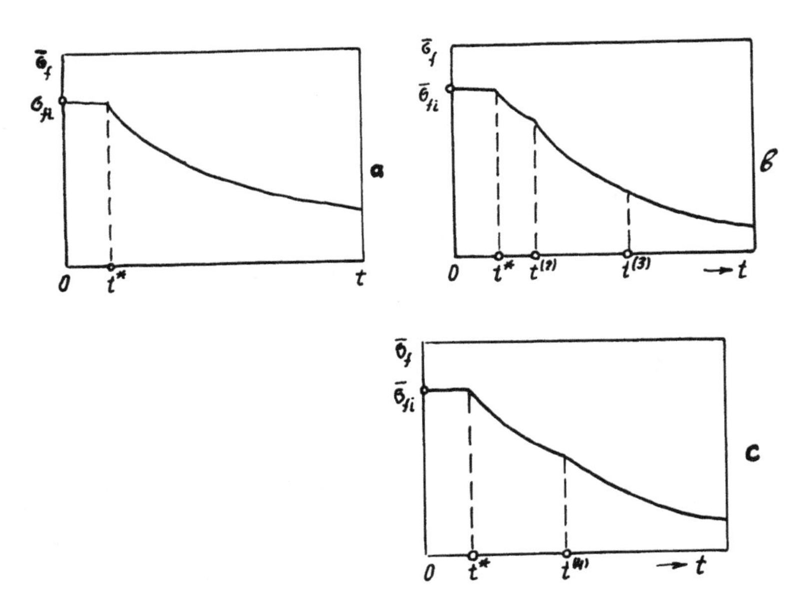

Fig. (9) - Typical curves showing the effect of the layer thickness t on the fracture stress in the fiber $\bar{\sigma}_f$: (a) for $t^*<t_s$ and $K_{Ic} \leq K_{Is}$; (b) for $t^*<t_s$ and $K_{Is} < K_{Ic} < K^{(II)}_{Imin}$; (c) for $t^*>t_s$ and the condition that with increasing of t at first $K^{(I)}_I$ is less but then is more than K_{Ic}

Experimental verification of the general results of the analysis is provided by tensile testing of borsic fibers with the deposit (layer) thickness of which is 1.5, 3.5 and 8.5 μm. The diameter of the fiber is 0.1 μm. The deposit was produced by VCD process. Temperature and time of the process were not so much to change (reduce) the strength of boron component of the borsic fibers. Fibers were tested on a "Shemadzu" tensile machine with strain rate 0.1 s^{-I} and gage length of 25 mm.

The results of this test are shown in Figure 10. They show that a drastic decrease in strength of the fibers occurs when the thickness of the deposit is greater than ~1.5 μm. Figure 10 shows the experimental and theoretical effects of layer thickness on the UTS of the borsic fibers. The calculated results are shown for different values of β_ℓ (β_ℓ = 4, 5 and 6). These results were produced according to additivity (law of mixture) considering the influence of each component to UTS of the borsic fibers. The strength of the layer of each thickness was calculated by equation (13) assuming $\sigma_{\ell 1} \approx 2000$ MPa for the thickness 0.02 mm which is equivalent to silicone carbide with a diameter 0.1 μm. The values used in the calculation were Young's moduli of the boron and silicon carbide fibers equal to 380000 and 470000 MPa, respectively, primary strength of the boron fibers (without deposit) of 2980 MPa. The critical thickness of the deposit t^* was calculated by equation (12). It equals 3.88; 2.15; 1.18 and 0.65 μm for β_ℓ = 3,

4, 5 and 6, respectively.

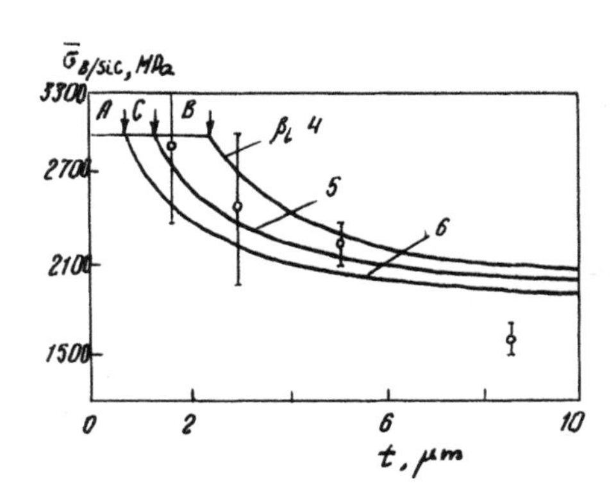

Fig. (10) - The effect of the deposit thickness t on the UTS of the borsic fibers. The curves are theoretically calculated results. Points correspond to the experimental data

It can be seen that the experimental and theoretical data for β_ℓ = 4 to 5 coincide quite well. These values of β_ℓ were most typical for SiC fibers. Neve theless, there is quite a large discrepancy between experimental and theoretical data at a thickness of 8.5 μm due probably to an increasing influence of the sec ondary effects at larger thicknesses of the deposit. The proper nature of these effects has not been stated. The value of the critical thickness of the silicon carbide deposit, from Figure 10, is somewhere between points B (for β_ℓ = 4) and C (for β_ℓ = 5). As a first approximation, it can be assumed to be about 1.5 μm.

CONCLUSIONS

1. The ultimate tensile strength of aluminum/boron fiber composites is affected by the density of bonding zones so that at the tensile value of it, i.e., at the middle interface strength, maximal longitudinal strength of the composite is attained in some cases.

2. The critical thickness of the brittle layer in fiber composite materials or brittle deposit on the brittle fibers is proportional to the fiber's diameter and increases with increasing Young's modulus of the fibers and the strength of the layer.

3. It has been shown by metallographic analysis and acoustic emission that the fiber/layer system is fractured by two processes.

4. From the position of fracture mechanics, these processes of fracture of the fiber/layer system are analyzed. The analysis results in three probable variants of the effect of the layer thickness on the fracture stress of the fibe

5. The critical thickness of the silicon carbide deposit on the boron component of the borsic fibers is about 1.5 μm.

EFERENCES

[1] Shorshorov, M. Kh., Ustinov, L. M. and Gukasjan, L. E., "Phys. Chim. Obrab. Mater.", No. 3, 132, 1979.

[2] Jackson, P. W., Baker, A. A. and Braddick, D. M., J. Mater. Sci., Vol. 6, No. 5, 427, 1971.

[3] Kelly, A., "Strong Solids", Clarendon Press, Oxford, 1973.

[4] Kelly, A. and Davies, G. J., Metallurgical Review, Vol. 10, No. 37, 1, 1965.

[5] Shorshorov, M. Kh., Kolesnichenko, V. A., Usupov, R. S. and Ustinov, L. M., "Phys. Chim. Obrab. Mater.", No. 4, 117, 1978.

[6] Shorshorov, M. Kh., Gukasjan, L. E. and Ustinov, L. M., "Phys. i. Chim. Obrab. Mater.", No. 1, 128, 1980.

[7] Ustinov, L. M., "Phys. i. Chim. Obrab. Mater.", No. 2, 122, 1980.

[8] Cooper, J. A. and Kelly, A., J. Mech. and Phys. Solids", Vol. 15, No. 4, 279, 1967.

[9] Mileiko, S. T., Sorokin, N. M. and Zirlin, A. M., "Mechanika Polimerov", No. 5, 840, 1973.

[10] Rosen, B. W., Fiber Composite Materials, ASM, 37, 1965.

[11] Ustinov, L. M., Shorshorov, M. Kh. and Kuznetzov, Yu. G., Proceedings 1974 and 1975 International Conference on Composite Materials", Ed. E. Scala, E. Anderson, J. Toth and B. R. Noton, AIME, 644, 1976.

[12] Weibull, W. A., "A Statistical Theory of the Strength of Materials", Ing. Vetenskaps. Acad. Handl. (Roy. Swedish Inst. Eng. Research Proc.), No. 151, 1939.

[13] Metcalf, A. G., J. Compos. Mater., No. 1, 356, 1967.

[14] Friedrich, E., Pompe, W. and Kopjev, I. M., J. Mater. Sci., Vol. 9, 1911, 1974.

[15] Shorshorov, M. Kh., Zirlin, A. M., Ustinov, L. M., Katinova, L. V., Gamnova, V. I., Mitkin, A. S. and Moguchi, L. N., Phys. i. Chim Obrab. Mater., No. 1, 119, 1976.

[16] Ustinov, L. M., Gamnova, V. I. and Shorshorov, M. Kh., Proceedings of Fifth International Conference for Powder Metallurgy, Dresden, Part II, 36-1, 1973.

[17] Shorshorov, M. Kh., Ustinov, L. M. and Gamnova, V. I., Phys. i Chim. Obrab. Mater., No. 2, 112, 1974.

[18] Zak, A. R., and Williams, M. L., Fract. AIME, Ser. E, J. Appl. Mech., No. 1, 142, 1963.

[19] Erdogan, F., Engng. and Fract. Mech., Vol. 4, 811, 1972.

[20] Swenson, D. O. and Ram, C. A., Int. J. Fract. Mech., Vol. 6, No. 4, 377, 1970.

[21] Shorshorov, M. Kh., Vinogradov, L. V. and Ustinov, L. M., Mechanika Kompositnych Materialov, No. 6, 982, 1979.

[22] Shorshorov, M. Kh., Ustinov, L. M., Zirlin, A. M., Olefirenko, V. I., and Vinogradov, L. V., J. Mater. Sci., Vol. 14, No. 8, 1850, 1979.

[23] Sereda, V. E. and Finkel, V. M., Problemy Prochnosti, No. 12, 24, 1977.

[24] Morozov, E. M., Izvestia Vuzov, Stroitelstvo i arhitektura, No. 12, 57, 1969.

[25] Paris, P. C. and Sih, G. C., Fracture Toughness Testing and its Application ASTM STP 381, 30, 1964.

[26] Brown, W. F. and Srawley, J. E., Plane Strain Crack Toughness Testing of High-Strength Metallic Materials, ASTM STP 410, 1968.

[27] Morin, D., Verre Textile Plastiques Reinforce, No. 3, 16, 1974.

AN ESTIMATION OF THE COMPRESSIVE STRENGTH OF A FIBROUS COMPOSITE

M. Arcisz

Institute of Fundamental Technological Research
Warsaw, Poland

INTRODUCTION

We consider a fibrous composite composed of straight parallel fibres subjected to a compressive load in the filament direction. Various failure modes of a fibrous composite are observed under the compressive load like filament micro-buckling, matrix yielding, constituents debonding, shear failure, etc. Which of these failure modes is likely the first to appear in a composite depends not only on the constituents properties but on many other factors including a fabrication process. A theoretical prediction of a failure mode, even is possible, is beyond the framework of this paper. We confine our attention to an a priori assumed failure mode.

We assume that a fibrous composite fails under compressive load by micro-buckling of fibers and we propose to estimate the compressive strength of such composite. Moreover, we assume that the geometry and the density of the fibres are such that the composite can be identified with a transversely isotropic continuum, which is further assumed to be elastic and incompressible.

Under these assumptions, we consider a rectangular block subjected to the uniform compressive load in the filament direction. Applying the procedure of superposition of a small deformation on finite deformations [1], we determine an adjacent equilibrium state. The boundary conditions are chosen in such a way that the plane boundaries of the block remain plane in the adjacent equilibrium state; then the derived solution describes the internal buckling [2].

This perturbed state of the equivalent homogeneous medium is to model the micro-buckling of fibres in a real composite. The load for which a non-zero superposed deformation first becomes possible is the upper bound of the composite compressive strength. This load can be identified with the composite strength only in the case of composites for which the micro-buckling of fibres is the only possible failure mode.

BASIC EQUATIONS

The fibrous composite is identified with a locally transversely isotropic elastic incompressible continuum with the isotropy axis determined by the fibres direction.

A deformation of a continuum in which a material particle initially at X_α in the rectangular Cartesian coordinate system x moves at time τ to x_i in the same system x, is described by the functions

$$x_i = x_i(X_\alpha, \tau) \tag{1}$$

Let us denote by $\underset{\sim}{A}$ and $\underset{\sim}{a}$, a fibre direction in the initial and current instant of time, respectively. The components of $\underset{\sim}{A}$ and $\underset{\sim}{a}$ referred to the system x are related as follows

$$a_i = A_\alpha x_{i,\alpha} \tag{2}$$

where $(\ )_{,\alpha}$ denotes the partial derivative with respect to X_α. The usual summation convention applies to any repeated subscripts.

For an elastic transversely isotropic continuum, the strain energy is a function of five invariants of the strain tensor and a vector tangent to the isotropy axis [3]. Choosing the Cauchy strain tensor as the strain measure and identifying the isotropy axis with the fibre direction, we have the following full set of independent invariants

$$\begin{aligned} &I_1 = C_{\alpha\alpha}, \qquad I_2 = \frac{1}{2}[(C_{\alpha\alpha})^2 - C_{\alpha\beta}C_{\beta\alpha}] \\ &I_3 = \det(C_{\alpha\beta}) \\ &I_4 = A_\alpha C_{\alpha\gamma}C_{\gamma\beta}A_\beta, \qquad I_5 = A_\alpha C_{\alpha\beta}A_\beta \end{aligned} \tag{3}$$

where

$$C_{\alpha\beta} = x_{i,\alpha}x_{i,\beta} \tag{4}$$

The assumption of incompressibility implies

$$I_3 = 1 \tag{5}$$

The constitutive equation for an elastic incompressible continuum has the form

$$t_{ij} = 2x_{i,\alpha}x_{j,\beta}\frac{\partial W}{\partial C_{\alpha\beta}} \tag{6}$$

where t_{ij} are the components of the Cauchy stress tensor and W is the strain energy per unit volume. In view of the condition (5), W can be written in the form

$$W(I_K) = \overline{W}(I_1, I_2, I_4, I_5) - \frac{1}{2}p \tag{7}$$

where p is an arbitrary scalar function. Then we have

$$t_{ij} = \Phi_K \overset{K}{F}_{ij} - p\delta_{ij} \tag{8}$$

where

$$\Phi_K = \frac{\partial \overline{W}}{\partial I_K}, \quad \overset{K}{F}_{ij} = x_{i,\alpha} x_{j,\beta} \frac{\partial I_K}{\partial C_{\alpha\beta}} \tag{9}$$

Substituting equation (3) into equation (9), we find

$$\begin{aligned} &\overset{1}{F}_{ij} = \overline{C}_{ij}, \qquad \overset{2}{F}_{ij} = \overline{C}_{kk}\overline{C}_{ij} - \overline{C}_{ik}\overline{C}_{kj} \\ &\overset{3}{F}_{ij} = \delta_{ij}, \qquad \overset{4}{F}_{ij} = a_i \overline{C}_{jk} a_k + a_j \overline{C}_{ik} a_k \\ &\overset{5}{F}_{ij} = a_i a_j \end{aligned} \tag{10}$$

where

$$\overline{C}_{ij} = x_{i,\alpha} x_{j,\alpha} \tag{11}$$

is the Finger strain tensor.

Suppose that a deformation of body consists of a finite static deformation $x_i(X_\alpha)$ and a superposed small deformation $\varepsilon u_i(x_j)$, where ε is a small constant such that ε^2 and higher powers can be neglected as compared with ε. The equilibrium state of a body produced by the deformation $x_i(X_\alpha)$ will be called the state B, while the equilibrium state of the body produced by the deformation

$$\overset{*}{x}_i(X_\alpha) = x_i(X_\alpha) + \varepsilon u_i(x_j) \tag{12}$$

will be named the state $\overset{*}{B}$.

The deformation gradient in the state $\overset{*}{B}$, referred to the system x, is

$$\overset{*}{x}_{i,\alpha} = x_{i,\alpha} + \varepsilon u_{i,j} x_{j,\alpha} \tag{13}$$

and the vector tangent to a fibre

$$\overset{*}{a}_i \equiv A_\alpha \overset{*}{x}_{i,\alpha} = a_i + \varepsilon a_j u_{i,j} \tag{14}$$

The Cauchy and Finger strain tensors at the state $\overset{*}{B}$ are

$$\begin{aligned} \overset{*}{C}_{\alpha\beta} &\equiv \overset{*}{x}_{i,\alpha} \overset{*}{x}_{i,\beta} = C_{\alpha\beta} + \varepsilon x_{i,\alpha} x_{j,\beta} (u_{i,j} + u_{j,i}) \\ \overset{*}{\overline{C}}_{ij} &\equiv \overset{*}{x}_{i,\alpha} \overset{*}{x}_{j,\alpha} = \overline{C}_{ij} + \varepsilon (\overline{C}_{ik} u_{j,k} + \overline{C}_{jk} u_{i,k}) \end{aligned} \tag{15}$$

respectively, while the stresses associated with the deformation (13) are

$$\overset{*}{t}_{ij} = \overset{*}{\Phi}_K (\overset{K}{F}_{ij})^* - \overset{*}{p} \delta_{ij} \tag{16}$$

where

$$\overset{*}{\Phi}_K \equiv \Phi_K(\overset{*}{I}_M),\ (\overset{K}{F}_{ij})^* \equiv \overset{K}{F}_{ij}(\overline{C}_{k\ell}, \overset{*}{a}_m),\ \overset{*}{p} = p+\varepsilon p' \tag{17}$$

Substituting (14) and (15) into (3) and (10), we obtain

$$\overset{*}{I}_K = I_K + \varepsilon i_K \tag{18}$$

where

$$i_K = 2\, \overset{K}{F}_{ij} u_{i,j} \tag{19}$$

and

$$(\overset{K}{F}_{ij})^* = \overset{K}{F}_{ij} + \varepsilon \overset{K}{f}_{ij} \tag{20}$$

where

$$\begin{aligned}
\overset{1}{f}_{ij} &= \overset{1}{F}_{ik} u_{j,k} + \overset{1}{F}_{jk} u_{i,k} \\
\overset{2}{f}_{ij} &= \overset{2}{F}_{ik} u_{j,k} + \overset{2}{F}_{jk} u_{i,k} + (\overline{C}_{ij}\overline{C}_{mk} - \overline{C}_{mi}\overline{C}_{jk})(u_{m,k} + u_{k,m}) \\
\overset{4}{f}_{ij} &= \overset{4}{F}_{ik} u_{j,k} + \overset{4}{F}_{jk} u_{i,k} + (a_i\overline{C}_{kj} + a_j\overline{C}_{ki})(u_{m,k} + u_{k,m})a_m \\
\overset{5}{f}_{ij} &= \overset{5}{F}_{ik} u_{j,k} + \overset{5}{F}_{jk} u_{i,k}
\end{aligned} \tag{21}$$

Expanding the functions $\Phi_K(I_K + \varepsilon i_K)$ in a Taylor series about I_K, we find

$$\overset{*}{\Phi}_K = \Phi_K + \varepsilon \Phi_{KM} i_M \tag{22}$$

where

$$\Phi_{KM} = 2\,\frac{\partial^2 \overline{W}}{\partial I_K \partial I_M} \tag{23}$$

Substituting equations $(17)_3$, (20), (22) into (16), we obtain

$$\overset{*}{t}_{ij} = t_{ij} + \varepsilon t'_{ij} \tag{24}$$

where

$$t'_{ij} = \Phi_K \overset{K}{f}_{ij} + 2\Phi_{KM} \overset{K}{F}_{ij} \overset{M}{F}_{k\ell} u_{k,\ell} - p'\delta_{ij}. \tag{25}$$

The assumption of incompressibility in the state $\overset{*}{B}$, $\overset{*}{I}_3 = 1$, implies $i_3 = 0$ nd from equation (19), it follows that the superposed deformation is restricted y the constraints

$$u_{i,i} = 0 \tag{26}$$

RECTANGULAR BLOCK SUBJECTED TO A COMPRESSIVE LOAD IN THE FIBRE DIRECTION

Consider a rectangular block, initially occupying a region $-L_\alpha \leq X_\alpha \leq L_\alpha$, reinorced by straight fibres parallel to x_1-axis and subjected to a pure homogeeous deformation

$$x_1 = \lambda_1 X_1,\ x_2 = \lambda_2 X_2,\ x_3 = \lambda_3 X_3 \tag{27}$$

In the state B, the block occupies the region

$$-\ell_i \leq x_i \leq \ell_i \tag{28}$$

The initial and current fibre directions are

$$A_\alpha = (1,0,0),\ a_i = (\lambda_1,0,0) \tag{29}$$

espectively.

From the incompressibility condition (5), we obtain

$$\lambda_1\lambda_2\lambda_3 = 1 \tag{30}$$

The stresses, (8), associated with the deformation (27) are

$$\begin{aligned} t_{11} &= \lambda_1^2[\Phi_1 + (\lambda_2^2+\lambda_3^2)\Phi_2 + 2\lambda_1^2\Phi_4 + \Phi_5] - p \\ t_{22} &= \lambda_2^2[\Phi_1 + (\lambda_1^2+\lambda_3^2)\Phi_2] - p \\ t_{33} &= \lambda_3^2[\Phi_1 + (\lambda_1^2+\lambda_2^2)\Phi_2] - p \end{aligned} \tag{31}$$

$$t_{12} = t_{13} = t_{23} = 0$$

From the equilibrium equations $t_{ij,j} = 0$, we obtain

$$p = \text{const.} \tag{32}$$

We consider the rectangular block subjected to a compressive load $t>0$ acting n the fibre direction, with the surfaces $x_2 = \pm\ell_2$ free of load, i.e.,

$$t_{11} = -t,\ t_{22} = 0 \tag{33}$$

The latter condition allows to determine the pressure p. If, additionally, $t_{33} = 0$, then $\lambda_2=\lambda_3$ and in view of the incompressibility condition (30), the deormation (27) is characterized by one parameter λ_1.

On the equilibrium state B, determined by (27) and (31), we superpose an arbitrary small static deformation $\varepsilon u_i(x_j)$ and investigate the existence of an adjacent equilibrium state.

The stresses associated with the deformation

$$\overset{*}{x}_i = x_i + \varepsilon u_i(x_j) \tag{34}$$

are determined by (24) and (25) and for the deformation (27) and the fibre direction (29), we obtain

$$\begin{aligned}
t'_{11} &= 2[\lambda_1^2\Phi_1 + \lambda_1^2(\lambda_2^2+\lambda_3^2)\Phi_2 + 4\lambda_1^4\Phi_4 + \Phi_5\lambda_1^2 + \Phi_{KM}\overset{K}{\xi}_1\overset{M}{\xi}_1]u_{1,1} \\
&\quad + 2(\lambda_1^2\lambda_2^2\Phi_2 + \Phi_{KM}\overset{K}{\xi}_1\overset{M}{\xi}_2)u_{2,2} + 2(\lambda_1^2\lambda_3^2\Phi_2 + \Phi_{KM}\overset{K}{\xi}_1\overset{M}{\xi}_3)u_{3,3} - p' \\
t'_{22} &= 2(\lambda_1^2\lambda_2^2\Phi_2 + \Phi_{KM}\overset{K}{\xi}_2\overset{M}{\xi}_1)u_{1,1} + 2[\lambda_2^2\Phi_1 + \lambda_2^2(\lambda_1^2+\lambda_3^2)\Phi_2 + \Phi_{KM}\overset{K}{\xi}_2\overset{M}{\xi}_2]u_{2,2} \\
&\quad + 2(\lambda_2^2\lambda_3^2\Phi_2 + \Phi_{KM}\overset{K}{\xi}_2\overset{M}{\xi}_3)u_{3,3} - p' \\
t'_{33} &= 2(\lambda_1^2\lambda_3^2\Phi_2 + \Phi_{KM}\overset{K}{\xi}_3\overset{M}{\xi}_1)u_{1,1} + 2(\lambda_2^2\lambda_3^2\Phi_2 + \Phi_{KM}\overset{K}{\xi}_3\overset{M}{\xi}_2)u_{2,2} \\
&\quad + 2[\lambda_3^2\Phi_1 + \lambda_3^2(\lambda_1^2+\lambda_2^2)\Phi_2 + \Phi_{KM}\overset{K}{\xi}_3\overset{M}{\xi}_3]u_{3,3} - p' \\
t'_{12} &= (\lambda_2^2\Phi_1 + \lambda_2^2\lambda_3^2\Phi_2 + \lambda_1^2\lambda_2^2\Phi_4)u_{1,2} \\
&\quad + [\lambda_1^2\Phi_1 + \lambda_1^2\lambda_3^2\Phi_2 + (2\lambda_1^4+\lambda_1^2\lambda_2^2)\Phi_4 + \lambda_1^2\Phi_5]u_{2,1} \\
t'_{13} &= (\lambda_3^2\Phi_1 + \lambda_2^2\lambda_3^2\Phi_2 + \lambda_1^2\lambda_3^2\Phi_4)u_{1,3} \\
&\quad + [\lambda_1^2\Phi_1 + \lambda_1^2\lambda_2^2\Phi_2 + (2\lambda_1^4+\lambda_1^2\lambda_3^2)\Phi_4 + \lambda_1^2\Phi_5]u_{3,1} \\
t'_{23} &= (\lambda_3^2\Phi_1 + \lambda_1^2\lambda_3^2\Phi_2)u_{2,3} + (\lambda_2^2\Phi_1 + \lambda_1^2\lambda_2^2\Phi_2)u_{3,2}
\end{aligned} \tag{35}$$

where

$$\begin{aligned}
\overset{1}{\xi}_i &= (\lambda_1^2,\lambda_2^2,\lambda_3^2) \\
\overset{2}{\xi}_i &= [\lambda_1^2(\lambda_2^2+\lambda_3^2),\ \lambda_2^2(\lambda_1^2+\lambda_3^2),\ \lambda_3^2(\lambda_1^2+\lambda_2^2)] \\
\overset{4}{\xi}_i &= (2\lambda_1^4,0,0),\quad \overset{5}{\xi}_i = (\lambda_1^2,0,0)
\end{aligned} \tag{36}$$

From the equilibrium equations for the state $\overset{*}{B}$

$$(t_{ij} + \varepsilon t'_{i,j})_{,j} = 0$$

In view of $t_{ij,j} = 0$ and (26), we obtain

$$\begin{aligned}
&b_{11}u_{1,11} + b_{12}u_{1,22} + b_{13}u_{1,33} + b_{14}u_{3,13} - p'_{,1} = 0 \\
&b_{21}u_{2,11} + b_{22}u_{2,22} + b_{23}u_{2,33} + b_{24}u_{3,23} - p'_{,2} = 0 \\
&b_{31}u_{3,11} + b_{32}u_{3,22} + b_{33}u_{3,33} + b_{34}u_{2,23} - p'_{,3} = 0
\end{aligned} \tag{38}$$

where

$$\begin{aligned}
b_{11} &= \lambda_1^2(\Phi_1 + \lambda_3^2\Phi_2 - \lambda_2^2\Phi_4 + 6\lambda_1^2\Phi_4 + \Phi_5) + 2\Phi_{KM}\overset{K}{\xi}_1(\overset{M}{\xi}_1 - \overset{M}{\xi}_2) \\
b_{12} &= \lambda_2^2(\Phi_1 + \lambda_3^2\Phi_2 + \lambda_1^2\Phi_4) \\
b_{13} &= \lambda_3^2(\Phi_1 + \lambda_2^2\Phi_2 + \lambda_1^2\Phi_4) \\
b_{14} &= \lambda_1^2(\lambda_3^2 - \lambda_2^2)(\Phi_2 + \Phi_4) + 2\Phi_{KM}\overset{K}{\xi}_1(\overset{M}{\xi}_3 - \overset{M}{\xi}_2) \\
b_{21} &= \lambda_1^2[\Phi_1 + \lambda_3^2\Phi_2 + (2\lambda_1^2 + \lambda_2^2)\Phi_4 + \Phi_5] \\
b_{22} &= \lambda_2^2(\Phi_1 + \lambda_3^2\Phi_2 - \lambda_1^2\Phi_4) + 2\Phi_{KM}\overset{K}{\xi}_2(\overset{M}{\xi}_2 - \overset{M}{\xi}_1) \\
b_{23} &= \lambda_3^2(\Phi_1 + \lambda_1^2\Phi_2) \\
b_{24} &= \lambda_2^2[(\lambda_3^2 - \lambda_1^2)\Phi_2 - \lambda_1^2\Phi_4] + 2\Phi_{KM}\overset{K}{\xi}_2(\overset{M}{\xi}_3 - \overset{M}{\xi}_1) \\
b_{31} &= \lambda_1^2[\Phi_1 + \lambda_2^2\Phi_2 + (2\lambda_1^2 + \lambda_3^2)\Phi_4 + \Phi_5] \\
b_{32} &= \lambda_2^2(\Phi_1 + \lambda_1^2\Phi_2) \\
b_{33} &= \lambda_3^2(\Phi_1 + \lambda_2^2\Phi_2 - \lambda_1^2\Phi_4) + 2\Phi_{KM}\overset{K}{\xi}_3(\overset{M}{\xi}_3 - \overset{M}{\xi}_1) \\
b_{34} &= \lambda_3^2[(\lambda_2^2 - \lambda_1^2)\Phi_2 - \lambda_1^2\Phi_4] + 2\Phi_{KM}\overset{K}{\xi}_3(\overset{M}{\xi}_2 - \overset{M}{\xi}_1)
\end{aligned} \tag{39}$$

Equations (38) completed by the incompressibility condition (26) constitute a set of four linear partial differential equations of second order for four unknown functions, u_i, p' of three variables x_i and the parameters λ_A.

We proceed now to establish the range of loads t for which a non-zero superposed deformation u_i governed by equations (26) and (38) are admissible for every specimen dimension $\overset{*}{\ell}_i$. To achieve this result, we have to formulate such boundary conditions for the state B which exclude the possibility of the specimen buckling since the latter phenomenon depends on the specimen dimensions, then characterizes rather stability of a specimen than material properties. To prevent the buckling, we require that the surfaces $\overset{*}{x}_i = \pm\ell_i$, plane in the state B, remain plane in the state B, i.e., we require the vanishing of the normal displacements on the whole boundary of the block. Denoting by n_i the normal vector to the boundary, we have

$$\text{on } x_i = \pm\ell_i \quad u_i n_i = 0 \tag{40}$$

Moreover, we require that the boundary be free of tractions, i.e.,

$$\text{on } x_i = \pm\ell_i \quad \overset{*}{t}_{ij} = 0 \text{ for } i\neq j \tag{41}$$

From equations (40) and (41) by substituting (31) and (35), we obtain

$$\begin{aligned}
&\text{on } x_1 = \pm\ell_1 \quad u_1 = 0,\ u_{2,1} = 0,\ u_{3,1} = 0\\
&\text{on } x_2 = \pm\ell_2 \quad u_2 = 0,\ u_{1,2} = 0,\ u_{3,2} = 0\\
&\text{on } x_3 = \pm\ell_3 \quad u_3 = 0,\ u_{1,3} = 0,\ u_{2,3} = 0
\end{aligned} \tag{42}$$

To solve the boundary value problem determined by (26), (38), (42), we expand the functions $u_i(x_j)$, $p'(x_j)$ in the orthogonal basis of trigonometric functions. Substituting this expansion into (42), we find that some terms vanish and the remaining part, already satisfying the boundary conditions, is the following

$$\begin{aligned}
u_1 &= \sum_k \sum_m \sum_n U_{1kmn} \sin \frac{k\pi}{\ell_1} x_1 \cos \frac{m\pi}{\ell_2} x_2 \cos \frac{n\pi}{\ell_3} x_3\\
u_2 &= \sum_k \sum_m \sum_n U_{2kmn} \cos \frac{k\pi}{\ell_1} x_1 \sin \frac{m\pi}{\ell_2} x_2 \cos \frac{n\pi}{\ell_3} x_3\\
u_3 &= \sum_k \sum_m \sum_n U_{3kmn} \cos \frac{k\pi}{\ell_1} x_1 \cos \frac{m\pi}{\ell_2} x_2 \sin \frac{n\pi}{\ell_3} x_3\\
p' &= \sum_k \sum_m \sum_n P_{kmn} \cos \frac{k\pi}{\ell_1} x_1 \cos \frac{m\pi}{\ell_2} x_2 \cos \frac{n\pi}{\ell_3} x_3
\end{aligned} \tag{43}$$

where the coefficients U_{ikmn}, P_{kmn} are constant.

Substituting (43) into equations (38), we obtain, for every k,m,n, the fol-
owing set of homogeneous algebraic equations

$$
\begin{aligned}
&\nu_1 U_{1kmn} + \nu_2 U_{2kmn} + \nu_3 U_{3kmn} = 0 \\
&(b_{11}\nu_1^2 + b_{12}\nu_2^2 + b_{13}\nu_3^2)U_{1kmn} + b_{14}\nu_1\nu_3 U_{3kmn} - \nu_1 P_{kmn} = 0 \\
&(b_{21}\nu_1^2 + b_{22}\nu_2^2 + b_{23}\nu_3^2)U_{2kmn} + b_{24}\nu_2\nu_3 U_{3kmn} - \nu_2 P_{kmn} = 0 \\
&b_{34}\nu_2\nu_3 U_{2kmn} + (b_{31}\nu_1^2 + b_{32}\nu_2^2 + b_{33}\nu_3^2)U_{3kmn} - \nu_3 P_{kmn} = 0
\end{aligned}
\tag{44}
$$

where

$$\nu_1 = \frac{k\pi}{\ell_1} \quad \nu_2 = \frac{m\pi}{\ell_2} \quad \nu_3 = \frac{n\pi}{\ell_3} \tag{45}$$

A non-zero solution of equations (44) exists if the determinant of this set is equal to zero:

$$\Delta(\nu_i, \lambda_A) = 0 \tag{46}$$

A general discussion of equation (46) is difficult to perform and we restrict ourselves to two special cases:

$$
\begin{aligned}
&(i) \quad \lambda_2 = \lambda_3 \\
&(ii) \quad \nu_3 = 0
\end{aligned}
\tag{47}
$$

The case (i) corresponds to the experiment in which $t_{22} = t_{33}$; in particular, to such that the surfaces $x_2 = \pm\ell_2$, $x_3 = \pm\ell_3$ are free of load.

The case (ii) corresponds to the plane superposed deformation: $u_i = u_i(x_1, x_2)$ for $i = 1,2$, $u_3 = 0$.

Under the assumptions (47), the condition (46) has the form

$$b_{12}\eta^2 + (b_{11}+b_{22})\eta + b_{21} = 0 \tag{48}$$

where

$$\eta = \frac{\nu_2^2+\nu_3^2}{\nu_1^2} \tag{49}$$

A non-zero solution u_i exists if the equation (48) has at least one positive root. The coefficients in (48) depend on λ_A and $\Phi_K(\lambda_A)$, $\Phi_{KM}(\lambda_A)$; however, without specifying the functions $\Phi_K(\lambda_A)$, $\Phi_{KM}(\lambda_A)$, we are able to estimate at least some of these coefficients. For this purpose, let us consider a deformation consisting of a pure homogeneous deformation and a shearing in the fibre direction:

$$x_1 = \lambda_1 X_1 + \beta\lambda_2 X_2, \; x_2 = \lambda_2 X_2, \; x_3 = \lambda_3 X_3 \tag{50}$$

The non-vanishing tangent stress is

$$t_{12} = \beta\lambda_2^2(\Phi_1 + \lambda_3^2\Phi_2 + \lambda_1^2\Phi_4) \tag{51}$$

From the requirement that the sign of the shearing stress be compatible with the sign of the shear angle, it follows that

$$b_{12} = \lambda_2^2(\Phi_1 + \lambda_3^2\Phi_2 + \lambda_1^2\Phi_4) > 0 \tag{52}$$

Considering the deformation

$$x_1 = \lambda_1 X_1, \; x_2 = \lambda_2 X_2 + \beta\lambda_3 X_3, \; x_3 = \lambda_3 X_3 \tag{53}$$

i.e., a shearing in the plane perpendicular to the fibres, we have

$$t_{23} = \beta\lambda_3^2(\Phi_1 + \lambda_1^2\Phi_2) \tag{54}$$

and from the same arguments, it follows that

$$\Phi_1 + \lambda_1^2\Phi_2 > 0 \tag{55}$$

Moreover, for $\lambda_2 = \lambda_3$, the coefficient of η can be written in the form

$$b_{11} + b_{22} = \lambda_1^2 \frac{d\sigma_{11}}{d\lambda_1} + \lambda_2^2(\Phi_1 + \lambda_1^2\Phi_2) - 2b_{12} \tag{56}$$

where $\sigma_{11} = \lambda_2\lambda_3 t_{11}$ is the nominal stress. Since the first two terms on the r.h.s. are positive

$$b_{11} + b_{22} > -2b_{12} \tag{57}$$

Similarly, for the plane underlying strain with $\lambda_3 = 1$, we obtain

$$b_{11} + b_{22} = \lambda_1^2 \frac{d\sigma_{11}}{d\lambda_1} - 2b_{12} \tag{58}$$

and again inequality (57) holds.

Finally, we observe that $b_{21} = b_{12} + t_{11} - t_{22}$ and substituting (33), we obtain

$$b_{21} = b_{12} - t \tag{59}$$

The equation (48), in view of (52), has one positive root η for $b_{21} < 0$, i.e., for

$$t > b_{12} \tag{60}$$

We find that

$$t \to b_{12} \text{ as } \eta \to 0 \tag{61}$$

Since for any compressive load t satisfying (60) an internal buckling is dmissible, the compressive strength of a fibrous composite has the following estriction:

$$t_c \leq \lambda_2^2(\Phi_1 + \lambda_3^2\Phi_2 + \lambda_1^2\Phi_4) \tag{62}$$

For $b_{21} \geq 0$, equation (48) has positive roots if

$$b_{11} + b_{22} < 0 \text{ and } t \geq b_{12} \left[1 - \left(\frac{b_{11}+b_{22}}{2b_{12}}\right)^2\right] \tag{63}$$

In view of inequalities (52) and (57), the r.h.s. in $(63)_2$ is positive. We ind that

$$t = b_{12}\left[1 - \left(\frac{b_{11}+b_{22}}{2b_{12}}\right)^2\right] \text{ for } \eta = -\frac{b_{11}+b_{22}}{2b_{12}} \tag{64}$$

The inequality $(63)_2$ is relevant only for the materials having the property hat in some interval of $\lambda_1 \leq 1$, $b_{11} + b_{22} < 0$ and $b_{21} \geq 0$. If such a material xists, then its compressive strength has a restriction stronger than equation 62), namely

$$t_c < b_{12}\left[1 - \left(\frac{b_{11}+b_{22}}{2b_{12}}\right)^2\right] \tag{65}$$

EFERENCES

1] Green, A. E., Rivlin, R. S. and Shield, R. T., "General Theory of Small Elastic Deformations Superposed on Large Elastic Deformations", Proc. R. Soc. A, 211, pp. 128-154, 1952.

2] Biot, M. A., "Internal Buckling under Initial Stress in Finite Elasticity", Proc. R. Soc. A, 273, pp. 306-328, 1963.

3] Green, A. E. and Adkins, J. E., "Large Elastic Deformations and Non-Linear Continuum Mechanics", Clarendon Press, 1960.

EQUATION OF STATE FOR REINFORCED PLASTIC MATERIALS SUBJECTED TO MECHANICAL AND THERMAL LOADING WITH THE ACCOUNT TAKEN OF DAMAGE AND PHYSICAL-CHEMICAL TRANSFORMATIONS

G. S. Pisarenko and V. S. Dzyuba

Academy of Sciences of the Ukr. SSR
Kiev, USSR

Wide application of reinforced plastics in modern technology has become possible due to a convenient combination in them of high strength and low heat conduction which makes it possible to use them in structural components subjected simultaneously to heating and mechanical loading. Irreversible physical-chemical transformations associated with destruction and pyrolysis, as well as local loosening of the material causing failure, occur in the material during the processes of heating and mechanical loading.

The models of a solid subjected to deformation, for instance: thermo-elastic [1], thermo-visco-elastic [2], the model of ultimate state [3], have found wide application in the investigations into the behavior of materials, reinforced plastics in particular, and the results obtained are used in practice. These models, however, do not take into account all the variety of phenomena taking place in reinforced plastics at high temperatures and under mechanical loading conditions.

It is necessary to consider all the irreversibilities that occur in the material during destruction which are accompanied by different chemical reactions, damage caused by mechanical loads and filtration processes taking place due to the migration of gaseous components of the material. All these phenomena contribute to softening and, eventually, to failure of the material. Their contributions to failure are not similar.

In some cases, reinforced plastics can be considered as a uniform anisotropic medium. Consider an element of such a medium infinitely small from the standpoint of physics. Assume that the state of the element is characterized by a system of independent variables S (or T), E_{ij}, L_{ij}, C_{ij}, where S is the entropy per unit of medium mass, T is the absolute temperature, E_{ij} are the components of the geometrically small strain tensor, L_{ij} are the components of the damage tensor and C_{ij} are the components of the chemical potential tensor. Here, a material model parameter which varies in the range $0 \leq L \leq 1$ during the process of thermal and mechanical loading, is assumed to represent the accumulated damage. The value L=0 will correspond to a nondamaged material, while L=1 will correspond to the material that failed. Thus, the above parameter reflects the processes that occur in the material which are associated with irreversible changes in the material. The accumulation of these changes characterizes, unambiguously, the damage summation process taking place in the material. On one hand, these changes testify to the increase in loosening of the material and in the formation of pores, cracks and other imperfections which weaken the adhesive forces in microvolumes of the material. On the other hand, at a certain stage they characterize strengthening of the material (caused by filtration processes, for example). However, in both cases, the irreversibility of the process is a common feature.

One of the most acceptable parameters for the evaluation of this process is the variation of the system entropy. Therefore, the accumulated damage is assumed to be expressed by the following formula

$$L_\tau^n = \sum_{\tau=0}^{\tau} \frac{S_\tau^n}{S^n}$$

where L_τ^n is the damage accumulated in the material under "n"-effect (effect of any kind) during a certain period of time τ; S_τ^n is the entropy of the material under similar conditions; S^n is the limiting value of the entropy (at the moment of failure).

On the basis of the first and the second laws of thermodynamics, the increment of the internal energy E per unit mass can be written in the following way:

$$dE = TdS + \sigma_{ij} d\varepsilon_{ij} + \Sigma\mu_i dn_i \tag{1}$$

where σ_{ij} are the stress tensor components; μ_i is a potential characterizing the progression of a certain process in the material; n_i is the concentration of the given process in the material.

Equation (1) is a generalized Gibb's equation. It should be assumed that the above equation is valid not only for reversible processes but for irreversible ones as well, providing the deviation of the system from the equilibrium state is small.

For the type of materials under investigation, the last term in equation (1) can be presented as a sum:

$$\mu_i dn_i = R_{ij} dL_{ij} + M_{\alpha\beta} dC_{\alpha\beta} \tag{2}$$

where R_{ij} are the components of the damage potential tensor; $M_{\alpha\beta}$ are the components of the chemical potential tensor.

When considering the processes occurring in the material subjected to loading and heating, damage accumulation and chemical transformations should be regarded as the main accompanying factors which are irreversible.

The increment of free energy, F, per unit mass with the account taken of (1) and (2) will be

$$dF = - SdT + \sigma_{ij} d\varepsilon_{ij} + R_{ij} dL_{ij} + M_{\alpha\beta} dC_{\alpha\beta} \tag{3}$$

Here, it should be pointed out that both internal energy E and free energy F are state functions, therefore, infinitesimal increments of these functions will be complete differentials. Variation in free energy can be presented in a different form:

$$dF = \left(\frac{\partial F}{\partial T}\right)_{E,L,C} dT + \left(\frac{\partial F}{\partial \varepsilon}\right)_{T,L,C} dE_{ij} + \left(\frac{\partial F}{\partial L}\right)_{\varepsilon,T,C} dL_{ij} + \left(\frac{\partial F}{\partial C}\right)_{T,\varepsilon,C} dC_{\alpha\beta} \tag{4}$$

Comparison of the right term of equations (3) and (4) gives the following equations:

$$\sigma_{ij} = \left(\frac{\partial F}{\partial \varepsilon}\right)_{T,L,C}; \; S = - \left(\frac{\partial F}{\partial T}\right)_{\varepsilon,L,C}; \; R_{ij} = \left(\frac{\partial F}{\partial L}\right)_{T,\varepsilon,C}; \; M_{\alpha\beta} = \left(\frac{\partial F}{\partial C}\right)_{T,\varepsilon,L} \tag{5}$$

Assume that the function F in an analytical function of its variables, i.e., continuous and differentiable in the neighborhood of its natural state F_0 and converging at the point F_0. At the initial instant $\tau=0$, the thermodynamic system is in the state of thermodynamic equilibrium. When the system deviation from the equilibrium state is small, the function F can be presented as an exponential series and written as

$$\begin{aligned} F_{(\varepsilon,L,T,C)} &= F_{(o,o,o,T_0)} + \frac{\partial F}{\partial \varepsilon_{ij}} \varepsilon_{ij} + \frac{\partial F}{\partial L_{ij}} L_{ij} + \frac{\partial F}{\partial C_{\alpha\beta}} C_{\alpha\beta} + \frac{\partial F}{\partial T} (T-T_0) \\ &+ \frac{1}{2!} \Big[\frac{\partial^2 F}{\partial \varepsilon_{ij} \partial \varepsilon_{k\ell}} \varepsilon_{ij} \varepsilon_{k\ell} + 2 \frac{\partial^2 F}{\partial \varepsilon_{ij} \partial L_{k\ell}} \varepsilon_{ij} L_{k\ell} + 2 \frac{\partial^2 F}{\partial \varepsilon_{ij} \partial C_{\alpha\beta}} \varepsilon_{ij} C_{\alpha\beta} \\ &+ 2 \frac{\partial^2 F}{\partial \varepsilon_{ij} \partial T} \varepsilon_{ij} (T-T_0) + \frac{\partial^2 F}{\partial T^2} (T-T_0)^2 + 2 \frac{\partial^2 F}{\partial L_{ij} \partial T} L_{ij} (T-T_0) \\ &+ 2 \frac{\partial^2 F}{\partial C_{\alpha\beta} \partial T} (T-T_0) C_{ij} + \frac{\partial^2 F}{\partial L_{ij} \partial L_{k\ell}} L_{ij} L_{k\ell} + \frac{\partial^2 F}{\partial C_{ij} \partial C_{k\ell}} C_{ij} C_{k\ell} \\ &+ 2 \frac{\partial^2 F}{\partial L_{ij} \partial C_{k\ell}} L_{ij} C_{k\ell} \Big] + \ldots \end{aligned} \tag{6}$$

including terms of second order.

Assume that

$$\frac{\partial^2 F}{\partial \varepsilon_{ij} \partial \varepsilon_{k\ell}} = A_{ijk\ell}; \quad \frac{\partial^2 F}{\partial T^2} = m; \quad \frac{\partial^2 F}{\partial L_{ij} \partial L_{k\ell}} = D_{ijk\ell};$$

$$\frac{\partial^2 F}{\partial C_{ij} \partial C_{\alpha\beta}} = \mu_{\alpha\beta ij}; \quad \frac{\partial^2 F}{\partial \varepsilon_{ij} \partial C_{\alpha\beta}} = \ell_{ij\alpha\beta}; \quad \frac{\partial^2 F}{\partial \varepsilon_{ij} \partial L_{k\ell}} = \gamma_{ijk\ell}; \quad \frac{\partial^2 F}{\partial \varepsilon_{ij} \partial T} = -\beta_{ij}; \tag{7}$$

$$\frac{\partial^2 F}{\partial L_{ij} \partial T} = \eta_{ij}; \quad \frac{\partial^2 F}{\partial C_{ij} \partial T} = \mu_{ij}; \quad \frac{\partial^2 F}{\partial L_{ij} \partial C_{\alpha\beta}} = g_{ij\alpha\beta}; \quad \theta = T - T_o; \quad F_o = 0$$

Substituting from (7) into the representation of the function in (6)

$$F_{(\varepsilon,T,L,C)} = \frac{1}{2} [A_{ijk\ell}\varepsilon_{ij}\varepsilon_{k\ell} - 2\beta_{ij}\varepsilon_{ij}\theta + m\theta^2 + R_{ijk\ell}L_{ij}L_{k\ell} + \mu_{ij\alpha\beta}C_{ij}C_{\alpha\beta}$$

$$= 2(\gamma_{ijk\ell}\varepsilon_{ij}L_{k\ell} + \eta_{ij}L_{ij}\theta + \mu_{ij}C_{ij}\theta + g_{ij\alpha\beta}L_{ij}C_{\alpha\beta}$$

$$+ \ell_{ij\alpha\beta}\varepsilon_{ij}C_{\alpha\beta})] \tag{8}$$

Using the relationship in (5), the values of σ_{ij}, S, R_{ij}, $M_{\alpha\beta}$ are

$$\sigma_{ij} = \left(\frac{\partial F}{\partial \varepsilon}\right)_{T,L,C} = A_{ijk\ell}\varepsilon_{k\ell} - B_{ij}\theta + \gamma_{ijk\ell}L_{k\ell} + \ell_{ij\alpha\beta}C_{\alpha\beta}$$

$$S = -\left(\frac{\partial F}{\partial T}\right)_{\varepsilon,L,C} = \beta_{ij}\varepsilon_{ij} - m\theta - \eta_{ij}L_{ij} - \mu_{ij}C_{ij}$$

$$R_{ij} = \left(\frac{\partial F}{\partial L}\right)_{T,\varepsilon,C} = D_{ijk\ell}L_{k\ell} + \eta_{ij}\theta + \gamma_{ijk\ell}\varepsilon_{k\ell} + g_{ij\alpha\beta}C_{\alpha\beta} \tag{9}$$

$$M_{\alpha\beta} = \left(\frac{\partial F}{\partial C}\right)_{T,\varepsilon,L} = \mu_{\alpha\beta ij}C_{ij} + \mu_{\alpha\beta}\theta + g_{\alpha\beta ij}L_{ij} + \ell_{\alpha\beta ij}e_{ij}$$

It can be shown that

$$\beta_{ij} = \left(\frac{\partial\sigma_{ij}}{\partial T}\right)_{\varepsilon,L,C} = \frac{1}{T}\Sigma\left[\left(\frac{\partial E}{\partial\varepsilon}\right)_{T,L,C} + \sigma_{ij}\right]$$

$$\eta_{ij} = \left(\frac{\partial R_{ij}}{\partial T}\right)_{\varepsilon,L,C} = \frac{1}{T}\Sigma\left[\left(\frac{\partial E}{\partial L}\right)_{T,\varepsilon,C} + R_{ij}\right] \qquad (10)$$

$$\mu_{\alpha\beta} = \left(\frac{\partial M_{\alpha\beta}}{\partial T}\right)_{\varepsilon,S,L} = \frac{1}{T}\Sigma\left[\left(\frac{\partial E}{\partial C}\right)_{T,\varepsilon,L} + M_{\alpha\beta}\right]$$

To evaluate the thermal state of reinforced plastics under conditions of high temperature, it is necessary to determine the rate of specific thermal flux passing through the material taking into account equation (2):

$$\dot{q}_{i,i} = C\rho_0 \frac{\partial T}{\partial\tau} \qquad (11a)$$

$$\dot{q}_{i,i} = T\dot{S} = T(C_\varepsilon\dot{t} + \beta_{ij}\dot{\varepsilon}_{ij} + R_{ij},_T\dot{L}_{ij} + \mu_{ij,T}\dot{C}_{ij}) \qquad (11b)$$

$$C\rho_0 \frac{\partial T}{\partial\tau} = div(\lambda\nabla T) + \Sigma h_i i_i - \Sigma(j_{gucp} + \Pi\rho_i b_i v_i)C\nabla T \qquad (11c)$$

Equating the right term of equation (11b) to the right term of equation (11c), we shall obtain

$$div(\lambda\nabla T) + \Sigma h_i i_i - \Sigma(j_{gucp} + \Pi\rho_i b_i v_i)C\nabla T - T(C_\varepsilon\dot{t} + \beta_{ij}\dot{\varepsilon}_{ij} + R_{ij,T}\dot{L}_{ij} + \mu_{ij,T}\dot{C}_{ij}) = 0 \qquad (12)$$

This equation describes the main processes in the material such as heat transfer, occurring due to heat conduction (the first term), internal sources ($\Sigma h_i i_i$); diffusion and filtration processes (the third expression); the rates of heating, deformation, damage accumulation and chemical reactions proceeding (in the last brackets).

Consider the limiting values of the entropy, which correspond to failure of the material. This limiting value is assumed to consist of two parts: the first part is responsible for the damage associated with the heating of the material and with chemical reactions taking place in the material during its pyrolysis. The second part is responsible for the damage that occurs in the material due to mechanical loading. Since in the first and second cases damage is accumulated in different volumes, the magnitudes of damage are not similar when the accumulation of the entropy is the same.

In the general case, the limiting value of the entropy that corresponds to the material failure caused by loading can be represented as a functional depending on temperature, pressure and three stress invariants.

$$S_n = f(\Pi_1, \Pi_2, \Pi_3, T, P); \; \Pi_1 = \sigma_{ij}\delta_{ij}; \; \Pi_2 = \sigma_{ij}\sigma_{ij}; \; \Pi_3 = \sigma_{ij}\sigma_{nm}\sigma_{pq}$$

Assuming that the limiting value of the entropy depends on the stress invariants in different ways, this dependence can be written in the following way

$$S_n = S_{o(P,T)} + (A_{ij}\sigma_{ij})^{\alpha} + (B_{ijnm}\sigma_{ij}\sigma_{nm})^{\beta} + (C_{ijnmpq}\sigma_{ij}\sigma_{nm}\sigma_{pq})^{\gamma} + \dots$$

For the materials under investigation, this criterion of the ultimate state is characterized by the fact that its magnitude depends on the loading directio

To be more concrete, take $\alpha = 1$; $\beta = \frac{1}{2}$; $\gamma = \frac{1}{3}$. Then this criterion will acquire the following form:

$$S_n = \begin{cases} S_{o(P,T)} + A_{ij}\sigma_{ij} + (B_{ijnm}\sigma_{ij}\sigma_{nm})^{1/2} + (C_{ijnmpq}\sigma_{ij}\sigma_{nm}\sigma_{pq})^{1/3} + \dots \\ S_{T(P,T)} \end{cases} \quad (13$$

Here, the first expression characterizes the fact that the system entropy reaches the limiting value in the given direction of loading depending on the mode of mechanical loading. The second expression corresponds to the limiting value of the system entropy in the case when the body is subjected only to thermal loads.

It should be noted that both expressions depend on temperature and the gaseous medium pressure.

REFERENCES

[1] Kovalenko, A. D., Fundamentals of Thermoelasticity, Kiev, Naukova Dumka, p. 307, 1970.

[2] Il'yushin, A. A. and Pobedrya, B. E., Fundamentals of the Mathematical Theory of Thermoviscoelasticity, M., Nauka, p. 280, 1970.

[3] Gol'denblatt, I. I., Bazhanov, V. L. and Kopnov, V. A., The Entropy Principle in the Theory of Creep and Rupture Strength of Polymeric Materials, Mekhanika Polimerov, No. 2, pp. 251-261, 1971.

STRENGTH OF LAYERED COMPOSITE CYLINDRICAL SHELLS UNDER DYNAMIC LOADING

A. Bogdanovich and V. Tamuzs

L.S.S.R. Academy of Sciences
Riga, USSR

Under actual conditions of use, thin-walled shell constructions are subjected to the action of various static, vibrational and impulsive loads applied simultaneously for finite time periods. The loss of a construction's load-carrying capacity may occur at any stage of the loading; therefore, it is necessary to solve a wide range of problems. The shell must be analyzed for both the case of the loading modes applied separately and the case of several modes of loading applied jointly. This work presents the results obtained in recent years by solving two classes of dynamic problems concerning layered composite cylindrical shells:

1. the analysis of the stress-strain state and the calculation of the time of the first ply failure under sufficiently long impulsive loads,

2. the analysis of the stress-strain state and the calculation of the fatigue longevity of a shell under axial vibrational loads.

The methods described below also allow the preliminary loading of a shell by static forces to be taken into account.

CALCULATION OF THE STRESS-STRAIN STATE OF CYLINDRICAL SHELLS UNDER DYNAMIC AXIAL COMPRESSION AND DYNAMIC EXTERNAL PRESSURE

The solution in a physically linear but geometrically nonlinear formulation of the problem of nonaxisymmetric buckling of imperfect cylindrical shells under dynamic loads (axial compression or external pressure) has been of interest to many authors. In addition to the works described in the monograph [1], the works concerning axial loading [2,3] and external pressure [4,5] should be noted. A detailed analysis of the most valuable data obtained in solving the problem of the buckling of cylindrical shells under the action of axial-compressive load impulses is given in [6]. In the latter, several still unsolved problems (concerning both axial dynamic compression and dynamic external pressure) were formulated:

1. It is necessary to prove the choice of the deflection approximation for solving the nonlinear problems of dynamic buckling by means of variational methods. (Such an approximation must provide for the required accuracy in determining the displacement and stress fields at an arbitrary point of a shell throughout the loading process).

2. It is necessary to investigate thoroughly the influence of the initial imperfection field characteristics on the load-carrying capacity of shells;

3. The load-carrying capacity criterion must be connected with the peculiar ities of the composite material and the strength properties of the layered composite;

4. The problem of the application of the linear-elastic-material model to the calculation of the time till the first ply failure arises if of considerable interest.

It should also be noted that the first ply failure in composite material constru tions is by no means identical to the total loss of the load-carrying capacity; hence, there is considerable interest in the investigation of the failure proces following the failure of the first ply.

Additionally, a new principle for the deflection approximation in the Galjorkin method applied to geometrically nonlinear problems of buckling of layere orthotropic cylindrical shells will be considered [7].

It is assumed that for a layered medium considered as a whole, the Kirchoff-Love hypotheses are valid; this allows the use of the equations of motion of homogeneous anisotropic shells. The stiffness characteristics are calculated ac cording to [8]. Only such layered structures in which "nonorthotropic" members in the equations of motion can be neglected will be examined. These considerations, with the usual admissions of a technical theory of cylindrical shells, allow the use of the well-known Donnell's type nonlinear equations for the shell deflection $w(x,y,t)$ and the stress function $\phi(x,y,t)$:

$$A_{22}\frac{\partial^4\phi}{\partial x^4} + (A_{66}+2A_{12})\frac{\partial^4\phi}{\partial x^2\partial y^2} + A_{11}\frac{\partial^4\phi}{\partial y^4} + \frac{1}{R}\frac{\partial^2(w-w_o)}{\partial x^2} = \left(\frac{\partial^2 w}{\partial x\partial y}\right)^2 - \frac{\partial^2 w}{\partial x^2}\frac{\partial^2 w}{\partial y^2} - \left(\frac{\partial^2 w_o}{\partial x\partial y}\right)^2 + \frac{\partial^2 w_o}{\partial x^2}\frac{\partial^2 w_o}{\partial y^2} \quad (1)$$

$$D_{11}\frac{\partial^4(w-w_o)}{\partial x^4} + 2(D_{12}+2D_{66})\frac{\partial^4(w-w_o)}{\partial x^2\partial y^2} + D_{22}\frac{\partial^4(w-w_o)}{\partial y^4} = \frac{1}{R}\frac{\partial^2\phi}{\partial x^2} + \frac{\partial^2 w}{\partial x^2}\frac{\partial^2\phi}{\partial y^2} + \frac{\partial^2 w}{\partial y^2}\frac{\partial^2\phi}{\partial x^2} - 2\frac{\partial^2 w}{\partial x\partial y}\frac{\partial^2\phi}{\partial x\partial y} - P(t)\frac{\partial^2 w}{\partial x^2} - \frac{q(t)}{R}\frac{\partial^2 w}{\partial y^2} - \mu\frac{\partial^2 w}{\partial t^2} \quad (2)$$

where $w_o(x,y)$ is the given initial deflection; x and y are the longitudinal and circumferential coordinates; μ is the weight of a shell surface unit; D_{ij} and

A_{ij}, the components of bending stiffness and the compliancy tensors for a multilayered orthotropic medium [8]; P(t), the uniformly distributed axial load applied to the edges; and q(t), the uniformly distributed external pressure.

If the geometrically nonlinear terms in (1) and (2) are neglected, we can satisfy these equations and the simply supported edge conditions can be satisfied by substituting each member of the following series:

$$w(x,y,t) = \sum_{m=1}^{\infty} \sum_{n=0}^{\infty} W_{mn}(t) \sin\alpha_m x \cos\beta_n y \tag{3}$$

$$\phi(x,y,t) = \sum_{m=1}^{\infty} \sum_{n=0}^{\infty} F_{mn}(t) \sin\alpha_m x \cos\beta_n y \tag{4}$$

where $\alpha_m = \frac{\pi m}{L}$ and $\beta_n = \frac{n}{R}$. As a result by solving the geometrically linear problem for each fixed pair of wave numbers (m,n) an ordinary differential equation is obtained from which the function $W_{mn}(t)$ can be found. An attempt to use the expansions (3) and (4) to solve a geometrically nonlinear problem leads to a calculation difficulty: all the modes in (3) and (4) are mutually connected, as a result of which it is impossible to obtain equations in an acceptable form for determining $W_{mn}(t)$ and $F_{mn}(t)$.

To investigate the possibility of the application of the Galjorkin method to these problems, a special analysis of the effect of the connectivity of circumferential and axial modes of buckling was carried out. Without going into detail, this effect causes a decrease in the value of $|W_{mn}(t)|$ for any m and n until this function reaches the first maximum. Accounting for the connectivity of the axial modes affects the results much less than accounting for the connectivity of the circumferential ones. Thus, sufficiently precise results may be obtained by taking into account the connectivity of the finite number of circumferential modes*.

In accordance with this for the solution of the geometrically nonlinear problems of dynamic buckling, it is suggested that the deflection and initial imperfections be approximated in the following way:

$$\begin{aligned} w_m(x,y,t) &= \sin\alpha_m x \sum_{n=n_0}^{N} W_{mn}(t) \cos\beta_n y \\ w_m^o(x,y) &= \sin\alpha_m x \sum_{n=n_0}^{N} W_{mn}^o \cos\beta_n y \end{aligned} \tag{5}$$

---

*It should be noted that in the case of external pressure, the connectivity of the various axial modes does not appear at all, since the dominating modes of buckling are m=1.

The approximation (5) with two additional insignificant terms was used in [5] to solve the problem of buckling of an isotropic cylindrical shell under dynamic external pressure. A detailed mathematical solution of the problem is given in [7]. The solution of the problem is finally reduced to the numerical integration of a system of $N-n_o+1$ ordinary nonlinear differential equations. If the calculation $W_{mk}(t)$ for axial modes with numbers $m_o,\ldots,M$ is desired, then such an integration should be carried out $M-m_o+1$ times. The total deflection can the be written in the form:

$$w(x,y,t) = \sum_{m=m_o}^{M} \sin\alpha_m x \sum_{n=n_o}^{N} W_{mn}(t)\cos\beta_n y \tag{6}$$

An expression in the form of a double trigonometric series can also be written for the total stress function.

In the case where not only the connectivity of the axial modes of buckling, but also the circumferential ones is neglected, the solution of the problems is greatly simplified. When approximating deflection as one term of the series (3) with fixed m and n:

$$w_{mn}(x,y,t) = \tilde{W}_{mn}(t)\sin\alpha_m x\cos\beta_n y \tag{7}$$

in order to determine each of the functions $\tilde{W}_{mn}(t)$, an ordinary nonlinear differential equation is obtained:

$$\frac{d^2\tilde{W}_{mn}}{dt^2} + \omega_{mn}^2 \left[\tilde{W}_{mn}\left(1 - \frac{P(t)}{P_{mn}^*} - \frac{q(t)}{q_{mn}^*}\right) - W_{mn}^o\right] + d_{mn}\tilde{W}_{mn}\left(\tilde{W}_{mn}^2 - W_{mn}^{o2}\right) = 0 \tag{8}$$

where ω_{mn}, P_{mn}^*, and q_{mn}^* are the frequency of natural bending vibrations, the critical static axial load, and the critical static external pressure, corresponding to the modes of buckling (m,n):

$$\alpha_{mn} = \frac{1}{16\mu}\left(\frac{\alpha_m^4}{A_{11}} + \frac{\beta_n^4}{A_{22}}\right)$$

The total deflection in this case can be presented in the form:

$$w(x,y,t) = \sum_{m=m_o}^{M}\sum_{n=n_o}^{N} \tilde{W}_{mn}(t)\sin\alpha_m x\cos\beta_n y \tag{9}$$

According to the assumed Kirchoff-Love hypotheses, the stresses at an arbitrary point of a shell are connected with the deflection and the stress functio by the relations

$$\sigma_{xx}(x,y,z,t) = \frac{1}{h}\frac{\partial^2\phi}{\partial y^2} - z\,[B_{11}\frac{\partial^2(w-w_o)}{\partial x^2} + B_{12}\frac{\partial^2(w-w_o)}{\partial y^2}] + \sigma^o_{xx}$$

$$\sigma_{yy}(x,y,z,t) = \frac{1}{h}\frac{\partial^2\phi}{\partial x^2} - z\,[B_{12}\frac{\partial^2(w-w_o)}{\partial x^2} + B_{22}\frac{\partial^2(w-w_o)}{\partial y^2}] + \sigma^o_{yy} \qquad (10)$$

$$\sigma_{xy}(x,y,z,t) = -\frac{1}{h}\frac{\partial^2\phi}{\partial x\partial y} - 2zB_{66}\frac{\partial^2(w-w_o)}{\partial x\partial y}$$

here B_{11}, B_{12}, B_{22}, B_{66} are the components of an orthotropic material stiffness atrix; σ^o_{xx} and σ^o_{yy} are the momentless stresses.

The formulas (6) and (10) or (9) and (10), together with the corresponding quations for calculating $W_{mn}(t)$ or $\tilde{W}_{mn}(t)$, allow the determination of the stress-train state at an arbitrary point of a shell.

The numerical solution of the problems was carried out in the following way:

a) The initial conditions were assumed in the form:

$$W_{mn}\Big|_{t=0} = W^o_{mn}; \quad \frac{dW_{mn}}{dt}\Big|_{t=0} = 0 \qquad (11)$$

he system of nonlinear differential equations for $W_{mn}(t)$ with fixed m and initial conditions (11) was numerically integrated using the fourth order Runge-Kutta method (it was repeated several times to establish the values of n_o and N, providing sufficient accuracy for practical purposes).

b) The summation of the series (6) for deflection was carried out (the limits of summation m_o and M were determined in accordance with the required accuracy and depend on the given initial imperfections, P(t) or q(t) dependencies, as well as on the geometric parameters of the shell). As a result, the deformed surface of a shell can be determined at any moment in time. The most dangerous points of this surface were then ascertained (usually they coincided with the points where deflection has an extremum).

c) At these points, the time dependencies of the stresses at two values of the coordinate z normal to the middle surface of a shell were calculated. As the expressions (10) are linear functions of z, these time dependencies are sufficient for a stress analysis through a shell thickness.

d) The time t^* at which at least at one point of a shell stresses go out of the failure surface was determined. The load values $P(t^*)$ or $q(t^*)$, as well as impulse values $I_p(t^*) = \int_0^{t^*} P(t)dt$ or $I_q(t^*) = \int_0^{t^*} q(t)dt$ were designated critical.

In this manner, the critical load (impulse) values are determined at the moment of the first ply failure in a shell. This first ply failure might be the breaking of fibers, the cracking of the matrix, the formation of local plastic zones in the matrix, or the disjunction of the matrix from the fibers.

As an illustration, several results of the calculation of a deformed shell surface are presented. An epoxy carbon composite shell consisting of six identical unidirectional layers is considered. The angles between the reinforcement direction of a layer and the axis of the shell are: +45, -45, 90, 90, -45, +45 (packet A) and +45, -45, 0, 0, -45, +45 (packet B). It is assumed that both the axial load and the external pressure linearly increase in time. When solving the axial compression problem the distribution of the amplitudes of the initial imperfections is taken in the form:

$$W^o_{mn} = 0.2\ h\ \frac{(-1)^{\ell+n}}{m^2}\ e^{-|n-3|} \tag{12}$$

and when the external pressure problem is under consideration, the imperfections are assumed in the form:

$$W^o_{mn} = 0.02\ h\ (-1)^{\ell+n}\ e^{-\frac{(m-1)^2}{10}-\frac{(n-4)^2}{4}} \tag{13}$$

where h is the total thickness of a shell, $\ell = m/2$ with even m values and $\ell = (m+1)/2$ with odd m values.

The deflection dependencies on the coordinate x in cross-section $y = \pi R$ are presented in Figure 1 for the case of axial dynamic compression. The dependencies

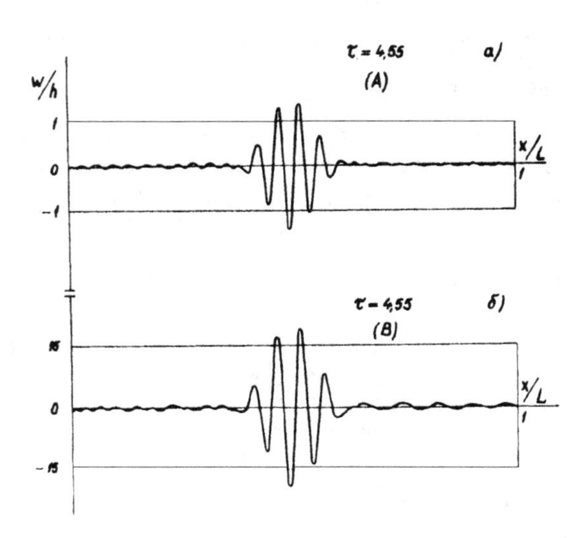

Fig. (1) - Deflection dependencies on axial coordinate for two epoxy carbon composite packets (axial compression)

for packet A (Figure 1a) and packet B (Figure 1b) are shown for the same moment in time. As shown in the figure, a region of intensive buckling propagates to the central part of the shell and consists of eight zones of dents and bulges.

This can be explained by the chosen distribution (12). Note that here a turn of two middle layers in the packet by 90° has caused the maximum value of the deflection to increase approximately ten times under the same load. The distributions of the deflection along the circumferential coordinate in cross-section x = 0.49L corresponding to the same moment in time are presented in Figure 2.

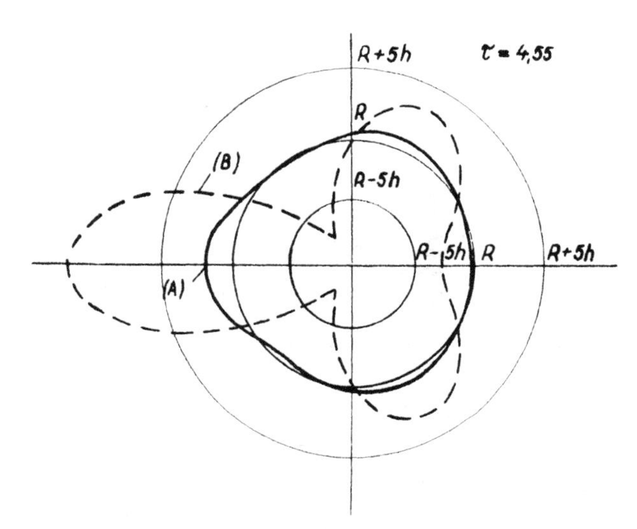

Fig. (2) - Deflection dependencies on circumferential coordinates for two epoxy carbon composite packets (axial compression)

An example of the calculation of the deflection dependence on the circumferential coordinate is given in Figure 3 for the case of a shell subjected to dynamic external pressure. For the packet A in cross-section x = 0.5L, the

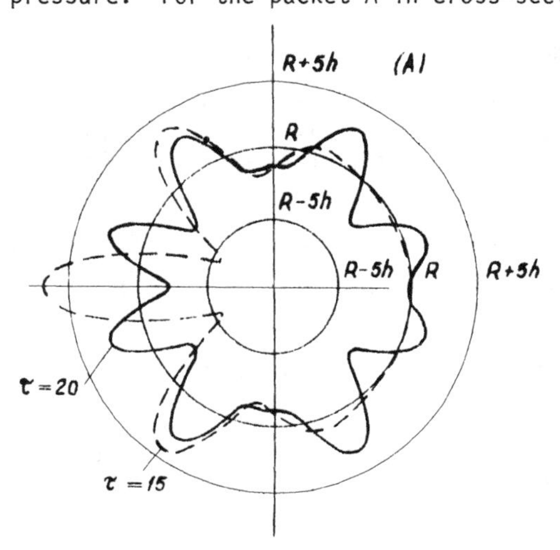

Fig. (3) - Deflection dependencies on circumferential coordinates for packet A at two time moments (external pressure)

solid line represents the dependencies calculated with the multi-term approximation of the deflection, the dash line - with the one-term approximation. As can be seen, the maximum deflection value under considerably greater pressure (at $\tau = 20$) with the multi-term approximation is lower than that with the one-term approximation at $\tau = 15$. This shows that in the case under consideration, there is a strong nonlinear effect.

The dependencies of stresses σ_{xx} (solid lines) and σ_{yy} (dashed lines) on time in cross-section $y = \pi R$ under the axial loading of a shell for packets A and B are presented in Figure 4a and 4b. It can be seen that an intensive increase of

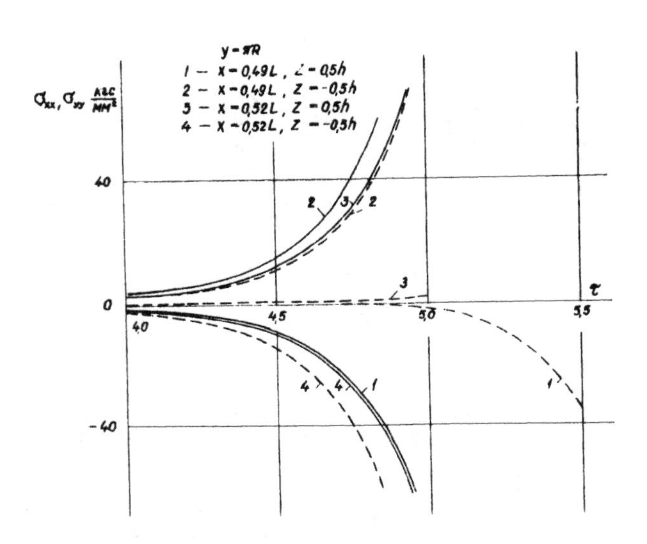

Fig. (4a) - Stress dependencies on time for packet A (axial compression)

stresses for packet A begins at $\tau \approx 4{,}5$, but begins for packet B at $\tau \approx 3{,}9$. Hence, it can be concluded that under this type of dynamic loading the structure of a layered shell with circumferential reinforcement of the two middle layers is preferable.

The dependencies of stresses σ_{xx} and σ_{yy} on time ($x = 0{,}5L$) under dynamic external pressure are presented in Figure 5. The solid lines refer to the multi-term deflection approximation results, the dashed lines to the single-term ones. Note that if for packet B (Figure 5b) the difference between the two approximations lies only in the decrease of the first maximum of the stresses, then for packet A (Figure 5a) the time till a rapid increase in the stresses is considerably greater thwn using multi-term deflection approximation. The results presented suggest that in order to obtain a reliable solution to the problem of finding an optimum layered shell structure under conditions of dynamic external pressure, it is necessary to use the multi-term deflection approximation.

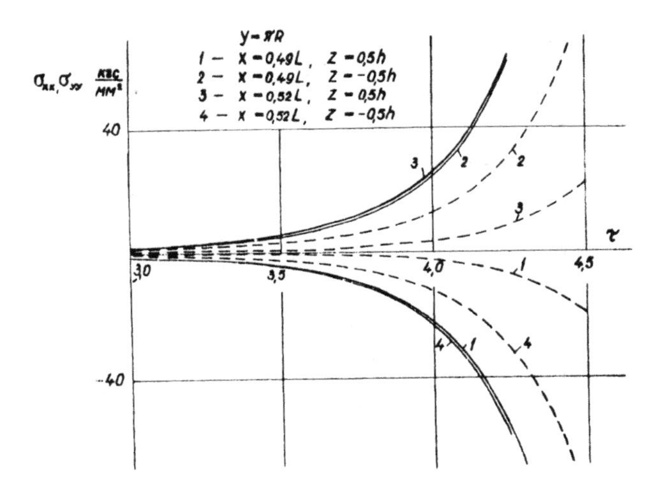

Fig. (4b) - Stress dependencies on time for packet B (axial compression)

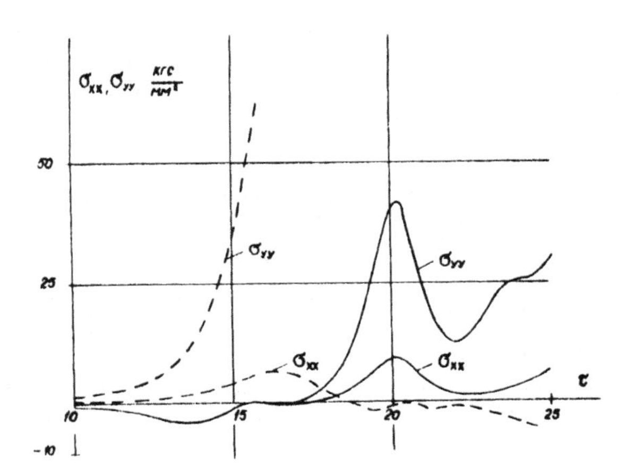

Fig. (5a) - Stress dependencies on time for packet A (external pressure)

In concluding this section, it should be noted that the method of calculation of the stress state of a shell (worked out in [7]) can be developed for the case of combined dynamic loading by both an axial force and external pressure, as well as for the case of the joint static and dynamic loading. This question has been considered in detail in [9].

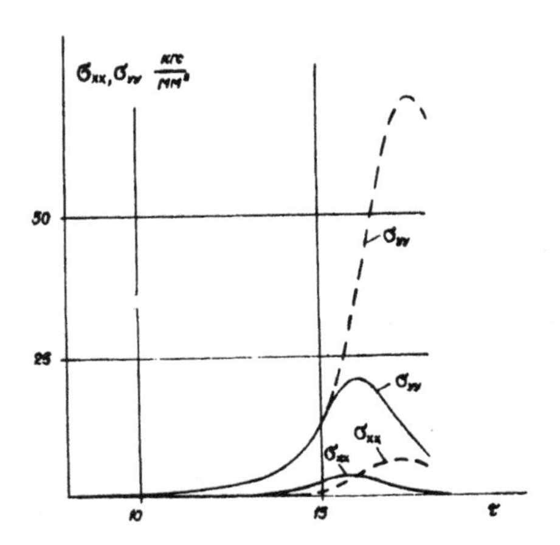

Fig. (5b) - Stress dependencies on time for packet B (external pressure)

CALCULATION OF STRESS-STRAIN STATE OF CYLINDRICAL SHELLS UNDER AXIAL VIBRATIONAL LOADS

This problem, similar to those already considered, arises when the reaction of a shell to the axial load of the type $P(t) = P_o + P_t \cos\theta t$ is investigated. For the solution of this problem, the same nonlinear equations of motion (1) - (2) as in part 1 can be used. However, in this case, the shell deformation process is of a different nature. If there are initial geometrical imperfections or transversal disturbances, the action of axial harmonic forces can cause bending resonance vibrations. They are of beating character with amplitudes several orders bigger than initial values. In the situation when the construction works in such a resonance regime the problem of calculating its longevity arises.

The first task in the calculation consists of dividing the regimes in resonance and nonresonance vibrations. It is sufficiently elaborated for various mechanical systems with distributed parameters. The relative ease of the solution is due to the possibility of using a geometrically linear approach. As in the problems of part 1, the deflection and the stress function can be presented in the form of expansions (3) - (4). Finally, it is necessary to investigate the stability of the Mathieu equation solutions for each form of vibration (m,n).

Regions of stable and unstable solutions are usually represented on the plane $\{\delta,\theta\}$, where $\delta = P_t/2(P^*-P_o)$ and P^* is the Euler critical value of an axial load. A typical initial part of the spectrum of the dynamic instability regions is presented in Figure 6*.

---

*All the results mentioned in this part are obtained for a unidirectional epoxy carbon composite shell with reinforcement along the shell axis; R/h = 200, L/R = 2.

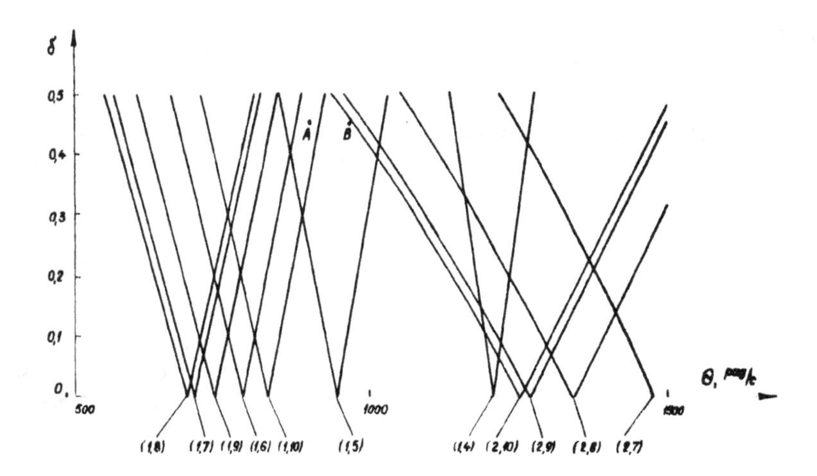

Fig. (6) - Initial part of the main dynamic instability region spectrum for the unidirectional epoxy carbon composite shell

In order to calculate the displacements and stresses in a shell vibrating
resonance regimes, it is necessary to solve a geometrically nonlinear problem.

The application of the Galjorkin method requires, as in part 1, the estab-
shment of a principle which can be applied in choosing the necessary term of
ries (3). Such a principle was suggested in [10]; it leads to deflection ap-
oximations that differ basically from all those previously used (e.g., in [11,
]). The essence lies in the following. If the given parameters of the ex-
rnal load correspond to the point of the plane $\{\delta,\theta\}$ which belongs only to one
gion of dynamic instability (say, for the vibration mode (m_1,n_1)), then the
flection approximation with the term of series (3) with numbers m_1, n_1 is quite
fficient. If at this point of a parameter plane two regions of dynamic insta-
lity (say, with numbers m_1, n_1 and m_2, n_2) are intersecting, then for the de-
ection approximation the two corresponding terms of series (3)

$$w(x,y,t) = W_{m_1n_1}(t)\sin\alpha_{m_1}x\cos\beta_{n_1}y + W_{m_2n_2}(t)\sin\alpha_{m_2}x\cos\beta_{n_2}y \qquad (14)$$

e necessary. Under vibrations in such a resonance regime, all other terms of
ries (3) will be several orders smaller than those in (14). When the external
ad corresponds to the intersection of three instability regions on a plane
$\delta,\theta\}$ the three term deflection approximation is formed in a similar manner.
en using multi-term deflection approximations to find the function $Wm_in_i(t)$,
in part 1, ordinary nonlinear differential equation systems are obtained. In
similar way, when summing all the modes taken account of for a fixed time, there
ist not only deflection dependencies on coordinates, but using formulas (10) -

stress dependencies on coordinates, too. However, the basic principle of deflection approximation formation is essentially different from the case of impulsive loading. The method described above was developed in [13] for the problem of viscoelastic shell nonlinear parametric vibrations.

Several results of the stress calculation in a shell where the load corresponds to points in one region or in the intersection of two regions of dynamic instability will be considered. The results presented in Figures 7 and 8 belon

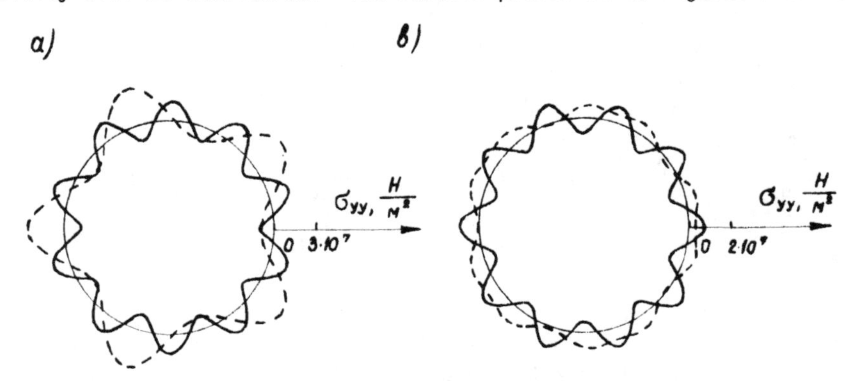

Fig. (7) - Circumferential stress dependencies on circumferential coordinate (a - outer surface, b - inner surface) at P_o = 0, P_t = 0.9P*, θ = 900 rad/s, solid line - t = 0.210s, dash line t = 0.398s

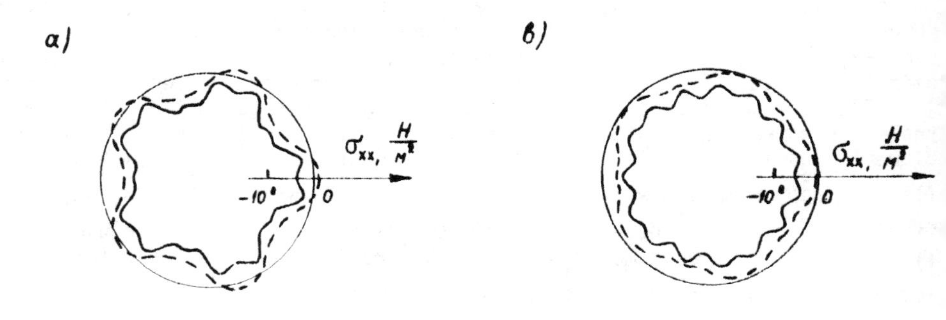

Fig. (8) - Axial stress dependencies on circumferential coordinate (a - outer surface, b - inner surface) at P_o = 0, P_t = 0.9P*, θ = 900 rad/s, solid line - t = 0.210s, dash line - t = 0.398s

to point A on the parameter plane, Figure 6, situated in the intersection of tw regions for vibration modes (1,5) and (1,10). The dependency of $\sigma_{xx}(y)$ and σ_{y} on the circumferential coordinate are presented in cross-section x = 0,5L for values of time: t = 0,21s when the function $W_{1,10}(t)$ has a maximum and at t = 0,398s when the function $W_{1,5}(t)$ has a maximum. At the first of these times (solid line) ten maximums and ten minimums exist, but at the second moment, only five maximums and five minimums are seen for each of the stresses. At the times when the values of the functions $W_{1,5}(t)$ and $W_{1,10}(t)$ are proximate, in cross section x = const, there are rather complicated stress disturbances corresponding neither to fifth nor tenth vibration modes. Thus, the most interesting features of parametric vibrations is the periodic change of both the number of node lines and their location on the shell surface when the load corresponds to the crossing of several dynamic instability regions.

From the results presented in Figure 8, we note that momentless stresses $\sigma_{xx} = -P(t)/h$ which do not depend on coordinates x and y are of great importance.

Figure 9 presents the stress dependence on the circumferential coordinate for the case when point B on the parameter plane, Figure 6, remains within the

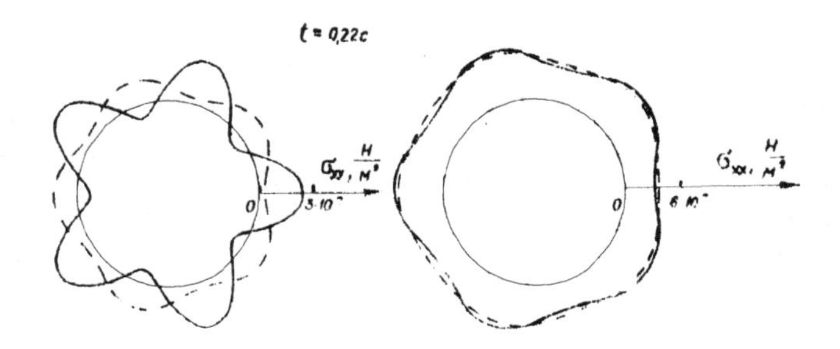

Fig. (9) - Circumferential and axial stress dependencies on circumferential coordinate at $P_0 = 0$, $P_t = 0.9\ P^*$, $\theta = 970$ rad/s, solid line - outer surface, dash line - inner surface

dynamic instability region for vibration mode (1,5), but has gone out of the region for mode (1,10); the load amplitude here is the same as in the previous case; however, the load frequency is higher.

The coordinates (both on the surface and through the thickness of a shell) at which the stresses reach maximum values, as well as the stress dependencies on time at any point on a shell can be determined using the methods of stress analysis developed for layered cylindrical shells working in different parametric vibration regimes.

THE FAILURE ANALYSIS OF MULTI-LAYERED CYLINDRICAL SHELLS UNDER DYNAMIC LOADS

Methods described above allow the calculation of the dependencies $\sigma_{xx}(x,y,z,t)$, $\sigma_{yy}(x,y,z,t)$, $\sigma_{xy}(x,y,z,t)$ at an arbitrary time t. The task now under consideration consists of determining the time $t=t^*$ and the corresponding value of dynamic load when the first ply failure of a shell takes place. Let us underline that in accordance with the formulation of the problem described in part 1 (assumptions of linear elastic continuous material) the calculation of stresses is valid till the moment t^* only.

In order to calculate t^*, it is necessary to use a strength criterion assuming an anisotropic body and a state of plane stress, checking the criterion applicability using fixed points $z=z_0$ and $t=t_0$. Such an analysis permits the determination of the coordinates on the shell surface (x_0,y_0) where the first ply failure took place, the coordinate z_0 (i.e., the layer in which the first ply failure occurred), and the time t^* at which it occurred.

For the case of impulsive loading of a shell by axial compression or external pressure this method of strength analysis was carried out in [7] by using the maximum strength criterion and in [9] by using the tensor polynomial anisotropic body strength criterion suggested by Malmeisters in [14]:

$$F(\sigma_{ij}) = P_{11}\sigma_{11} + P_{22}\sigma_{22} + P_{1111}\sigma_{11}^2 + P_{2222}\sigma_{22}^2 + 2P_{1122}\sigma_{11}\sigma_{22} + 4P_{1212}\sigma_{12}^2 = 1 \qquad (15)$$

where the stresses σ_{11}, σ_{22}, σ_{12} are related to the principal axes of a layer. The strength surface tensor components p_{11}, p_{22}, p_{1111}, p_{1111}, p_{2222}, p_{1212} can be calculated by means of the strength values obtained from the experiments at various simple ways of loading*. There is a certain difficulty in finding the value p_{1122} (pointed out, e.g., in [15-17]) because for this purpose, it is nece sary to examine the strength characteristics of an anisotropic material under bi axial loading.

To analyze the sensitivity of both a critical time t^* calculation and the location of the first ply failure to the value of p_{1122} in criteria (18) in the problems under consideration, a special investigation was carried out. Unidirectional, epoxy carbon, layered, cylindrical shells having layers related to the x direction at various angles were examined. The main results will be presented here. Under axial, dynamic compression, the value of p_{1122} only slightl influences the results of calculation of the time of the first ply failure and the first defect location in a shell for all kinds of laminate structures excep in the case of reinforcement at small angles (<5°) to the x direction. Under dy namic external pressure, the value of p_{1122} can influence the calculation resul in a small range of reinforcement angles (≈ 75° ÷ 85° to the x direction). Thi preliminary analysis, apparently, is needed in every concrete case because it lends greater reliability to the construction strength analysis results when there is a lack of reliable data of the value p_{1122} for the material.

According to the results [7,9] of the first ply failure moment calculation in a shell, the following main conclusions can be drawn. Under axial dynamic compression the first ply failure usually takes place when in the zones of the most intensive buckling the deflection is 1-1.5 times greater than the shell thickness. At the same time, even for thin shells, a marked distinction in the solutions of linear and nonlinear problems arises with deflections 2.5 times greater than the shell thickness. Thus, a reliable value of the first ply failure moment gives a solution to the problem in a geometrically linear approach.

---

*As the dynamic process of deformation and failure of a construction is investigated, the values p_{ij}, $p_{ijk\ell}$ are the deformation rate functions.

A basically different situation develops when a shell is subjected to dy-
amic external pressure. For certain packet structures, failure conditions are
ot satisfied even with deflections 4-5 times greater than the shell thickness.
onsequently, not only is solving the geometrically nonlinear problem necessary,
ut also reliable results cannot be obtained when the connectivity of circum-
erential modes of buckling is neglected (this is obvious, e.g., from the re-
ults presented in Figure 5). However, we must say that such laminate structures
naturally, unsuccessful ones) are possible where fracture under dynamic external
ressure takes place at small deflections. The results presented in Table 1 (in
ases "L" fracture started in the region of geometrical linearity, in cases "N" -

TABLE 1

o.	ϕ_1	ϕ_2	ϕ_3	ϕ_4	ϕ_5	ϕ_6	τ^*	$q(\tau^*)\cdot 10^{-6}\cdot N$	L/N	$\sigma_{11}(\tau^*)/F_{11}$
1	90°	90°	90°	90°	90°	90°	>8	> 5.53	N	
2	90°	90°	+80°	-80°	90°	90°	7.8	5.43	N	0.28
3	90°	90°	+75°	-75°	90°	90°	6.0	4.16	N	0.17
4	90°	90°	+54°	-54°	90°	90°	6.0	4.75	L	0.06
5	90°	90°	+45°	-45°	90°	90°	6.4	5.27	L	0.05
6	+80°	90°	90°	90°	90°	-80°	5.9	4.05	L	0.28

n the region of geometrical nonlinearity) illustrate what has been stated above.
he angles of reinforcement (counting layers from the inner surface of a shell)
re denoted as ϕ_i; τ^* is the nondimensional time of the first ply failure in a
hell.

As can be seen from the results presented, a shell with a circumferential
einforcement of the layers is optimal (the strength criterion was not satisfied
or any layer at $\tau \leq 8$). In other cases, the value of the critical dynamic load
s considerably lower; this decrease is especially obvious when the fracture
tarts in the region of geometrical linearity. It should be noted that the best
tiffness characteristics of those presented occur in a shell with two middle
ayers situated at angles of ±45°; this explains the relatively high value of
(τ^*) for the corresponding shell. However, from the point of strength properties,
shell with circumferential reinforcement is superior.

In the last graph of Table 1, a ratio of the stress value σ_{11} in the direction
f reinforcement for the layer that failed first (at the first ply failure moment)
o the strength limit F_{11} in the same direction is presented. For all the cases
nder consideration, this ratio is less than unity and it is still smaller in the
roblem of axial dynamic compression. Thus, the first ply failure is not con-
ected with the breakage of fibers. Such a result arouses interest in the in-
estigation of the shell deformation and the fracture process after the first ply
ailure moment.

At present, several approaches to the analysis of layered composite material
ehavior after first ply failure are known (some of them are described in [17-
9]). They were applied to very simple static problems only: the investigation
f layered plates under uniaxial tension or layered cylindrical shells under in-

ternal pressure. Certain qualitative data about the fracture process in layerec cylindrical shells subjected to the dynamic forces considered in this paper after the first ply failure can be obtained when using the model described in [19]. According to this model, if the strength criterion for some layer is satisfied, but the ratio σ_{11}/F_{11} in this layer is less than unity, it is assume that a breakage of the matrix has taken place. For this condition, it is assume that the layer is unloaded from shearing stresses and stresses normal to fibers, but keeps its load-carrying capacity in the fiber direction. In accordance wit this, new effective values of the stiffness and compliance matrices of the laye packet are determined. If the load continues to increase the second ply failur moment can be determined. If, in the same layer, the condition $\sigma_{11}/F_{11} \geq 1$ is satisfied, it is assumed that the layer has failed completely. Consequently, evaluation of the time till the complete loss of the load-carrying capacity of a shell can be obtained.

The numerical realization of the described method for the case of axial dynamic cylindrical shell compression has shown that the failure process has an avalanche-like character: the moment of the loss of total load-carrying capaci hardly exceeds the first ply failure moment. However, in the case of dynamic e ternal pressure, the load at which the total loss of the load-carrying capacity takes place can be considerably greater than the load corresponding to the firs ply failure.

Consider the results of a multi-layered cylindrical shell failure calculati for external pressure linearly increasing in time. The values of failure momen τ_i^* with the number of a layer in which the failure took place and the type of t failure (in cases "m" there was a condition $\sigma_{11}/F_{11} < 1$, in cases "f" - a condition $\sigma_{11}/F_{11} \geq 1$) for four packets are presented in Table 2. Once again, an essential difference in the first ply failure moments calculated with one-term an multi-term deflection approximations for the first packet should be noted. For other three packets, not only the first ply failure moments, but also several other failure moments coincide when calculating with one-term and multi-term deflection approximations.

From the results presented, it follows that in the problems of layered composite shell buckling, the beginning of fiber breakage in at least one layer pr cedes the failure of the matrix (or the disjunction of fibers from matrix) in almost all shell layers. This is especially evident for the last two packets: the first fiber failure takes place a considerable time period after the matrix failure in the last layer. The result of such a failure peculiarity is a considerably greater load value corresponding to the loss of the total load-carrying capacity of a shell in comparison with its value corresponding to the first ply failure.

It has to be mentioned that for a more precise evaluation of layered structures load-carrying capacity subjected to loading regimes that cause buckling, it is necessary to use models that take into account the local failure nature in each layer.

In conclusion, the strong analysis of cylindrical shells subjected to axial vibration loads will be considered briefly. Making use of the calculation meth described in part 2, it is possible at each fixed value of coordinate $z=z_0$ for

Packet Structures $(\phi_1,\phi_2,\phi_3,\phi_4,\phi_5,\phi_6)$		One-Term Approximation												Multi-Term Approximation			
90, 90, +80,-80, 90, 90	τ_i^*	6,0	6,1	6,2	6,4	6,5	6,7	6,8	7,1	7,7				7,8	7,9		
	Layer No.	4	3	6	5	1	6	2	5	4				3	4		
	Failure Type	m	m	m	m	m	f	f	f	f				m	m		
90, 90, +45,-45, 90, 90	τ_i^*	6,4	6,5	7,1	7,3	7,6	7,7	7,9	8,1	8,2	9,4			6,4	6,5	7,1	7,2
	Layer No.	4	3	6	5	6	1	5	2	2	3			4	3	6	1
	Failure Type	m	m	m	m	f	f	f	m	f	f			m	m	m	m
90,+45, -45,+45, -45, 90	τ_i^*	9,1	9,2	9,3	9,5	9,6	10,3	11,0	11,5	12,4	12,7	13,4	13,5				
	Layer No.	5	2	6	4	3	1	6	1	2	5	3	4				
	Failure Type	m	m	m	m	m	m	f	f	f	f	f	f				
+45,-45, 90, 90, +45,-45	τ_i^*	8,4	8,6	8,8	8,9	9,6	10,3	10,9	11,0	11,3	12,0	12,5	12,6				
	Layer No.	6	1	5	2	4	3	6	4	3	1	2	5				
	Failure Type	m	m	m	m	m	m	f	f	f	f	f	f				

each fixed shell surface point (x_0,y_0) to obtain dependencies $\sigma_{xx}(t)$, $\sigma_{yy}(t)$, $\sigma_{xy}(t)$. Knowing the resonance vibration frequency and the stress amplitude values $\tilde{\sigma}_{xx}$, $\tilde{\sigma}_{yy}$, $\tilde{\sigma}_{xy}$, it is possible to determine the time (or the number of cycles) till the beginning of the fatigue fracture of a construction at the given parameters θ, P_t, P_0 of the external load.

Thus, considering the results presented in Figure 7÷9, we find that the greatest values of stresses at θ = 900 rad/s are: $\sigma_{xx} = -7.5 \cdot 10^7 \frac{N}{M^2}$ (at t = 0.21s) and $\sigma_{yy} = 2.7 \cdot 10^7 \frac{N}{M^2}$ (at t = 0.398s). At frequency, θ = 970 rad/s at the moment t = 0.22s the maximum values of stresses are somewhat lower: $\sigma_{xx} = 5.6 \cdot 10^7 \frac{N}{M^2}$, $\sigma_{yy} = 2.5 \cdot 10^7 \frac{N}{M^2}$. Using a special program for finding the maximum stress values in a shell at the initial (several seconds long) time period, it was determined that these values are a bit greater than the ones given above. Typical values of the static strength limit for uniaxial compression of unidirectional epoxy carbon composites in the fiber direction exceed by an order the obtained maximum stress value in this direction. However, in the direction normal to the fibers, typical values of the static strength limit for uniaxial tension less than twice exceed the maximum stress value in this direction. Thus in the example under consideration, circumferential stresses arising the course of the parametric vibration process can lead to a rise in fatigue defects.

CONCLUSIONS

The described methods for solving some dynamic problems of layered composite cylindrical shells allow:

1. calculation of the stress-strain state at any point of a shell at an arbitrary time moment;

2. determination, using the strength criterion of anisotropic body for the case of a plane stress state, of the first ply failure moment (including fatigue failure, too), as well as the location of the first failure in a shell; this allows the value of applied impulse (or amplitudes or impulse load) or the value of vibrational load amplitude at which the first seats of destruction appear to be found;

3. investigation of the multi-layered shell fracture process after the first ply failure moment under the dynamic loads mentioned above.

REFERENCES

[1] Vol'mir, A. S., Nonlinear Dynamics of Plates and Shells, Moscow, Nauka, 432 p., 1972 (in Russian).

[2] Roth, R. S. and Klosner, I. M., "Nonlinear Response of Cylindrical Shells Subjected to Dynamic Axial Loads", AIAA Journal, 10, pp. 1788-1794, 1964.

[3] Bazhenov, V. G. and Igonicheva, E. V., "Dynamic Buckling and Post-Critical Behaviour of a Thin Cylindrical Shell with Initial Imperfections under Axial Impact Loading", Prikladnie Problemi Prochnosti i Plastichnosti, 6, pp. 98-106, 1977 (in Russian).

[4] Kadashevich, Yu. P. and Pertsev, A. K., "On the Stability of Cylindrical Shell under Dynamic Loading", Izv. Akad. Nauk SSSR, OTN, Mekhanika i Mashinostroenie, 3, pp. 30-33, 1960 (in Russian).

[5] Grigoljuk, E. I. and Srebovskiy, A. I., "Thin Cylindrical Shells under External Pressure Impulse", Inzhenernij Zhurnal MTT, 3, pp. 110-118, 1968 (in Russian).

[6] Bogdanovich, A. E., "Review of the Works on Cylindrical Shell Stability under Dynamic Loading", Elektrodinamika i Mekhanika Sploshnikh Sred, pp. 68-105, 1980 (in Russian).

[7] Bogdanovich, A. E. and Feldmane, E. G., "Load-Carrying Capacity Calculation of Composite Cylindrical Shells under Dynamic Loading", Mekhanika Kompozitnikh Materialov, 3, pp. 476-484, 1980 (in Russian).

[8] Ambartsumjan, S. A., The General Theory of Anisotropic Shells, Moscow, Nauka, 448 p., 1974 (in Russian).

[9] Bogdanovich, A. E. and Feldmane, E. G., "Deformation of Composite Shells under Combined Dynamic Loading", Abstracts of the IY USSR Conference on Mechanics of Polymer and Composite Materials, Riga, p. 14, 1980 (in Russian).

[10] Bogdanovich, A. E. and Feldmane, E. G., "On the Solving of Nonlinear Parametric Vibration Problem for Cylindrical Shells", Akad. Nauk SSSR, Mekhanika Tverdogo Tela, 1, pp. 171-177, 1979 (in Russian).

[11] Yao, I. C., "Nonlinear Elastic Buckling and Parametric Excitation of a Cylinder under Axial Loads", Trans. ASME, Ser. E, Journal Appl. Mech., 1, pp. 109-115, 1965.

[12] Vol'mir, A. S. and Ponomarjov, A. T., "Nonlinear Parametric Vibrations of Composite Material Cylindrical Shells", Mekhanika Polimerov, 3, pp. 531-539, 1973 (in Russian).

[13] Bogdanovich, A. E. and Feldmane, E. G., "Nonlinear Parametric Vibrations of Visco-Elastic Orthotropic Cylindrical Shells", Prikladnaya Mekhanika, 4, pp. 49-55, 1980 (in Russian).

[14] Malmeisters, A. K., "Geometry of Strength Theories", Mekhanika Polimerov, 4, pp. 519-534, 1966 (in Russian).

[15] Wu, E. M., "Optimal Experimental Measurements of Anisotropic Failure Tensors", Journal Composite Materials, Vol. 6, pp. 472-489, 1972.

[16] Wu, E. M., "Phenomenological Anisotropic Failure Criterion", Composite Materials, Academic Press, Vol. 2, pp. 353-431, G. P. Sendecky, ed., 1974.

[17] Tsai, S. W. and Hahn, H. T., "Failure Analysis of Composite Materials", Inelastic Behaviour of Composite Materials, AMD, Vol. 13, pp. 73-96, C. T. Herakovich, ed., 1975.

[18] Protasov, V. D., Ermoljenko, A. F., Filipenko, A. A. and Dimitrijenko, I. P "Investigation of Load-Carrying Capacity of Layered Cylindrical Shells by Means of the Fracture Process Computer Modelling", Mekhanika Kompozitnikh Materialov, 2, pp. 254-261, 1980 (in Russian).

[19] Rowlands, R. E., "Flow and Failure of Biaxially Loaded Composites", Inelastic Behaviour of Composite Materials, AMD, Vol. 13, pp. 97-125, C. T. Herakovich, ed., 1975.

SECTION VII

EXPERIMENTS AND TESTS

MAGEABILITY EVALUATION OF ORGANIC AND CARBON FIBER PLASTICS BY NONDESTRUCTIVE CHNIQUE

V. A. Latishenko and I. G. Matiss

Latvian SSR Academy of Sciences
Riga, USSR

STRACT

Organic and carbon fiber composites have a number of specific test condi-
ons which must be taken into account if the damageability of these materials
investigated by nondestructive testing. Solution of the problem led to the
velopment of a modified testing technique employing acoustic and electrical
ectrometry.

TRODUCTION

In the broadest sense, damageability of composite materials is determined
the variation in their structure due to mechanical loading, aging, moisture
sorption, technological imperfection, etc. Such structural changes can be
aluated by nondestructive testing techniques, particularly ultrasonic, thermal,
tical, acoustic emission, electrical, etc. These considerations were the sub-
ct of our report at the First Symposium of Fracture Mechanics of Composite Ma-
rials [1]. Damage studies of organic and carbon fiber composites by nondestruc-
ve testing however have a number of characteristic features which must be taken
to account and are discussed in this report.

A. Electrical Properties of Organic and Carbon Fiber Composites

From the point of view of electrical properties, organic and carbon fi-
r plastics have specific test conditions. Regenerated cellulose, polyester,
lyvinyl alcohol, aramide, polyimide as a filament and epoxy, polyester, poly-
ide as a matrix of organic composites are low loss dielectrics with slight in-
insic conductivity. These considerations provide an obvious disadvantage to
e eddy-current method for structural study of organic plastics and indicate an
significant information capability for the dissipation factor as a characteris-
c response from capacitance method.

On the contrary, the same analysis for carbon fiber composites with a
ther high intrinsic conductivity of the filament encourages an opposite conclu-
on and makes the eddy current technique very promising for structural studies
these materials. Similar considerations were expressed by other investigators
].

The dielectric constant of organic plastics filaments lies within a range of 2.5 - 3.0 relative units, whereas the same parameter of the matrix is within the limits of 3.0 - 3.8. Hence, the dielectric constant of both components of the composite - filament and matrix - adopts comparatively close value and determination of the structural imperfections, particularly the reinforcement ratio variation becomes rather problematic.

Damage accumulation caused by crack propagation, bulk porosity and othe similar imperfections of organic plastics refers to the problem of detection of air inclusions (a dielectric constant close to one relative unit) in a media with comparatively low dielectric constant. Consequently, solution of this pro lem by capacitance method deals with designing high resolution capacitance measuring devices.

In the most advantageous situation is a problem of controlling the mois ture absorption in organic fiber plastics by capacitance technique. In this case, the volume of material under examination contains inclusions with a high dielectric constant, which can be detected with high reliability. In addition, the moisture absorption detection can serve as an efficient facility for the damage evaluation due to microcrack propagation, since this process is accompan by a considerable increase of the moisture absorption ability. The quantitativ sensitivity comparison of the capacitance method with all the mentioned structural characteristics; reinforcement ratio, bulk porosity and moisture absorpti is given in Figure 1. The results were obtained from mathematical models of capacitance transducers and test structures [3]. The relationships exposed in Fi ure 1 enable to compare a sensitivity to structural characteristics of organic fiber plastics (ε_R = 2.5 + 3.0), fiber glass plastics (ε_R = 5.5 + 6.0) and othe composite structures depending on resolution of the measuring equipment and dielectric constant of the components or the entire composite. An apparatus (typ 7211) for determination of the dielectric constant with a resolution of 0.02 relative units was designed at the Institute of Polymer Mechanics. The technic data of this apparatus are as follows: dielectric constant range is within 1 - relative units; frequency - 1 MHz; compensation of surface unevenness influence up to 0.3 mm with maximal error not exceeding 5%. The frontal view of the apparatus is given in Figure 2.

An indirect study of the electrical characteristics and the interrelate structural changes of carbon fiber composites was carried out by eddy current technique. The measuring device designed for this purpose comprises a radio fr quency oscillator tuned inductively by an eddy current transducer and a frequen measuring unit. The experimental relationships between the frequency deviation as a characteristic response and a number of structural parameters (reinforceme ratio, velocity of ultrasonic oscillations and modulus of elasticity) are exhib ited in Figure 3. All the parameters are given in relative units. A character istic response of the eddy current technique is

$$\overline{F} = \frac{\Delta f}{f_o}$$

where Δf is a frequency deviation of the radio frequency oscillator and f_o is a initial value of the frequency.

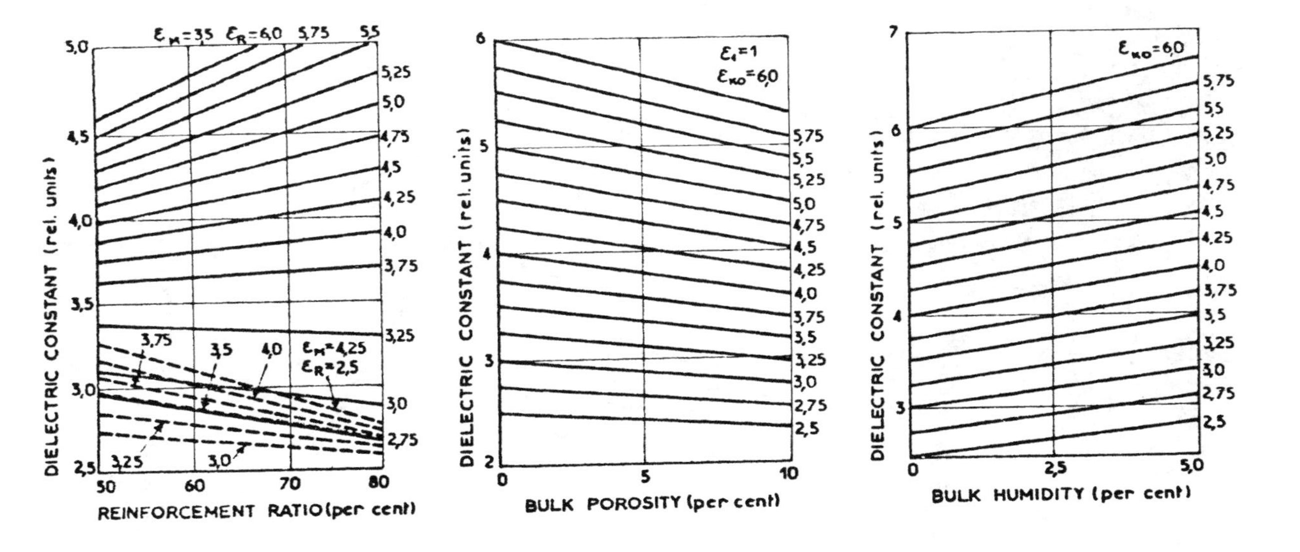

Fig. (1) - Dependence of dielectric constant on: (a) reinforcement ratio for different dielectric constant values of the filament (solid lines) and for different dielectric constant values of the matrix (dashed lines); (b) bulk porosity for different initial dielectric constant values of the composite; (c) bulk humidity for different initial dielectric constant values of the composite

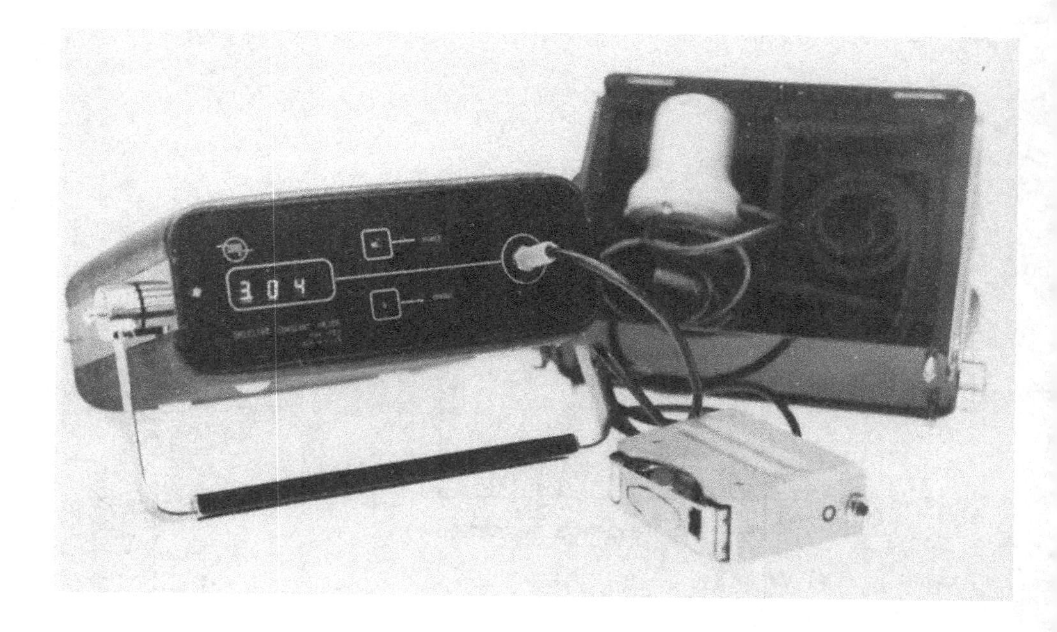

Fig. (2) - Frontal view of type 7211 apparatus for dielectric constant measurement

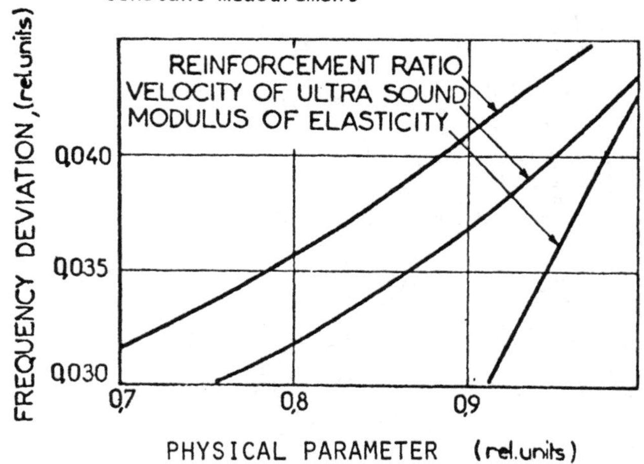

Fig. (3) - Relationships of frequency deviation from reinforcement ratio of carbon fiber plastic, velocity of ultrasonic waves and modulus of elasticity

The structural parameters of the carbon fiber plastic are

$$\overline{M} = \frac{M}{M_{max}}$$

where M_{max} is a maximal value of the parameter and M is a running value of the parameter.

B. Acoustic and Electrical Spectrometry

Damage evaluation by the ultrasonic method is also connected with some specific test conditions. The wave propagation is determined by the properties of material, while in carbon fiber composites this process is specified by a remarkable signal distortion due to great differences in elastic properties of the components. Both components of organic fiber plastics are elastic viscous materials and therefore cause significant attenuation of the ultrasonic waves. In order to take into account these measurement singularities as well as the electrical properties of organic and carbon composites, spectrometric investigations of the electrical and acoustic characteristics (frequency dependence determination) were carried out. For this purpose, a computer assisted measurement system was designed. Frontal view of the system is given in Figure 4. The technical details of this complicated measurement equipment will not be discussed here, but the basic principles involved will be presented. Frequency dependence of the

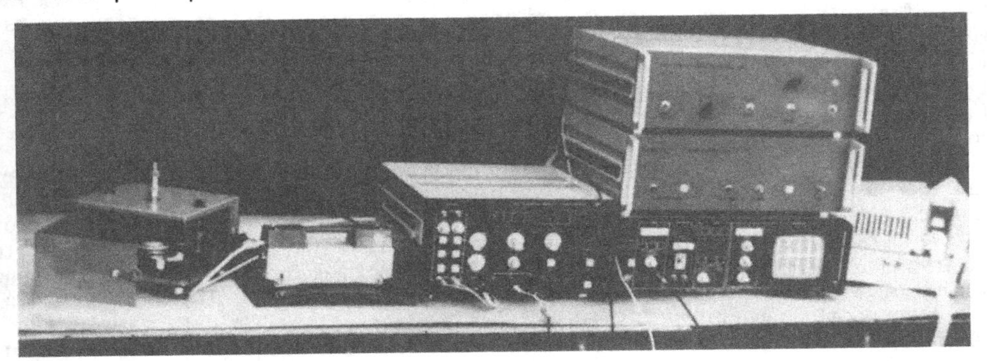

Fig. (4) - Frontal view of measurement system for spectrometric investigations

electrical characteristics is determined by recording and Fourier frequency analysis on the dielectric polarization current of the capacitance transducer in the material under surveillance. On the other hand, frequency dependence of the acoustic characteristics is evaluated by recording and Fourier frequency analysis of a distorted acoustic signal after having passed through a certain region of the test material. The main technical data of the measurement system are the following:

- minimal value of the relaxed real component of the complex dielectric constant 0.2;
- minimal value of the imaginary component of the complex dielectric constant 0.1;
- frequency range of the dielectric analysis 10^{-4} + 3Hz;
- measurement range of the phase velocity of ultrasonic waves 1, 500 - 12,000 m/sec;

- measurement range of the ultrasonic waves attenuation 0.005 - 0.6 cm^{-1}

- frequency range of the acoustic analysis 100 kHz - 2 MHz.

Some experimental outcomes of the moisture absorption studies in organic fiber composites [4] are in Figure 5.

The frequency dependence of both the relaxed real and imaginary components of the complex dielectric constant at different gravimetric humidities is shown in Figure 5. This frequency dependence (especially of the dissipation factor) provides additional information about moisture absorption in organic fiber plastics. For example, a specimen in the initial stage is characterized by one unpronounced relaxation process, whereas at even low values of humidity, a significant increase in the dielectric loss is observed (see curve 3 in Figure 5) and two clearly defined relaxation peaks are noticeable.

Consequently, spectrometry widens the range and utility of physical characteristics to be involved in damage evaluation. In addition, spectrometric studies are of interest in the study of transducers and test condition optimization of single frequency electric and acoustic measuring devices such as apparatus type 7211.

In Figure 6, the time dependence of gravimetric humidity is represented. Intrinsic conductivity and both components of the complex dielectric constant for organic fiber plastics in absolute humitity are shown. It can be seen in Figure 6 that the most intensive absorption process occurs from 130 to 350 hours and then saturation is taking place. Quantitative evaluation of this process can be performed by the measurement of intrinsic conductivity, while other informative parameters continue to change after saturation. This feature demonstrates an invaluable characteristic of the capacitance method for studying the influence of one of the most dangerous composites degradation factors. The variations in the electrical properties of organic fiber plastics after humidity saturation demonstrates the existence of structural changes and consequently damage accumulation independent of the equilibrium moisture content of the material. This stage of moisture absorption cannot be investigated by other conventional techniques.

C. Thermal Method

Finally, there is one more method - a thermal method which looks promising for the damage evaluation. Nondestructive testing of a number of parameters indirectly characterizing the strength of material illustrates experimental results obtained by an apparatus from thermal activity measurement [5]. In these investigations, the thermal activity as a characteristic response was used. A relationship between the thermal activity and the density of organic fiber plastics was established, Figure 7. Analogous results were acquired for a carbon fiber composite. In addition, the dependence of thermal activity, density and tensile strength on the molding pressure were plotted. In Figure 8, the relationships are given in relative units:

$$\overline{A} = \frac{\Delta A}{A}; \ \overline{\gamma} = \frac{\Delta \gamma}{\gamma}; \ \overline{\sigma} = \frac{\Delta \sigma_p}{\sigma}$$

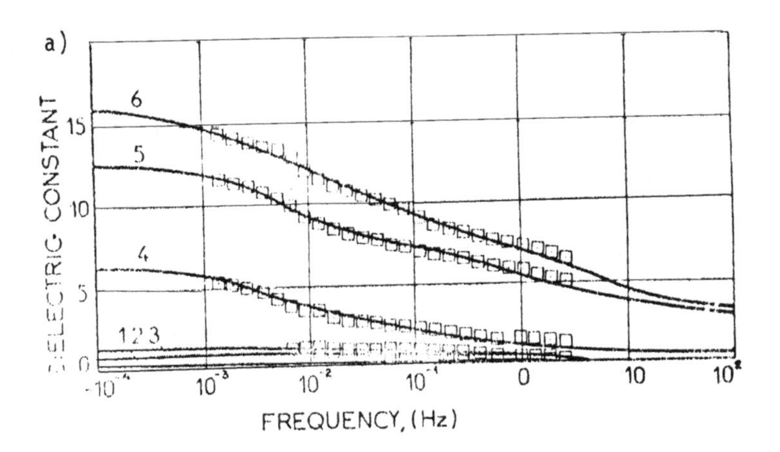

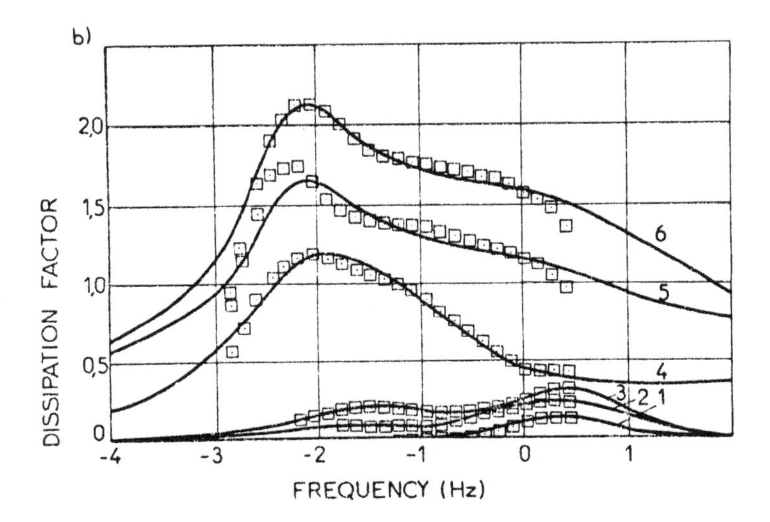

Fig. (5) - Frequency dependence of dielectric constant (a) and dissipation factor (b) for different gravimetric humidities of organic plastic: 1 - 0%; 2 - 0.5%; 3 - 1.8%; 4 - 5.2%; 5 - 5.35% and 6 - 5.4%

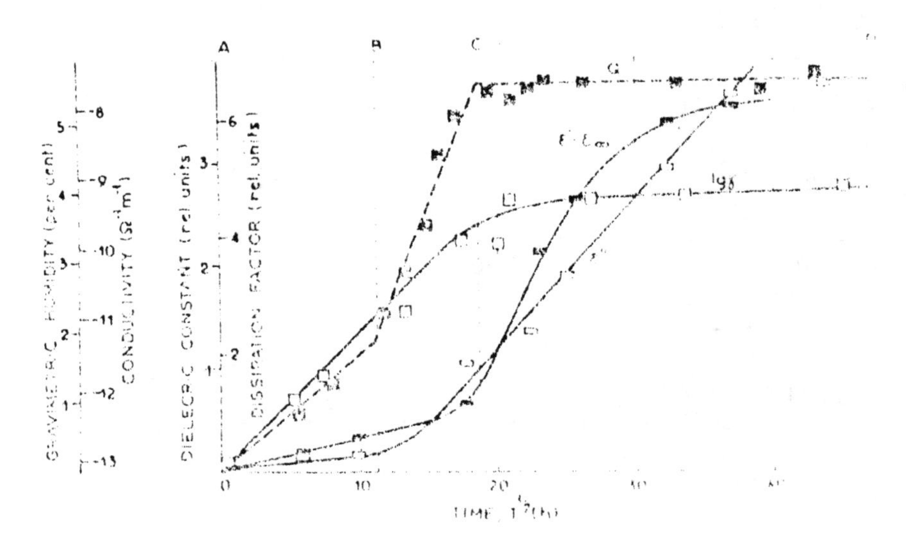

Fig. (6) - Time dependence of intrinsic conductivity, gravimetric humidity and components of complex dielectric constant for organic plastic in absolute humidity

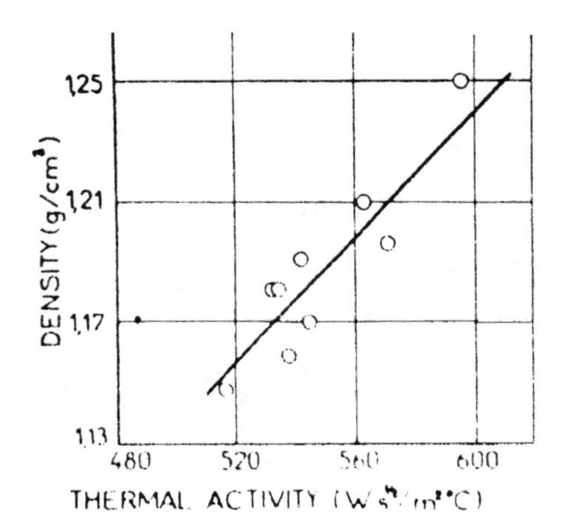

Fig. (7) - Relationship of thermal activity and density of organic plastic

where A, γ and σ are initial values of the thermal activity, density and tensi strength; ΔA, $\Delta\gamma$, $\Delta\sigma$ are the running values of the same parameters.

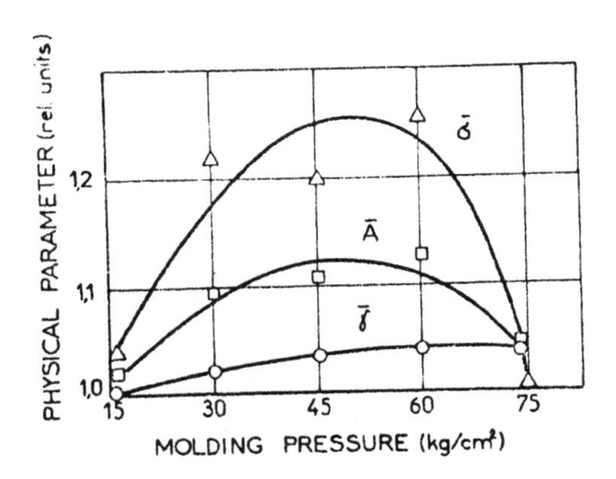

Fig. (8) - Dependence of thermal activity, density and tensile strength on molding pressure

CONCLUSION

The considered examples of damage evaluation by nondestructive testing illustrate the growing potential of this technique. For the study of the specific test conditions required for organic and carbon fiber composites, modified testing techniques involving eddy-currents, acoustic and electrical spectrometry are preferred. Further work is needed to investigate the informative capabilities of nondestructive testing for different kinds of damage.

REFERENCES

[1] Latishenko, V. A. and Matiss, I. G., "Methods and Means of Damageability of Composite Materials", Fracture of Composite Materials, Sijthoff and Noordhoff, Alphen aan den Rijn, The Netherlands, 1979.

[2] Owston, C. N., "Eddy Current Methods for the Examination of Carbon Fibre Reinforced Epoxy Resins", Materials Evaluation, No. 11, pp. 237-244, 1976.

[3] Matiss, I. G., Yemkostnije Preobrazovateli Dlja Nerazrushyuchego Kontrolja, Izdatelstvo "Zinatne", Riga, 255 p., 1977.

[4] Pone, D. A. and Shtrauss, V. D., Issledovanije Dielektricheskih Svoistv v Diapazone Infranizkih Chastot Nekotorih Polimerov v Processe Vodopogloschenija. Tezisi Dokladov 1-oi Konf. Molodih Specialistov po Mehanike Polimerov, "Zinatne", pp. 102-104, 1977.

[5] Zinchenko, V. F. and Negrejeva, S. N., Metod Teplovovo Kontrolja Pokazatelei Strukturi i Svoistv Kompozitnih Materialov, "Promishlennaja Teplotehnika", No. 1, 1981.

MECHANICAL LUMINESCENCE STUDY OF COMPOSITE FRACTURE IN A PLANE-STRESSED STATE

G. A. Teters, U. E. Krauja, R. B. Rikards and Z. T. Upitis

Academy of Sciences of the Latvian SSR
Riga, USSR

The mechanical luminescence method (photoemission) is a comparatively new method of studying fracture processes in polymeric materials. The first publications along these lines appeared in the early sixties [1-3]. In those early works, a relationship between the photoemission and the destruction processes was found. However, a serious study of fracture developments in polymer-based composites was undertaken quite recently [4,5].

The possibilities of employing the mechanoluminescence method for studying structures made of composite materials are of practical interest. In this respect, it is important to detect the initial stages of material destruction under complex states of stress. The problem has been most correctly solved in the testing of tubular samples; therefore, in the present work, attention has been focussed on the regularities of the luminescence excitation in helically wound (fiber orientation at an angle ±45°) and cross-ply glass fiber reinforced tubular samples. In the interest of understanding specific features of the fracture process the experimental data from rectangular flat samples of unidirectionally reinforced GFRP and OFRP, obtained by the mechanical luminescence method, have been presented in the work. Samples were tested in uniaxial tension.

"Blade"-type samples were considered to be flat. Their cross sectional dimensions over the working zone were 5 x 0.5 mm. The length of samples was 130 mm. In total, 25 such samples were tested. Geometrical shapes of tubular samples are shown in Figure 1. The working zone of cross-ply tubular samples consisted of three layers. The layer orientation was as follows: the first layer was oriented along the axis 2, the second layer along axis 1, the third layer along axis 2. The wall thickness of the sample was 1.1 mm. Helically would tubular samples were reinforced as follows: the first layer at an angle +45° to the longitudinal material axis 1, the second at an angle -45°. The layers were periodically repeated. The wall thickness of the samples was 1.4 mm. On the whole, 41 cross-ply tubes and 13 helically wound tubes were tested.

Tubular samples were tested under tension, compression, internal pressure and under their various combinations. Loading was applied in the testing machine by computer control. Stresses σ_{11}, σ_{22} and σ_{12} were calculated according to the formulae:

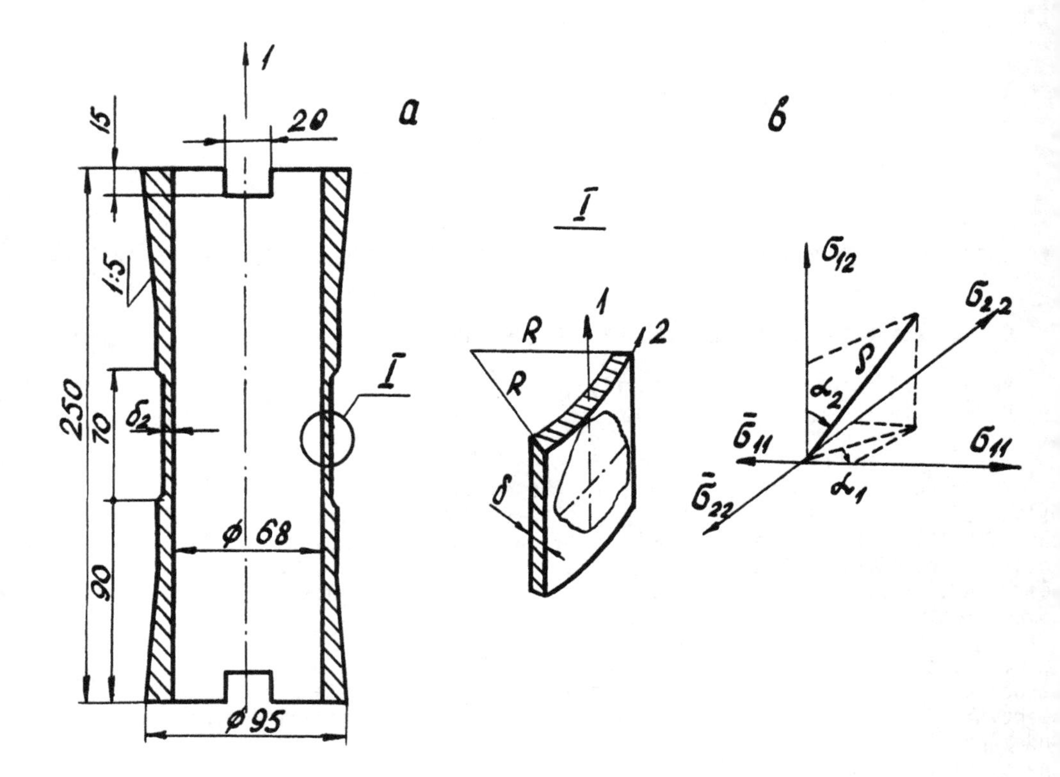

Fig. (1) - Schematic representation of a tubular sample made of glass fiber reinforced plastic ; (a) and plane stress state in spherical coordinates (b)

$$\sigma_{11} = \frac{N}{F} + \frac{p \cdot D_a}{4\delta}; \quad \sigma_{22} = \frac{p \cdot D_a}{2\delta}; \quad \sigma_{12} = \frac{M}{W_p}$$

where N is the axial load; p is the pressure of a liquid; M is the torque, W_p is the polar resistance moment; F is the cross-sectional area of the tube; D_a is the average diameter of the tube; δ is the wall thickness of the tubular sample.

Apart from assessing the regularities of the luminescence excitation and strength characteristics, the elastic behavior of tubular samples was also studied. The modulus of elasticity E_{11} and Poisson's ratio μ_{12} were determined in tension along the axis 1, and E_{22} and μ_{21} were determined under internal pressure. It should be noted that the necessary data were recorded over the linear portion of the relations σ_{11} (ε_{11}), σ_{22} (ε_{22}), σ_{12} (γ_{12}). Components of the stress tensor coincide with the material axis 1, σ_{22} - 2 (along the circumferen - see Figure 1.

The intensity of the mechanoluminescence was measured by a photon-counting chnique. A block diagram of the apparatus for recording sample photoemission given in Figure 2. Samples were placed in a light-opaque chamber. The hous-

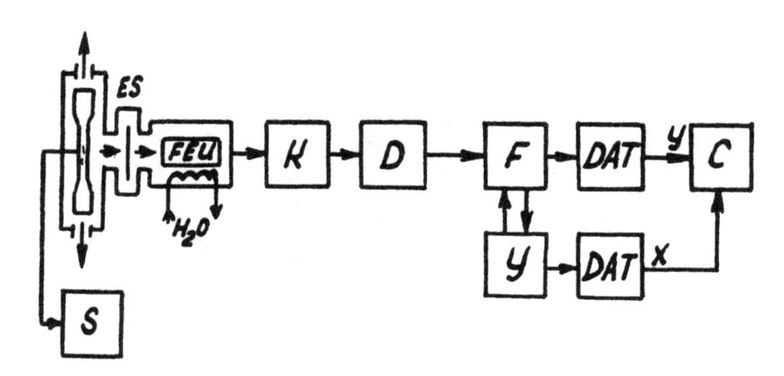

Fig. (2) - A block diagram of apparatus for recording sample photoemission: 0 is the sample; ESis the electromechanical shield; FEU is the photoelectronic multiplier of the type FEU-79, K is the pulse amplifier, F are frequency meters of the type C3-38, y is a control block, DAT are digital-analog transformers, C is the self-recorder PDS-021, H_2O is the cooling system for the housing incorporating FEU, S is the strain recorder

g of the photon detector FEU was placed as near as possible to the sample, and s cooled in running water at the temperature of +15°C, the dark background was ± 0,4 pulses/sec. Between the chamber and the housing, an electromechanical ield,ES, was placed. A selected unit of the photoelectronic multiplier FEU- served as a photon-receiver. The output pulses FEU were intensified by a lse amplifier K. Isolation and specification of the amplified photoelectronic lses were accomplished by means of an integral discriminator D . Pulses were unted by an electronic frequency counter of the type C3-38. Control block y gistered the counting cycles and at the end of each cycle, the frequency coun- r was put into operation. Pulse integration took place in a second. The data re recorded by a two-coordinate self-recorder PDS-021 through data transforma- on from digital into analog form by two digital-analog transformers DAT.

The photoemission was measured in arbitrary units of light emission (a.u.ℓ.e). nce the spectral distribution of light was not studied, it was impossible to ve precise data on the actual number of photons.

Consider the results obtained on flat samples. For unidirectionally rein- rced GFRP samples in tension along the fiber direction, the onset of the photo- ission was detected at (58,75% ± 5,4%) of the destructive stress (σ^*_{11}). The agram of the change in the luminescence intensity with the stress σ_{11} has two ctions, Figure 3a. The first section is characterized by low intensity of the otoemission (5-20 arbitrary units of light emission per second) the second by werful flash-ups of the photoemission. It has been established that the first ash-up, of total power approximately 1000 arbitrary units, was accompanied by mple damage in the form of a longitudinal crack along the reinforcing fibers.

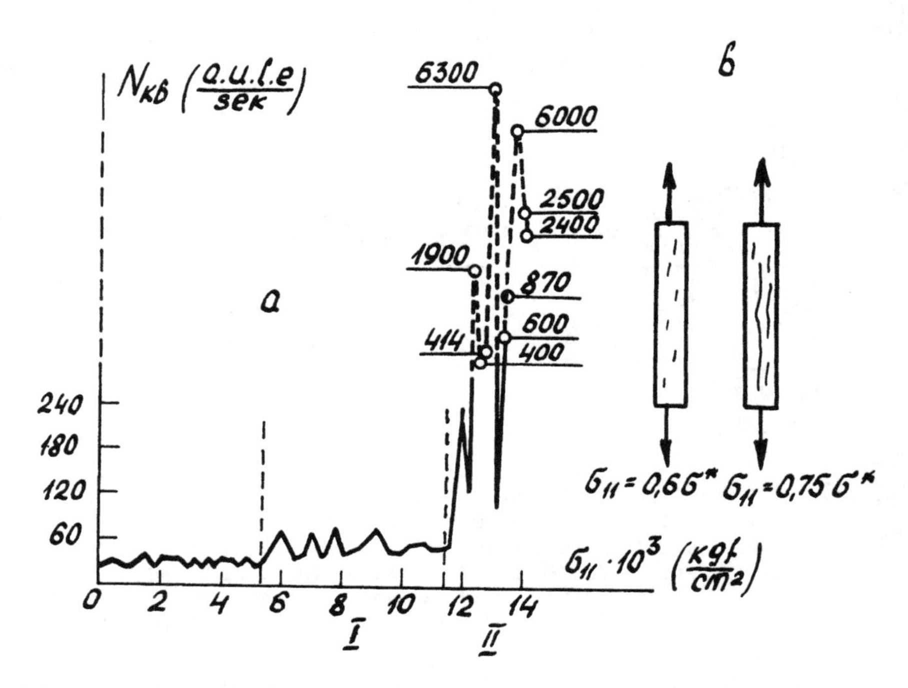

Fig. (3) - The photoemission intensity versus the stress σ_{11} for unidirectional reinforced fiberglass. I, II are the stages of destruction; b is th scheme for visual observation of the photoemission

It is of interest to note that the photoemission in the unidirectionally re inforced GFRP may very well be observed visually after the eye has been accustomed to darkness, Figure 3b. Thus, at $\sigma_{11} \sim 0,6\ \sigma^*_{11}$ (where σ^*_{11} is the destruc tive stress) a few light flashes were observed in different sites of the sample at $\sigma_{11} \sim 0,75\ \sigma^*_{11}$, long lines of light emission parallel to the fibers over the entire working zone of the samplê were observed. It must be noted that the lig emission was observed visually only for this type of material.

It is clear that the experimental data at the onset of the photoemission (mechanoluminescence) confirm the fracture model of the material proposed elsewhere [6,7]. The above experiments demonstrate that at σ_{11} = (58,75% ± 5,4%)σ^*_1 the destruction process of the material starts, accompanied by a low level of the photoemission intensity, stage 1. The first flashes of emitted photons, of total power approximately 1000 arbitrary units of light emission, testify to th crack formation in some localized sample sites. On further loading of the samp the fracture at the interface along the fibers - fiber pullout from the matrix takes place. It is important that the first stage of fracture may be foreseen from the values of photoemission intensity when the fiber breakage is not likel to occur as yet. This assumption allows the establishment of the allowable ran of loads for the material.

The study of mechanical behavior of GFRP components yields additional information on the destruction process of composites. Upon loading of an epoxy binder ED-10, the photoemission starts at σ_{11} = 0,35 of the destructive stress for the binder and reaches 1485 ± 366 arbitrary units, Figure 4a. Upon loading glass fi-

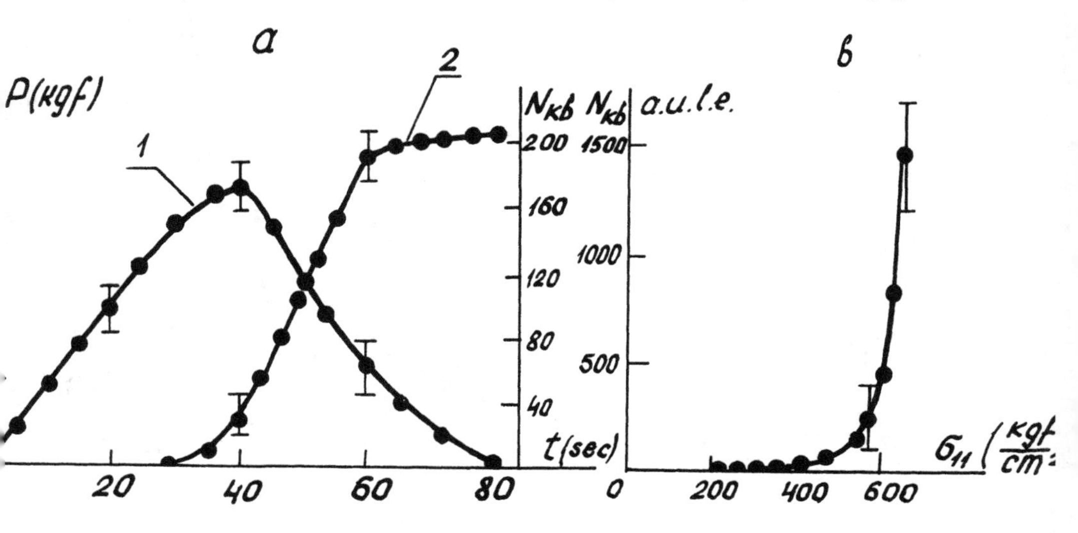

Fig. (4) - Change in the applied load P (1) and the total amount of the photoemission N_{kb} (2) with time for glass fibers; (a) change in the total amount of the photoemission with the stress σ_{11} for epoxy binder ED-10 (b)

bers, the onset of the photoemission was registered at the instant the load had reached its maximal magnitude, see Figure 4b. The mechanoluminescence in GFRP components is characterized by a gradual accumulation of a small number of emitted photons (236 ± 32 arbitrary units).

It is difficult to explain the photoemission excitation in unidirectional GFRP in terms of the character of mechanoluminescence in its components. The total amount of emitted photons in GFRP components during the whole deformation process did not exceed 2000 arbitrary units, while only one flash-up of the photoemission in GFRP reaches 1000-6000 arbitrary units (the overall photoemission in the unidirectionally reinforced GFRP reaches 21618 ± 2002 arbitrary units). Evidently, the photoemission in a composite is associated not only with the rupture in the binder and fibers, but also with the degradation of the interface. As was previously demonstrated, during this deformation process, a long crack parallel to the fibers develops after the first flash-up of the photoemission. This suggests that flash-ups of 1000 arbitrary units and more occur at the instant of breaking of long portions of glass fibers away from the matrix, after the fiber breakage has already taken place.

In stretching the unidirectional OFRP along the fibers, the onset of photoemission, Figure 5, curve 1, was detected much earlier (σ_{11} = 0,36 σ^*_{11}) than in

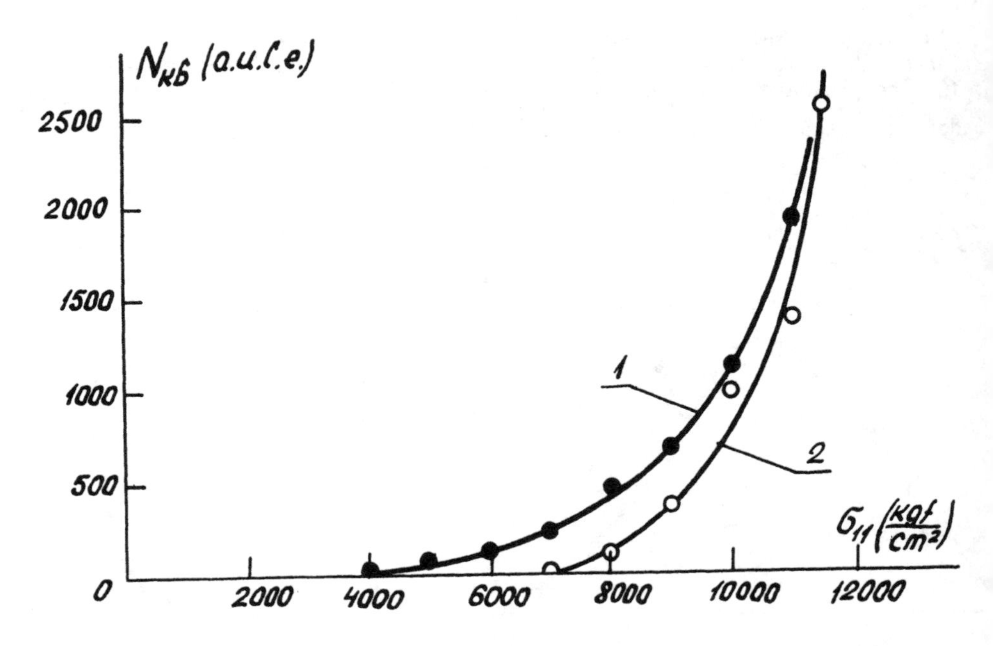

Fig. (5) - Dependence of the amount of arbitrary units of light emission N_{kb} on tensile stresses σ_{11} for unidirectionally reinforced organic plastics (curve 1), the curve 2 for the same material having loading prehistory

GFRP of the same reinforcement structure, see Figure 3. Under these conditions, the photoemission in OFRP is accumulated gradually, without visible jumps, and the total power of light emission is an order of magnitude less than in the unidirectional GFRP. Mechanoluminescence in GFRP and OFRP develops differently. The fact is evidently associated with different fracture mechanisms. In the testing of OFRP, having load prehistory, the photoemission starts later and the accumulation of arbitrary units of light emission proceeds according to another rule, see Figure 5, curve 2.

The magnitudes of the stress σ^{**}_{ij} at which the photoemission starts and the destructive stresses σ^{*}_{ij} in a complex state of stress obtained in the testing of cross-ply tubular samples are presented in Table 1. The fracture behavior has revealed, Figures 6 and 7, that under all kinds of loading the loss of stability of cylindrical tubular samples had not occurred. Various light emissions in the initial and final stages of destruction in a three-dimensional space of a plane state of stress are shown in Figures 8-13. In the figures, the ultimate state in failure is marked by a broken line; the stressed state, in which the onset of the photoemission was detected - by a chain-dotted line; the theoretical prediction of the initial stage of material destruction, obtained by the method of finite elements - by a solid line. As seen in Figure 8, the inner surface (2) lies nonsymmetrically relative to the surface 1 of the ultimate state of the material. Under tension and internal pressure, early stages of the state of stress have been detected, in which the onset of mechanoluminescence commences (0.12 σ^{*}_{11} and

TABLE 1 - MAGNITUDES OF STRESSES σ_{ij}^{**}, AT WHICH THE PHOTOEMISSION COMMENCES, AND THE DESTRUCTIVE STRESSES σ_{ij}^{*} FOR CROSS-PLY GFRP TUBES IN A COMPLEX STATE OF STRESS

No.	σ_{11}^{**} $\frac{kgf}{cm^2}$	σ_{22}^{**} $\frac{kgf}{cm^2}$	σ_{12}^{**} $\frac{kgf}{cm^2}$	σ_{11}^{*} $\frac{kgf}{cm^2}$	σ_{22}^{*} $\frac{kgf}{cm^2}$	σ_{12}^{*} $\frac{kgf}{cm^2}$	α_1 deg.	α_2 deg.
1	647	-	-	5217	-	-	0	90
2	662	-	-	-	-	-	0	90
3	486	250	-	3758	2145	-	29	90
4	647	375	-	4851	2794	-	29	90
5	577	995	-	3312	6623	-	63	90
6	632	1085	-	3785	7570	-	63	90
7	-		-	1285	7038	-	80	90
8	-	1537	-	-	7584	-	90	90
9	-	1708	-	-	-	-	90	90
10	- 633	1565	-	-	-	-	112	90
11	- 659	1642	-	-	-	-	112	90
12	-1520	1520	-	-1712	1712	-	135	90
13	-2027	2027	-	-2096	2096	-	135	90
14	-2019	821	-	-2019	821	-	158	90
15	-1687	-	-	-1763	-	-	180	90
16	-1795	-	-	-1816	-	-	180	90
17	-	-	-	-1916	-	-	180	90
18	-	-2350	-	-2350	-	-	270	90
19	-	-2474	-	-2680	-	-	270	90
20	533	- 631	-	1521	-1642	-	313	90
21	-	-	-	1381	-1540	-	313	90
22	466	-	52	-	-	-	0	83
23	537	-	62	-	-	-	0	83
24	344	-	344	477	-	477	0	45
25	352	-	352	548	-	548	0	45
26	-	-	379	-	-	504	0	0
27	-	-	422	-	-	631	0	0
28	- 324	-	324	- 553	-	553	180	45
29	- 426	-	-426	- 580	-	580	180	45
30	-1257	-	260	-1384	-	290	180	78
31	-1353	-	282	-1401	-	290	180	78
32	-	1007	324	-	1399	434	90	72
33	-	1027	330	-	2190	685	90	72
34	-	219	304	-	236	323	90	36
35	-	295	395	-	372	502	90	36
36	331	331	276	501	501	412	45	60
37	368	368	307	707	707	586	45	60
38	163	163	409	205	205	503	45	30
39	- 324	324	450	- 417	417	566	135	46
40	- 343	343	483	- 369	369	516	135	46
41	- 721	721	370	- 860	860	440	135	70

Elastic characteristics of the material: $E_{11} = 2,2.10^5 \pm 4,1.10^4$ (kgf/cm²), $\mu_{12} = 0,148 \pm 0,032$, $E_{22} = 4.15.10^5 \pm 1,8.10^4$ (kgf/cm²), $G_{12} = 0,305.10^5 \pm 0,53.10^4$ (kgf/cm²), angles α_1 and α_2. (See in Figure 1(b).

Fig. (6) - Failure modes of tubular cross-ply GFRP samples in tension along the axis 1 - No. 1, under internal pressure - No. 2, in compression along the axis 1 - No. 3, in torsion - No. 4

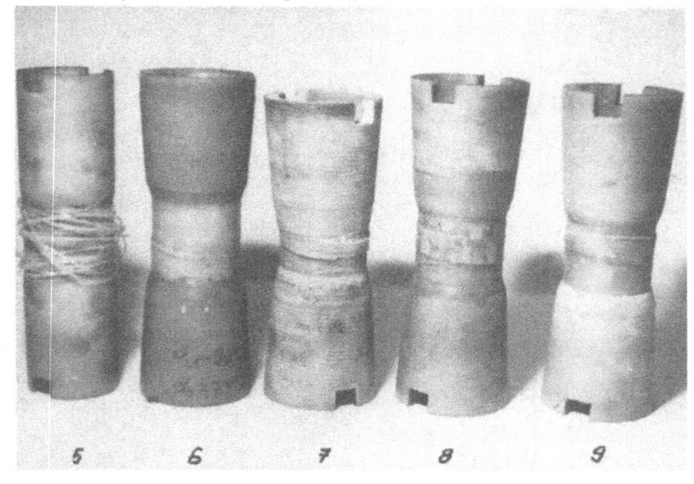

Fig. (7) - Failure modes of cross-ply tubular GFRP samples under the combined effect of tension and internal pressure, No. 5, torsion and internal pressure, No. 6, torsion and compression, No. 7, torsion, tension and internal pressure, No. 8, torsion, compression and internal pressure, No. 9

0,21 σ^*_{11}, respectively). This may be attributed to the fact that the layers perpendicular to the loading direction fail first. However, the photoemission in compression was observed only in the final stage of deformation (0,95 σ^*_{11} and 0,96 σ^*_{11}, respectively). This demonstrates the existence of different fracture

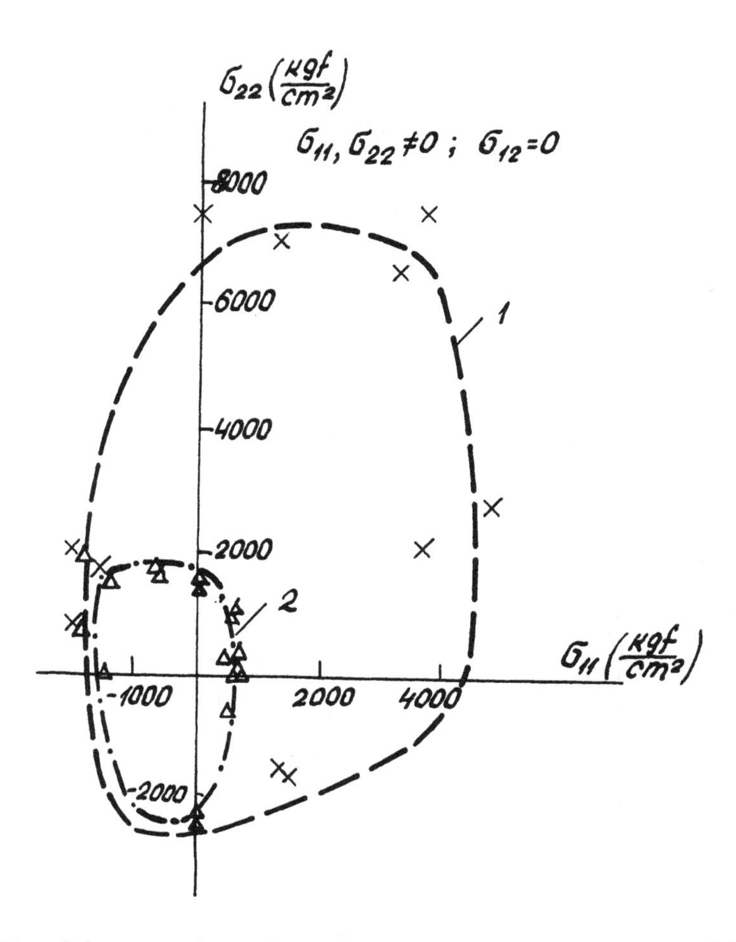

Fig. (8) - Experimental surfaces of destructive stresses (1) and stresses, at which the photoemission (2) commences at $\sigma_{12} = 0$

chanisms in tension and compression. In pure shear (see Figure 10), the me-anoluminescence starts at 0,705 σ^{*}_{12}.

Magnitudes of stresses σ^{**}_{ij}, at which the photoemission starts, and the de-ructive stresses σ^{*}_{ij} in a complex stressed state in the testing of helically und tubular samples of the reinforcement orientation at an angle $\phi \pm 45°$, are esented in Table 2. The failure modes have revealed that upon loading, loss stability did not take place. In Figure 14, the magnitudes of destructive resses (x) and the state of stress, corresponding to the onset of the photo-ission (Δ) at $\sigma_{12} = 0$ are presented.

In both cases, Malmeisters' strength criterion [8] has been employed in the proximation. The values of the tensor components of the strength surface were

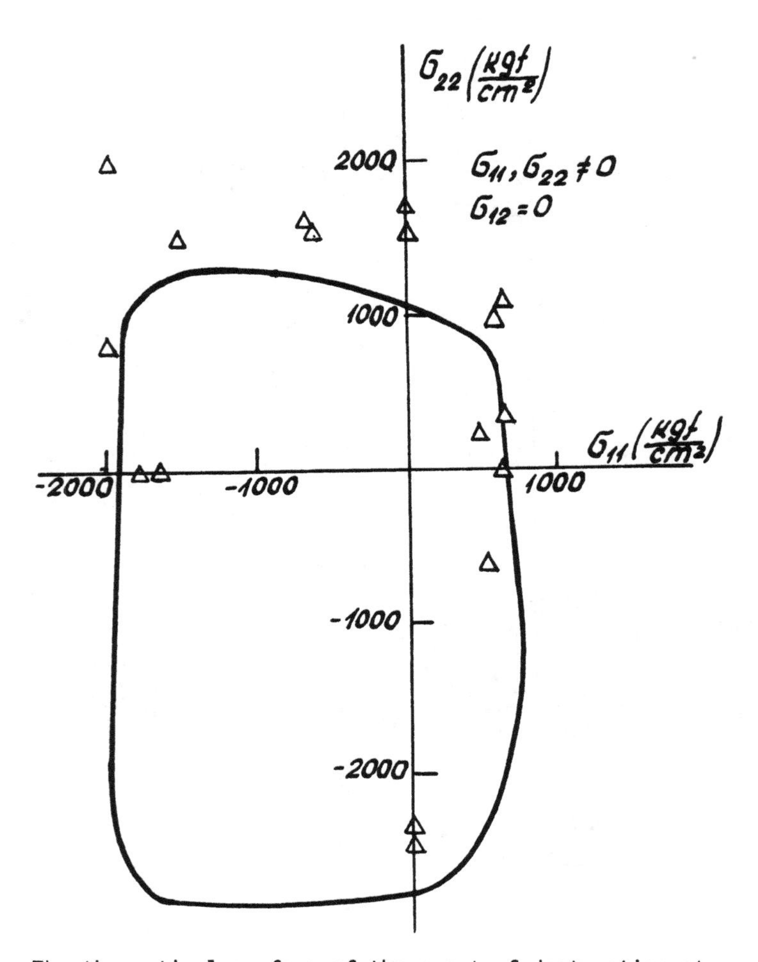

Fig. (9) - The theoretical surface of the onset of destruction at $\sigma_{12} = 0$

calculated by the method of least squares and the relative error is given in Table 3.

It is of interest to note that under the combined effect of tensile force and internal pressure (see Figure 12), the composite material sustains higher loads. Thus, in uniaxial tension GFRP sustain $\sigma_{11}^{*} = 1128$ kgf/cm^2, under internal pressure - $\sigma_{22}^{*} = 1572$ kgf/cm^2, however, under the combined effect $\sigma_{11}^{*} = \sigma_{22}^{*}$ the magnitude of stresses reaches $\sigma_{11}^{*} = \sigma_{22}^{*} = 4546$ kgf/cm^2, i.e., the composite sustains the load 3,37-fold higher. It follows from above that GFRP reinforced at an angle ±45+ are applicable in thin-walled structures, subjected to biaxial tension of the same magnitude.

An ellipsoid, characterizing the state of stress in helically wound reinforced composite when the onset of the mechanoluminescence commences (see Figure 14, curve 2) in contrast to the initial fracture surface of cross-ply GFRP

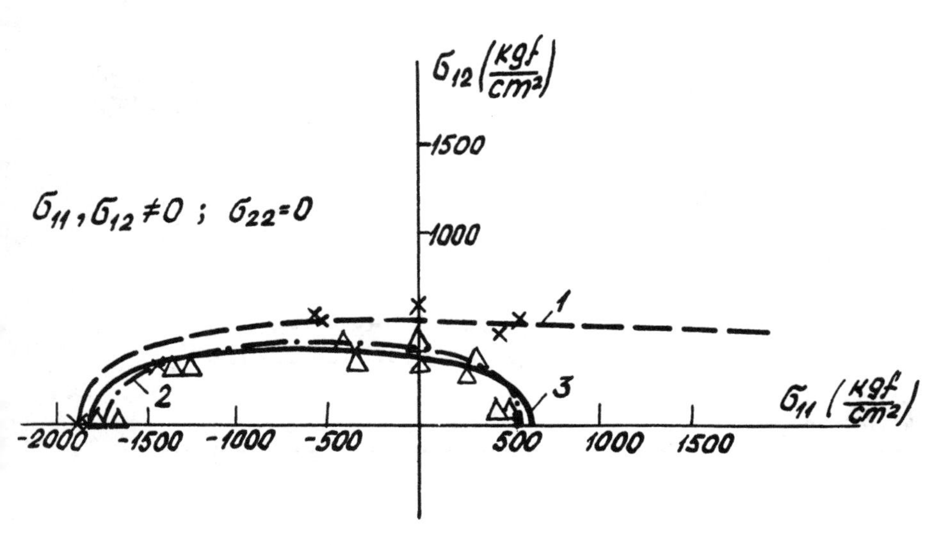

Fig. (10) - Experimental surfaces of the destructive stresses (1), the stresses at which the photoemission (2) commences, the theoretical curve of the onset of destruction (3) at $\sigma_{22} = 0$

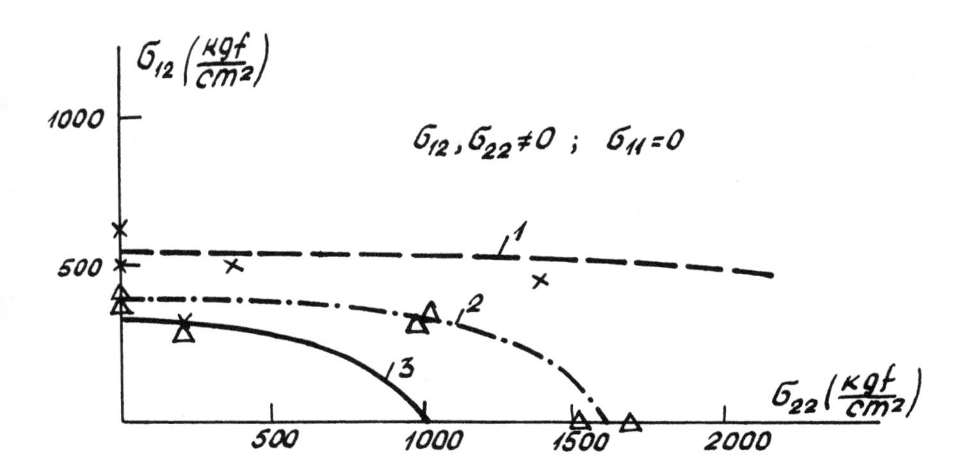

Fig. (11) - Experimental surfaces of the destructive stresses (1), the stresses at which the photoemission (2) commences, the theoretical curve of the onset of destruction (3) at $\sigma_{11} = 0$

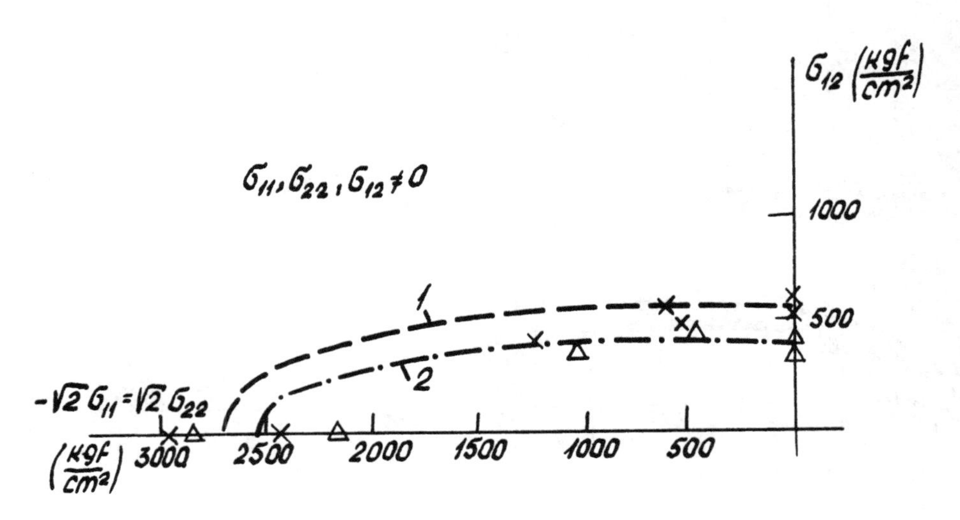

Fig. (12) - Experimental surfaces of the destructive stresses (1) and stresses, at which the onset of the photo-emission (2) at $-\sqrt{2}\sigma_{11} = \sqrt{2}\sigma_{22}$, $\sigma_{12} \neq 0$ starts

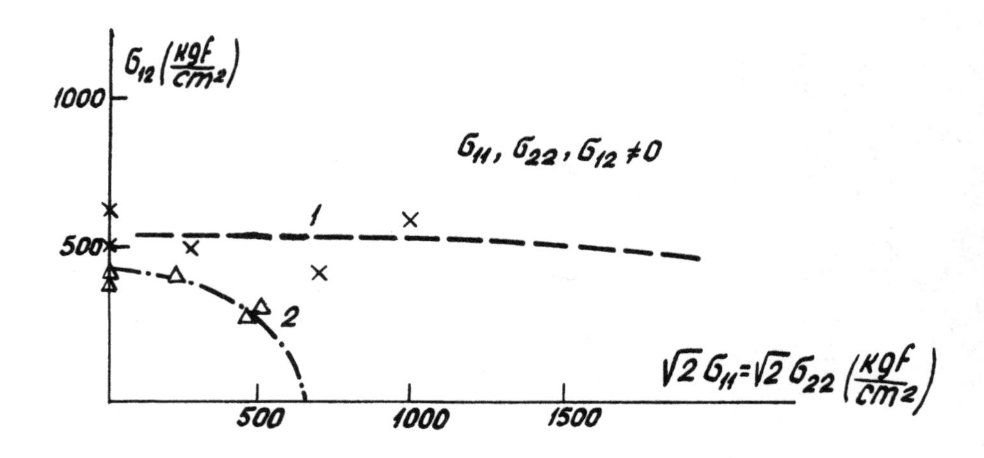

Fig. (13) - Experimental surfaces of the destructive stresses (1) and stresses, at which the photoemission (2) starts at $\sqrt{2}\sigma_{11} = \sqrt{2}\sigma_{22}$, $\sigma_{12} \neq 0$

TABLE 2 - MAGNITUDES OF STRESSES σ^{**}, AT WHICH THE PHOTOEMISSION COMMENCES, AND THE DESTRUCTIVE STRESS σ^{*}_{ij} OF HELICALLY WOUND GFRP TUBES IN A COMPLEX STATE OF STRESS, ϕ = ±45° (σ_{12} = 0)

o.	σ^{**}_{11} (kgf/cm²)	σ^{**}_{22} (kgf/cm²)	σ^{*}_{11} (kgf/cm²)	σ^{*}_{22} (kgf/cm²)	α_1 (deg.)	α_2 (deg.)
1	595	-	1204	-	0	90
2	559	-	1052	-	0	90
3	1350	887	4284	2877	33	90
4	1691	1691	4546	4546	45	90
5	556	948	1756	3006	60	90
6	617	1048	2670	4615	60	90
7	-	606	-	1598	90	90
8	-	502	-	1546	90	90
9	- 315	264	- 607	547	137	90
0	- 427	373	- 956	873	137	90
1	- 771	-	-1383	-	180	90
2	- 776	-	-1316	-	180	90
3	- 586	-	-1141	-	180	90

TABLE 3 - COMPONENTS OF STRENGTH SURFACE TENSOR FOR GFRP (ϕ ±45°)

inal and nitial tages of racture Fig. 14)	Components of Strength Surface Tensor						$\Omega(\vec{x})$
	$p_{11}\cdot 10^3$ $\frac{cm^2}{kgf}$	$p_{22}\cdot 10^3$ $\frac{cm^2}{kgf}$	$p_{1111}\cdot 10^6$ $\frac{cm^4}{kgf}$	$p_{2222}\cdot 10^6$ $\frac{cm^4}{kgf}$	$2p_{1122}\cdot 10^6$ $\frac{cm^4}{kgf}$	$4p_{1212}\cdot 10^6$ $\frac{cm^4}{kgf}$	%
1	-0,0116	-0,164	0,631	0.54	+0.54	+0,55	16,4
2	-0,0099	-0.25	2,22	2,96	+2,96	+2,25	10,6

see Figures 8 and 9), is located symmetrically to the ultimate strength surface. hus, under internal pressure, the light emission starts, on the average, at 0,35 $^{*}_{22}$. In stretching and compression along the axis 1, the mechanoluminescence was etected at 0,51 σ^{*}_{11} and 0,55 $\bar{\sigma}^{*}_{11}$, respectively. This confirms the assumption hat the fracture mechanisms causing the photoemission in tension and compression f the composite reinforced at an angle ±45° are similar.

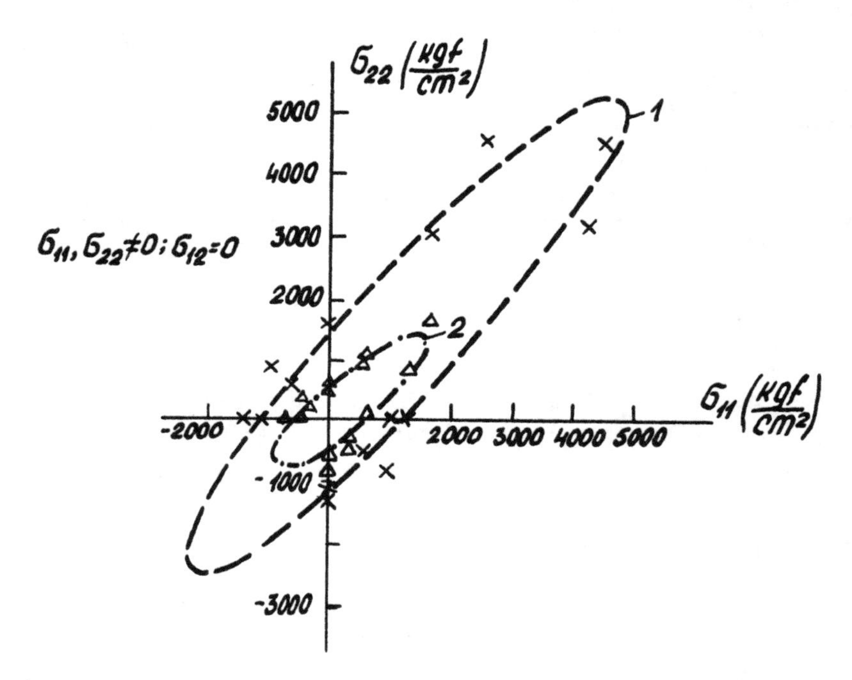

Fig. (14) - The surface of strength in an ultimate state of the material 1 (X), and the surface, describing the state of stress, in which light emission - 2 (Δ) starts

The experimental results at the onset of the photoemission in GFRP cross-ply samples were compared with the theoretically predicted surfaces of initial destruction. Methods, based on the micromechanical analysis of the onset of destruction in components of the composite material, was described in the works [9, 10]. In this case, the field of microstresses in a composite is determined from the solution of the three-dimensional problem by the method of finite elements. The method is employed for assessment of the initial destruction surfaces of cross-ply GFRP having linearly elastic isotropic components. The composite with the given fiber packing and prescribed deformation characteristics is divided into finite elements and the microstresses in the finite elements in a preassumed macrostressed state are determined. By employing the conditions of strength for separate components of the composite, the finite element is defined in which the ultimate microstresses have been built up. This defines the onset of fracture of the composite.

A polynomial is used as a criterion of destruction in the binder. Its shape being established by expansion of the function into spheres by means of tensors [11]. The strength surface in the space of principal stresses is described by the equation:

$$\Psi(I_1, I_2, I_3) \equiv K_1 I_2^{1/2} + K_2 I_1 + K_3 I_2^{-1/2} I_1^2 + K_4 I_2^{-1} I_1^3 + K_5 I_2^{-1} I_3 + K_6 I_2^{-3/2} I_1 I_3 + K_7 I_2^{-3/2} I_1^4 - 1 = 0$$

re, $I_1 = \sigma_{ii}$; $I_2 = \sigma_{ij}\sigma_{ij}$; $I_3 = \sigma_{ij}\sigma_{jk}\sigma_{ki}$ - are invariants of the stress ten-
r; k_i are constants determined from respective strengths r_{ijk}.

The criterion of Mises is used as the condition of fiber breakage: $\frac{3}{2} I_{2D}$ $(\tau_{100})^2$; here r_{100} is the tensile strength of the fiber and I_{2D} is the second variant of the stress deviator.

The methods involves the following initial data (data for cross-ply GFRP are acketed): intensity of the reinforcement of the composite $\theta_1 = 0{,}33$, $\theta_2 = 0{,}67$ $_1 = 0{,}33$, $\theta_2 = 0{,}67$); fiber fraction by volume $\mu = 60\%$ ($\mu = 59{,}9\%$);modulus of asticity of the fiber $E_f = 10^6$ kgf/cm² ($E_f = 0.95 \cdot 10^6$ kgf/cm²); Poisson's tio of the fiber $\mu_f = 0{,}21$ ($\mu_f = 0{,}21$); tensile strength of the fiber $\sigma_{11}^f = 3{,}7$ 10^4 kgf/cm² ($\sigma_{11}^f = 3{,}85 \cdot 10^4$ kgf/cm²); the modulus of elasticity of the binder $= 3{,}5 \cdot 10^4$ kgf/cm² ($E_b = 3{,}5 \cdot 10^4$ kgf/cm²); Poisson's ratio of the binder $= 0.35$ ($\mu_b = 0{,}35$); tensile strength of the binder $\sigma_{11}^b = 0{,}6 \cdot 10^3$ kgf/cm²) $^b_{11} = 0{,}57 \cdot 10^3$ kgf/cm²); compression strength of the binder $\bar{\sigma}_{11}^b = 1{,}2 \cdot 10^3$ f/cm² ($\bar{\sigma}_{11}^n = 1{,}3 \cdot 10^3$ kgf/cm²).

The theoretical surface of initial destruction is shown in Figure 9. As is en, the theoretical surface of initial destruction agrees well with the surface the onset of the photoemission in the material. The exception is the state of ress under internal pressure in combination with compression along the axis 1. is may be explained by the fact that under internal pressure the second layer, inforced along the axis 1, fails first. Additionally, the second layer is lo-ted between the other two layers, as a result the light pulses should pass rough the outer layer having thickness of 0,337 mm. Experiments performed on e unidirectional GFRP samples [12] have shown that on passing light pulses rough the layer of thickness 0,337 mm about 50% of the total amount of arbi-ary units of light emission are absorbed. Accordingly, the photoelectronic ltiplier does not fix the initial stages of light emission and pulses are reg-tered with a time-lag. This accounts for the fact that under internal pres-re the experimental points lie beyond the theoretical curve.

The comparison of the theoretical and experimental curves was made also in ctions in a plane state of stress with $\sigma_{22} = 0$ (Figure 10) and $\sigma_{11} = 0$ (Fig-e 11). A good theory-experiment agreement was found.

NCLUSIONS

1. The experimental results have shown that the mechanolumiscence method may employed for studying the structural composite materials. The initial stage the material destruction as well as the regularities of destruction kinetics ve been determined by the method.

2. In comparing the theoretical calculations of the onset of crack forma-on in components of the composite with experimental data, it has been estab-shed for the first time that the mechanoluminescence reflects the initial pro-sses of destruction in GFRP under a plane as well as a complex state of stress.

REFERENCES

[1] Butyagin, P. Yu.,"Active Intermediate States in Mechanical Fracture of Poly mers", (in Russian), Dikladi Akademiyi Nauk SSSR, Vol. 140, No. 1, pp. 145-148, 1961.

[2] Kisl'uk, M. U. and Butyagin, P. Yu., "On the Possibilities of Studying Mechanical Destruction of Polymers by the Hemiluminescence Method", (in Russian), Vysokomol. Soed., 59, pp. 612-615, 1967.

[3] Butyagin, P. Yu., Erofeev, V. S., Musaelyan, I. N., Patrikeev, G. A., Streletskii, A. N. and Shulyak, A. D., "Luminescence Accompanying Mechanical Deformation and Destruction of Polymers", (in Russian), Vysokomol. Soed Ser. A, No. 2, pp. 290-299, 1970.

[4] Krauja, U. E., Laizan, Ya. B., Upitis, Z. T. and Tutan, M. Ya., "Mechanoluminescence in the Stretching of Fiberglass", Polymer Mechanics, Vol. 13, No. 2, pp. 283-287, 1977.

[5] Rikards, R. B., Teters, G. A. and Upitis, Z. T., "Fracture Models of Composites with Different Fiber Orientation", Mechanics of Composite Materials No. 2, pp. 222-227, 1979.

[6] Broutman, L. J. and Krock, R. H., Modern Composite Materials, 672 pages, 1967.

[7] Regel, V. R., "Studies in Physics of Strength for Composite Materials. Review", Mechanics of Composite Materials, No. 6, pp. 999-1020, 1979.

[8] Malmeisters, A. K., "Geometry of Strength Theories", Polymer Mechanics, No. 4, pp. 519-534, 1966.

[9] Rikards, R. B. and Chate, A. K., "Initial Yield Surface of Unidirectionally Reinforced Composite Materials in the Plane Stressed State", Polymer Mechanics, Vol. 12, No. 4, pp. 633-639, 1976.

[10] Rikards, R. B. and Chate, A. K., Initial Fracture Surfaces of Orthogonally Reinforced Composites, in: Mechanics of Deformable Media, Issue 4, pp. 97-107, Kuibishev, 1979.

[11] Lagzdynsh, A. Zh., "On the Expansion of Scalar Function on a Unit Sphere by Means of Tensor Components", Polymer Mechanics, No. 1, pp. 30-36, 1974.

[12] Upitis, Z. T., Study of Fracture Processes in Multilayered Composites as a Function of Reinforcement Structure. C.Sc. Thesis, Riga, 20 pages, 1980.

COMPARATIVE EVALUATION OF SHEAR TEST METHODS FOR COMPOSITES

Yu. M. Tarnopol'skii and T. Kincis

L.S.S.R. Academy of Sciences
Riga, USSR

INTRODUCTION

The possibility of obtaining reliable numerical estimates by solving the problems of fracture mechanics is based on exact and accurate information about strength and stiffness of composites. To this end, new test methods are continuously being developed, and already existing ones checked and reconsidered. Regardless of the achieved progress in the field, separate test methods have been mastered to various extents. As before, the study of stiffness and, particularly, shear strength presents difficulties. It is exactly fracture in shear, which in many respects, limits the load carrying capacity of structures made from advanced types of composites [1,2]. An attempt at selecting and evaluating the most promising shear test methods for advanced fibrous composites (fiberglass-, carbon-, boron- and organic plastics), conducted on flat, tubular and ring specimens has been made in this survey. The obtained information has been presented in summary tables, containing the loading scheme, determinable characteristics, formulae for calculation, physical, structural and geometrical limitations. The survey contains some 20 loading schemes; the total number of loading schemes described in various sources is considerably greater (see, for instance, [3]). In general, test methods differ among themselves by the way and degree of minimization of "ballast" stresses and strains; it is practically impossible to generate the state of pure shear in specimens. The survey is based on the analysis of experimental data and experience of its authors over many years.

IN-PLANE SHEAR

Historically, the first loading scheme was that of panel shear (picture-frame) type test in a four-link test rig. Initially, this type of test was developed for testing plywood. Regardless of a number of modifications pertaining to the mode of fixation of a specimen and load transfer [4,5], the method has not gained wide acceptance for composites. The main disadvantages are nonuniform stress and strain distribution in the specimen gage section and large specimen dimensions.

The panel shear test was followed by the rail shear test (schemes I-1 and I-2, Table 1) [6,7]. The rail shear test is a simple and economical technique, though it has several limitations. At the free edges of the specimen, the state of stress differs from the state of pure shear. Fixed specimen edges undergo

TABLE 1 - METHODS OF INVESTIGATION IN-PLANE SHEAR CHARACTERISTICS

		Rail Shear	Tension of a Strip		Square Plate Twist
Loading Scheme		I-1, I-2	I-3	I-4	I-5
Determinable Characteristics		G_{12}, τ^{u}_{12}	G_{12}, τ^{u}_{12}	G_{12}	G_{12}
Measurable Values		P, P_u, ε_{45}	$P; \varepsilon_0, \varepsilon_{45}, \varepsilon_{90}$ or $\varepsilon_0, \varepsilon_{120}, \varepsilon_{240}$	$P, \varepsilon_1, \varepsilon_2, \varepsilon_{\pm 45}$	P, w_p
Limitations: Structural	Layup	0°;90°;0/90°	0°	0°;90°;0/90°	0°;90°;0/90°
	Orientation	0°,90°	10-15°	0°,90°,45°	0°,90°
Physical		-	-	-	a linear range of the curve $P_r w_p$
Geometrical		$\ell/b > 10$	$\ell/b \approx 14$	-	$\frac{1}{25} > \frac{h}{\ell} > \frac{1}{100}$

Notes: Relations for calculations, see [3] (schemes I-1, I-2, I-4 and I-5) and [9] (scheme I-3). P_u is load at failure; w_p is the deflection at the loading corner.

lamping in rails. The effect of edge zones and uniformity of tangential stress istribution through the specimen width depends on the length-to-specimen width atio of the gage section ℓ/b and on the relation of elastic constants G_{12}/E_2 of he material under investigation. For composites, the effect of edge zones is egligible at $\ell/b > 10$, except for materials having a high Poisson's ratio.

Estimates of elastic constants by the rail shear type test are insensitive o relative dimensions ℓ/b, because measurements are taken in the middle of the pecimen gage section, where the state of stress is the most uniform. Edge lamping and the loading type appreciably affect the measured strength τ^u_{12}. In imple fixtures (scheme I-1), the specimen is in a biaxial stress state. This ay introduce inaccuracies into the measurements of τ^u_{12}. Therefore, it is more easonable to employ double fixtures (scheme I-2). In comparison to testing in four-link rig, the rail shear test yields higher values of shear modulus and omewhat lower strength values, Table 2.

ABLE 2 - IN-PLANE SHEAR CHARACTERISTICS DETERMINED BY THE TWO TECHNIQUES [8]

Material	G_{12}, GPa		τ^u_{12}, MPa	
	Panel Shear	Rail Shear	Panel Shear	Rail Shear
GRP, 0°	5,24	5,48	48,0	44,4
GRP, 0/90°	5,47	5,53	50,2	48,2
FRP, 0/90°	10,40	11,00	49,5	48,2
FRP, 0/90°	3,92	4,10	44,2	43,5

The method of tension on an anisotropic strip with various fiber layups schemes I-3 and I-4) is outstanding for its apparent simplicity. However, nisotropic strips are not employed for determination of the shear strength in ne fiber layup plane, since the method yields lower values. In the testing of strip with a fiber layup ±45° in tension, the state of pure shear is not generated and normal stresses also act in shear areas. This results in somewhat ower stresses on the curve τ-γ.

In the testing of a strip from a unidirectional material with a fiber lay- at an angle θ, the numerical estimate of the angle, at which relative shear trains $\gamma_{12}/\varepsilon_x$ are maximal and tangential stress τ_{12} reaches its ultimate value, epends on the anisotropy of elastic and strength characteristics of the material 9]. For advanced composites, the optimal magnitude of the angle θ is equal to)-15°. Owing to high sensitivity of stress relations to a change in the angle , an allowance for the angle of specimen cutting, angle of strain gage application and load direction has been specified as equal to ±1°. In order to generate uniform state of stress, relatively narrow specimens have been employed (ℓ/b 14).

For assessment of the in-plane shear modulus, the square plate twist test as been widely used, firstly substantiated for composites by Tsai [10,11] scheme I-5). Its popularity is ascribed to a simple formula for calculations

$G_{12} = \frac{3P\ell^2}{h^3 W_p}$. However, experiments must be performed with utmost care, because in practice, the method has been greatly simplified. The formula was obtained from the linear theory and it is applicable only at small deflections of plates from homogeneous through the thickness and orthotropic in specimen axes materia Calculations according to refined formulae, allowing for large plate deflection [12,13], show that the linearity of the relationship $P \sim W_p$ is maintained appro mately up to $W_p \approx h$. However, due to possible loss of stability of the specime it is advised to confine the deflections to $W_p \leq 0{,}5\ h$ [14]. The relative plat thickness h/ℓ is defined by two conditions: the effect of transverse shear on deflection (at large ratios h/ℓ) and the possible loss of stability (at low rat h/ℓ). The boundaries of ratios h/ℓ, presented in Table 1, are for boron plasti [15]. However, in [16], it has been established from test results for fibergla carbon and boron plastics with various fiber layups that consistent data may be obtained already at $h/\ell < 1/15$. The specimen should be strictly planar, becaus the value h^3 enters the formula for calculations. The distance from support or load application point to the apex of plate angles must not exceed 2h; the accu racy is higher nearer to the plate angles. Deflection W_p is measured in the in tial linear range of the diagram $P \sim W_p$.

In the work [17], the method has been extended for the study of the shear stiffness of multi-layered nonhomogeneous angle-ply plates. For plates with anti-symmetrical fiber layups, it is possible to exclude bending. The formulae allowing in terms of the shear modulus and layup geometry to assess the effecti shear characteristics of a package, have been derived. The method has been experimentally verified for carbon plastics with 13 different layups; numerical estimates are presented in [17]. Disregard for nonhomogeneity of layups may lead to errors higher than 200%.

Generalized experimental data from the estimation of the various methods of determination of the strength and stiffness in the fiber layup plane are presented in Table 3, but their comparative evaluation in Figures 1, 2a and 2b; th method of torsion a thin-walled tube is of reference; the curves 1 and 2 in Fig ure 1 have been plotted under conditions of elimination the restraint of axial strains.

Comparison of the panel shear, rail shear tests, tension of an anisotropic strip and a square plate twist shows that in the determination of the in-plane shear modulus, all the methods yield comparative results (see, the initial rang of strain curves in Figure 1). In the assessment of the shear strength, the panel shear test and tension of an anisotropic strip are quantitatively compara ble, but the estimates of the shear strengths obtained in the rail shear test and in three-point bending are outstanding. Evaluation of the mode of failure presents difficulties [5,20].

INTERLAMINAR SHEAR

In experimental practice, the three-point bending of relatively short bars or ring segments (Table 4, scheme 4-1 and 4-2) is the most widespread method fo the determination of the interlaminar shear strength τ_{13}^u. A refined solution of the bending problem for a relatively short bar from an anisotropic material [20] has shown that the state of stress essentially differs from the one pre-

Method	Modulus, GPa				Strength, MPa			
	GFRP (ν_f=50-54%) [8,16]	BFRP (ν_f=50%) [8,16]	CFRP (ν_f=58%) [8,16,19]	OFRP (ν_f=65% [18]	GFRP (ν_f=50-54%) [8]	BFRP (ν_f=58%) [8]	CFRP (ν_f=50%) [8]	OFRP (ν_f=65%) [18]
Rail shear	5,50	11,0	4,10		46,3	48,2	43,5	
10° off-axis tension				1,99				19,3
±45° off-axis tension	5,50	3,95	3,20	1,90				29,8
Square plate twist	5,10	4,00	2,90					
Three-point bending	4,35	2,80	2,90					36,8
Panel shear in a four-link rig	5,35	10,40	3,92		49,1	49,5	44,2	
Torsion of a rod	5,20	4,60	2,70	1,86				31,9
Torsion of a thin-walled tube			4,30	1,74			47,8	31,3

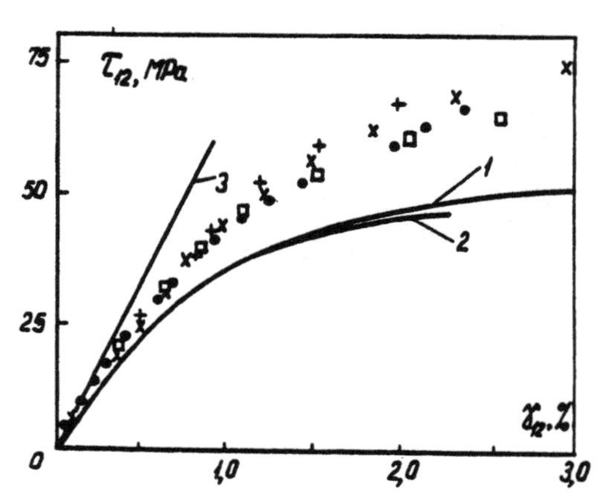

Fig. (1) - Stress-strain curves for carbon-epoxy composites in shear. Methods and layups: panel shear, four-link rig, • - (0°), □- (0°/90°), rail shear, + - 0/90°; tension of a strip, x - ±45°. Torsion: 1 - a tube of circular and 2 - quadratic cross section with a layup ($0°/90_2°/0°/90_2°/0°/90_2°/0°$); 3 - a square plate

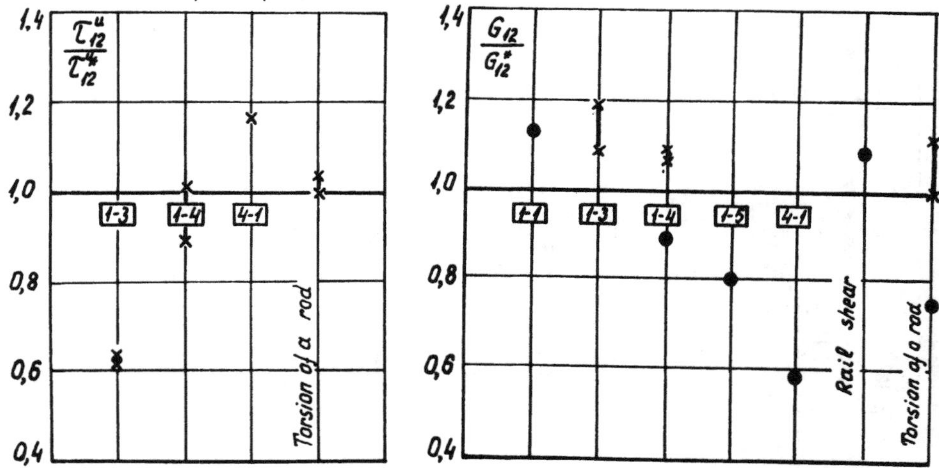

Fig. (2) - A comparative evaluation of different methods for determination of shear resistance. G_{12}^{*}, τ_{12}^{u*} are the shear modulus and shear strength, determined by torsion test of a thin-walled tube. • - CFRP, x - OFRP

scribed by the technical theory of bending. Distribution of tangential stresse throughout the height of a relatively short bar from an anisotropic material o in the center of a halfspan approximately agrees with the technical theory of bending: near the points of concentrated load application the distribution of tangential stresses along the bar height shows a distinct maximum near the loa bar surface. For relatively short bars from anisotropic materials, the sectior with a constant ordinate of maximal tangential stresses are absent. Besides, compressive transverse stresses act along the entire length of a short bar and near the contact areas high compressive contact stresses are observed. Due to the deviations, the experimentally determined interlaminar shear strength τ_{13}^{u} decreases with an increase in the relative span ℓ/h and, therefore, the test results for short bars under three-point bending may serve only for a qualitative comparison of mechanical behavior of different composites.

One must treat with utmost care the problem of employing the three-point bending test for determination of the interlaminar shear modulus G_{13} from the test results of two series of specimens with various ratios ℓ/h; in practice (Table 3), lower values of G_{13} are commonly obtained. Transverse shear w_{τ} contribute only partly to the entire specimen deflection w. In order to reliably determine the shear modulus by properly selecting the relative span ℓ/h, depending on the degree of anisotropy of the material under investigation E_1/G_{13}, a sufficiently high ratio of deflections $K_{\tau} = \frac{w_{\tau}}{w}$ (no less than 0.3) must be ensured.

For calculation of the interlaminar shear strength on ring segments, the formula for prismatic bars is used. However, in so doing, it should be remembered that in segments, in contrast to prismatic bars, normal stresses also act along the entire span length, the direction of which depends on the loading scheme. In the case of loading segments with convexity upward (scheme 4-2), the stresses are tensile (σ_r^+), with convexity downward - compressive (σ_r^-); in the latter case, the stresses σ_r^- interfere with the opening of the delamination crack due to tangential stresses $\tau_{\theta r}$ and in such a way "increase" the interlaminar shear resistance of the material. The value of the shearing force Q depends on the method of specimen fixation: in the case of a specimen supported on a plate or sharp edge of a prism $Q = \frac{P}{2}$; in the case of cylindrical supports $Q = \frac{P}{2} \cdot \frac{1}{\cos\theta}$. The dimensions of the segments must be selected so that the normal hoop stresses σ_{θ} and normal radial stresses σ_r be negligibly small compared to the value of tangential stresses $\tau_{\theta r}$. The conditions for selecting the ratio of dimensions ℓ/R and ℓ/h are listed in Table 4.

The restricted range of application of the three-point bending has resulted in the quest for other techniques. The method of torsion of specimens with a circular notch yields good results, especially for spatially reinforced materials. Two types of specimens have been employed - hollowed (scheme 4-4) and non-hollowed (scheme 4-3). It is important that the geometrical parameters of a notch - its relative width ℓ/d, diameter d or the wall thickness h be properly selected.

Investigations [22] show that within an ℓ/d ratio of 0.2 to 1.0, the length of the specimen gage section ℓ does not affect the value of the measured strength τ_{rz}^{u}. The increase in the gage section diameter d from 5 to 15mm does not affect the value of the strength, but on further increase of the diameter (d > 15 mm),

TABLE 4 - METHODS OF INVESTIGATION INTERLAMINAR SHEAR CHARACTERISTICS

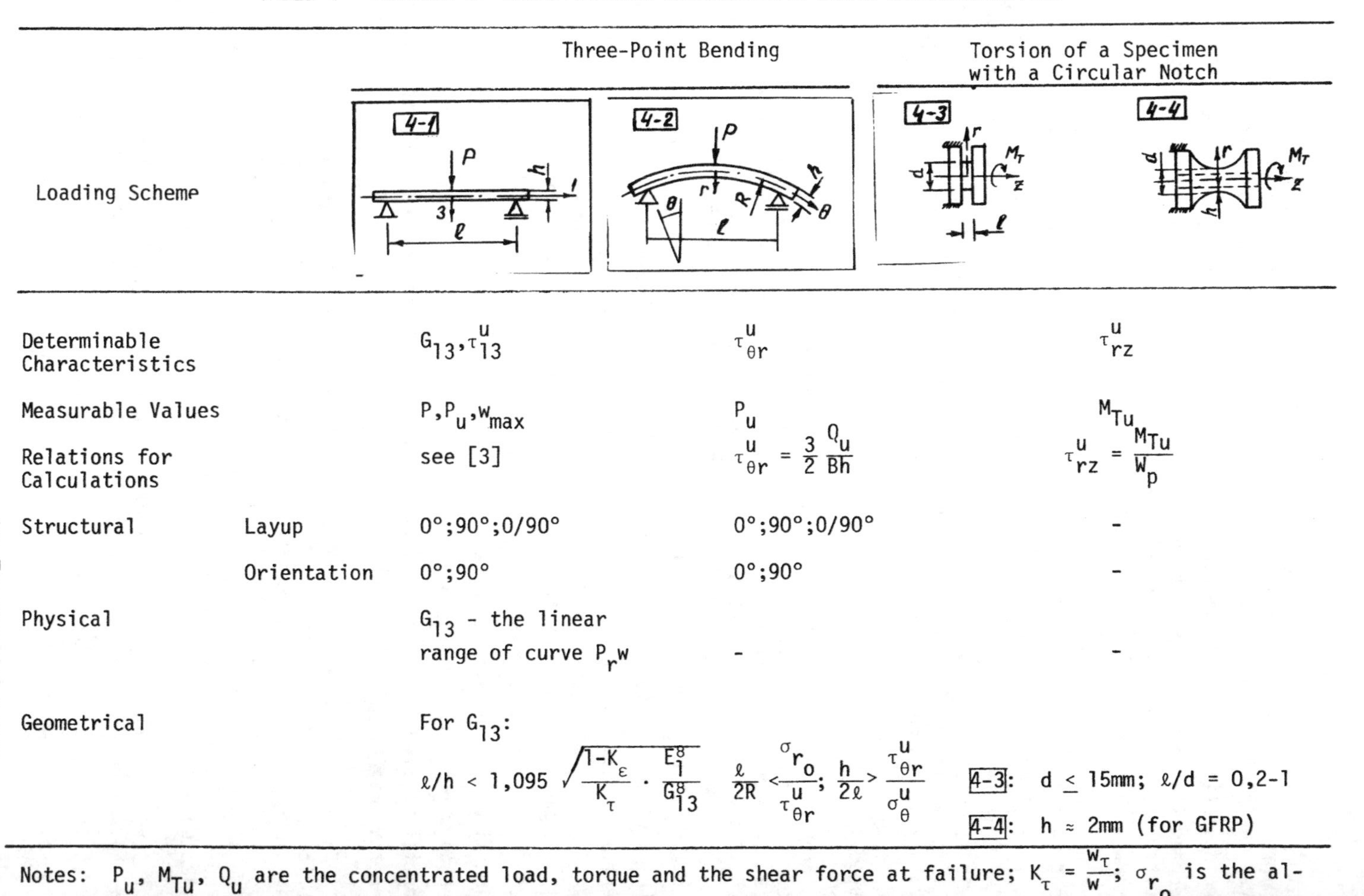

		Three-Point Bending		Torsion of a Specimen with a Circular Notch
Loading Scheme		4-1	4-2	4-3, 4-4
Determinable Characteristics		G_{13}, τ^u_{13}	$\tau^u_{\theta r}$	τ^u_{rz}
Measurable Values		P, P_u, w_{max}	P_u	M_{Tu}
Relations for Calculations		see [3]	$\tau^u_{\theta r} = \frac{3}{2}\frac{Q_u}{Bh}$	$\tau^u_{rz} = \frac{M_{Tu}}{W_p}$
Limitations: Structural	Layup	0°;90°;0/90°	0°;90°;0/90°	-
	Orientation	0°;90°	0°;90°	-
Physical		G_{13} - the linear range of curve $P_r w$	-	-
Geometrical		For G_{13}: $\ell/h < 1{,}095 \sqrt{\frac{1-K_\varepsilon}{K_\tau} \cdot \frac{E^8_1}{G^8_{13}}}$	$\frac{\ell}{2R} < \frac{\sigma_{r_o}}{\tau^u_{\theta r}}$; $\frac{h}{2\ell} > \frac{\tau^u_{\theta r}}{\sigma^u_\theta}$	4-3: $d \leq 15$mm; $\ell/d = 0{,}2-1$ 4-4: $h \approx 2$mm (for GFRP)

Notes: P_u, M_{Tu}, Q_u are the concentrated load, torque and the shear force at failure; $K_\tau = \frac{w_\tau}{w}$; σ_{r_o} is the al-

sharp fall in the measured strength τ_{rz}^u occurs. The wall thickness h for GRP specimens is equal to 2 mm [23].

By employing different methods for determination of τ_{13}^u, good results may be obtained provided the level of test technique is high; data for FGRP are presented in Table 5. With an increase in the degree of anisotropy, the error for different methods is greater.

TABLE 5 - INTERLAMINAR SHEAR STRENGTH (IN MPa) DETERMINED BY VARIOUS TECHNIQUES [22]

Method	Glass Textolite (V_f = 45%)	Epoxy Resin
Square plate twist	71	75
Torsion of a bar of rectangular cross section	84	66
Torsion of a bar of circular cross section	73	-
Torsion of a specimen with a circular notch	71	73

Referring to Table 3, in which generalized characteristics of FGRP, BFRP, CFRP and OFRP with various reinforcement patterns, obtained on flat specimens, are presented. It follows from the data comparison that in the determination of the shear modulus, all the methods yield results within the data scatter for the experiment. In determination of the shear strength, the methods of tension of an anisotropic strip and three-point bending must be sharply distinguished. There are several reasons for such a deviation. In the case of tension of an anisotropic strip, either the method itself may be inapplicable for the determination of the shear strength or the angle θ = 10° may be improperly selected. The three-point bending test may be affected by the deficiencies of the method or by the specific features of the material under investigation (for instance, OFRP in compression often do not exhibit a linear-elastic behavior; in this case, the formulae of the technical bending theory are inapplicable). The analysis shows that the methods of tension of an anisotropic strip, torsion of a square plate and torsion of a bar of rectangular cross section yield the most reliable readings, the three-point bending has the widest, practically inadmissible data scatter.

RING-TYPE SPECIMENS

During the last years, the information obtained in the testing of ring-type specimens has greatly broadened. The most widespread methods for determination of the shear characteristics on ring specimens are presented in Table 6. The bending of intact rings with concentrated forces (scheme 6-1) has been experimentally well verified, and provided the relative specimen thickness had been properly selected, the methods yields reliable results. The interlaminar shear modulus $G_{\theta r}$ in terms of the experimentally measured load P and the change in the vertical diameter W_0 is determined analogously to the case of three-point bending of prismatic bars [24].

To ensure accuracy of the method of ring tension with concentrated forces in determination of the interlaminar shear modulus, the relative specimen thickness h/R must be selected with due regard for the degree of anisotropy of the material $E_\theta/G_{\theta r}$ and the contribution of the tangential stresses $W_{0\tau}/W_0$ to the change in the vertical diameter, which must not be lower than $(0.25\text{-}0.30)W_0$. In assessing the interlaminar shear strength $\tau^u_{\theta r}$ by the method of bending intact rings, the failure by shear - delamination starting in the midplane - must be ensured by the proper selection of the relative specimen thickness h/R.

In the case of assessment of the interlaminar shear strength $\tau^u_{\theta r}$ on split rings, the specimens are loaded with concentrated forces P, applied to the rigid cantilevers in such a way that their line of action passed through the ring center (scheme 6-2) [25].

The loading scheme 6-3 [26] is the best for the determination of the shear moduli from torsion tests on ring-type specimens from the viewpoint of the uniformity of the state of stress. In this case, the effect of bending and transverse shear is negligible. Technically, the loading scheme 6-4 [27] is simpler in its realization, however, in this case, the effect of bending is negligible only under the conditions indicated in Table 6. Application of intact rings for determination of the shear moduli, due to extremely stringent geometrical limitations, is reasonable only in the cases when elastic constants and strength are determined on the same specimens [28]. The shear moduli in terms of the torsional rigidity C, determined in torsion tests, have been calculated in the same way as in the case of torsion of straight bars.

Experimental data from torsion tests of rings are presented in [28]; their comparative evaluation is given in Figure 3.

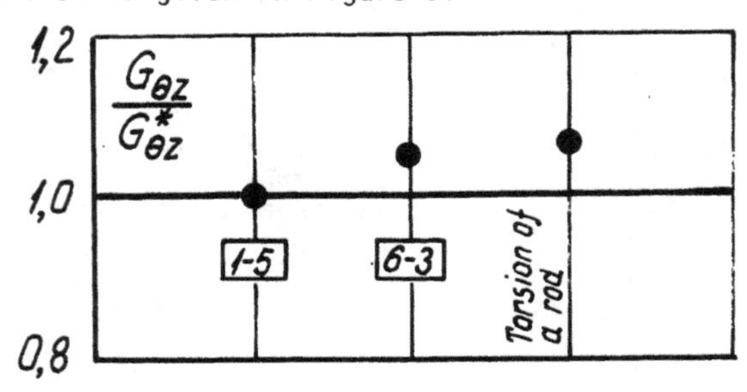

Fig. (3) - A comparative evaluation of the methods of shear modulus determination for CFRP (V_f = 64%). $G^*_{\theta r}$ is the shear modulus determined by torsion of a thin-walled tube

It follows from the comparison of experimental data that the numerical estimates $G_{\theta r}$ and $G_{\theta z}$, determined by various techniques, shows a good agreement, regardless of the different fabrication technology for flat, circular and tubular specimens.

TABLE 6 - BENDING AND TORSION OF RING SPECIMENS

		Bending of an Intact Ring	Bending of a Split Ring	Torsion of a Split Ring	
Loading Scheme					
Determinable Characteristics		$G_{\theta r}, \tau_{\theta r}^{u}$	$\tau_{\theta r}^{u}$	$G_{\theta r}, G_{\theta z}$	$G_{\theta r}, G_{\theta z}$
Measurable Values		P, P_{u}, w_{o}	P_{u}	P, w_{p}	P, w_{p}
Relations for Calculations		see [3,24]	$\tau_{\theta r}^{u} = \frac{3}{2}\frac{P_{u}}{Bh}$	$C = \frac{PR^{3}}{w_{p}}\theta$	$C = \frac{w_{p}}{\pi PR^{3}} - \frac{1}{E_{\theta}]}$
Limitations: Structural	Layup	0°;90°;0/90°			
	Orientation	0°,90°			
Physical			-	-	-
Geometrical		For $G_{\theta r}$: $\frac{R}{h} < 0{,}727 \sqrt{\frac{1-K_{\tau}}{K_{\tau}}\frac{E_{\theta}}{G_{\theta r}}}$ For $\tau_{\theta r}^{u}$: R/h = 5-12	-	-	$\frac{1}{E_{\theta}]} \to 0$ at B/h > 3

Notes: R - is the mean radius; w_{p} is the deflection at the point of loading (scheme 6-3) or the mutual displacement of load application points (scheme 6-4).

CONCLUDING REMARKS

The survey allows us to state that over the last decade, great advances hav been made in the development of shear test methods and evaluation of their effi cacy. This allows us to better understand and evaluate the structural capabili of advanced types of composites. Regardless of the progress, the study of the shear strength presents difficulties. A series of methods, in particular those which appeared during the last years, are in need of further corrections along with the accumulation of experience. It is necessary to note that the material is based on the experience of testing composites of the first generation based on a polymeric matrix. Research practice of the last years has revealed that a number of methods are also applicable to composites of next generations, based on carbon, ceramic and metallic matrices. However, the experience of the testi of composites of the first generation cannot be directly extended to a wider cl of composites without an additional consideration of the matrix properties.

REFERENCES

[1] Tarnopol'skii, Yu. M. and Roze, A. V., Osobennosti raschota detaljei iz armirovannikh plastikov. (Peculiarities of Structural Design from Reinforced Plastics), Zinātne Press, Riga, 1969.

[2] Zhigun, I. G. and Polyakov, V. A., Svoystva prostranstvenno armirovannikh plastikov. (Properties of Spatially Reinforced Plastics), Zinātne Press, Riga, 1978.

[3] Tarnopol'skii, Yu. M. and Kincis, T. Ya., Metodi staticheskikh ispitaniy armirovannikh plastikov (Methods for Static Testing of Reinforced Plastic Khimiya Press, 3rd ed., Moscow, 1981. (Translation of the 2nd ed: Tarnopol'skiy, Yu. M. and Kintsis, T. Ya. Static Methods of Testing Reinforc Plastics, NASA TT F-16669, April 1976).

[4] Skudra, A. M., Bulavs, F. J. and Rocens, K. A., Kriechen und Zeitstandver von verstärkten Plasten. VEB Deutscher Verlag für Grundstoffindustrie, L zig, 1975.

[5] Terry, G. A., Comparative Investigation of Some Methods of Unidirectional In-Plane Shear Characteristics of Composite Materials. Composites, Vol. 10, No. 4, pp. 233-237, 1979.

[6] Whitney, J. M., Stansbarger, D. L. and Howell, H. B., "Analysis of the Ra Shear Test", J. Composite Materials, pp. 24-34, January 1971.

[7] Sims, D. F., "In-Plane Shear Stress-Strain Response of Unidirectional Composite Materials", J. Composite Materials, pp. 124-128, January 1973.

[8] Auzukalns, Ya. V., Birze, A. N. and Bulavs, F. Ya., Prochnostniye svoystv armirovannikh plastikov pri szhatiyi pod uglom k napravleniyu armirovaniy (Strength Properties of Reinforced Plastics in Compression at an Angle to the Fiber Direction), in: Nerazrushayustiye metodi ispytaniya stroyiteln materialov, Issue 2, Riga Polytechnic, Riga, pp. 86-95, 1976.

[9] Chamis, C. C., Sinclair, J. H., "Ten-Day Off-Axis Test for Shear Properti in Fiber Composites", Exp. Mechanics, Vol. 17, No. 9, pp. 339-346, 1977.

10] Tsai, S. W., "Experimental Determination of the Elastic Behavior of Orthotropic Plates", Trans. ASME, Ser. B, Vol. 87, No. 3, pp. 315-318, 1965.

11] Tsai, S. W. and Springer, G. S., "The Determination of Moduli of Anisotropic Plates", Trans. ASME, Ser. E, Vol. 30, No. 3, pp. 467-468, 1963.

12] Foye, R. L., "Deflection Limits on the Plate-Twisting Test", J. Composite Materials, pp. 194-198, April 1967.

13] Chandra, R., "On Twisting of Orthotropic Plates in Large Deflection Regime", AIAA Journal, Vol. 14, No. 8, pp. 1130-1131, 1976.

14] Purslow, D., "The Shear Properties of Unidirectional Carbon Fibre Reinforced Plastics and Their Experimental Determination", Aeronautical Research Current Paper No. 1381, 1977.

15] Handbook of Fiberglass and Advanced Plastics Composites, G. Lubin, ed., New York, 1969.

16] Zhigun, I. G., Yakushin, V. A., Tanevskii, V. V. and Mikhailov, V. V., "Analysis of Certain Methods of Determining Shear Moduli. I. Testing Composites Uniform over the Thickness", Polymer Mechanics, Vol. 12, No. 1, pp. 112-119, 1976.

17] Polyakov, V. A. and Tanevskii, V. V., Eksperimental'naya otsenka sdvigovoi zhestkosti kompozitov s peremennim zakonom ukladki armaturi (Experimental Evaluation of Shear Stiffness of Composites with Varying Fiber Layup). Mekhanika kompozitnikh materialov, No. 5, pp. 912-918, 1980.

18] Chiao, C. C., Moore, R. L. and Chiao, T. T., "Measurement of Shear Properties of Fibre Composites. Part I. Evaluation of Test Methods", Composites, Vol. 8, No. 3, pp. 161-169, 1977.

19] Rabotnov, Yu. N., Danilova, I. N., Polilov, A. N., Sokolova, T. V., Karpeikin, I. S. and Vainberg, M. V., "Strength of Winding Epoxide Carbon- and Fiberglass-Plastics in Twisting, Stretching and Transverse Bending, Polymer Mechanics, Vol. 14, No. 2, pp. 174-179, 1978.

20] Hua, P. H., "Measuring In-Plane Shear Properties of Glass-Fiber/Epoxy Laminates", Composites Technology Review, Vol. 2, No. 2, pp. 3-13, 1980.

21] Tarnopol'skii, Yu. M., Zhigun, I. G. and Polyakov, V. A., "Distribution of Shearing Stresses under Three-Point Flexure in Beams Made of Composite Materials", Polymer Mechanics, No. 1, 13, pp. 52-58, 1977.

22] Zhigun, I. G., Yakushin, V. A. and Ivonin, Yu. N., "Analysis of Methods of Determining the Interlaminar Shear Strength of Composite", Polymer Mechanics, Vol. 12, No. 4, pp. 573-580, 1976.

23] McKenna, G. B., Mandell, J. F. and McGarry, F. J., "Interlaminar Strength and Toughness of Fiberglass Laminates", in: SPI Reinforced Plastics/Composites Institute, Proc. 29th Annual Conference, Section 13-C, pp. 1-8, Washington, 1974.

[24] Tarnopol'skii, Yu. M., Roze, A. V. and Shlitsa, R. P., "Testing of Wound Rings under Concentrated Forces", Polymer Mechanics, No. 4, 5, pp. 719-727, 1969.

[25] Nikolaev, V. P., O metodike ispitaniy kolec iz armirovannikh materialov na prochnostj pri mezhsloinom sdvige. (On the Test Methodics for Determination the Interlaminar Shear Strength on RP Rings), in: Trudy MEI, Dinamika i prochnost' mashin, Iss. 164 (MEI, Moscow), pp. 92-96, 1972.

[26] Nikolaev, V. P., "Determination of the Shear Moduli of Glass-Reinforced Plastics Using Ring Specimens", Polymer Mechanics, Vol. 7, No. 6, pp. 984-986, 1971.

[27] Greszczuk, L. B., "Shear-Modulus Determination of Isotropic and Composite Materials", in: Composite Materials: Testing and Design, ASTM STP 460, pp. 140-149, 1969.

[28] Kintsis, T. Ya. and Shlitsa, R. P., "Determination of the Shear Modulus of Composites from Experiments on the Twisting of Circular Specimens", Polymer Mechanics, Vol. 14, No. 5, pp. 764-767, 1978.

KPERIMENTAL DETERMINATION OF FLAW SHAPES AND STRESS INTENSITY DISTRIBUTIONS; ONDITIONS FOR APPLICATION TO COMPOSITE MATERIALS

C. W. Smith

Virginia Polytechnic Institute and State University
Blacksburg, Virginia 24061

BSTRACT

An extension of studies described in the proceedings of the first USA-SSR Symposium concerning the application of the "frozen stress" method to the etermination of both flaw shapes and stress intensity distributions in three imensional (3D) crack problems in single phase materials is presented. Analy-ical foundations are extended to include shear modes of fracture and its' po-ential use in angle ply laminates is discussed.

NTRODUCTION

One of the most frequently encountered types of in-service fractures in-olves the stable growth of a sub-critical flaw under fatigue loading to a ritical size at which quasi-brittle fracture occurs. Such flaw growth is sually not self-similar, and the problem geometry often involves curved crack ronts, non-planar cracks and varying stress intensity factors (SIFs) along ne flaw borders. Continued improvements in both the soft and hardware of igital computers has led to significant advances in the realm of numerical nalysis and improved precision has recently led to close correlation between number of these solutions for semi-elliptic surface flaws in flat plates nder both tension and bending [1]. Progress also continues in evaluating ingular fields for a wide variety of 3D crack problems using various analy-ical approaches [2]. Despite these significant advances, complications such s those noted above still hamper the analysts efforts to render a number of roblems in this class tractable.

Due to the mathematical complications in many such problems and their echnological importance, the author and his students began, over a decade go, to study ways of obtaining information about such problems experimen-ally, with the goal of achieving sufficient accuracy that the experimental ethod might be used for computer code verification. The approach chosen is nown as the "frozen stress" method [3] and it was combined with the equations f linear elastic fracture mechanics in order to develop a technique for stimating SIF distributions along the border of a flaw in a three dimensional cracked body problem. The early form of the method for estimating Mode I SIF distributions is found in [4]. Moreover, it has recently been extended to include estimates of Mode II and Mode III SIF distributions [5] as well as Mode I.

In applying the technique to a variety of 3D cracked body problems, both real and artificial (machined) cracks have been used. Recently, in connection with a study of nozzle corner cracks in plates [6][7] the author noticed that, under certain conditions, the shapes of the real cracks grown in photoelastic models under monotonic load above critical temperature were identical to those produced by tension-tension fatigue loading of A508 reactor vessel steel models. Subsequently efforts have been made to clarify the conditions which must be met so that the real cracks in the photoelastic models will accurately model fatigue cracks in metals. If this can be achieved, an experimental modelling method will be available with the potential for predicting both flaw shapes and SIF distributions in 3D cracked body problems where neither are known a-priori.

These ideas were introduced at the first USA-USSR Symposium on Fracture [8]. Since that time, further studies have been conducted towards clarifying the conditions under which sub-critical flaw growth occurs, and how fatigue cracks grown in metals can be simulated in photoelastic models. After briefly reviewing the features of the experimental method, the present paper describes details of fatigue flaw simulation in the photoelastic models, and the constraints which must be met in order for the method to successfully predict crack growth in composite materials.

ANALYTICAL FOUNDATIONS - A SUMMARY

The general problem geometry and notation for studying cracks in single phase materials is given in Figure 1. For Mode I loading, where the photoelastic fringes are symmetric with respect to the crack plane (Figure 2), we may write, along $\theta = \pi/2$, [8]

$$\frac{K_{AP}}{q(\pi a)^{1/2}} = \frac{K_I}{q(\pi a)^{1/2}} + \frac{f(\sigma^0_{ij})}{q} \{\frac{r}{a}\}^{1/2} \quad (1)$$

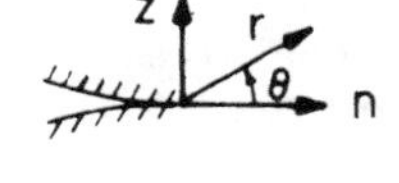

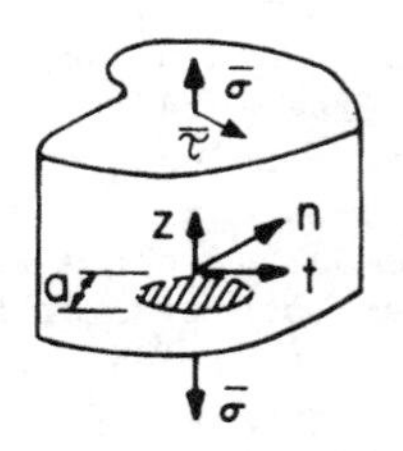

Fig. 1 General Problem Geometry and Notation.

where:

$K_{AP} = \tau_{max}(8\pi r)^{1/2}$, the Apparent SIF and $\tau_{max} = Nf/2t'$ where N is the stress fringe order, f is the material fringe value and t' is the thickness of a thin slice of material taken mutually orthogonal to the crack surface and the crack border (nz plane of Fig. 1)

σ^0_{ij} represents the contribution of the regular stresses to the fringe order in the measurement zone and produces "folding" and change in eccentricity of the fringe loops as r increases from zero.

q is a load parameter, such as remote stress or internal essure.

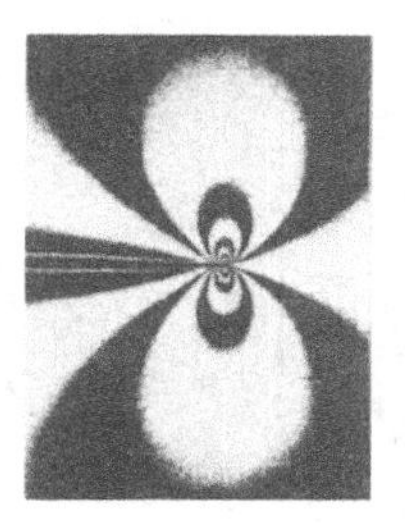

Fig. 2 Local Mode I Stress Fringe Pattern

Equation (1) reveals that, in the zone dominated by the ress singularity, $K_{AP}/q(\pi a)^{1/2}$ varies linearly with $(r/a)^{1/2}$. us, by establishing the linear zone, we may extrapolate ac ss a non-linear zone near the crack tip in order to measure $/q(\pi a)^{1/2}$. An example for a slice removed from the geometry ich we will study parallel to the nz plane is shown in Fig e 3.

A similar approach can be applied when Mode II is also esent. However, in this case, the fringe loops will be ro- ted from the symmetric Mode I position, and, as in the Mode case, there may be some folding and change in eccentricity of the fringe ops due to the influence of σ^0_{ij}. This requires a determination of θ^0_m at e crack tip by extrapolation but it is often a horizontal extrapolation as ctured in Figure 4. The equation for Modes I and II corresponding to Equa- on (1) is [9], for fringe loops approaching the geometry of Fig. 4.

$$\frac{K^*_{AP}}{q(\pi a)^{1/2}} = \frac{K^*}{q(\pi a)^{1/2}} + \frac{f(\sigma^0_{ij})}{q} \{\frac{r}{a}\}^{1/2} \tag{2}$$

ere

$$K^*_{AP} = [(K_{1AP}\sin\theta_m + 2K_{2AP}\cos\theta_m)^2 + (K_{2AP}\sin\theta_m)^2]^{1/2} = \tau_{max}(8\pi r)^{1/2}$$

$$K^* = [(K_1\sin\theta^0_m + 2K_2\cos\theta^0_m)^2 + (K_2\sin\theta^0_m)^2]^{1/2} \text{ which is solved with}$$

$$[\frac{K_2}{K_1}]^2 - \frac{4}{3}[\frac{K_2}{K_1}] \cot 2\theta^0_m - \frac{1}{3} = 0 \text{ for } K_2/K_1$$

d K^* can be obtained by extrapolating from a graph of $K^*_{AP}/q(\pi a)^{1/2}$ vs $/a)^{1/2}$ in a fashion similar to Figure 3.

In order to evaluate K_{III}, a sub-slice is taken from the nz slice and ewed along the n direction. For this case, the equation corresponding to uation (1) is, along $\theta = \pi/2$ [5]

$$\frac{K^{**}_{AP}}{q(\pi a)^{1/2}} = \frac{K^{**}}{q(\pi a)^{1/2}} + \frac{f(E\bar{\varepsilon},\sigma^0_{ij})}{q} \{\frac{r}{a}\}^{1/2} \tag{3}$$

ere

$$K^{**}_{AP} = \sqrt{2}\,[\frac{1}{16}(K_{1AP} + K_{2AP})^2 + K^2_{3AP}]^{1/2} = \tau_{max}(8\pi r)^{1/2}$$

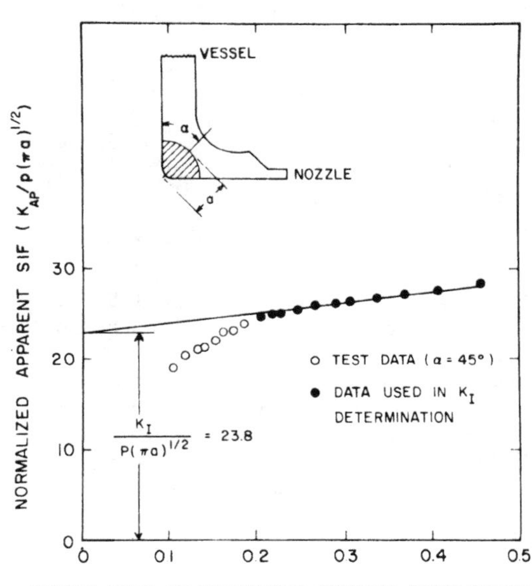

Fig. 3 Estimating SIF Value from Typical Test Data

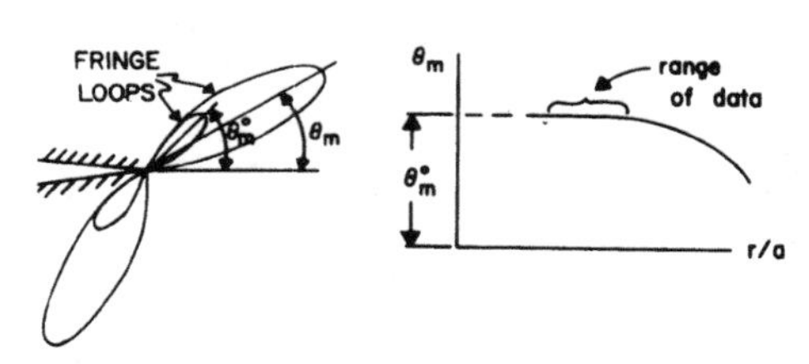

Fig. 4 Determination of θ_m^0 for Mixed Mode Case

$$K^{**} = \sqrt{2}\ [\frac{1}{16}\ (K_1 + K_2)^2 + K_3^{\ 2}]^{1/2}$$

$E\bar{\varepsilon}$ = Young's Modulus times the average normal strain through slice thickness in n direction

Again, one extrapolates from a graph of $K^{**}_{AP}/q(\pi a)^{1/2}$ vs $\{r/a\}^{1/2}$ in order to obtain $K^{**}/q(\pi a)^{1/2}$.

Thus by the photoelastic analysis of a slice parallel to the nz plane and a sub-slice from that slice using a white light field and Tardy Compensation, it is possible to estimate K_1, K_2 and K_3 when all three are acting simultaneously. The accuracy of the estimation of K_1 in the absence of K_2 is estimated to be $\pm$ 5% for 3D problems. However, K_2 estimates in the presence of Mode I are about half as precise, and K_3 estimates in the presence of Modes I and II are somewhat less precise than those for K_2. Moreover, while the presence of Mode II produces visual dissymetry in the local fringe loops, Mode III may not be discernable.

EXPERIMENTS AND OBSERVATIONS

The Experimental Method

It was desired to study the subcritical growth of a flaw under conditions of substantial constraint and stress gradients. Prior experience [10] suggested that cracks emanating from the reentrant corners of reactor vessel nozzle junctures provided a sufficiently complex combination of constraints and stress gradients for the study.

Scale models of boiling water reactors (BWRs) were cast in parts from a suitable stress freezing photoelastic material. Starter cracks were inserted t three different locations around the circular reentrant nozzle corner as hown in Fig. 5a and the model parts were then glued together forming vessel odels each of which contained two diametrically opposite nozzles (Fig. 5b). ach model was then heated to critical temperature and sufficient internal ressure was applied to grow the starter cracks. When the cracks reached the esired size as viewed through a glass port in the oven wall, the pressure was educed sufficiently to terminate flaw growth, and the cracked model was ooled to room temperature, freezing in the stress fringes and deformation ields produced above critical temperature. The models were then unloaded ith negligible recovery and thin slices were extracted from the models arallel to the nz plane (Fig. 1).

Stress analysis of the uncracked model geometry showed that the maximum rincipal tensile stress direction changed from the hoop direction in the ozzle to the hoop direction in the vessel and that the vessel value was oughly an order of magnitude greater than the nozzle value.

rack Shapes and SIF Distributions

i) Crack orientation A, (Fig. 5a.) - Crack shapes and SIF distributions or this case were presented at the first USA-USSR Conference [8]. Figure 6 ummarizes the variations in flaw shapes and corresponding SIF distributions

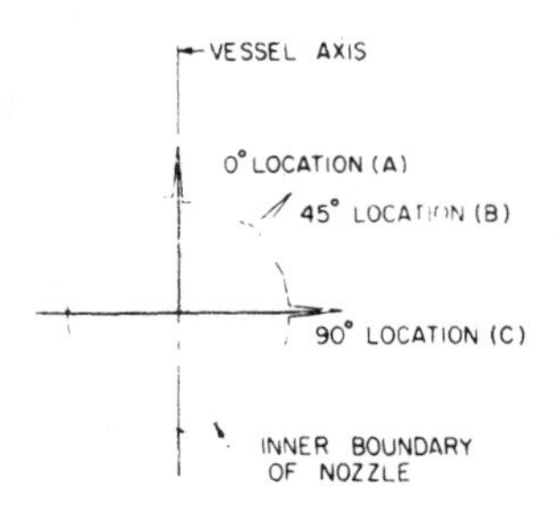

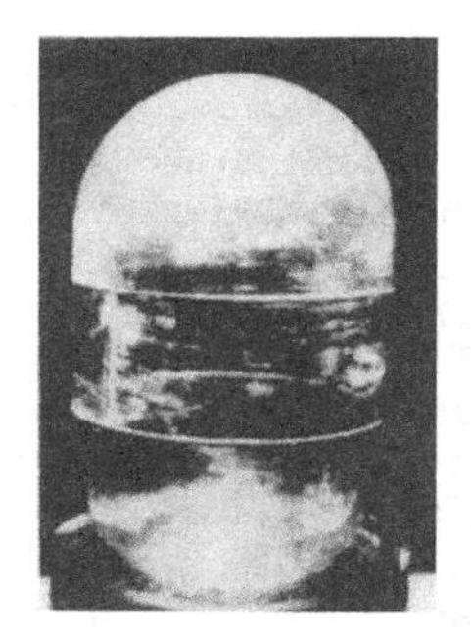

ig. 5 (a) Flaw locations around nozzle axis (b) Photo of a photoelastic model of a BWR.

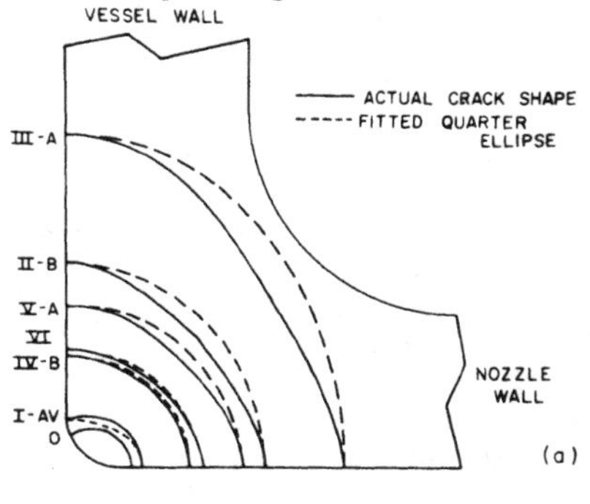

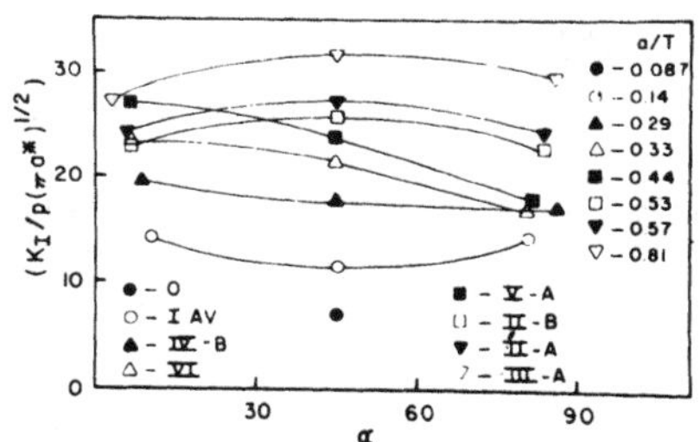

Fig. 6 (a) Flaw shapes and (b) Corresponding SIF distributions for Case A of Fig. 5(a).

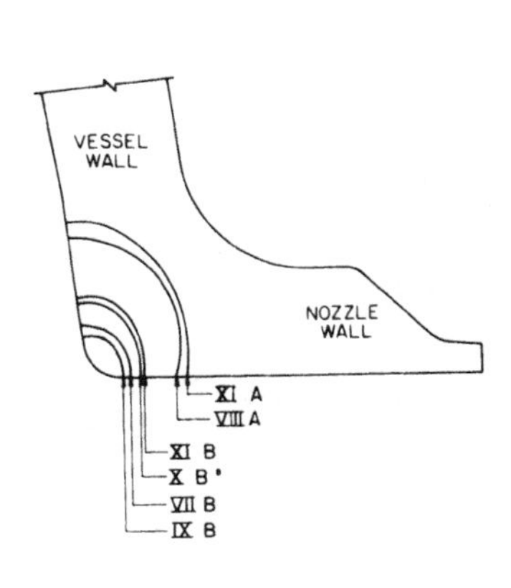

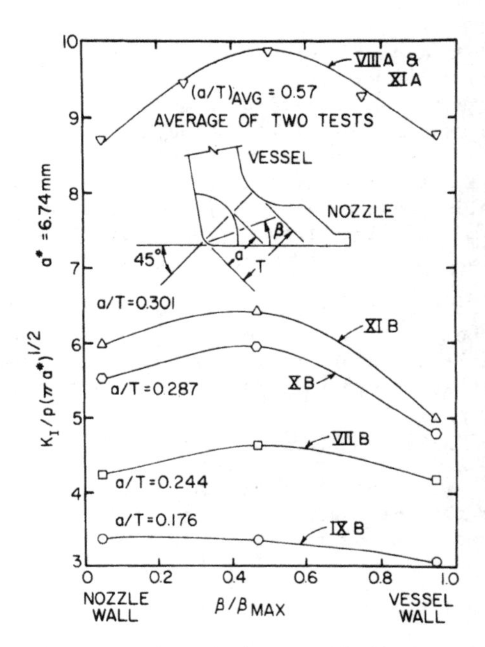

Fig. 7 (a) Flaw Shapes and (b) Corresponding SIF Distributions for Case C of Fig. 5(a).

for this case. The crack growth was clearly non-self-similar. All flaw growth was in the plane of the vessel hoop stress and when a starter crack was slightly misoriented from this plane, it quickly grew back into that plane. However, in so doing the shape of the crack front changed. This change is believed due to local boundary and stress gradient effects rather than a material history effect.

ii) Crack Orientation C (Fig. 5a) - Crack shapes and SIF distributions for this case are given in Fig. 7. These results still reveal non-self-similar flaw growth but with less variation in flaw shape and SIF distribution than in Case A. All flaw growth was in the same plane again. However, for slightly misoriented starter cracks, the cracks turned towards the plane of the vessel hoop stress and did not return to Orientation C.

iii) Crack Orientation B (Fig. 5a) - The starter flaw geometry for this case was initially a planar nearly quarter elliptic flaw under pure Mode I loading at n in Fig. 8(a) and Mixed Mode loading at the vessel wall. When the crack began to grow, it remained in the plane of the hoop stress near n, but it turned immediately away from the 45° direction at the vessel wall and gradually approached the hoop stress plane of the vessel. Photoelastic analysis showed that the presence of Mode II at the vessel wall caused an immediate change in direction of flaw growth. Up to $a/T \simeq 0.3$ the flaw grew as two separate flaws joined at P (Fig. 8a) but, at this a/T level, the two parts coalesced, eliminating P and leaving a single non-planar flaw with a smooth crack front nv. The radical change in the SIF distribution accompanying the elimination of P is recorded in Fig. 8(b).

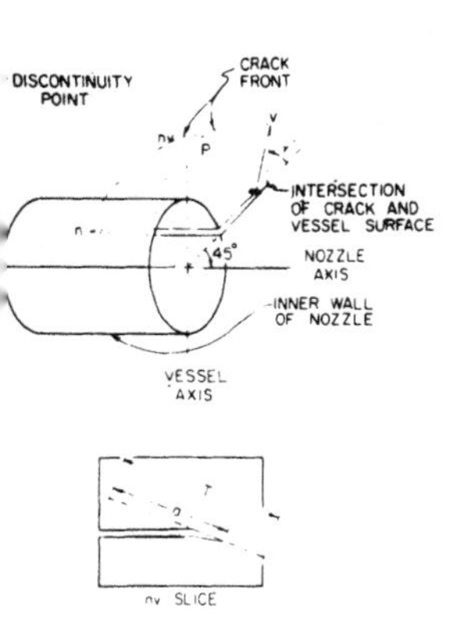

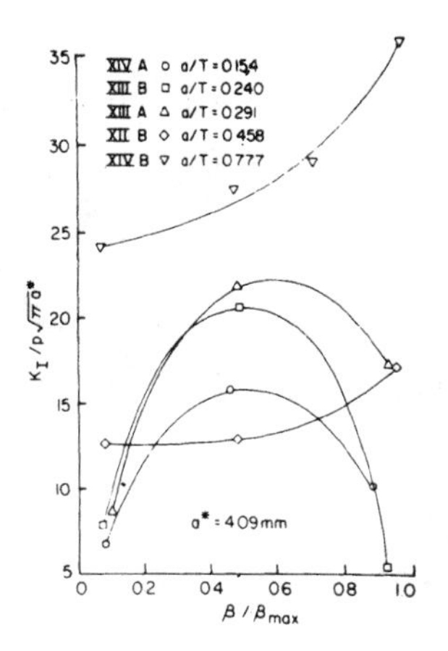

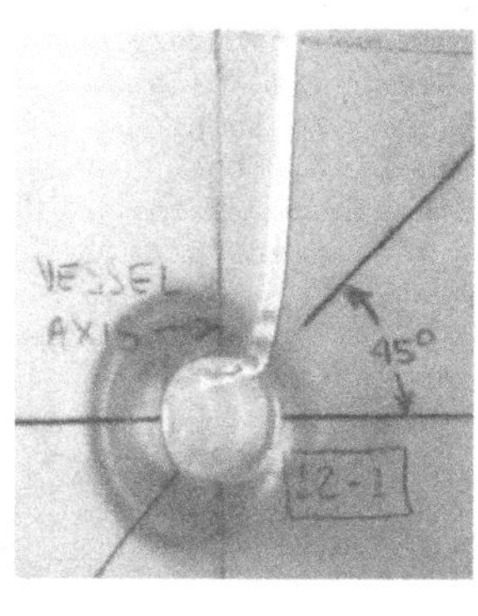

ig. 8 (a) Flaw Shapes and (b) Corresponding SIF Distributions for Case B of Fig. 5a. β is the angle turned in the plane of the initial flaw from the nozzle wall.

Fig. 9 Photo of Through Thickness Flaw

Fig. 9 shows a photograph of one of the cracks which was driven com-
letely through the wall of the juncture and clearly shows how the crack pur-
ues the principal (hoop stress) planes in the walls of the vessel and nozzle
espectively.

Photoelastic stress fringe loops revealed a very small amount of dysym-
etry throughout flaw growth near the vessel wall.

ONDITIONS FOR APPLICATION TO COMPOSITES

A qualitative description of the cracking in an off-axis graphite-epoxy
aminate containing a stress raiser with a fiber to matrix modulus ratio ≈
00 was given at the first USA-USSR Conference with the aid of Fig. 10. This
escription indicated that cracks initiate
t the stress raiser and extend parallel
o the fibers in each layer. This is ac-
ompanied by delaminations (and some fi-
er breakage); the collective result be-
ng the development of a damage zone.
ig. 11 shows one of a series of cracks
hich initiated in the polyimide matrix
f a cross-ply graphite polyimide laminate
gain with fiber to matrix modulus ratio
100. These cracks are usually equally
paced and, after propagating through the
pper layer of the laminate in Fig. 11
arallel to the fibers, they arrest in
he resin rich region between the upper and lower

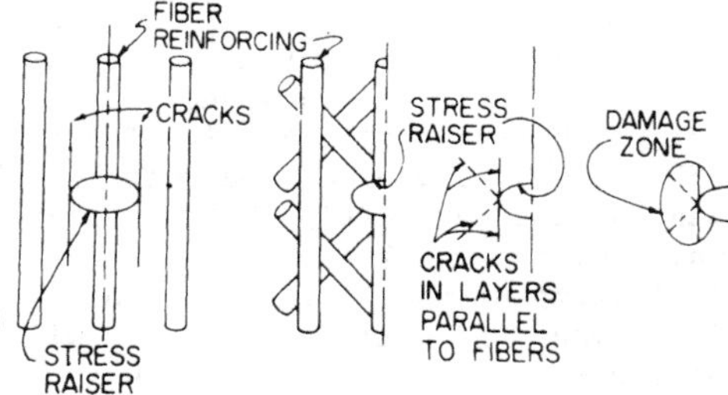

Fig. 10 Damage Zone in a Composite Laminate

laminate layers. Analysis [11] has shown that, when cracks cross interfaces between dissimilar materials or are located along bond lines, the order of the stress singularity is weaker than $r^{-1/2}$.

From our observations in the preceding section, we expect a crack in a single phase material in a field of strong gradients and constraints to change its direction in order to pursue the planes of maximum principal stress. It may be conjectured that the continuously changing direction of the crack as it grows through the upper layer in Fig. 11 results from a combination of it's attempt to follow a principal plane of stress created by triaxial constraints induced in the matrix by thermally or mechanically induced fiber tension and the crack's effort to find its way around the reinforcing fibers along paths of highest singularity order. When the crack reaches the resin rich interface between adjacent cross plies, it arrests upon emanating from the zone of high constraint in the upper layer. However, additional loading may cause the crack to interact with defects in the resin and/or other cracks to generate delamination.

Fig. 11 Transverse Crack in a Composite Laminate [ref. 13].

Efforts [12] to study crack growth in composites optically by producing fibers with refractive indices which match that of the matrix and treating the material as homogeneously anisotropic have met with only limited success to date. On the other hand, if one foregoes optical analysis of the fibers, one might consider substituting a stress freezing material for the matrix, and obtain a SIF distribution along the tip of a crack after it penetrates into the resin rich region of Fig. 11. The thickness of this layer, however, is of the order of only 0.025 mm, and this is an order of magnitude smaller than the range of most crack tip photoelastic work. Moreover, for accurate slice analysis, the crack front would have to be straight for at least one or two millimeters normal to the plane of Fig. 11. Because of these rather severe limitations, it would seem preferable to model the layers as homogeneous anisotropic materials but the resin-rich interface could still be an isotropic stress freezing material. One could then scale the thickness of the resin rich region upward by at least two orders of magnitude. Two dimensional mathematical models of this class are already available [11].

When the frozen stress method is used to predict both flaw shapes and SIF stributions for sub-critical flaw growth, it is subject to the following strictions:

i) The model material is elastic and incompressible.

ii) The model geometry and starter crack should be geometrically similar those of the prototype.

iii) Small scale yield conditions must prevail in the prototype.

iv) The fatigue loading of the prototype should be tension-tension.

To these must be added the requirement that the thermal coefficients of pansion between the laminate layers and the photoelastic material be matched , for thermal stress simulation, controlled. This is a severe restriction. fact, while these modifications are simple to describe, they may be very fficult to carry out in the laboratory with consistent quantitative accur- y.

Despite these drawbacks, the complexity of the problem is such that the sual features associated with the frozen stress approach may lend new in- ght into the three dimensional effects upon damage zone development in com- site materials.

KNOWLEDGEMENTS

The author wishes to acknowledge the contributions of his former students I. Jolles, W. H. Peters, W. T. Hardrath, T. S. Fleischman and A. Andonian d the stimulating discussions with his colleague, Prof. Robert Plunkett. He so acknowledges the support of the Oak Ridge National Laboratory under REG/CR-0640 ORNL/SUB/7015-2 and The National Science Foundation under Grant . Eng. 76-20824 for parts of this study.

FERENCES

] McGowan, J. J., Ed., "A Critical Evaluation of Numerical Solutions to the 'Benchmark' Surface Flaw Problem" (In Press), Experimental Mechanics, August 1980.

] Sih, G. C., Ed., "Three Dimensional Crack Problems", Mechanics of Fracture, Vol. 2, Noordhoff International Publishing Co., Leyden, 1975.

] Smith, C. W., "Use of Three Dimensional Photoelasticity in Fracture Mechanics and Progress in Related Areas", Chapter 1, Experimental Techniques in Fracture Mechanics - No. 2, A. S. Kobayashi, Ed., pp. 2-58, 1975.

] Smith, C. W., "Stress Intensity Estimates by a Computer Assisted Photoelastic Method", Fracture Mechanics and Technology, Vol. 1, G. C. Sih and C. L. Chow, Eds., Sitjhoff and Noordhoff, pp. 591-605, 1977.

[5] Smith, C. W., "Use of Photoelasticity in Fracture Mechanics" (In Press), Mechanics of Fracture, Vol. VI, G. C. Sih, Ed., Sitjhoff-Noordhoff International, 1980.

[6] Smith, C. W. and Peters, W. H., "Experimental Observations of 3D Geometric Effects in Cracked Bodies", Developments in Theoretical and Applied Mechanics, Vol. 9, pp. 225-234, 1978.

[7] Smith, C. W. and Peters, W. H., "Prediction of Flaw Shapes and Stress Intensity Distributions in 3D Problems by the Frozen Stress Method", Preprints of Sixth International Conference on Experimental Stress Analysis, pp. 861-864, Sept. 1978.

[8] Smith, C. W., "Observations of Three Dimensional Geometric Effects in Crack Growth and Implications for Composite Material Structures", Fracture of Composite Materials, G. C. Sih and V. P. Tamuzs, Eds., Sitjhoff and Noordhoff, pp. 145-158, 1978 (see also Journal of Mechanics of Composite Materials, No. 2, 1979, pp. 201-210 (In Russian).

[9] Smith, C. W., Jolles, M. I. and Peters, W. H., "Stress Intensities for Cracks Emanating from Pin-Loaded Holes", Flaw Growth and Fracture, ASTM STP 631, pp. 190-201, 1977.

[10] Smith, C. W., Peters, W. H., Hardrath, W. T. and Fleischman, T. S., "Stress Intensity Distributions in Nozzle Corner Cracks of Complex Geometry", NUREG/CR-0640 ORNL/SUB/7015-2, Oak Ridge National Laboratory, Oak Ridge, TN, 1979.

[11] Erdogan, F., "Fracture Problems in Composite Materials", J. of Engineering Fracture Mechanics, Vol. 4, pp. 811-840, 1972.

[12] Rowlands, R. E., Dudderar, T. D., Prabhakaran, R. and Daniel, I. M., "Holographically Determined Isopachies and Isochromatics in the Neighborhood of a Crack in a Glass Composite", Experimental Mechanics, Vol. 20, No. 2, pp. 55-56, Feb. 1980.

[13] Herakovich, C. T., Davis, J. G., Jr. and Mills, J. S., "Thermal Micro-Cracking in Celion 6000/PMR-15 Graphite/Polyimide (In Press), Proc. of Int. Conference on Thermal Stresses in Materials and Structures in Severe Thermal Environments, March 1980.

ECULIARITIES OF THE FRACTURE OF DRY AND WET COMPACT BONE TISSUE

I. V. Knets and A. E. Melnis

Academy of Sciences of the Latvian SSR
Riga, USSR

Determination of the character of compact bone tissue fracture has attracted he attention of many investigators for two reasons. First, compact bone tissue s a material the structure of which has been improved over many thousands of ears, and this fact ensures the optical mechanical functioning of bone in vi-al activities. Secondly, along with a very intensive development of new com-osite materials, the problem of increasing the load-carrying capacity by main-aining the same low specific weight is significant. A study of bone tissue omposite structure might give some new ideas for development of highly effec-ive structural composite materials.

Bone tissue has been considered as a composite material in a number of pa-ers. For instance, in [1-4], it was assumed to be a two-phase material con-isting of high modulus mineral fibers and soft protein matrix. The elastic haracteristics of a composite material were calculated on the basis of the inear mixture law taking into account the given moduli and volume fractions of oth components. The analysis of applicability of the different structural mod-ls for the two-phase material was presented in [5,6]. It was pointed out that he Voigt and Reuss models provide the lower and upper bounds of composite ma-erial elastic characteristics respectively, but a more reasonable agreement ith real bone tissue moduli could be reached by using the Hashin-Shtrikman nd Hirsch models. Two structure levels in bone tissue were pointed out in 7,8]. On the first level, the composite material is formed by viscoelastic ollagen fibers fulfilling the role of a matrix, with interconnected dispersed ineral crystals serving as the reinforcement. On the second level, this com-osite material is formed by a bonding matrix (amorphous ground substance) and steons, considered as large hollow reinforcing fibers.

In [9,10], the compact bone tissue is regarded as a three-phase (three-omponent) composite material with five different structure levels. The third tructure level has been chosen as the basic one upon viewing the compact bone issue as a solid anisotropic composite material in the phenomenological deter-ination of its elastic and deformation properties. On this level, the com-osite material is formed by a physically nonlinear interfibrillar substance erforming the role of a matrix, and collagen-mineral fibers, in which collagen icrofibrils are closely combined with mineral crystals by stereochemical bonds, etting up the united reinforcing elements.

However, during the investigation of the process of compact bone tissue fracture, the analysis has to be carried out on the fourth and fifth structure levels as well. On the fourth level, the composite material is formed by the bonding matrix and reinforcing structural elements - lamellae, which have the configuration of thin-walled plates, curved panels or cylindrical shells. Each lamella contain many collagen-mineral fibers with a certain orientation. On the fifth level, compact bone tissue is built up by the bonding substance and the embedded osteons - hollow cylinders formed by many concentrically located lamellae. Collagen-mineral fibers in the adjacent lamellae form the angles from 45 to 90°.

Compact bone tissue has specific fracture characteristics because of its peculiar structure. The survey of early publications (till 1967) on this subject was presented in [11]. However, during the last ten years, many new paper have appeared in which the bone tissue fracture is examined from the point of view of the fracture mechanics of composite materials.

For example, the fracture toughness of bone tissue in tension along the longitudinal axis of bone was investigated in [12]. The specimens had a transversally machined edge crack of length c and the tip of this artificial crack was specially formed to give it a constant radius of curvature r. It was determined that for a constant crack length c, a variation in the radius of curvature r does not affect the fracture stress σ^*_{11} significantly, but for a given radius of curvature r, the fracture stress σ^*_{11} is a linear function of $c^{-1/2}$. This allowed the critical stress intensity factor K_{Ic} = 4.6 MPam$^{1/2}$, the critical energy release rate (in fact, the released energy per unit of the created crack surface) G_c = 560 J/m^2, and the intrinsic crack length c_o = 340 mkm responsible for the fracture in a "perfect" (unnotched) bone section to be calculated.

The longitudinal shear of compact bone tissue was analyzed in [13]. It was determined that the fracture stress $\sigma^*_{13} = 0.75 \times \sigma^*_{23}$ = 50.40 ± 14.08 MPa, and the critical energy release rate G_c = 20720 ± 9310 J/m^2. The sum of fracture stresses, determined on the decalcified and deproteinized specimens, is significantly less than σ^*_{13} of normal bone tissue. The formation of cracks on the interface between osteons and the bonding matrix, and between lamellae in osteons are considered to be the basic mechanism of the fracture of bone tissue

The propagation of a crack in bone tissue along the longitudinal axis of bone during transversal tension was investigated in [14]. The specimens had a machined edge crack with a constant length along the osteons. It is found out that the critical stress intensity factor determined on the basis of linear elastic fracture mechanics is not affected by the specimen thickness (in the range from 1.85 to 3.82 mm), but it changes significantly (by 30%) even at a small change in material density (by 5%). The value of K_{Ic} = 4 MPam$^{1/2}$ at the density ρ = 2 g/cm^3. The value of G_c is determined from the experimental compliance calibration curve of specimens and it is found to be proportional to the corresponding K_{Ic}, i.e., $K_{Ic} = E^* G_c$, where E* is the effective modulus proosed for anisotropic materials in [15]. The experimentally determined value f E^* = 9420 MPa corresponds quite well with the theoretical modulus based on he transversally isotropic model of compact bone tissue.

In [16], the propagation of a crack along the longitudinal axis of bone in ransversal tension was also analyzed. But here, the length of an artificial rack was not kept constant and the crack propagation during loading was meaured microscopically. It is determined that an increase in the cross-head speed rom 1.7×10^{-6} to 33×10^{-6} m/sec produces an increase in the average crack veocity from 2.1×10^{-5} to 27×10^{-5} m/sec, the critical stress intensity factor $_{Ic}$ from 2.4 to 5.2 $MPam^{1/2}$, and G_c from 920 to 2780 J/m^2. The fracture occurs y slow propagation of a stable crack.

The fracture toughness of fresh compact bone tissue in bending was investiated in [17]. The specimens had a V-notch on the tension side. On some speciens, a pre-crack was introduced at the root of the notch. It is found out that ver the range of strain rates from 7×10^{-6} to 10^{-2} sec^{-1}, the average value of $_{Ic} = 5.7 \pm 1.4$ $MPam^{1/2}$, and is independent of the existence of a pre-crack. It ndicates, that in this case, the major effort in bone tissue fracturing is in he crack propagation rather than initiation. The average critical crack length s 0.36 mm, but the plastic zone size is much less - 0.0168 mm. This difference ustifies the applicability of linear elastic fracture mechanics to describe the racture toughness of bone.

In [18], the work required to fracture a compact bone tissue specimen in ending was determined. The specimens in the middle of span had a special tringular cross-section. It is determined that the average work-to-fracture (toughess) is 9030 ± 3270 J/m^2. Catastrophic crack propagation occurs in specimens ith large osteons and it causes a smooth fracture surface.

The propagation of a crack along the longitudinal axis of bone in transverse ension was investigated also in [19]. The specimens had a machined edge crack. t is found out that an increase of cross-head speed from 1.7×10^{-7} to 84 10^{-7} m/sec produces an increase in the crack velocity from 1.75×10^{-5} to 23.6 10^{-5} m/sec, K_{Ic} from 4.46 to 5.38 $MPam^{1/2}$, and G_c from 1736 to 2796 J/m^2. or this slow crack propagation, a relatively rough fracture surface is produced nd some osteon pull-out is observed. The fracture energy, being evaluated from he area under the load-deflection curve, for the range of cross-head speed menioned above also increases from 764 to 2125 J/m^2. But at the next investigated ross-head speed (170×10^{-7} m/sec), the transition from slow to catastrophic rack propagation occurs, resulting in a sudden decrease of fracture energy to nly 125 J/m^2.

In [20], it was pointed out that the compact bone tissue had three different odes of failure on the microstructural level. In the first case, interosteonic racture takes place forming the fracture surface around osteons through the cement lines. In the second case, osteonic fracture occurs in which the fracture urface propagates through single osteons and microsamples containing several osteons. In the third case, intraosteonic fracture is seen - the fracture surface goes through the interlamellar region between the collagen rich lamellae of single osteons.

As is generally known, the characteristics of the mechanical properties of compact bone tissue and, especially, the fracture behavior of bone depend significantly upon its moisture content. However, there are available data for tw ultimate moisture conditions of bone only - the relatively dry and very wet. I should be mentioned that these conditions are often interpreted differently by investigators. Some of them assume the bone tissue to be dry when it is kept a the room temperature, but others - when dried in the thermostat at 50 or even 105°C. Neglecting to evaluate the real moisture condition of bone samples has brought a wide experimental data scatter and even the controversial conclusions about the influence of moisture on the mechanical behavior of bone tissue.

The task of this paper is an experimental investigation of the mechanical b havior of compact bone tissue over a wide range of fixed values of moisture con tent.

The specimens for the investigation were taken from the left tibia mid-diaphysis of five men having died in accidents at the age from 20 to 30. The bon and the specimens of bone tissue were kept in polyethylene packages at the temperature from -4 to -7°C up to the time of testing. The period of storage from the moment of autopsy until the experiment did not exceed 16 days. The specime with a rectangular cross-section were cut along the longitudinal axis of bone. Their length was 100 ± 1 mm, width - 6.0 ± 0.2 mm and thickness - 1.0 ± 0.1 mm. Testing of specimens in uniaxial tension was carried out at a constant strain rate (0.0007 s^{-1}) in a loading equipment [21] developed in the Institute of Pol mer Mechanics of the Latvian SSR Academy of Sciences. Measuring of strains was performed by a special electro-mechanical strain gauge [22].

The experimental bone tissue was divided into 5 groups depending on moistur content. The moisture of the specimens (in percent) was determined as a relative content of water in the material: $W = (P-P_o)100/P_o$, where P_o is a weight of specimen after drying at the temperature T = 50°C for 48 hours. The first moisture group was formed by dry specimens which were conditionally assumed to be completely dry (W=0%). The second moisture group was formed by specimens which were kept after drying in a special chamber with a relative humidity of 65 ± 2%. After reaching equilibrium, the moisture content of the specimens was measured at W = 2.5%. The specimens of the third moisture group after drying were kept at a relative humidity of 80%. Their moisture W at equilibrium was 5%. The fourth moisture group included those specimens which were kept after drying at the relative humidity of 95%. Their moisture was W = 8.5%. The fift moisture group was composed of specimens kept in distilled water until full wat saturation was reached (W = 10.5%). In each group from 10 to 12 specimens were tested. Each experiment was performed at the temperature of 21 ± 1°C and relative air moisture of 65 ± 2%. After testing in tension, the fracture surface o specimens was investigated in a scanning electron microscope.

The average experimental stress σ_{11} - strain ε_{11} curves for five moisture contents are shown in Figure 1. It is seen that the stress-strain behavior sig nificantly depends upon the bone tissue moisture. If the σ_{11}-ε_{11} curve for dry

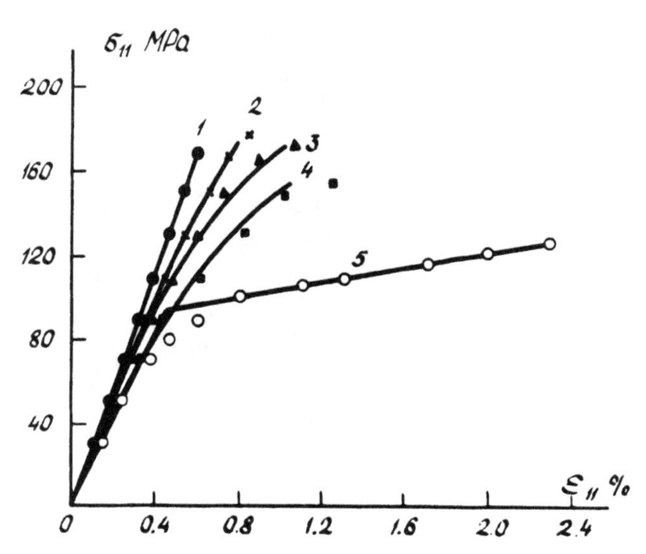

Fig. (1) - Dependence of stress σ_{11} - strain ε_{11} curve of compact bone tissue upon its moisture W: 0(1), 2.5(2), 5(3), 8.5(4) and 10.5%(5). Solid lines - approximation curves

ne tissue (curve 1) is linear till the very moment of fracture, then in other ses the nonlinearity of those curves is quite reasonable. With an increase of isture of the material, the stress level at which the nonlinearity appears de- eases. It should be mentioned that the knowledge of the region of linear σ_{11} $_{11}$ behavior for bone tissue is important because according to the opinion of ny investigators, the bone works under physiological loads in this region. The rve 5 (W = 10.5%) on Figure 1 presents special interest illustrating the sig- ficant effect of moisture on the mechanical properties of bone tissue. Its aracter is similar to the typical stress-strain curve of elastic-plastic mate- al with a strongly marked yield point. Such a behavior of wet bone tissue is ntioned in [23,24]. The determined in those investigations plastic strain is ual to 70% of the total fracture strain of compact bone.

The experimentally obtained stress-strain curves of compact bone tissue for e first four moisture groups are approximated by a cubic polynomial series:

$$\varepsilon_{11} = a_1(1+\alpha W)\sigma_{11} + a_2 W\sigma_{11}^3$$

ere the coefficients of approximation $a_1 = 3.55 \times 10^{-5}$ MPa^{-1}, $a_2 = 1.16 \times 10^{-8}$ a^{-3} and $\alpha = 2.77$. The average square deviation of the approximation for the ole series of curves is 4.9%.

It should be mentioned that the value of fracture stress σ_{11}^{*} also depends on the moisture of bone tissue which has to be taken into account in determination of the fracture strain ε^*_{11}. For the fifth moisture group (curve 5, Figure 1) the stress-strain curve was approximated by two straight lines. The stress-strain curves show that the content of moisture in bone tissue determines its modulus of elasticity, fracture stress and fracture strain, Figure 2. It is

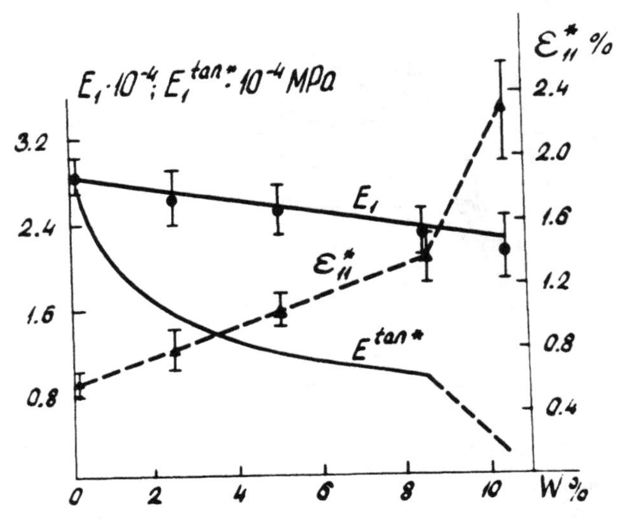

Fig. (2) - Dependence of initial modulus of elasticity E_1, tangent modulus of elasticity in fracture E_1^{tan*} and fracture strain ε^*_{11} upon the moisture W of compact bone tissue

seen that the increase in W causes the increase of ε^*_{11} and decrease of both E_1 and E_1^{tan*}. Especially interesting is the character of the change of E_1^{tan*}, ha ing three different regions of moisture influence. In the first region ($W \leq 5$ the small increase in W causes significant decrease in E_1^{tan*}. In the second r gion ($5\% < W \leq 8.5\%$), this modulus is practically constant, but in the third r gion ($W > 8.5\%$) the increase in W again causes a sharp decrease in E_1^{tan*}. The value of E_1^{tan*} for wet bone tissue decreases 21 times in comparison with the d tissue. The change of initial modulus of elasticity E_1 is more uniform and 28 less in absolute value - for a completely wet bone than for dry bone.

The experimental dependence of both fracture stress σ^*_{11} and yield stress σ on the moisture of bone tissue is shown in Figure 3. It is seen, that at smal values of moisture the stress σ^*_{11} does not depend on it, but the increase in m ture content causes sharp decrease of σ^*_{11}. For example, at the value of W = 1 the fracture stress σ^*_{11} is 30% lower than at W = 2.5%.

The yield stress of bone tissue decreases significantly at small values of moisture, but with the increase of W the intensity of the change of σ_{11e} decre

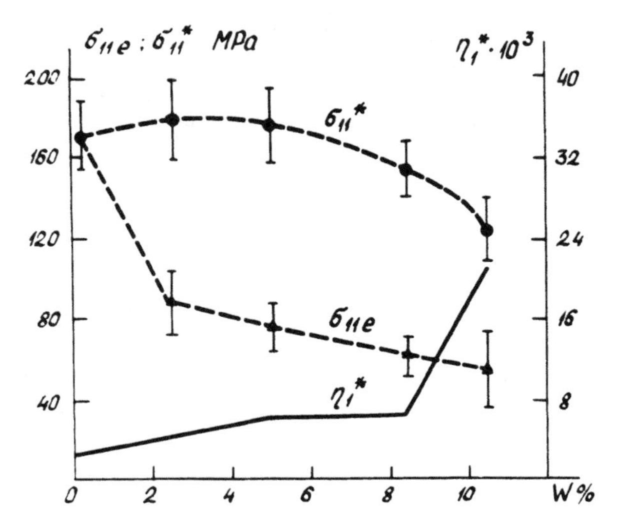

Fig. (3) - Experimentally determined dependence of fracture stress σ_{11}^*, yield stress σ_{11e} and parameter of resistibility to fracture η_1^* upon the moisture W of compact bone tissue

nd the relationship σ_{11e}-W becomes practically linear. Hence, the change of W t its small values affect the yield stress σ_{11e}, but at its large values, it ffects the fracture stress σ_{11}^*.

The important parameter of the deformation and strength properties of a maerial is the specific strain energy U_1. As it does not characterize the reistibility of the material to the mechanical loading in its full extent, we use he parameter of resistibility η, proposed in [4]. The maximum of this paramter determines the restibility to fracture and is determined as follows:

$$\eta_1^* = \frac{U_1^*}{\sigma_{11}^*}$$

here U_1^* is the specific strain energy at the moment of fracture.

The dependence of η_1^* upon the moisture W in bone tissue is presented on Figre 3 also. There are three regions where relationship η_1^*-W is different. In he first region, there is a uniform increase of η_1^* from its minimum value (0.003) t W = 0% to 0.0058 at W = 5%. In the second region, $(5 < W \leq 8.5\%)$ the value of η_1^* is practically constant (0.006), but in the third region, η_1^* increases signifiantly. The maximum of $\eta_1^* = 0.0215$ is found at W = 10.5%. The analysis of the tress-strain curve at this moisture shows that such an increase in the resistbility to fracture takes place because of plastic deformation of bone tissue.

Experimental data show that a change in moisture content simultaneously changes the bone tissue mechanical characteristics and causes some changes in the material structure. This is evidenced by an increase in both linear dimensions and, consequently, the volume of specimens. The average values of the relative change in linear dimensions along longitudinal (ε_{11w}), transversal (ε_{22w}) and radial (ε_{33w}) axis, and the volume ($\Delta V/V_0$) at five moisture levels are presented in Table 1. They have been calculated with respect to dimensions and volume of dry specimens, i.e., at W = 0%.

TABLE 1

Specimen Moisture %	ε_{11w}, %	ε_{22w}, %	ε_{33w}, %	$\Delta V/V_0$, %
2.5	0.17 ± 0.06	0.62 ± 0.15	0.53 ± 0.11	1.33 ± 0.20
5.0	0.35 ± 0.10	1.50 ± 0.24	1.04 ± 0.19	2.54 ± 0.36
8.5	0.59 ± 0.13	2.20 ± 0.33	2.06 ± 0.27	4.77 ± 0.60
10.5	0.69 ± 0.14	3.43 ± 0.63	3.06 ± 0.36	7.32 ± 0.71

The analysis of those data show that the swelling of bone tissue in the transversal and radial directions is 4-5 times higher than in the longitudinal direction. Such an anisotropy of bone tissue swelling could be explained by th orientation of collagen fibers mainly along the longitudinal axis of bone. Therefore, the resistibility of bone tissue in transversal and radial direction is lower than in the longitudinal direction. The parameter of resistibility to fracture, in accordance with [25], in the longitudinal direction is also 4-5 ti higher than in other directions. Consequently, the water in the bone tissue cr ates the condition of bulk tension and causes a larger increase in dimensions along the transversal and radial directions.

The water which acts on the organic components of bone tissue - collagen an interfibrillar substance - is changing not only the volume and spatial ratio be tween organic and mineral components of bone. It seems that the water affects the strength of adhesion between the components also and consequently, in its turn, significantly changes the mechanical behavior of bone tissue as a whole.

Upon complete extraction of water from the bone tissue, its organic components become brittle and due to shrinkage, there is a possibility of the develc ment of microcracks. Such a material might be considered as a linear elastic composite in which the brittle organic matrix is reinforced by disperse particles of mineral substance - the crystals of hydroxyapatites. The high stiffness of dry bone tissue until the very moment of fracture is attributed to a strong interaction between crystals and organic matrix. Therefore, the dry bor tissue fractures like a solid monolithic material. The angle between the fracture surface and longitudinal axis of bone is 90° approximately. The analysis of photomicrographs, Figure 4, shows that the fracture occurs due to the accumu lation and propagation of cracks.

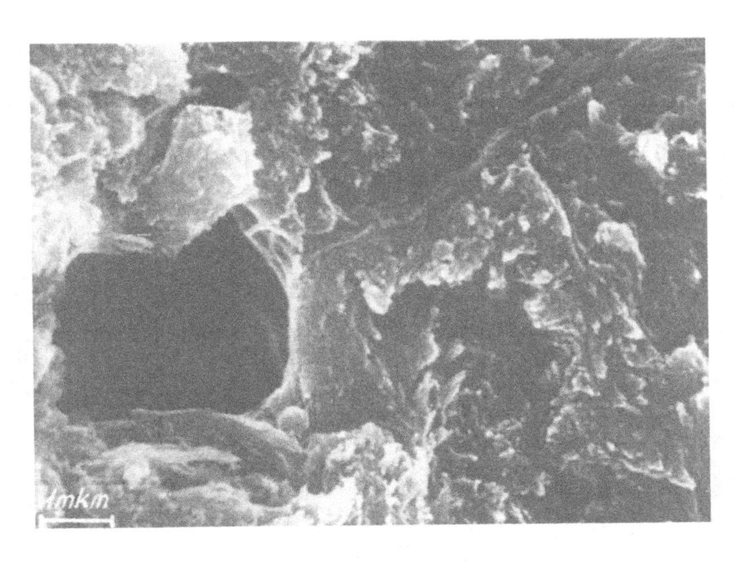

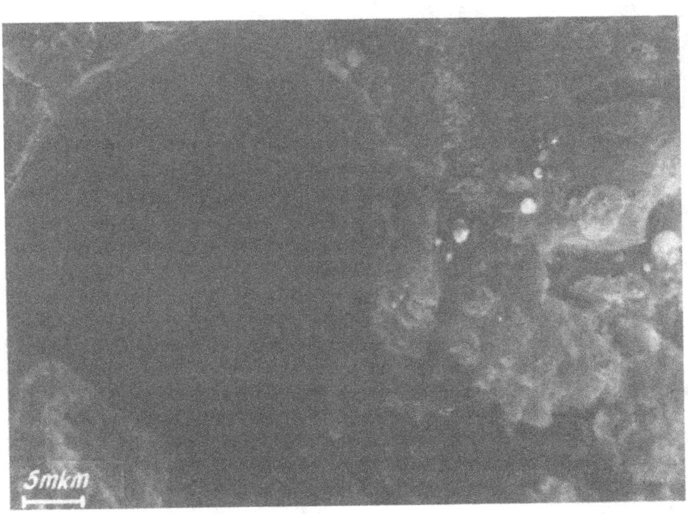

Fig. (4) - Fracture surface of dry compact bone tissue (W = 0%)

The stress-strain curves in loading and unloading indicate that the increase in bone tissue moisture W causes more expressed viscous behavior of bone. If for the first two moisture groups the hysteresis loop is not obtained, then for the next three groups it becomes obvious. Upon complete unloading, there is some permanent strain which reaches its maximum at W = 10.5%.

The stress-strain curve of the completely wet bone tissue, Figure 1, has a distinct yield point which indicates that a loss of material solidity, and a marked change in the modulus of elasticity occur. It might be supposed that in the wet bone, the mineral component is deformed elastically, but the organic component - plastically. Apparently, the role of the crystals of hydroxyapatite is to retard the yielding of the matrix. The fracture surface which forms an angle of 45° to the direction of specimen axis indicates that the material fracture is caused by shear stresses and there exist some sliding between the miner and organic components. Such an assumption is confirmed by photomicrographs, Figure 5, which show the pull-out of osteons and their fragments from the matri

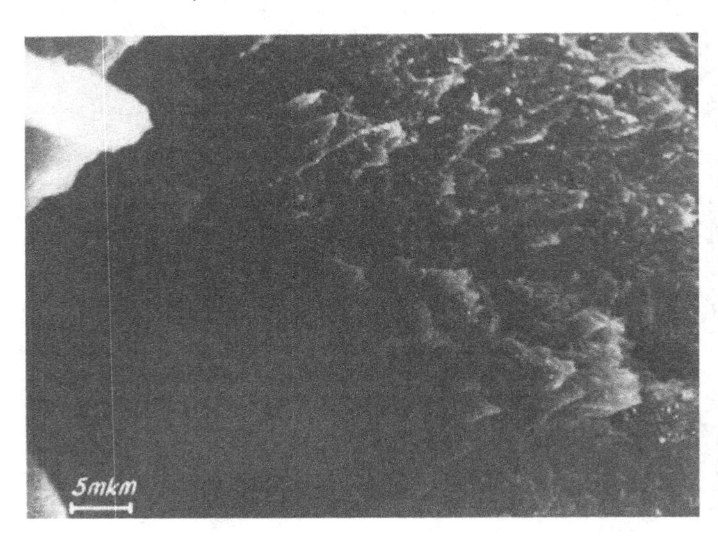

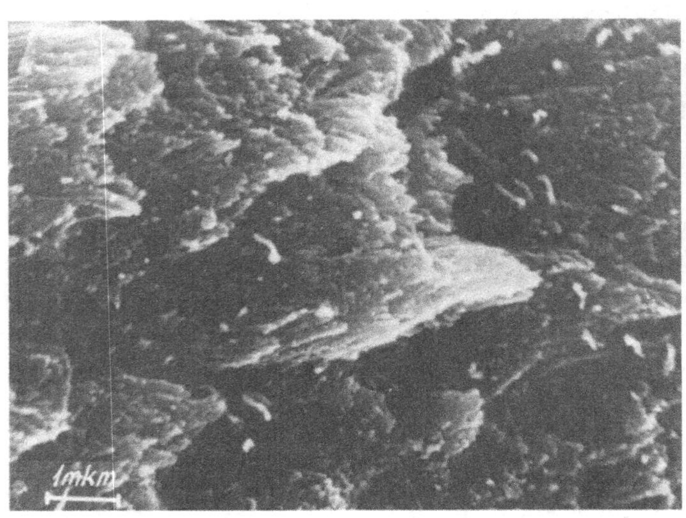

Fig. (5) - Fracture surface of completely wet compact bone tissue (W = 10.5%)

The question arises which moisture content of bone tissue is more characteristic of the normal physiological condition? It is known that the fresh compac bone tissue has the average water content of 10 to 20% and approximately a half of it is in a combined form. Therefore, the amount of combined water in the bo vivo is in the range of 5 to 10%. Consequently, the characteristics of the chanical properties of bone tissue, determined at the moisture W of 5 to 8.5%, rrespond to the parameters of bone tissue in vivo. The fracture surface of ecimens with such moisture content, Figure 6, is rougher in comparison to dry

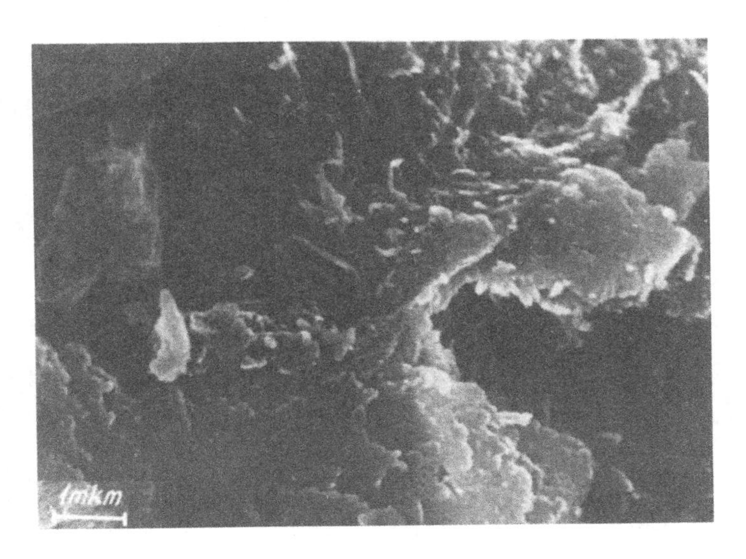

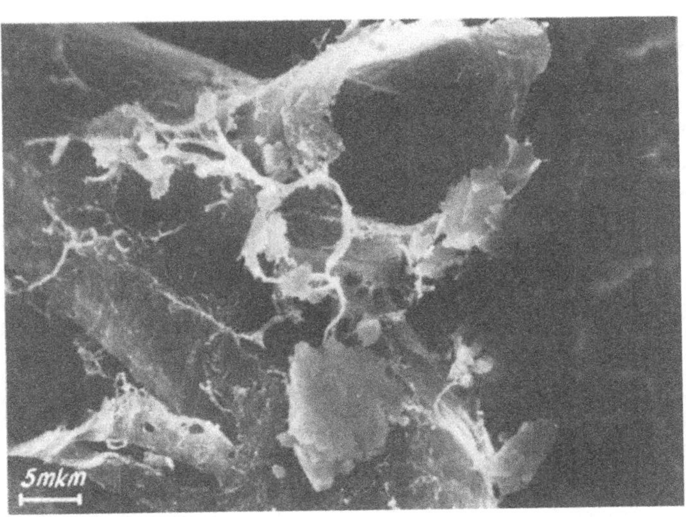

Fig. (6) - Fracture surface of compact bone tissue (W = 8.5%)

bone tissue, but the pull-out of osteons is expressed in a lesser extent as compared with a wet bone tissue. The fracture surface itself lies at an angle of 70 to 75° to the direction of the specimen axis.

CONCLUSIONS

The mechanical properties and fracture process of compact bone tissue significantly depend on its moisture. Bone tissue with a moisture of 5 to 8.5% corresponds to bone under normal physiological conditions. The largest strengt at rather high values of stiffness and resistibility to fracture of compact bon tissue is found to be exactly in this range of moisture. The fracture surface of such bone tissue is rough and the fracture occurs due to both normal and shear stresses.

REFERENCES

[1] Currey, J. D., "Three Analogies to Explain the Mechanical Properties of Bone", Biorheology, Vol. 2, No. 1, pp. 1-10, 1964.

[2] Piekarski, K., "Fracture of Bone", J. Appl. Physics, Vol. 41, No. 1, pp. 215-223, 1970.

[3] Welch, D. O., "The Composite Structure of Bone and its Response to Mechanical Stress", Recent Advances in Engineering Science, New York, Vol. 5, Par 1, pp. 245-262, 1970.

[4] Knets, I. V., Yanson, Kh. A., Saulgozis, Yu. Zh. and Pfafrod, G. O., "Resistance of Bone Tissue to Tensile Fracture", Polymer Mechanics, Vol. 7, No. 6, pp. 962-967, 1971. (Translation of "Mekhanika Polimerov" from Russian by Consultants Bureau, New York).

[5] Katz, J. L., "Hard Tissue as a Composite Material. I. Bounds on the Elastic Behavior", J. Biomech., Vol. 4, No. 5, pp. 455-473, 1971.

[6] Piekarski, K., "Analysis of Bone as a Composite Material", Int. J. Eng. Sci., Vol. II, No. 6, pp. 557-565, 1973.

[7] Cooke, F. W., Zeidman, H. and Scheifele, S. J., "The Fracture Mechanics of Bone - Another Look at Composite Modeling", J. Biomed. Mater. Res. Symposium, No. 4, pp. 383-399, 1973.

[8] Katz, J. L., "Hierarchal Modeling of Compact Haversian Bone as a Fiber Reinforced Material", 1976 Advances in Bioengineering, ASME, New York, pp. 17-18, 1976.

[9] Knets, I. V., "Mechanics of Biological Tissue. A Review", Polymer Mechani Vol. 13, No. 6, pp. 434-441, 1977. (Translation of "Mekhanika Polimerov" from Russian by Consultants Bureau, New York).

[10] Knets, I. V., "Fracture of Compact Bone Tissue", Fracture of Composite Materials, Sijthoff and Noordhoff, Alphen aan den Rijn, pp. 303-310, 1979.

[11] Herrmann, G. and Liebowitz, H., "Mechanics of Bone Fracture", Fracture, Academic Press, New York, London, Vol. 7, pp. 771-840, 1971.

[12] Bonfield, W. and Datta, P. K., "Fracture Toughness of Compact Bone", J. Biomech., Vol. 9, No. 3, pp. 131-134, 1976.

[13] Saha, S., "Longitudinal Shear Properties of Human Compact Bone and its Constituents, and the Associated Failure Mechanisms", J. Mater. Sci., Vol. 12, pp. 1798-1806, 1977.

[14] Wright, T. M. and Hayes, W. C., "Fracture Mechanics Parameters for Compact Bone - Effects of Density and Specimen Thickness", J. Biomech., Vol. 10, No. 7, pp. 419-430, 1977.

[15] Sih, G. C., Paris, P. C. and Irwin, G. R., "On Cracks in Rectilinearly Anisotropic Bodies", Int. J. Fracture Mech., Vol. 1, No. 3, pp. 189-202, 1965.

[16] Bonfield, W., Grynpas, M. D. and Young, R. J., "Crack Velocity and the Fracture of Bone", J. Biomech., Vol. 11, No. 10/12, pp. 473-479, 1978.

[17] Robertson, D. M., Robertson, D. and Barrett, C. R., "Fracture Toughness, Critical Crack Length and Plastic Zone Size in Bone", J. Biomech., Vol. 11, No. 8/9, pp. 359-364, 1978.

[18] Moyle, D. D., Welborne, J. W. III and Cooke, F. W., "Work to Fracture of Canine Femoral Bone", J. Biomech., Vol. 11, No. 10/12, pp. 435-440, 1978.

[19] Behiri, J. C. and Bonfield, W., "Crack Velocity Dependence of Longitudinal Fracture in Bone", J. Mater. Sci., Vol. 15, pp. 1841-1849, 1980.

[20] Frasca, P., Harper, R. A. and Katz, J. L., "Mechanical Failure on the Microstructural Level in Haversian Bone", Fracture 1977, Vol. 3, Waterloo (Canada), pp. 1167-1172, 1977.

[21] Knets, I. V., Krauya, U. E. and Vilks, Yu. K., "Acoustic Emission in Human Bone Tissue under Uniaxial Tension", Polymer Mechanics, Vol. II, No. 4, pp. 589-593, 1975. (Translation of "Mekhanika Polimerov" from Russian by Consultants Bureau, New York).

[22] Vilks, U. K., "The Strain Gauge", Patent USSR No. 355486, 1972.

[23] Burstein, A. H., Currey, J. D., Frankel, V. H. and Reilly, D. T., "The Ultimate Properties of Bone Tissue: The Effect of Yielding", J. Biomech., Vol. 5, No. 1, pp. 35-44, 1972.

[24] Reilly, D. T. and Burstein, A. H., "The Mechanical Properties of Cortical Bone", J. Bone Joint Surg., Vol. 56A, No. 5, pp. 1001-1022, 1974.

[25] Knets, I. V., Saulgozis, Yu. Zh. and Yanson, Kh. A., "Deformability and Strength of Compact Bone Tissue in Tension", Polymer Mechanics, Vol. 10, No. 3, pp. 419-423, 1974. (Translation of "Mekhanika Polimerov" from Russian by Consultants Bureau, New York).

[26] Knets, I. V., Pfafrod, G. O. and Saulgozis, Yu. Zh., "Deformation and Fracture of Hard Biological Tissue", Zinatne, Riga, 1980 (in Russian).

LAWS AND DEFECTS OF STRUCTURAL CARBON FIBERS

I. L. Kalnin

Celanese Research Company
Summit, New Jersey 07901

NTRODUCTION

Most of the so-called high performance fibers, such as carbon, boron, or lass, used as the reinforcement in structural composites are inherently brittle. evertheless, the utilization of brittle fibers in composites intended for long erm dynamic load-bearing applications is growing rapidly. Meanwhile, considerble efforts have been made to determine the prevailing mechanisms of the brittle fiber failure, particularly of carbon fibers (CF) and their composites.

The fracture of different types of CF has now been studied for over a decade since the advent of the technology for converting commercial polyacryonitrile (PAN) and rayon yarns into the corresponding CF assemblies. The accumulated evidence indicates clearly that the fracture of a CF originates at one of a number of structural defects (also called flaws) that are distributed throughout the CF, including the surface. According to the postulates of brittle fracture, the failure always takes place at the most severe flaw present in the test length of the stressed fiber. The purpose of this paper is to review the origin and the characteristics of the flaws found in the CF, the effect of these flaws on the CF strength and fracture strain, and the probable fracture mechanisms. The emphasis is on CF made by pyrolysis of PAN or rayon, since information on the flaws present in the CF made from pitch is still rather scarce. The present indications are that the ex-pitch CF contains the same types of flaw found in the ex-PAN fibers [1].

THE TYPES OF FLAWS

Extrinsic flaws are introduced into the fiber structure either during the fabrication of the polymeric precursor or during its conversion into the CF by controlled pyrolysis. The process steps involved in the fabrication of a typical "wet spun" PAN fiber precursor are: a) extrusion of an aqueous or non-aqueous solution containing 10-20 wt. % of the polymer through a multifilament spinneret into a suitable liquid nonsolvent; b) controlled coagulation of the extrudate to form the filaments, and c) washing, stretching, drying and collection of the processed filament bundles. Figure 1 gives a diagram of a typical laboratory spinning apparatus and the attendant processing steps.

The subsequent pyrolysis process of the precursor fiber, Figure 2, is comprised of (a) so-called "stabilization" step during which the fiber is rendered

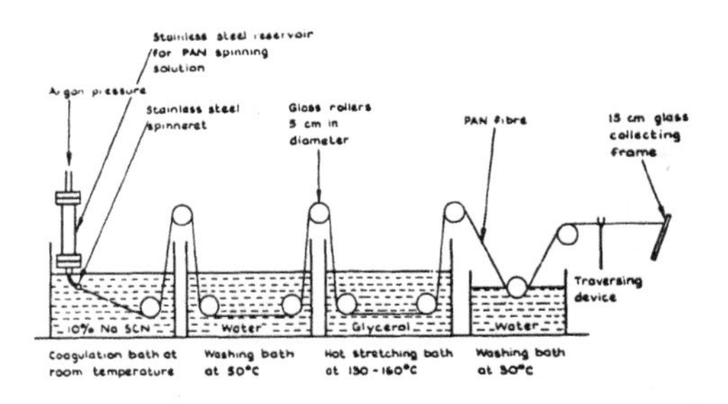

Fig. (1) - Diagram of wet spinning of polyacrylonitrile fibers on laboratory scale [13]

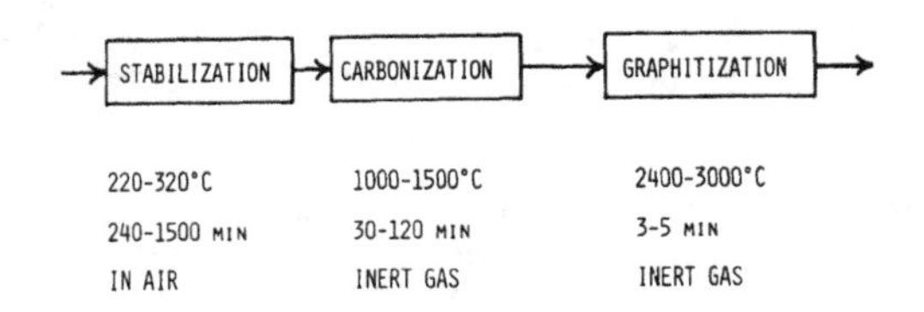

Fig. (2) - Diagram of the polyacrylonitrile fiber pyrolysis stages

infusible, followed by (b) "carbonization"-heating the stabilized fiber to 1100 -1500°C in an inert atmosphere as a result of which a "high strength" (or Type II) fiber containing ~90 wt. % carbon is formed. An additional heat treatment in the 2500-2900°C range is added if a "high modulus" (or Type I) CF product is desired. The published literature indicates that the fracture-inducing flaws may be introduced into the fiber during every one of the above mentioned proces sing steps. These flaws can be divided into two groups - internal flaws and su face flaws, depending on whether they are located inside or at the fiber surfac

INTERNAL FLAWS

The internal flaws are either cavities or particulate impurities (inclusion of various shapes and sizes. The largest cavities encountered are long prisms or needles, up to 4 μm in width and < 100 μm long. Having been observed in bot the CF and its precursor fiber, they are believed to originate from certain deficiencies in the spinning process, such as gas bubbles in the spinning solutio or local overstress in the spin line [2,3]. The observation that these defects are present in some fibers while missing altogether from others, supports this conclusion. Other long cylindrical cavities, located in the center of the CF a not seen in the precursor, are reportedly due to volatilization of an insuffici stabilized fiber core [4,5]. Most of the cavities, seen both before and after e fiber pyrolysis, are substantially smaller up to a few μm in width, 5-10 μm length and are either irregular or diconical in shape, as seen in Figure 3. e former are attributed to spinning process faults, such as irregular extrusion too rapid coagulation, while the latter have been shown to be microcracks

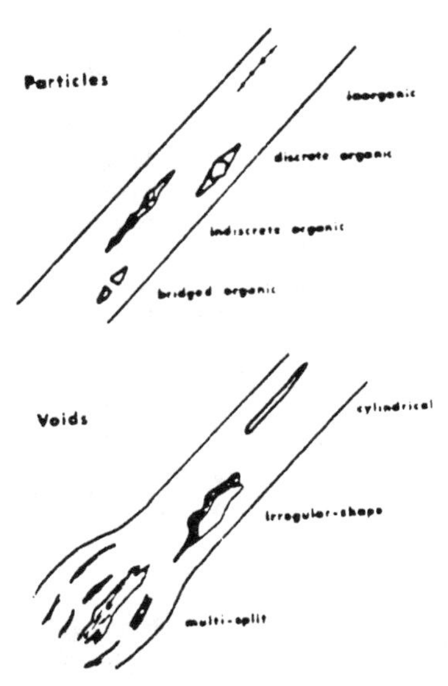

Fig. (3) - Schematic representation of types of internal defects observed in polyacrylonitrile precursor fibers [6]

enerated by stresses around the rigid impurity inclusion during the spinning. inally, CF heat treated above ca. 1800°C often contain elliptical or onion-haped cavities, 1-3 μm wide. These appear to be formed by inorganic impurity olatilization or thermal decomposition at high temperatures [7,8]. It should e noted that virtually all of the cavities are oriented with their longest di-ension closely parallel to the fiber axis; consequently, their critical dimen-ions are the transverse ones.

Particulate impurities can be organic as well as inorganic. The former are elatively large in size and are often surrounded by a diconical cavity. Little s known about their origin and properties except that they are birefringent and elt around 200°C [6]. The inorganic particles are generally smaller, <1 μm and o not release enough volatiles to create cavities in the high strength CF. At igher carbonization temperatures, >1800°C, however, they appear to generate bub-le-like cavities which contribute to fiber fracture [7].

The smaller cavities and inclusions present in the CF have been conveniently bserved in detail by means of a high voltage transmission electron microscope - n excellent tool for locating dispersed particles or holes >0.1 μm size [9]. rystallites of three-dimensional graphite found primarily in the high modulus CF constitute another type of internal flaw. Such crystallites have been reported to be present on the walls of the elliptical cavities, apparently having been formed by conversion of the underlying carbon matrix into graphite by catalytically active volatized impurities. Because of their large anisotropy and mis-orientation with respect to the fiber axis, they have been regarded as sources of critical microcracks [10].

SURFACE FLAWS

In the majority of cases, particularly for the high strength CF, the tensil fracture will initiate at the fiber surface and propagate transversely to the f ber axis. The most conspicuous and damaging flaw sites are those at which two or more fibers have fused together during the stabilization and broken apart later, leaving a jagged groove or pit in one fiber and a corresponding protrusi on the other one [11]. Other clearly noticeable but less damaging surface flaw are: etch pits arising from severe surface treatment [12], fibrillation caused by excessive stretching, and subsurface cavities [5]. The most common ones, ho ever, are granular impurities, <1 μm in size, deposited on the fiber from the n mally dusty ambient atmosphere during any or all of the fiber processing steps and fused to the surface during the pyrolysis [13]. With both the internal cav ties and the larger surface impurities eliminated, the fracture initiates at th small surface pits and granules, but at substantially higher strength levels th before [14]. Finally, when high strength CF is tested by a bending technique i which only a short length, ~0.05 μm is under high stress, the fracture may orig nate at a surface site where no flaw is seen by SEM [15].

INTRINSIC FLAWS

As the flaws associated with processing inadequacies are eliminated, the i herent irregularities of the fiber microstructure may limit the realizable frac ture stress levels. Several theories have been proposed to explain the fractu of flaw-free CF, each utilizing different microstructural models and focusing different crack initiation mechanisms. Cooper and Mayer [16], employing a mode put forward by Johnson and Tyson [17] according to which the CF is built-up of stacks of oriented graphitic crystallites, 1-10 μm in size, separated by twist tilt boundaries, suggest that the crystallites contain glissile dislocations wh under increasing shear stress, are driven toward the boundaries. There the re-sulting dislocation pile-ups generate progressively increasing stresses which p duce microcracks that eventually coalesce into a critical crack. According to theory, the fracture strength is inversely proportional to the square root of t basal plane crystallite size, Figure 4, which means that it decreases with inc ing Young modulus. Experiments utilizing relatively flaw-free CF indicate, how ever, that the fracture strength is nearly independent of the modulus [14].

Williams et al [18] carried out filament bending tests on ex-rayon CF obta ing stress-strain curves which became non-linear at >0.6% strain and assumed th shape of tensile yield curves. Upon release of the stress, most of the deform tion was recoverable. Utilizing the fibrillar microstructural model of Bacon a Silvaggi [19], the authors attribute the inelastic non-linearity to interfibri debonding and suggest that the intrinsic CF failure takes place progressively local yielding of the interfibrillar regions following the mechanism proposed Marsh [20] for unflawed glass. They do not, however, rule out the occurrence a purely brittle, (Griffith's type), fracture mechanism at this stage.

For ex-PAN fibers, Tysor [21] has proposed an intrinsic fracture mechanism fundamentally analogous to that of Williams et al, but based on a more detaile structural model which pictures the fiber as a two-phase composite material co

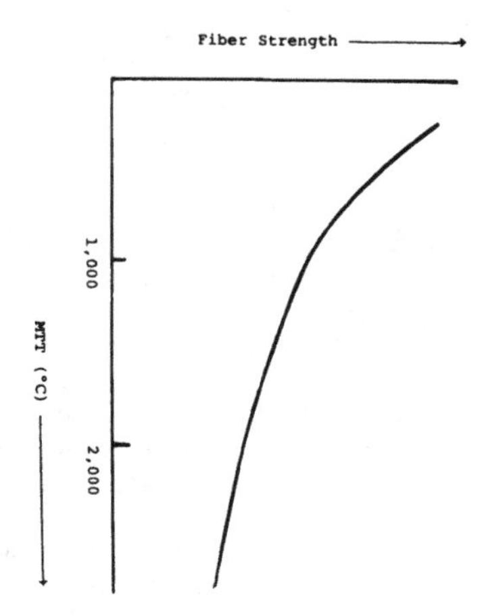

Fig. (4) - Proposed variation of CF strength with heat treatment temperature, assuming validity of the dislocation pile-up mechanism [16]

-ised of a highly oriented crystalline ribbon matrix enclosing more or less longated pockets which may be nearly empty, as in the case of high modulus CF, filled by highly disordered low density carbon, as in the case of high strength F, Figure 5. The disordered phase is the one undergoing the "yielding" during

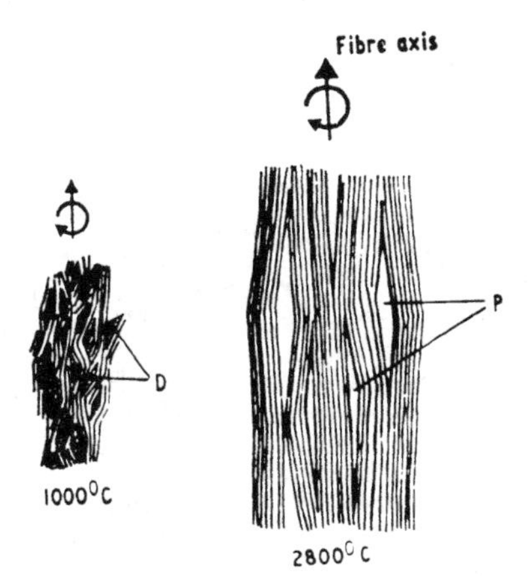

Fig. (5) - Schematic diagram of possible carbon fiber microstructure. The discontinuous phase consists of pores that may be empty (P) or filled with disordered carbon (D) of varying density [21]

the fiber loading as described by the theory of Marsh. From the ordered and dis-ordered phase dimensions, respectively, calculated from small angle x-ray scattering data, Tyson obtains a yield stress of ca. 4 GPa. The theory also shows that the fracture strain of CF heat treated at various temperatures increases linearly with increasing disordered phase density or, in other words, with decreasing heat treatment temperature (HTT). Data from other sources, [23], indicate, however, that the fracture strains may peak at HHT of 1200-1400°C.

Reynolds and Sharp [22] considered the ex-PAN CF as an assembly of undulati ribbons composed of graphite crystallites with their basal planes oriented gene ally along the fiber axis. Although the mean angle of misorientation, ϕ, is sh by x-ray diffraction to be fairly small, 5-20°, decreasing with increasing Youn modulus, the authors estimate that a significant portion of the crystallites wi be misoriented at much greater angles, ~2ϕ. As the fiber is strained in tensio the most misaligned crystallites are subjected to a greatly enhanced (10-20 fol shear strain so that they will progressively fail in shear, generating off-axis microcracks until a critical crack is nucleated and fracture occurs. From the available fracture strain data, the authors infer that the limiting shear strai for a contemporary CF is 20%. If such fracture mechanism applies, the fracture strength should increase linearly with increasing Young modulus [8].

Such behavior, however, has been noted only in hot stretched CF. In a late paper, Reynolds et al [23] provide evidence that the misorientation flaws can b quite detrimental by presenting calculations to show that the shear stress enhancement can generate failures at tensile stresses and strains as low as 1.3 - 1.5 GPa and 0.3 - 0.8%, respectively, if sufficiently many highly misoriented (ϕ ~ 50-70°) crystallites are present.

FLAW MODIFICATION

In the absence of definitive information as to the criticality of the observed flaws, the approach often taken was to implement changes in fiber processing that might somehow affect the flaws and determine the attendant change in the mean fracture stress, $\overline{S}_f$. Already in the early stages of the CF developmen Johnson et al observed that oxidative surface etching significantly increased the $\overline{S}_f$ of Type II CF and attributed the increase to elimination of the worst su face flaws leaving behind less severe internal as well as surface flaws [2,3]. To elucidate the effects of rigid inclusions and cavities found in the PAN fibe Moreton and Watt developed techniques for spinning under nearly dust-free and v free conditions and thus eliminated the optically observable internal cavities impurities as well as the surface impurities >1 μm in size. As a result, the m surface flaw density in the precursor decreased from ca. 0.7 mm^{-1} to 0.02 mm^{-1}, and the $\overline{S}_f$ of CF made from this precursor increased by 80% with the fractures n originating solely at the surface. Relatively high $\overline{S}_f$ values were obtained on heat treated even at high temperatures, >1600°C, and were independent of the ga length. Deliberate contamination of spun fiber surface by fine particles of si ica, ferric oxide or carbon black followed by pyrolysis at 800-2400°, on the ot hand, resulted in a progressively greater strength degradation above 1000°C unt at 2400°C the $\overline{S}_f$ was only ~1 GPa - one-third that of the uncontaminated CF. Th fracture was found to originate at granules or pits generated by the contaminan [23]. Finally, hot stretching at >2300°C increases both the modulus and the $\overline{S}_f$

f ex-rayon and ex-PAN CF [24,25]. Although such treatment obviously must alter he critical flaw characteristics, neither the nature of these flaws nor the echanism by which they are altered have been, as yet, identified.

TATISTICS OF FLAWS AND THEIR RELATION TO FRACTURE STRENGTH

A characterization of both the observed flaws and the fiber fracture strengths y their distribution parameters is a prerequisite before correlations between hem can be established quantitatively. In addition to size, measured transverse-y to the fiber, the flaws have been characterized by their mean linear density, .e., number of flaws per unit length (usually per mm) or, alternately, by the requency of the flaw-free segment length, determined over a range of test engths (gauge lengths).

The most complete flaw analyses have been done by Thorne on a number of com-nerical PAN precursors as well as Type II CF prepared from some of these [6,26] nd by Sharp et al on both Type I and II CF [7-9]. Considering all of the opti-cally observable flaws, both cavities and impurities, as one population, Thorne found that the mean flaw density (MFD) a) varies over a wide range, 0.03 - 2.3 mm^{-1}, for the different precursors and b) may decrease substantially with increasing gauge length as considerable flaw-free lengths are encountered. For the majority of the precursors, the MFD distribution at a given gauge length could be repre-sented by a Poisson type frequency distribution. In cases in which this was not possible, the probability of obtaining a flaw-free gauge length was calculated empirically from the measured flaw-free segment length distribution. Subsequent-ly, the same kind of statistical analysis was applied to three different PAN lots (MFD of 0.01, 0.07 and 1.06 mm^{-1}, respectively) followed by stretching and pyrol-ysis at 1000°C to CF. The resulting probabilities of either the flaw or flaw-free length distributions of the precursor were compared to previously published fracture probabilities for the CF at 0.05 - 100 mm gauge length [27]. For one precursor lot, the agreement was good, whereas for the others, the probability of fracture at a given gauge length was much smaller than that predicted by the in-ternal flaw distributions. Reasonable agreement could be obtained by assuming that only ca. 10% of the supposedly critical internal flaws would induce fracture, with the rest of the fractures being initiated by some unidentified surface flaws.

Sharp et al [8] measured internal flaw distributions by high voltage TEM over unspecified gauge lengths on both Type I and Type II CF made from two different commercial precursors. Although they saw substantial differences in MFD and maximum flaw widths, these did not correlate with the fracture stresses. In an earlier paper [7], the authors broke some 35 fibers in tension and examined the fractured ends, finding that ca. two-thirds of the fibers fractured across a cav-ity the diameter, d, of which could be measured. A plot of the d against $\overline{S}_f$ (at 20 mm gauge length) showed that a Griffith type equation, $S_f \alpha d^{-1/2}$, does not hold, Figure 6. Assuming the validity of Griffith's relationship, nevertheless, and the graphite surface fracture energy of 4.2 J/m^2 [28], the calculated critical crack size is only about 0.1 - 0.3 μm. The authors propose that such cracks are present in the walls of cavities, having originated from the fracture of highly graphitic lamellae formed by catalytic graphitization during the fabrication of the Type I CF at HTT > 2000°C. An effort to relate the observed fracture flaws, both inter-nal and surface, to the fracture strength was also made by Whitney and Kimmel [5]. They conclude that most of the critical flaws are indeed 1-2 μm in size. Conse-

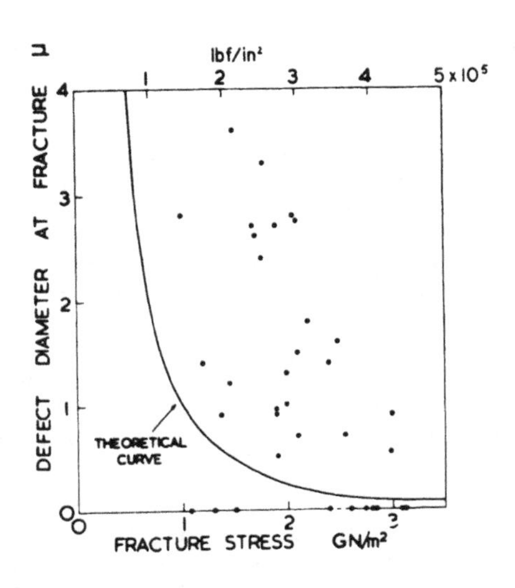

Fig. (6) - Fracture stress versus the measured internal defect diameter. The curve gives the theoretical Griffith relationship assuming basal surface fracture energy of graphite, 4.2 J/m^2 [7]

quently, assuming the validity of the Griffith relationship, the surface fractu energy is much higher, 15 - 55 J/m^2, rather than the 4.2 J/m^2 pertaining to per fect graphite. This surface energy enhancement is attributed to plastic deform tion taking place during the fracture process.

A more popular approach to flaw characterization is through determination o the fracture strength distributions over a range of gauge lengths. Most CF ten sile strength distributions can be satisfactorily represented by a two or three parameter Weibull distribution or the conventional Gaussian one [29]. In eithe case, if the dispersion of the fracture strength values in the given gauge leng range is similar, a plot of the $\overline{S}_f$ against the gauge length, L, will be a strai line on a log-log scale. A change in slope, on the other hand, may indicate th presence of more than one kind of flaw population. The first such study on exp mental CF prepared from the same precursor by pyrolysis at 1000 - 3000°C was re ported by Moreton [27], who found that the $\overline{S}_f$ decreases exponentially with L (5-50 mm range); the slope being approximately the same in all cases, ~30% stre drop per decade of L. The coefficients of variation (CV) calculated assuming a Gaussian distribution were rather large, 21-38%. Concurrent attempts to fit th measured distributions to a "weak link" model [30] were not particularly succes ful. McMahon measured $\overline{S}_f$ distributions of ten commercial CF over the L range o 2.5 - 76 mm [31]. His data, like Moreton's were fitted to an exponential $\overline{S}_f$ ve sus L relationship although both could have been linearized also on a log-log plot [32]. The $\Delta\overline{S}_f/\Delta$ log L values vary widely - from ca. 10% to 33% per decade

L, and the CV scatter is also large, 10 - 27%. Both of these studies indicate e existence of one type of critical flaw distribution but with significantly ffering frequencies and severities. The strongest fibers are characterized by ther widely and unevenly scattered small flaws, while the weakest ones possess relatively evenly distributed population of larger critical flaws. A more re- nt study [32] investigating the $\overline{S}_f$ of Courtaulds Type I and II CF over a L nge of 0.5 - 50 mm used a two parameter Weibull distribution to analyze the ta collected from two sets of samples of each CF type. Even so, the decrease strength with length varied from 16% to 33% per decade of L for the different mples, and the CV values calculated from the Weibull parameter, m, showed sub- antial scatter - 13 to 24%. In particular, it was observed that the Weibull m lues were different at different L and that two of the four sets of CF tested owed much lower $\overline{S}_f$ at L = 0.05 mm than predicted by the Weibull relation at rger L. This was attributed to a possible existence of a bimodal flaw distri- ition consisting of a small number of large flaws and a large number of small osely-spaced flaws. On the other hand, the decrease of the $\overline{S}_f$, occurring only t the shortest L value, might be due to a longer effective gauge length arising rom the so-called "clamp effect" [33].

JMMARY AND CONCLUSION

The flaws initiating CF fracture may be found in the fiber interior (inter- al flaws), at the surface (surface flaws) or may be an integral part of the fi- er microstructure (intrinsic flaws). The internal flaws, which can originate uring all stages of the fiber processing, are associated with a) cavities of arious shapes and sizes, more or less elongated in the fiber direction, b) solid nclusions containing organic or inorganic matter, and c) three-dimensional highly raphitic crystallites. The most damaging surface flaws-gaps or protrusions ormed by breaking apart of previously fused fiber are comparatively rare. The ost common ones are etch pits and irregular particles that can vary in size rom a few μm to below the limit of observability by SEM. These mostly originate rom dust particles deposited onto the fiber surface during the processing. In ddition, microstructural defects such as dislocations, disordered interfibril- ar carbon and misoriented graphite crystallites have been proposed as intrinsic ources of fracture-initiating flaws. Useful qualitative information regarding he nature of flaws and their effect on fracture strength has been obtained by leliberate process alterations such as application of various surface treatments, limination of ambient dust during the spinning and pyrolysis or, conversely, ontamination of the precursor with dust-like particles prior to pyrolysis.

Attempts to correlate the statistical features of the observed flaws with the fracture stresses lead to the following conclusions: a) the internal flaws are much less effective in initiating fracture than their statistics indicate; b) internal flaws up to 3 μm in size become critical at larger fracture stresses than expected from the Griffith relation as applied to graphite; c) internal flaws may act synergistically with the surface flaws to initiate fracture; and d) the Griffith surface fracture energy term may be many times larger than that pertain- ing to fracture of pure graphite because of a large contribution by the plastic deformation in front of the fracture crack.

Studies of the fracture statistics at different test lengths indicate that a) there is a large variability in the distribution parameters, not only for dif- ferent CF brands, but also between samples of the same kind of fiber, indicating a large fluctuation in size and density of the critical, presumably unimodal flaws, and b) in some fibers, the flaw distribution may be bimodal, composed of a small number of widely-spaced large flaws and a large number of closely-space small flaws. When the larger flaws are eliminated, the mean fracture stress be comes independent of the test length. Finally, the experimental results indica that the CF strength is presently limited at ca. 3 GPa by small surface flaws. If these can be eliminated, fracture strengths in the range of 4.5 - 6 GPa are predicted.

REFERENCES

[1] Bacon, R., Phil. Trans. Royal Soc. London, Ser. A, 294, p. 437, 1980.

[2] Johnson, J. W., J. Appl. Poly. Sci., Symposia, 9, p. 229, 1969.

[3] Johnson, J. W. and Thorne, D. J., Carbon, 7, p. 659, 1969.

[4] Johnson, J. W., Rose, P. G. et al, 3rd Conf. on Ind. Carbon and Graphite, Soc. Chem. Ind., London, p. 443, 1970.

[5] Whitney, W. and Kimmel, R. M., Nature Phys. Sci., 237, p. 93, 1972.

[6] Thorne, D. J., J. Appl. Poly. Sci., 14, p. 103, 1970.

[7] Sharp, J. V. and Burnay, S. G. in Carbon Fibers, The Plastics Institute, London, p. 68, 1971.

[8] Sharp, J. V., Burnay, S. G. et al, in Carbon Fibers, The Plastics Institut London, p. 25, 1974.

[9] Sharp, J. V. and Burnay, S. G., in Proc. 7th Int. Congress on Electron Microscopy, 1, p. 49, 1970.

[10] Burnay, S. G. and Sharp, J. V., J. Microscopy, 97, p. 153, 1973.

[11] Murphy, E. V. and Jones, B. F., Carbon, 9, p. 91, 1971.

[12] Mimeault, V. J., Fibre Sci. Technol., 3, p. 273, 1971.

[13] Moreton, R. in Carbon Fibers, The Plastic Institute, London, p. 73, 1971.

[14] Moreton, R. and Watt, W., Carbon, 12, p. 543, 1974.

[15] Jones, W. R. and Johnson, J. W., Carbon, 9, p. 645, 1971.

[16] Cooper, G. A. and Mayer, R. M., J. Mater. Sci., 6, p. 60, 1971.

[17] Johnson, D. J. and Tyson, C. N., Brit. J. Appl. Phys. 2, p. 787, 1969.

[18] Williams, W. S., Steffens, D. A. et al, J. Appl. Phys., 41, p. 4893, 1970.

[19] Bacon, R. and Silvaggi, A. F., Carbon, 6, p. 231, 1968.

[20] Marsh, D. M., Proc. Royal Soc. London, A282, p. 33, 1964.

[21] Tyson, C. N., J. Phys. D.; Appl. Phys., 8, p. 749, 1975.

[22] Reynolds, W. N. and Sharp, J. V., Carbon, 12, p. 103, 1974.

[23] Reynolds, W. N. and Moreton, R., Phil. Trans. Royal Soc. London, Ser. A, 294, p. 451, 1980.

[24] Bacon, R. and Schalamon, W. A., J. Appl. Poly. Sci., Symposia, 9, p. 285, 1969.

[25] Johnson, J. W., Marjoram, J. R. et al, Nature, 221, p. 357, 1969.

[26] Thorne, D. J., 3rd Conf. on Ind. Carbon and Graphite, Soc. Chem Ind. London, p. 463, 1970.

[27] Moreton, R., Fibre Sci. Technol., 1, p. 273, 1969.

[28] Cottrell, A. H., Proc. Royal Soc. London, A276, p. 1, 1963.

[29] Metcalfe, A. G. and Schmitz, G. K., Proc. ASTM, 64, p. 1075, 1964.

[30] Tippett, L. H. C., Biometrika, 17, p. 364, 1925.

[31] McMahon, P. E., SAMPE Quart., 6, October, 1974.

[32] Hitchon, J. W. and Phillips, D. C., Fibre Sci. Technol., 12, p. 217, 1979.

[33] Phoenix, S. L. and Sexsmith, R. G., J. Composite Mater., 6, p. 322, 1972.

ATIGUE CRACK PROPAGATION OF SHEET MOLDING COMPOUNDS IN VARIOUS ENVIRONMENTS

S. V. Hoa, A. D. Ngo and T. S. Sankar

Concordia University
Montreal, Quebec H3G 1M8

ABSTRACT

Effect of the absorption of water and isooctane on the rate of fatigue crack propagation of SMC-R30 and SMC-R65 is investigated. Crack extension gage is used to measure the crack length. Results show that the absorption of water increases the rate of fatigue crack propagation of the two sheet molding compound materials considerably whereas the absorption of isooctane into SMC-R65 does not.

INTRODUCTION

The need to reduce weight in structures such as aircrafts, automobiles, satellites and in recreational equipment has led to the rapid development of fiber reinforced plastics. Some of the recently developed fiber reinforced plastics that have great potential usage in the automotive industry are sheet molding compounds. These materials are composed of randomly distributed fibers in a matrix of polyester resin and various fillers. One of the problems with fiber reinforced plastics is that they tend to absorb water and their mechanical behavior can be adversely affected by this absorption. Most of the investigations done on the effect of environments on fiber reinforced plastics have focussed mainly on continuous graphite fiber reinforced epoxy materials for aircraft applications. Some examples of these works are the special technical publication of ASTM [1] and the AGARD Conference Proceedings [2]. These works show that graphite reinforced epoxy materials absorb water and water moisture and their strengths are reduced as a result. The absorption of moisture into polyester-E glass sheet molding compounds has recently been investigated by Loos et al [3]. In this paper, the effect of water and isooctane on the rate of crack propagation of sheet molding compounds is investigated.

EXPERIMENTS

The materials studied were sheet molding compounds SMC-R30 and SMC-R65. Details for these materials are given in Table 1. Specimens of nominal thicknesses of 2.54 mm were cut into dimensions as shown in Figure 1. The cracks were machined using a low speed saw with a diamond impregnated slitting wheel 0.20 mm thick. The liquids used were demineralized water and reagent grade isooctane.

TABLE 1 - MATERIALS SPECIFICATION

	SMC-R30	SMC-R65
Polyester resin	27 w/o	35.8 w/o
Calcium carbonate filler	41 w/o	0.35 w/o
E-glass	30 w/o 25.4 mm chopped	62.5 w/o
Balance (thickness and internal release and catalyst)	2 w/o	1.35 w/o
Tensile strength	70-100 MPa	175-210 MPa
Supplier	Somerville Industries Ltd. - Code No. G-1005-30	Budd-Code No. DSM-750

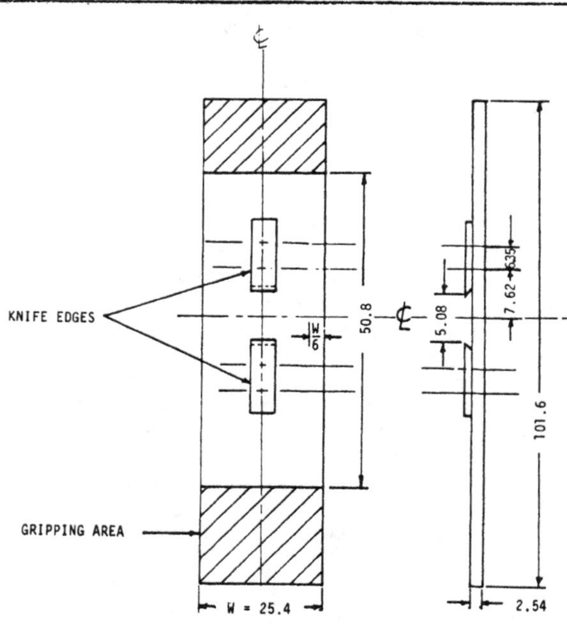

Fig. (1) - Double edge notched fatigue specimen

The extension of the cracks was measured using a crack extension gage MTS-632-02 supplied from MTS Corporation. Knife edges are attached to the specimen for mounting on the gage and are shown in Figure 1. Under fatigue loading, the material was completely immersed within the liquid as shown in Figure 2. The compliance measurement with the crack extension gage is performed after the liquid has been drained away.

Fig. (2) - Immersion of the specimen within the liquid under fatigue loading

The specimens were cycled from zero load to a maximum stress range of 50% f the ultimate tensile strength of the material. This low stress range is used o simulate the real life application where the loads applied are not very high. n LFE-150 fatigue machine fabricated by Fatigue Dynamics, Inc., with an axial oading attachment was used. A testing frequency of 5 cps was used.

The cycle crack growth tests of the two composites in air, water and isooc-ane were performed on specimens that had been immersed unstressed in these liq-ids for a period of 80 days. This time period is taken to ensure complete sat-ration of the specimens since saturation requires a shorter period as is shown n [3]. The percent weight uptake after 80 days immersing in the two liquids re shown in Table 2.

TABLE 2 - PERCENT WEIGHT UPTAKE AT 80 DAYS

Liquids	SMC-R30	SMC-R65
Water	2.17	2.28
Isooctane	0.16	1.10

EFFECT OF LIQUIDS ON THE COMPLIANCE OF THE SPECIMENS

The compliance of the specimens with various crack lengths is shown in Figure 3. These calibration curves are obtained by increasing the crack lengths b

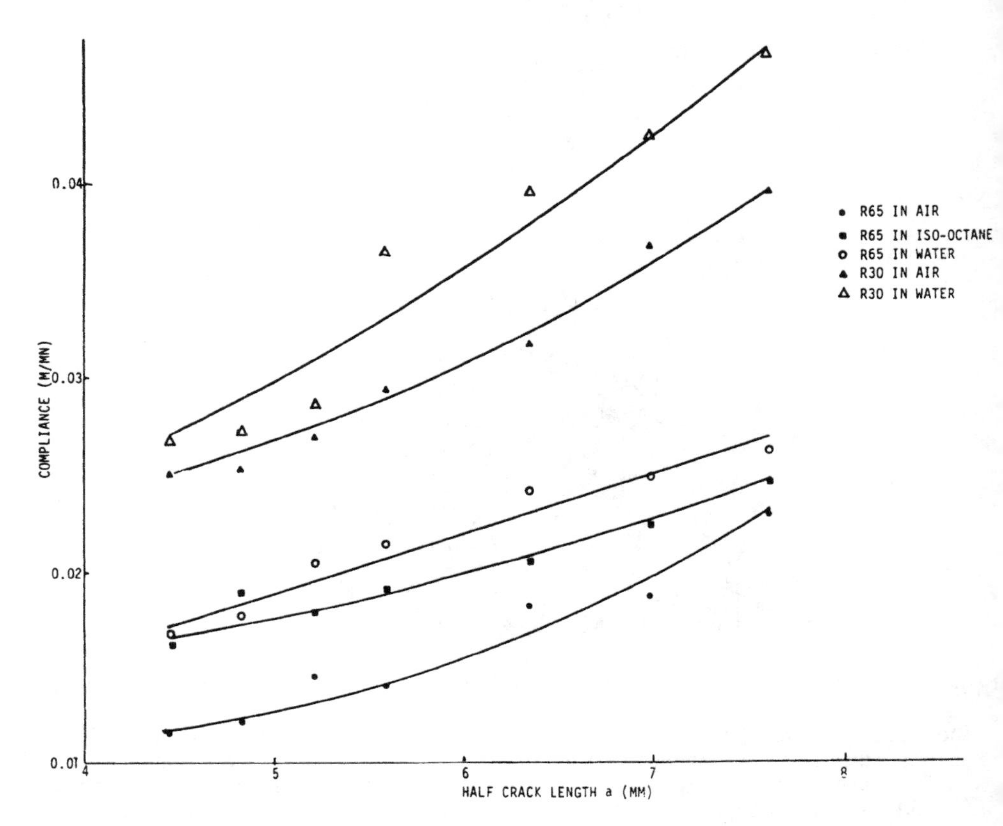

Fig. (3) - Compliance calibration curves

the same slitting wheel in small increments and by applying a small load to obtain a load displacement curve. For those specimens in water and in isooctane care is taken to ensure that the liquid surrounds the specimen at all times by wrapping liquid soaked sponges around the specimens once they are taken away f the liquid bath. It is evident from Figure 3 that the absorption of the liqui increases the compliance of the materials. For SMC-R65, water absorption increases the compliance more than isooctane absorption.

FATIGUE CRACK PROPAGATION RATE

The fatigue crack propagation is obtained by fatigue loading a specimen wi a precut crack of a certain length either in air or in the liquid. Different methods were tried out to measure the crack length as a function of the number of cycles. These include measurement using a travelling microscope, using penetrating dyes and using the calibration curves in Figure 3. The method using t ravelling microscope is not successful since the crack does not propagate along neat line as in metal. The presence of the fibers on the surface orients the rack to propagate along the direction of the fibers as will be discussed in ater sections. Use of penetrating dyes is not successful either since the dye ay run along surface cracks which indicate erroneously large crack length. sually, these surface cracks may not be through the thickness. The method using xtension gage was found most suitable and was used to obtain the results. This ethod was also suggested by Brown [9] and Owen [8]. However, the crack extension gage does not measure the crack length directly but correlation using the ompliance curves in Figure 3 has to be used. This means that at specified instants during the experiment, the fatigue test was stopped, the liquid evacuated in case the specimen was immersed in the liquid) and the compliance measured sing the crack extension gage. Value of the crack length was then obtained from he curve in Figure 3 knowing the compliance. The crack length versus the fatigue ycle curves for SMC-R30 are shown in Figure 4 and for SMC-R65 are shown in Figure . Three specimens are used for each test and the curves are drawn from the re-

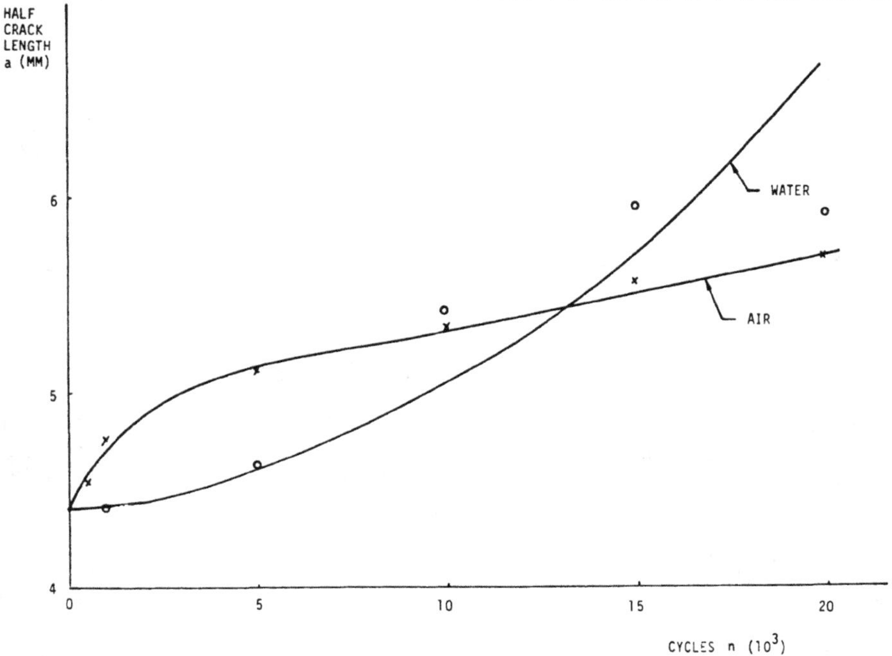

Fig. (4) - Fatigue crack growth in SMC-R30

ult of the most typical specimen. For SMC-R30, in air, there is a rapid rate f crack propagation in the initial cycles but the propagation rate comes to a onstant for subsequent cycles and the crack does not seem to run away at 2×10^4 ycles. However, in water, the rate of crack propagation increases rapidly after 000 cycles and the crack is running away at 2×10^4 cycles. For SMC-R65, the ehavior of fatigue crack propagation in air is similar to that of SMC-R30. ater also has similar effect on this material. The effect of isooctane on the atigue crack of SMC-R65, however, is not as pronounced. The half crack length oes not increase as much as compared to in air and the rate of crack propagation s even less than the rate of crack propagation in air. It is interesting to

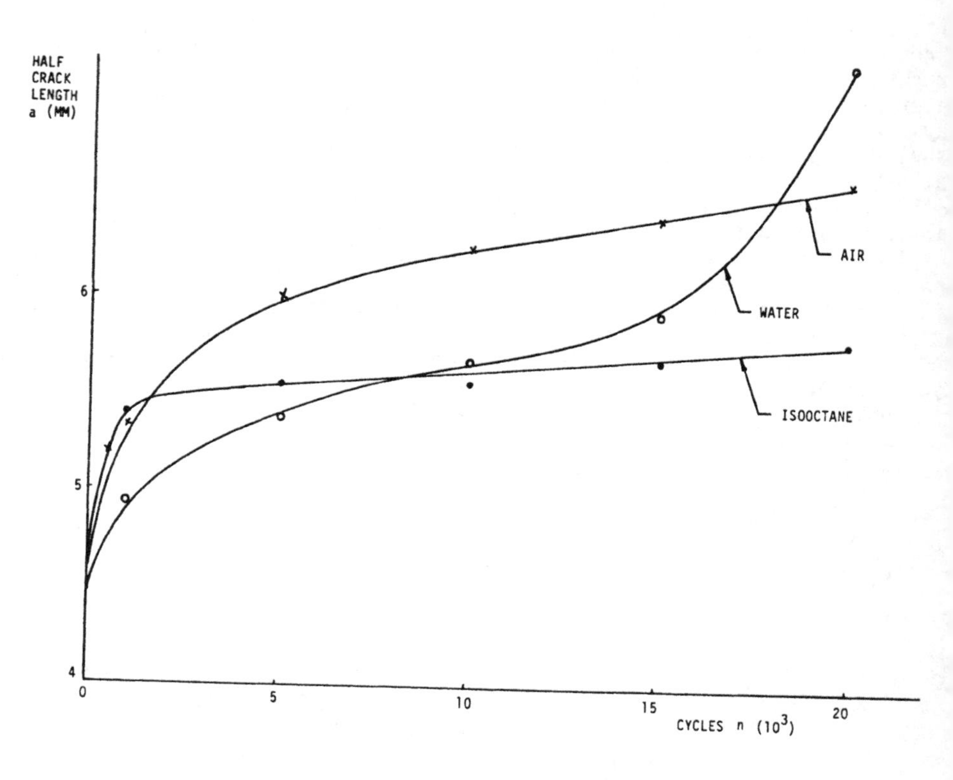

Fig. (5) - Fatigue crack growth in SMC-R65

note that work by Loos et al [3] shows that the saturation level of water into SMC-R65 is more than 3% whereas the saturation level of oil base liquids such as No. 2 diesel fuel, jet fuel, aviation oil and gasoline in SMC-R65 is much less than 1%.

There is certainly some correlation between the degree of absorption of the liquid and the rule of crack propagation. However, this correlation is not monotonic since absorption of isooctane seems to reduce the rate of crack propagation rather than increasing it. Results from creep experiments on SMC-R30 [5] also show that at 41 MPa, specimens tested in isooctane may last longer than specimens tested in air. Liquids are usually labeled plasticizers in their effect upon polymers [6]. The plasticizing effect softens the polymer matrix. This softening may or may not weaken the polymer. If the softening is only limited to the tip of the notches (due to irregularities on the surface), the notches can be blunted and the material is strengthened. However, if the softening is so extensive as to cover a significant thickness layer of the specimen, then the specimen is weakened. Some aspects of the complexity of the combined effect of liquid and stress on the mechanical behavior of polymers has been indicated in [7].

ISCUSSION AND CONCLUSION

The results of this investigation clearly show that the absorption of water as a detrimental effect upon the fatigue lives of sheet molding compounds. A horough understanding of the effect of liquids on the mechanical behavior of iber reinforced plastics materials is far from complete, however. Water absorbs nto and softens the matrix of the material but this softening effect may and may ot weaken the whole specimen or structure. Also, for two liquids, a liquid that bsorbs more into a polymer may have more detrimental effect under axial loading ut may have less detrimental effect under flexural loading. These different bservations have been made in the studies on the effect of liquids on the creep, lexural and axial fatigue loadings presented in [4,5] and in this paper. The xplanation for this variation in behavior remains to be investigated.

From this investigation, the following conclusions can be reached:

- Absorption of water increases the rate of crack propagation of both SMC-R30 and SMC-R65.
- Absorption of isooctane does not affect the rate of crack propagation of SMC-R65.
- Absorption of water causes swelling of the polyester resin and this facilitates crack propagation.
- Crack extension gage measurement of crack length is recommended since surface cracks do not represent through the thickness cracks.

ACKNOWLEDGEMENT

The financial assistance from the Natural Sciences and Engineering Research Council through Grant No. A-4013 is appreciated.

REFERENCES

[1] "Advanced Composite Material, Environmental Effect", ASTM STP 658, J. R. Vinson, ed., 1977.

[2] "Effect of Service Environment on Composite Materials", AGARD Conference Proceedings No. 288, Athens, Greece, April 1980.

[3] Loos, A. C., Springer, G. S., Sanders, B. A. and Tung, R. W., "Moisture Absorption of Polyester - E Glass Composites", J. Composite Materials, Vol. 14, p. 142, April 1980.

[4] Ngo, A. D., Hoa, S. V. and Sankar, T. S., "Effect of Environments on the Fatigue Behavior of Sheet Molding Compounds", Proceedings of the National Technical Conference, Society of Plastics Engineers, p. 75, November 1979.

[5] Hoa, S. V., "Creep of Fiber Glass Reinforced Plastics in Liquid Environments", Advances in Materials Technology in the Americas, ASME MD-1, p. 63, 1980.

[6] Bernier, G. A. and Kambour, R. P., "The Role of Organic Agents in the Stress Crazing and Cracking of Poly (2,6 Dimethyl - 1,4 Phenylene Oxide)", Macromolecules 1, 5, p. 393, 1968.

[7] Hoa, S. V., "Relative Influence of the Mobility and the Solubility Parameters of Fluids on the Mechanical Behavior of High Impact Polystyrene", J. Polymer Engineering and Science, Vol. 20, No. 17, p. 1157, November 1980.

[8] Owen, M. J. and Bishop, P. T., "Crack Growth Relationships for Glass Reinforced Plastics and Their Application to Design", J. Phys. D:Appl. Phys., Vol. 7, p. 1214, 1974.

[9] Brown, K. F. and Srawley, J. E., "Fracture Toughness Testing Methods", AST STP 410, p. 133, 1964.

LIST OF PARTICIPANTS

per, J.
val Air Development Center
rminster, Pennsylvania 18974

cisz, M.
stitute of Fundamental Technological
Research
rsaw, POLAND

bcock, C. D.
lifornia Institute of Technology
sadena, California 91125

daliance, R.
Donnell Aircraft Company
. Louis, Missouri 63166

gdanovich, A.
stitute of Polymer Mechanics
ademy of Sciences of the Latvian SSR
ga, USSR

lotin, V. V.
SR Academy of Sciences
scow, USSR

ton, R.
igh University
thlehem, Pennsylvania 18015

en, T.
val Air Development Center
minster, Pennsylvania 18974

iao, T. T.
wrence Livermore Laboratory
vermore, California 94550

u, S.-C.
AMMRC
tertown, Massachusetts 01272

u, D.
igh University
thlehem, Pennsylvania 18015

llins, R.
umman Aerospace Corporation
thpage, New York 11714

Crossman, F. W.
Lockheed Palo Alto Research Laboratory
Palo Alto, California 94304

Dunaevsky, V. V.
WABCO - Engineering Department
Wilmerding, Pennsylvania 15148

Feng, W. W.
Lawrence Livermore Laboratory
Livermore, California 94550

Gause, L.
Naval Air Development Center
Warminster, Pennsylvania 18974

Greszczuk, L. B.
McDonnell Douglas Astronautics Company
Huntington Beach, California 92647

Hahn, H. T.
Washington University
St. Louis, Missouri 53131

Harlow, D. G.
Drexel University
Philadelphia, Pennsylvania 19104

Hartranft, R. J.
Lehigh University
Bethlehem, Pennsylvania 18015

Hoa, S. V.
Concordia University
Montreal, Quebec H3G 1M8

Huang, S.
Naval Air Development Center
Warminster, Pennsylvania 18974

Ingraffea, A. R.
Cornell University
Ithaca, New York 14853

Kalnin, I.
Celanese Research Company
Summit, New Jersey 07901

Knets, I.
Institute of Polymer Mechanics
Riga 6, Latvian SSR, USSR

Kyanka, G. H.
Syracuse University
Syracuse, New York 13210

Lal, K. M.
NASA-Langley Research Center
Hampton, Virginia 23665

Lee, J. D.
The George Washington University
Washington, D.C. 20052

Lieu, M.
Lehigh University
Bethlehem, Pennsylvania 18015

Madenci, E.
Lehigh University
Bethlehem, Pennsylvania 18015

Matic, P.
Lehigh University
Bethlehem, Pennsylvania 18015

McAllister, L.
Fiber-Materials, Inc.
Biddeford, Maine 04005

Mileiko, S. T.
Academy of Sciences of the USSR
Moscow, USSR

Moyer, T.
Lehigh University
Bethlehem, Pennsylvania 18015

Nikitin, L. V.
USSR Academy of Sciences
Moscow, USSR

Nuismer, R. J.
University of Utah
Salt Lake City, Utah 84121

Ozen, M.
Lehigh University
Bethlehem, Pennsylvania 18015

Panaysuk, V. V.
SSR Academy of Sciences
Moscow, USSR

Parhizgar, S.
University of Wisconsin
Platteville, Wisconsin 53818

Phoenix, S. L.
Cornell University
Ithaca, New York 14853

Pindera, M. J.
Virginia Polytechnic Institute and St
University
Blacksburg, Virginia 24061

Reid, S.
Lehigh University
Bethlehem, Pennsylvania 18015

Reifsnider, K. L.
Virginia Polytechnic Institute and St
University
Blacksburg, Virginia 24061

Rivlin, R. S.
Lehigh University
Bethlehem, Pennsylvania 18015

Rowlands, R. E.
University of Wisconsin
Madison, Wisconsin 53706

Sendeckyj, G. P.
AFWAL/FIBE
Wright-Patterson Air Force Base, Ohio

Shih, S.
Lehigh University
Bethlehem, Pennsylvania 18015

Sih, G. C.
Lehigh University
Bethlehem, Pennsylvania 18015

Smith, C. W.
Virginia Polytechnic Institute and St
University
Blacksburg, Virginia 24061

Smith, G. F.
Lehigh University
Bethlehem, Pennsylvania 18015

Sun, C. T.
Purdue University
West Lafayette, Indiana 47907

Tabaddor, F.
B. F. Goodrich
Akron, Ohio 44313

amuzs, V.
nstitute of Polymer Mechanics
iga, USSR

ennyson, R. C.
niversity of Toronto
ownsview, Ontario, Canada

eters, G. A.
nstitute of Polymer Mechanics
iga, USSR

ng, T. C. T.
iversity of Illinois at Chicago
Circle
icago, Illinois 60680

inkle, J. C.
wrence Livermore Laboratory
vermore, California 94550

irk, A.
rns and Roe, Inc.
per Montclair, New Jersey 07643

siliev, V. V.
scow Institute of Aviation Technology
scow, USSR

loshin, A.
wa State University
nes, Iowa 50011

ng, A. S. D.
exel University
iladelphia, Pennsylvania 19104

, E. M.
wrence Livermore Laboratory
vermore, California 94550

, S.
Pont Company
lmington, Delaware 19898

rski, H.
stitute of Fundamental Technological
Research
rsaw, POLAND

eben, C.
neral Electric Company
iladelphia, Pennsylvania 19101

CPSIA information can be obtained
at www.ICGtesting.com
Printed in the USA
LVHW051033030520
654914LV00001B/100

9 789024 726998